T0224355

Communications
in Computer and Information Science 902

Commenced Publication in 2007
Founding and Former Series Editors:
Phoebe Chen, Alfredo Cuzzocrea, Xiaoyong Du, Orhun Kara, Ting Liu,
Dominik Ślęzak, and Xiaokang Yang

More information about this series at http://www.springer.com/series/7899

Qinglei Zhou · Qiguang Miao
Hongzhi Wang · Wei Xie
Yan Wang · Zeguang Lu (Eds.)

Data Science

4th International Conference
of Pioneering Computer Scientists,
Engineers and Educators, ICPCSEE 2018
Zhengzhou, China, September 21–23, 2018
Proceedings, Part II

 Springer

Editors
Qinglei Zhou
Zhengzhou University
Zhengzhou, Henan
China

Qiguang Miao
Xidian University
Xi'an, Shaanxi
China

Hongzhi Wang
Harbin Institute of Technology
Harbin, China

Wei Xie
Harbin University of Science
  and Technology
Harbin, China

Yan Wang
Zhengzhou Institute of Technology
Zhengzhou, China

Zeguang Lu
National Academy
  of Guo Ding Institute of Data Science
Beijing, China

ISSN 1865-0929   ISSN 1865-0937   (electronic)
Communications in Computer and Information Science
ISBN 978-981-13-2205-1   ISBN 978-981-13-2206-8   (eBook)
https://doi.org/10.1007/978-981-13-2206-8

Library of Congress Control Number: 2018951433

This Springer imprint is published by the registered company Springer Nature Singapore Pte Ltd.
The registered company address is: 152 Beach Road, #21-01/04 Gateway East, Singapore 189721, Singapore

Preface

As the general and program co-chairs of the 4th Interna
Computer Scientists, Engineers and Educators 2018
ICYCSEE), it is our great pleasure to welcome you to tl
ence, which was held in Zhengzhou, China, September 2
Computer Federation and Zhengzhou Computer Federati
and Henan Polytechnic University and National Acade
Data Science. The goal of this conference is to provide a
engineers, and educators.

The call for papers of this year's conference attracted
the hard work of the Program Committee, 125 papers v
conference proceedings, with an acceptance rate of 26
conference was data science. The accepted papers cover a
Basic Theory and Techniques for Data Science includin
Science, Computational Theory for Data Science, Big
cations, Data Quality and Data Preparation, Evaluati
Science, Data Visualization, Big Data Mining and Kn
tructure for Data Science, Machine Learning for Dat
Privacy, Applications of Data Science, Case Study of I
Management and Analysis, Data-Driven Scientific Re
matics, Data-Driven Healthcare, Data-Driven Managem
Data-Driven Smart City/Planet, Data Marketing and
Recommendation Systems, Data-Driven Security, Data
vation, Social and/or Organizational Impacts of Data S

We would like to thank all the Program Committee i
institutes, for their hard work in completing the reviev
made it possible to attain quality reviews for all the su
Their diverse expertise in each individual research a
exciting program for the conference. Their comments a
improve the quality of their papers and gain deeper ir

Great thanks should also go to the authors and p
support in making the conference a success. We thanl
from Springer, whose professional assistance was inv
proceedings.

Besides the technical program, this year ICPCSEE
the participants. We hope you enjoy the conference p

June 2018

Organization

The 4th International Conference of Pioneering Computer Scientists, Engineers and Educators (ICPCSEE, originally ICYCSEE) 2018 (http://2018.icpcsee.org) was held in Zhengzhou, China, during September 21–23 2018, hosted by Henan Computer Federation and Zhengzhou Computer Federation and Zhengzhou University and Henan Polytechnic University and National Academy of Guo Ding Institute of Data Science.

ICPCSEE 2018 General Chair

Qinglei Zhou   Zhengzhou University, China

Program Chairs

Yong Gan   Zhengzhou Institute of Technology, China
Qiguang Miao   Xidian University, China

Program Co-chairs

Qingxian Wang   Information Engineering University, China
Fengbin Zheng   Henan University, China
JiuCheng Xu   Henan Normal University, China
Jiexin Pu   Henan University of Science and Technology, China
ZongPu Jia   Henan Polytechnic University, China
Zhanbo Li   Zhengzhou University, China

Organization Chairs

Yangdong Ye   Zhengzhou University, China
WANG Yan   Zhengzhou Institute of Technology, China
Dong Liu   Henan Normal University, China
Junding Sun   Henan Polytechnic University, China
Zeguang Lu   National Academy of Guo Ding Institute of Data Science, China

Organization Co-chairs

Jianmin Wang   Zhengzhou University, China
Haitao Li   Zhengzhou University, China
Song Yu   Zhengzhou University, China
Song Wei   Zhengzhou University, China
Sun Yi   Zhengzhou University, China

Yan Gao   Henan Polytechnic University, China
Zhiheng Wang   Henan Polytechnic University, China
Fan Zhang   Zhengzhou Institute of Technology, China

Publication Chairs

Hongzhi Wang   Harbin Institute of Technology, China
Weipeng Jing   Northeast Forestry University, China

Publication Co-chairs

Xianhua Song   Harbin University of Science and Technology, China
Wei Xie   Harbin University of Science and Technology, China
Liuyuan Chen   Henan Normal University, China
Hui Li   Henan Polytechnic University, China
Xiaopeng Chang   Henan Finance University, China

Education Chairs

Shenyi Qian   Zhengzhou University of Light Industry, China
Miaolei Deng   Henan University of Technology, China

Industrial Chairs

Zheng Shan   Information Engineering University, China
Zhiyongng Zhang   Henan University of Science and Technology, China

Demo Chairs

Tianyang Zhou   Information Engineering University, China
Shuhong Li   Henan University of Economics and Law, China

Panel Chairs

Bing Xia   Zhongyuan University of Technology, China
Huaiguang Wu   Zhengzhou University of Light Industry, China

Poster Chairs

Guanglu Sun   Harbin University of Science and Technology, China
Liu Xia   Sanya Aviation and Tourism College, China

Expo Chairs

Shuaiyi Zhou   Henan Smart City Planning and Construction Specialized
                      Committee, China
Junhao Jia   Henan King Source Information Technology Co., Ltd., China

Expo Co-chairs

Liang Bing   Henan Smart City Planning and Construction Specialized
                      Committee, China
Dandan Jia   Henan Skylark Marketing Data Services Ltd., China

Registration/Financial Chair

Chunyan Hu   National Academy of Guo Ding Institute of Data Science,
                      China

ICPCSEE Steering Committee

Jiajun Bu   Zhejiang University, China
Wanxiang Che   Harbin Institute of Technology, China
Jian Chen   Paratera, China
Xuebin Chen   North China University of Science and Technology, China
Wenguang Chen   Tsinghua University, China
Xiaoju Dong   Shanghai Jiao Tong University, China
TIAN Feng   Institute of Software Chinese Academy of Sciences, China
Qilong Han   Harbin Engineering University, China
Yiliang Han   Engineering University of CAPF, China
Yinhe Han   Institute of Computing Technology, Chinese Academy
                      of Sciences, China
Hai Jin   Huazhong University of Science and Technology, China
Weipeng Jing   Northeast Forestry University, China
Wei Li   Central Queensland University, Australia
Min Li   Central South University, China
Junyu Lin   Institute of Information Engineering, Chinese Academy
                      of Sciences, China
Yunhao Liu   Michigan State University, America
Zeguang Lu   National Academy of Guo Ding Institute of Data Science,
                      China
Rui Mao   Shenzhen University, China
Qiguang Miao   Xidian University, China
Haiwei Pan   Harbin Engineering University, China
Pinle Qin   North University of China, China
Zhaowen Qiu   Northeast Forestry University, China
Zheng Shan   The PLA Information Engineering University, China
Guanglu Sun   Harbin University of Science and Technology, China

Jie Tang	Tsinghua University, China
Hongzhi Wang	Harbin Institute of Technology, China
Tao Wang	Peking University, China
Xiaohui Wei	Jilin University, China
Lifang Wen	Beijing Huazhang Graphics & Information Co., Ltd., China
Yu Yao	Northeastern University, China
Xiaoru Yuan	Peking University, China
Yingtao Zhang	Harbin Institute of Technology, China
Yunquan Zhang	Institute of Computing Technology, Chinese Academy of Sciences, China
Liehuang Zhu	Beijing Institute of Technology, China
Min Zhu	Sichuan University, China

ICPCSEE 2018 Program Committee Members

Chunyu Ai	University of South Carolina Upstate, America
Jiyao An	Hunan University, China
Xiaojing Bai	TsingHua University, China
Ran Bi	Dalian University of Technology, China
Yi Cai	South China University of Technology, China
Zhipeng Cai	Georgia State University, America
Cao Cao	State Key Laboratory of Mathematical Engineering and Advanced Computing, China
Zhao Cao	Beijing Institute of Technology, China
Baobao Chang	Peking University, China
Richard Chbeir	LIUPPA Laboratory, France
Che Nan	Harbin University of Science and Technology, China
Wanxiang Che	Harbin Institute of Technology, China
Bolin Chen	Northwestern Polytechnical University, China
Chunyi Chen	Changchun University of Science and Technology, China
Hao Chen	Hunan University, China
Quan Chen	Guangdong University of Technology, China
Shu Chen	Xiangtan University, China
Wei Chen	Beijing Jiaotong University, China
Wenliang Chen	Soochow University, China
Wenyu Chen	University of Electronic Science and Technology of China, China
Xuebin Chen	North China University of Science and Technology, China
Zhumin Chen	Shandong University, China
Ming Cheng	Zhengzhou University of Light Industry, China
Siyao Cheng	Harbin Institute of Technology, China
Byron Choi	Hong Kong Baptist University, China
Xinyu Dai	Nanjing University, China
Lei Deng	Central South University, China
Vincenzo Deufemia	University of Salerno, Italy
Jianrui Ding	Harbin Institute of Technology, China

Qun Ding	Heilongjiang University, China
Xiaofeng Ding	Huazhong University, China
Hongbin Dong	Harbin Engineering University, China
Xiaoju Dong	Shanghai Jiao Tong University, China
Zhicheng Dou	Renmin University of China, China
Jianyong Duan	North China University of Technology, China
Lei Duan	Sichuan University, China
Xiping Duan	Harbin Normal University, China
Junbin Fang	Jinan University, China
Xiaolin Fang	Southeast University, China
Guangsheng Feng	Harbin Engineering University, China
Jianlin Feng	Sun Yat-Sen University, China
Weisen Feng	Sichuan University, China
Guohong Fu	Heilongjiang University, China
Jianhou Gan	Yunnan Normal University, China
Jing Gao	Dalian University of Technology, China
Daohui Ge	Xidian University, China
Lin Ge	Zhengzhou University of Aeronautics, China
Dianxuan Gong	North China University of Science and Technology, China
Lila Gu	Xinjiang University
Yu Gu	Northeastern University, China
Hongjiao Guan	Harbin Institute of Technology, China
Tao Guan	Zhengzhou University of Aeronautics, China
Chunyi Guo	Zhengzhou University, China
Jiafeng Guo	Institute of Computing Technology, Chinese Academy of Sciences, China
Longjiang Guo	Heilongjiang University, China
Yibo Guo	Zhengzhou University, China
Yuhang Guo	Beijing Institute of Technology, China
Meng Han	Georgia State University, America
Meng Han	Kennesaw State University, America
Qi Han	Harbin Institute of Technology, China
Xianpei Han	Chinese Academy of Sciences, China
Yingjie Han	Zhengzhou University, China
Zhongyuan Han	Harbin Institute of Technology, China
Tianyong Hao	Guangdong University of Foreign Studies, China
Jia He	Chengdu University of Information Technology, China
Qinglai He	Arizona State University, America
Shizhu He	Chinese Academy of Sciences, China
Liang Hong	Wuhan University, China
Leong Hou	University of Macau, China
Yifan Hou	State Key Laboratory of Mathematical Engineering and Advanced Computing, China
Chengquan Hu	Jilin University, China
Wei Hu	Nanjing University, China
Zhang Hu	Shanxi University, China

Hao Huang   Wuhan University, China
Kuan Huang   Utah State University, America
Lan Huang   Jilin University, China
Shujian Huang   Nanjing University, China
Jian Ji   Xidian University, China
Ruoyu Jia   Sichuan University, China
Yuxiang Jia   Zhengzhou University, China
Bin Jiang   Hunan University, China
Feng Jiang   Harbin Institute of Technology, China
Hailin Jiang   Harbin Institute of Technology, China
Jiming Jiang   King Abdullah University of Science & Technology,
                       Saudi Arabia
Wenjun Jiang   Hunan University, China
Xiaoheng Jiang   Zhengzhou University, China
Peng Jin   Leshan Normal University, China
Weipeng Jing   Northeast Forestry University, China
Shenggen Ju   Sichuan University, China
Fang Kong   Soochow University, China
Hanjiang Lai   Sun Yat-Sen University, China
Wei Lan   Central South University, China
Yanyan Lan   Institute of Computing Technology, Chinese Academy
                       of Sciences, China
Chenliang Li   Wuhan University, China
Dawei Li   Nanjing Institute of Technology, China
Dun Li   Zhengzhou University, China
Faming Li   University of Electronic Science and Technology of China,
                       China
Guoqiang Li   Norwegian University of Science and Technology, Norway
Hua Li   Changchun University, China
Hui Li   Xidian University, China
Jianjun Li   Huazhong University of Science and Technology, China
Jie Li   Harbin Institute of Technology, China
Kai Li   Harbin Institute of Technology, China
Min Li   Central South University, China
Mingzhao Li   RMIT University, Australia
Mohan Li   Jinan University, China
Moses Li   Jiangxi Normal University, China
Peng Li   Shaanxi Normal University, China
Qingliang Li   Changchun University of Science and Technology, China
Qiong Li   Harbin Institute of Technology, China
Rong-Hua Li   Shenzhen University, China
Ru Li   Shanxi University, China
Sujian Li   Peking University, China
Wei Li   Georgia State University, America
Xiaofeng Li   Sichuan University, China
Xiaoyong Li   Beijing University of Posts and Telecommunications, China

Xuwei Li	Sichuan University, China
Yunan Li	Xidian University, China
Zheng Li	Sichuan University, China
Zhenghua Li	Soochow University, China
Zhijun Li	Harbin Institute of Technology, China
Zhixu Li	Soochow University, China
Zhixun Li	Nanchang University, China
Hongfei Lin	Dalian University of Technology, China
Bingqiang Liu	Shandong University, China
Fudong Liu	State Key Laboratory of Mathematical Engineering and Advanced Computing, China
Guanfeng Liu	Soochow University, China
Guojun Liu	Harbin Institute of Technology, China
Hailong Liu	Northwestern Polytechnical University, China
Ming Liu	Harbin Institute of Technology, China
Pengyuan Liu	Beijing Language and Culture University, China
Shengquan Liu	XinJiang University, China
Tiange Liu	Yanshan University, China
Yan Liu	Harbin Institute of Technology, China
Yang Liu	Peking University, China
Yang Liu	TsingHua University, China
Yanli Liu	Sichuan University, China
Yong Liu	Heilongjiang University, China
Binbin Lu	Sichuan University, China
Junling Lu	Shaanxi Normal University, China
Wei Lu	Renmin University of China, China
Zeguang Lu	Sciences of Country Tripod Institute of Data Science, China
Jianlu Luo	Officers College of PAP, China
Jiawei Luo	Hunan University, China
Jizhou Luo	Harbin Institute of Technology, China
Zhunchen Luo	China Defense Science and Technology Information Center, China
Huifang Ma	NorthWest Normal University, China
Jiquan Ma	Heilongjiang University, China
Yide Ma	Lanzhou University, China
Hua Mao	Sichuan University, China
Xian-Ling Mao	Beijing Institute of Technology, China
Jun Meng	Dalian University of Technology, China
Hongwei Mo	Harbin Engineering University, China
Lingling Mu	Zhengzhou University, China
Jiaofen Nan	Zhengzhou University of Light Industry, China
Tiezheng Nie	Northeastern University, China
Haiwei Pan	Harbin Engineering University, China
Fei Peng	Hunan University, China
Jialiang Peng	Norwegian University of Science and Technology, China
Wei Peng	Kunming University of Science and Technology, China

Xiaoqing Peng	Central South University, China
Yuwei Peng	Wuhan University, China
Jianzhong Qi	University of Melbourne, Australia
Yutao Qi	Xidian University, China
Shenyi Qian	Zhengzhou University of Light Industry, China
Shaojie Qiao	Southwest Jiaotong University, China
Hong Qu	University of Electronic Science and Technology of China, China
Weiguang Qu	Nanjing Normal University, China
Yining Quan	Xidian University, China
Zhe Quan	Hunan University, China
Shan Xiang	Harbin Institute of Technology, China
Zheng Shan	State Key Laboratory of Mathematical Engineering and Advanced Computing, China
Songtao Shang	Zhengzhou University of Light Industry, China
Yingxia Shao	Peking University, China
Qiaomu Shen	The Hong Kong University of Science and Technology, China
Hongwei Shi	Sichuan University, China
Jianting Shi	HeiLongjiang University of Science and Technology, China
Hongtao Song	Harbin Engineering University, China
Wei Song	North China University of Technology, China
Xianhua Song	Harbin Institute of Technology, China
Chengjie Sun	Harbin Institute of Technology, China
Guanglu Sun	Harbin University of Science and Technology, China
Minghui Sun	Jilin University, China
Penggang Sun	Xidian University, China
Tong Sun	Zhengzhou University of Light Industry, China
Xiao Sun	Hefei University of Technology, China
Yanan Sun	Sichuan University, China
Guanghua Tan	Hunan University, China
Wenrong Tan	Southwest University for Nationalities, China
Binbin Tang	Works Applications, China
Dang Tang	Chengdu University of Information Technology, China
Jintao Tang	National University of Defense Technology, China
Xing Tang	Huawei Technologies Co., Ltd., China
Hongwei Tao	Zhengzhou University of Light Industry, China
Lingling Tian	University of Electronic Science and Technology of China, China
Xifeng Tong	Northeast Petroleum University, China
Yongxin Tong	Beihang University, China
Vicenc Torra	Högskolan i Skövde, Sweden
Chaokun Wang	TsingHua University, China
Chunnan Wang	Harbin Institute of Technology, China
Dong Wang	Hunan University, China
Hongzhi Wang	Harbin Institute of Technology, China
Jinbao Wang	Harbin Institute of Technology, China

Suge Wang	Shanxi University, China
Xiao Wang	Zhengzhou University of Light Industry, China
Xin Wang	Tianjin University, China
Yingjie Wang	Yantai University, China
Yongheng Wang	Hunan University, China
Yunfeng Wang	Sichuan University, China
Zhenyu Wang	State Key Laboratory of Mathematical Engineering and Advanced Computing, China
Zhifang Wang	Heilongjiang University, China
Zhewei Wei	School of Information, Renming University, China
Zhongyu Wei	Fudan University, China
Bin Wen	Yunnan Normal University, China
Huaiguang Wu	Zhengzhou University of Light Industry, China
Huayu Wu	Institute for Infocomm Research, China
Rui Wu	Harbin Institute of Technology, China
Xiangqian Wu	Harbin Institute of Technology, China
Yan Wu	Changchun University, China
Yufang Wu	Peking University, China
Zhihong Wu	Sichuan University, China
Guangyong Xi	Zhengzhou University of Light Industry, China
Rui Xia	Nanjing University of Science and Technology, China
Min Xian	Utah State University, America
Degui Xiao	Hunan University, China
Sheng Xiao	Hunan University, China
Tong Xiao	Northeastern University, China
Yi Xiao	Hunan University, China
Minzhu Xie	Hunan Normal University, China
Deyi Xing	Soochow University, China
Dan Xu	University of Trento, Italy
Jianqiu Xu	Nanjing University of Aeronautics and Astronautics, China
Jing Xu	Changchun University of Science and Technology, China
Pengfei Xu	Xidian University, China
Ruifeng Xu	Harbin Institute of Technology, China
Ying Xu	Hunan University, China
Yaohong Xue	Changchun University of Science and Technology, China
Mingyuan Yan	University of North Georgia, America
Shaohong Yan	North China University of Science and Technology, China
Xuexiong Yan	State Key Laboratory of Mathematical Engineering and Advanced Computing, China
Bian Yang	Norwegian University of Science and Technology, Norway
Chunfang Yang	State Key Laboratory of Mathematical Engineering and Advanced Computing, China
Donghua Yang	Harbin Institute of Technology, China
Gaobo Yang	Hunan University, China
Lei Yang	Heilongjiang University, China
Ning Yang	Sichuan University, China

Yajun Yang	Tianjin University, China
Bin Yao	Shanghai Jiao Tong University, China
Yuxin Ye	Jilin University, China
Dan Yin	Harbin Engineering University, China
Meijuan Yin	State Key Laboratory of Mathematical Engineering and Advanced Computing, China
Minghao Yin	Northeast Normal University, China
Zhongxu Yin	State Key Laboratory of Mathematical Engineering and Advanced Computing, China
Zhou Yong	China University of Mining and Technology, China
Jinguo You	Kunming University of Science and Technology, China
Bo Yu	National University of Defense Technology, China
Dong Yu	Beijing Language and Culture University, China
Fei Yu	Harbin Institute of Technology, China
Haitao Yu	Harbin Institute of Technology, China
Lei Yu	Georgia Institute of Technology, America
Yonghao Yu	Harbin Institute of Technology, China
Zhengtao Yu	Kunming University of Science and Technology, China
Lingyun Yuan	Yunnan Normal University, China
Ye Yuan	Harbin Institute of Technology, China
Ye Yuan	Northeastern University, China
Kun Yue	Yunnan University, China
Yue Yue	SUTD, Singapore
Hongying Zan	Zhengzhou University, China
Boyu Zhang	Utah State University, America
Dongxiang Zhang	University of Electronic Science and Technology of China, China
Fan Zhang	Wuhan University of Light Industry, China
Haixian Zhang	Sichuan University, China
Huijie Zhang	Northeast Normal University, China
Jiajun Zhang	Institute of Automation, Chinese Academy of Sciences, China
Kejia Zhang	Harbin Engineering University, China
Keliang Zhang	PLAUFL, China
Kunli Zhang	Zhengzhou University, China
Liancheng Zhang	State Key Laboratory of Mathematical Engineering and Advanced Computing, China
Lichen Zhang	Shaanxi Normal University, China
Liguo Zhang	Harbin Engineering University, China
Meishan Zhang	Heilongjiang University, China
Meishan Zhang	Singapore University of Technology and Design, Singapore
Peipei Zhang	Xidian University, China
Ping Zhang	State Key Laboratory of Mathematical Engineering and Advanced Computing, China
Tiejun Zhang	Harbin University of Science and Technology, China
Wenjie Zhang	The University of New South Wales, Australia
Xiao Zhang	Renmin University of China, China

Xiaowang Zhang	Tianjin University, China
Yangsen Zhang	Beijing Information Science and Technology University, China
Yi Zhang	Sichuan University, China
Yingtao Zhang	Harbin Institute of Technology, China
Yonggang Zhang	Jilin University, China
Yongqing Zhang	Chengdu University of Information Technology, China
Yu Zhang	Harbin Institute of Technology, China
Yuhong Zhang	Henan University of Technology, China
Bihai Zhao	Changsha University, China
Hai Zhao	Shanghai Jiao Tong University, China
Jian Zhao	Changchun University, China
Qijun Zhao	Sichuan University, China
Xin Zhao	Renmin University of China, China
Xudong Zhao	Northeast Forestry University, China
Wenping Zheng	Shanxi University, China
Zezhi Zheng	Xiamen University, China
Jiancheng Zhong	Hunan Normal University, China
Changjian Zhou	Northeast Agricultural University, China
Fucai Zhou	Northeastern University, China
Juxiang Zhou	Yunnan Normal University, China
Tianyang Zhou	State Key Laboratory of Mathematical Engineering and Advanced Computing, China
Haodong Zhu	Zhengzhou University of Light Industry, China
Jinghua Zhu	Heilongjiang University, China
Min Zhu	Sichuan University, China
Ruijie Zhu	Zhengzhou University, China
Shaolin Zhu	Xinjiang Institute of Sciences and Chemistry of the Chinese Academy of Sciences, China
Yuanyuan Zhu	Wuhan University, China
Zede Zhu	Hefei Institutes of Physical Science, Chinese Academy of Sciences, China
Huibin Zhuang	Henan University, China
Quan Zou	Tianjin University, China
Wangmeng Zuo	Harbin Institute of Technology, China
Xingquan Zuo	Beijing University of Posts and Telecommunications, China

Contents – Part II

Classifying DNA Methylation Imbalance Data in Cancer Risk Prediction
Using SMOTE and Tomek Link Methods   1
   Chao Liu, Jia Wu, Labrador Mirador, Yang Song, and Weiyan Hou

Auxiliary Disease and Treatment System of Aortic Disease Based
on Mixed Reality 10
   Zishan Qiu, Jian Zhang, and Hui Gao

An Algorithm for Describing the Convex and Concave Shape of
Protein Surface 17
   Wei Wang, Keliang Li, Hehe Lv, Lin Sun, Hongjun Zhang, Jinling Shi,
   Shiguang Zhang, Yun Zhou, Yuan Zhao, and Jingjing Xv

Establish Evidence Chain Model on Chinese Criminal Judgment
Documents Using Text Similarity Measure 27
   Yixuan Dong, Yemao Zhou, Chuanyi Li, Jidong Ge, Yali Han,
   Mengting He, Dekuan Liu, Xiaoyu Zhou, and Bin Luo

Text Sentiment Analysis Based on Emotion Adjustment 41
   Mengjiao Song, Yepei Wang, Yong Liu, and Zhihong Zhao

Text Sentiment Analysis Based on Convolutional Neural Network
and Bidirectional LSTM Model 55
   Mengjiao Song, Xingyu Zhao, Yong Liu, and Zhihong Zhao

Research on Dynamic Discovery Model of User Interest Based on Time
and Space Vector 69
   Jinxiu Lin, Zhaoxin Zhang, Lejun Chi, and Yang Wang

Automatic Generation of Multiple-Choice Items for Prepositions
Based on Word2vec 81
   Wenyan Xiao, Mingwen Wang, Chenlin Zhang, Yiming Tan,
   and Zhiming Chen

ABPR– A New Way of Point-of-Interest Recommendation via
Geographical and Category Influence. 96
   Jingyuan Gao and Yan Yang

A Study on Corpus Content Display and IP Protection 108
   Jingyi Ma, Muyun Yang, Haoyong Wang, Conghui Zhu,
   and Bing Xu

Construction and Application of Diversified Knowledge Model for Paper
Reviewers Recommendation   120
  Hua Zhao, Wei Tao, Ruofei Zou, and Chunming Xu

Study on Chinese Term Extraction Method Based on Machine Learning   128
  Wen Zeng, Xiang Li, and Hui Li

Topic Detection for Post Bar Based on LDA Model   136
  Muzhen Sun and Haonan Zheng

EventGraph Based Events Detection in Social Media.   150
  Jianbiao He, Yongjiao Liu, and Yawei Jia

A Method of Chinese Named Entity Recognition Based
on CNN-BILSTM-CRF Model   161
  Sun Long, Rao Yuan, Lu Yi, and Li Xue

Research Progress of Knowledge Graph Based on Knowledge
Base Embedding   176
  Tang Caifang, Rao Yuan, Yu Hualei, and Cheng Jiamin

Dynamic Detection Method of Micro-blog Topic Based on Time Series.   192
  Deyang Zhang, Yiliang Han, and Xiaolong Li

An Evaluation Algorithm for Importance of Dynamic Nodes in Social
Networks Based on Three-Dimensional Grey Relational Degree   201
  Xiaolong Li, Yiliang Han, Deyang Zhang, and Xuguang Wu

Travel Attractions Recommendation with Travel Spatial-Temporal
Knowledge Graphs   213
  Weitao Zhang, Tianlong Gu, Wenping Sun, Yochum Phatpicha,
  Liang Chang, and Chenzhong Bin

Hierarchical RNN for Few-Shot Information Extraction Learning   227
  Shengpeng Liu, Ying Li, and Binbin Fan

An Improved Collaborative Filtering Algorithm and Application
in Scenic Spot Recommendation   240
  Wanhong Bian, Jintao Zhang, Jialin Li, and Lan Huang

The Algorithms of Weightening Based on DNA Sticker Model.   250
  Chunyan Zhang, Weijun Zhu, and Qinglei Zhou

Method and Evaluation Method of Ultra-Short-Load Forecasting
in Power System   263
  Jiaxiang Ou, Songling Li, Junwei Zhang, and Chao Ding

Ensemble of Deep Autoencoder Classifiers for Activity Recognition Based
on Sensor Modalities in Smart Homes............................. 273
  Serge Thomas, Mickala Bourobou, and Jie Li

An Anomaly Detection Method Based on Learning of "Scores Sequence" ... 296
  Dongsheng Li, Shengfei Shi, Yan Zhang, Hongzhi Wang, and Jizhou Luo

A Method of Improving the Tracking Method of CSI Personnel.......... 312
  Zhanjun Hao, Beibei Li, and Xiaochao Dang

Design and Implementation of the Forearm Rehabilitation System Based
on Gesture Recognition... 330
  Dexin Zhu, Zhiling Li, Kui Huang, and Sato Reika

Research on Traffic Passenger Volume Prediction of Sanya City Based
on ARIMA and Grey Markov Models.............................. 337
  Xia Liu, Fang Wan, Lei Chen, Zhao Qiu, and Ming-rui Chen

Predictive Simulation of Airline Passenger Volume Based
on Three Models.. 350
  Han-Tao Yang and Xia Liu

Research on Monitoring Methods for Electricity Hall Staff Based
on Autonomous Updating and Semi-supervising Model................ 359
  Yao Tang, Zhenjuan Qiao, Rui Zou, Xueming Qiao, Chenglin Liu,
  and Yiliang Wang

Research on Electricity Personnel Apparel Monitoring Model Based
on Auxiliary Categorical-Generative Adversarial Network............. 377
  Xueming Qiao, Yiping Rong, Yanhong Liu, and Ting Jiang

Optimization Method of Suspected Electricity Theft Topic Model Based
on Chi-square Test and Logistic Regression....................... 389
  Jian Dou and Ye Aliaosha

A Comparison Method of Massive Power Consumption Information
Collection Test Data Based on Improved Merkle Tree................ 401
  Enguo Zhu, Fangbin Ye, Jian Dou, and Chaoliang Wang

Towards Realizing Sign Language to Emotional Speech Conversion
by Deep Learning... 416
  Nan Song, Hongwu Yang, and Pengpeng Zhi

Noise-Immune Localization for Mobile Targets in Tunnels via Low-Rank
Matrix Decomposition... 431
  Hong Ji, Pengfei Xu, Jian Ling, Hu Xie, Junfeng Ding, and Qiejun Dai

Passenger Flow Forecast of Sanya Airport Based on ARIMA Model   442
    Yuan-hui Li, Hai-yun Han, Xia Liu, and Chao Li

Comparison of LVQ and BP Neural Network in the Diagnosis of Diabetes
and Retinopathy   455
    Jiarui Si, Yan Zhang, Shuaijun Hu, Li Sun, Shu Li, Hongxi Yang,
    Xiaopei Li, and Yaogang Wang

A Heuristic Indoor Path Planning Method Based on Hierarchical
Indoor Modelling   467
    Jingwen Li, Liqiang Zhang, Qian Zhao, Huiqiang Wang, Hongwu Lv,
    and Guangsheng Feng

Predicting Statutes Based on Causes of Action and Content of Statutes   477
    Zhongyue Li, Chuhan Zhuang, Jidong Ge, Chuanyi Li, Ting Lei,
    Peitang Ling, Mengting He, and Bin Luo

Adaptive Anomaly Detection Strategy Based on Reinforcement Learning   493
    Youchang Xu, Ningjiang Chen, Hanlin Zhang, and Birui Liang

Research on Country Fragility Assessment of Climate Change   505
    Yanwei Qi, Fang Zhang, and Zhizhong Wang

Data Analysis and Quality Management Research on the Integration
of Micro-Lecture-Oriented Design Theory Courses with Maker Practice   516
    Tiejun Zhu

A Cloud-Based Evaluation System for Science-and-Engineering Students   530
    Qian Huang, Feng Ye, Yong Chen, and Peiling Xu

From Small Scale Guerrilla Warfare to a Wide Range of Army Operations
the Development Direction of Software Production and Education   539
    Lei Xu, Huipeng Chen, Hongwei Liu, Yanhang Zhang, and Qing Wang

Gathering Ideas by Exploring Bursting into Sparks Through the Cross–To
Discuss Interdisciplinary Role in Cultivating Students' Innovation   545
    Lei Xu, Lili Zhang, Yanhang Zhang, Hongwei Liu, and Yu Wang

A High Precision and Realtime Physics-Based Hand Interaction
for Virtual Instrument Experiment   552
    Xu Han, Ning Zhou, Xinyan Gao, and Anping He

Application of Project Management in Undergraduates' Innovation
Experiment Teaching   564
    Qing Wang, Huipeng Chen, Hongwei Liu, Lei Xu, and Yanhang Zhang

Exploration and Research on the Training Mode of New Engineering
Talents Under the Background of Big Data   573
   Bing Zhao, Jie Yang, Dongxiang Ma, and Jie Zhu

An Empirical Study on the Influence Factors of Mobile Phone Dependence
of College Students Based on SPSS and AMOS   581
   Zhi-peng Ou and Xia Liu

The Reliability and Validity Analysis of Questionnaire Survey
on the Mathematics Teaching Quality in Higher Vocational Colleges.   594
   Yuan-hui Li, Xia Liu, and Hai-yun Han

Online Education Resource Evaluation Systems Based on MOOCs   605
   Yan Zhang and Han Cao

Analysis on Psychological Health Education of Graduate Students
from the Strengths Perspective   616
   Xiaoli Liu

Design and Implement of International Students' Management and Security
Warning System Based on B/S Architecture   623
   Yulu Zhang, Zhikun Li, Ya Wen, Jifu Wang, and Ruigai Li

Performance Prediction Based on Analysis of Learning Behavior   632
   Shaowei Sun, Xiaojie Qian, Lingling Mu, Hongying Zan,
   and Qing Zhang

Author Index   645

Contents – Part I

Development of Scientific Research Management in Big Data Era   1
 Bin Wang and Zhaowen Liu

The Competence of Volunteer Computing for MapReduce
Big Data Applications   8
 Wei Li and William Guo

Research on the Security Protection Scheme for Container-Based
Cloud Platform Node Based on BlockChain Technology   24
 Xiaolan Xie, Tao Huang, and Zhihong Guo

SeCEE: Edge Environment Data Sharing and Processing Framework
with Service Composition   33
 Yasu Zhang, Haiquan Wang, Jiejie Zhao, and Bo An

Research on Pricing Model of Offline Crowdsourcing Based
on Dynamic Quota   48
 Lu Yuan, Yan Zhou, Jia-run Fu, Ling-yu Yan, and Chun-zhi Wang

Research on Hybrid Data Verification Method for Educational Data   60
 Lin Dong, Xinhong Hei, Xiaojiao Liu, Ping He, and Bin Wang

Efficient User Preferences-Based Top-k Skyline Using MapReduce   74
 Linlin Ding, Xiao Zhang, Mingxin Sun, Aili Liu, and Baoyan Song

An Importance-and-Semantics-Aware Approach for Entity
Resolution Using MLP   88
 Yaoli Xu, Zhanhuai Li, and Wanhua Qi

Integration of Big Data: A Survey   101
 Jingya Hui, Lingli Li, and Zhaogong Zhang

Scene-Based Big Data Quality Management Framework   122
 Xinhua Dong, Heng He, Chao Li, Yongchuan Liu, and Houbo Xiong

The Construction Approach of Statutes Database   140
 Linxia Yao, Haojie Huang, Jidong Ge, Simeng Zhao, Peitang Ling,
 Ting Lei, Mengting He, and Bin Luo

Weighted Clustering Coefficients Based Feature Extraction and Selection
for Collaboration Relation Prediction   151
 Jiehua Wu

A Representation-Based Pseudo Nearest Neighbor Classifier   165
  Yanwei Qi

Research on Network Intrusion Data Based on KNN and Feature
Extraction Algorithm.   182
  Shuai Dong and Xingang Wang

PSHCAR: A Position-Irrelevant Scene-Aware Human Complex
Activities Recognizing Algorithm on Mobile Phones.   192
  Boxuan Jia, Jinbao Li, and Hui Xu

Visual-Based Character Embedding via Principal Component Analysis   212
  Linchao He, Dejun Zhang, Long Tian, Fei Han, Mengting Luo,
  Yilin Chen, and Yiqi Wu

A Novel Experience-Based Exploration Method for Q-Learning   225
  Bohong Yang, Hong Lu, Baogen Li, Zheng Zhang, and Wenqiang Zhang

Overlapping Community Detection Based on Community Connection
Similarity of Maximum Clique   241
  Xiaodong Qian, Lei Yang, and Jinhao Fang

Heterogeneous Network Community Detection Algorithm Based
on Maximum Bipartite Clique.   253
  Xiaodong Qian, Lei Yang, and Jinhao Fang

Novel Algorithm for Mining Frequent Patterns of Moving Objects Based
on Dictionary Tree Improvement.   269
  Yi Chen, Yulan Dong, and Dechang Pi

MalCommunity: A Graph-Based Evaluation Model for Malware
Family Clustering   279
  Yihang Chen, Fudong Liu, Zheng Shan, and Guanghui Liang

Negative Influence Maximization in Social Networks   298
  Jinghua Zhu, Bochong Li, Yuekai Zhang, and Yaqiong Li

Mining Correlation Relationship of Users from Trajectory Data   308
  Zi Yang and Bo Ning

Context-Aware Network Embedding via Variation Autoencoders
for Link Prediction   322
  Long Tian, Dejun Zhang, Fei Han, Mingbo Hong, Xiang Huang,
  Yilin Chen, and Yiqi Wu

SFSC: Segment Feature Sampling Classifier for Time Series Classification. . . .   332
  Fanshan Meng, Tianbai Yue, Hongzhi Wang, Hong Gao, and Yaping Li

Fuzzy C-Mean Clustering Based: LEO Satellite Handover   347
  Syed Umer Bukhari, Liwei Yu, Xiao qiang Di, Chunyi Chen, and Xu Liu

An Improved *Apriori* Algorithm Based on Matrix and Double Correlation
Profit Constraint   359
  Yuan Liu, Ya Li, Jian Yang, Yan Ren, Guoqiang Sun, and Quansheng Li

Mining and Ranking Important Nodes in Complex Network by K-Shell
and Degree Difference   371
  Jianpei Zhang, Hui Xu, Jing Yang, and Lijun Lun

Representation Learning for Knowledge Graph with Dynamic Step   382
  Yongfang Li, Liang Chang, Guanjun Rao, Phatpicha Yochum,
  Yiqin Luo, and Tianlong Gu

An Improved K-Means Parallel Algorithm Based on Cloud Computing   394
  Xiaofeng Li and Dong Li

Statistical Learning-Based Prediction of Execution Time of Data-Intensive
Program Under Hadoop2.0   403
  Haoran Zhang, Jianzhong Li, and Hongzhi Wang

Scheme of Cloud Desktop Based on Citrix   415
  Xia Liu, Xu-lun Huo, Zhao Qiu, and Ming-rui Chen

A Constraint-Based Model for Virtual Machine Data Access Control
in Cloud Platform   426
  Zhixin Li, Lei Liu, and Xin Wang

Improved DES on Heterogeneous Multi-core Architecture   444
  Zhenshan Bao, Chong Chen, and Wenbo Zhang

Task Scheduling of Data-Parallel Applications on HSA Platform   452
  Zhenshan Bao, Chong Chen, and Wenbo Zhang

Dual-Scheme Block Management to Trade Off Storage Overhead,
Performance and Reliability   462
  Ruini Xue, Zhongyang Guan, Zhibin Dong, and Wei Su

Cooperation Mechanism Design in Cloud Manufacturing Under
Information Asymmetry   477
  Haidong Yu and Qihua Tian

A Scheduling Algorithm Based on User Satisfaction Degree
in Cloud Environment   484
  Feng Ye, Yong Chen, and Qian Huang

E-CAT: Evaluating Crowdsourced Android Testing   493
  Hao Lian, Zemin Qin, Hangcheng Song, and Tieke He

Dual-Issue CGRA for DAG Acceleration   505
  Li Zhou, Jianfeng Zhang, and Hengzhu Liu

Interruptible Load Management Strategy Based on Chamberlain Model   512
  Zhaoyuan Xie, Xiujuan Li, Tao Xu, Minghao Li, Wendong Deng,
  and Bo Gu

A Method to Identify Spark Important Parameters Based
on Machine Learning   525
  Tianyu Li, Shengfei Shi, Jizhou Luo, and Hongzhi Wang

Design and Implementation of Dynamic Memory Allocation Algorithm
in Embedded Real-Time System   539
  Xiaohui Cheng, Yelei Guan, and Yi Zhang

A Heterogeneous Cluster Multi-resource Fair Scheduling Algorithm Based
on Machine Learning   548
  Wenbin Liu, Ningjiang Chen, Hua Li, Yusi Tang, and Birui Liang

A Network Visualization System for Anomaly Detection
and Attack Tracing   560
  Xin Fan, Wenjie Luo, Xiaoju Dong, and Rui Su

Opportunistic Concurrency Transmission MAC Protocol Based
on Geographic Location Information   575
  Jianfeng Wang, Dongjia Zhang, Haomin Zhan, Zhen Cao,
  and Hongbin Wang

Multi-channel Parallel Negotiation MAC Protocol Based on Geographic
Location Information   589
  Jianfeng Wang, Hongbin Wang, Haomin Zhan, Rouwen Dang,
  and Yang Bai

PBSVis: A Visual System for Studying Behavior Patterns of Pseudo
Base Stations   599
  Haocheng Zhang, Xiang Tang, Chenglu Li, Yiming Bian, Xiaoju Dong,
  and Xin Fan

C2C E-commerce Credit Model Research Based on IDS System   611
  Xiaotang Li

An Evolutionary Energy Prediction Model for Solar Energy-Harvesting
Wireless Sensor Networks   619
  Guangya Yang, Xue Hu, and Xiuying Chen

A Cooperative Indoor Localization Method Based on Spatial Analysis   628
  Qian Zhao, Yang Liu, Huiqiang Wang, Hongwu Lv, Guangsheng Feng,
  and Mao Tang

Phishing Detection Research Based on LSTM Recurrent Neural Network . . .   638
    Wenwu Chen, Wei Zhang, and Yang Su

Performance Evaluation of Queuing Management Algorithms in Hybrid
Wireless Ad-Hoc Network   646
    Ertshag Hamza, Honge Ren, Elmustafa Sayed, and Xiaolong Zhu

Selection of Wavelet Basis for Compression of Spatial Remote
Sensing Image   656
    Meishan Li, Jiamei Xue, and Hong Zhang

Recognition of Tunnel Cracks Based on Deep Convolutional Neural
Network Classifier.   666
    Min Yang, Qing Song, Xueshi Xin, and Lu Yang

Quality of Geographical Information Services Evaluation Based
on Order-Relation   679
    Yi Cheng, Wen Ge, and Li Xu

High Precision Self-learning Hashing for Image Retrieval   689
    Jia-run Fu, Ling-yu Yan, Lu Yuan, Yan Zhou, Hong-xin Zhang,
    and Chun-zhi Wang

Face Detection and Recognition Based on Deep Learning
in the Monitoring Environment.   698
    Chaoping Zhu and Yi Yang

Localization and Recognition of Single Particle Image in Microscopy
Micrographs Based on Region Based Convolutional Neural Networks   706
    Fang Zheng, FuChuan Ni, and Liang Zhao

A Novel Airplane Detection Algorithm Based on Deep CNN   721
    Ying Wang, Aili Wang, and Changyu Hu

Object Tracking Based on Hierarchical Convolutional Features.   729
    Aili Wang, Haiyang Liu, Yushi Chen, and Yuji Iwahori

A Volleyball Movement Trajectory Tracking Method Adapting
to Occlusion Scenes   738
    Ting Yu, Zeyu Hu, Xinyu Liu, Pengyuan Jiang, Jun Xie,
    and Tianlei Zang

Author Index   751

Classifying DNA Methylation Imbalance Data in Cancer Risk Prediction Using SMOTE and Tomek Link Methods

Chao Liu[1] , Jia Wu[1], Labrador Mirador[1], Yang Song[2],
and Weiyan Hou[1(✉)]

[1] School of Information Engineering,
Zhengzhou University, Zhengzhou 450001, China
houwy@zzu.edu.cn
[2] School of Mechatronic Engineering and Automation,
Shanghai University, Shanghai 200072, China

Abstract. Recent study shows that DNA methylation (DM) as a better bio-marker and help in improving the dichotomous outcome (tumor/normal) based on several features. Over the past years, rapid advances in next-generation sequence technology had led to the timely advent of The Cancer Genome Atlas (TCGA) project which provides the most comprehensive genomic data for various kinds of Cancer. However, TCGA data is faced with the problem of class imbalance and of high data dimensionality leading to an increase in the false negative rate. In this paper, uses Synthetic Minority Oversampling Technique (SMOTE) algorithm in the pre-processing phase as a method to maintain a balanced class distribution. SMOTE is combined with the Tomek Link (T-Link) under-sampling technique for data cleaning and removing noise. To reduce the feature space of the data only those genes for which mutations have been causally implicated in cancer is considered. These are obtained through resources like Catalogue of Somatic Mutations in Cancer (COSMIC) and Clinical Interpretation of Variants in Cancer (CIViC). Classification of patient samples is then performed utilizing several machine learning algorithms of Logistic Regression, Random Forest and Gaussian Naive Bayes. Each classifier performance is evaluated using appropriate performance measures. The methodology is applied on the TCGA DNA Methylation data for 28 various cancer types, which demonstrated a superior performance in class of patient samples.

Keywords: DNA methylation · TCGA · SMOTE · T-Link · Random Forest

1 Introduction

Classification modeling as being defined is process of learning a target function based on a trained data. It is one of the most important tasks in the machine learning and data-mining. Classification modeling generates less error when being applied to data's not previously seen. Imbalanced data are the most important issue in all applications in the real world. However, classification accuracy based on minority class often results to a

© Springer Nature Singapore Pte Ltd. 2018
Q. Zhou et al. (Eds.): ICPCSEE 2018, CCIS 902, pp. 1–9, 2018.
https://doi.org/10.1007/978-981-13-2206-8_1

classifier that is more biased towards the majority class. Therefore, enhancing the classification precision of a minority class is very significant to undertake. Imbalance class is the main challenge that influences the classification of the medical data. For this reason, new techniques and methods for dealing with class imbalance have been proposed. These techniques can be classified into three methods: those that amend the data distribution by resampling techniques (data level methods); those at the level of the learning algorithm which adapt a base class to deal with class imbalance (algorithm level methods); and those at the features selection level which find an optimal feature among the whole the features.

This paper proposes a combined solution to classify TCGA DM imbalanced data. The proposed solution is projected to successfully reduce dimensionality and balances the minority class using a combination of Synthetic Minority Oversampling Techniques (SMOTE) and Tomek Link (T-Link) sampling method. The combined used of the two sampling techniques is the key innovative methodological aspect of the study.

The remainder of this paper is organized as follows: Sect. 2 introduces the data being utilized, the proposed model, and the description of an algorithm training procedure as well as the performance measurements; Sect. 3 presents the demonstrated results; and Sect. 4 is the conclusion and future work plan.

2 Methodology

The paper uses Synthetic Minority Oversampling Technique (SMOTE) algorithm in the pre-processing phase as a method to maintain a balanced class distribution. SMOTE is combined with the T-Link under-sampling technique for data cleaning in order to remove noise. To reduce the feature space of the data, only those genes for which mutations have been causally implicated in cancer are considered. Data are obtained through the COSMIC and the CIViC online database resources. Classification of patient samples is undertaken with the use of several machine learning algorithms such as Logistic Regression, Random Forest and Gaussian Naive Bayes. Performance of each classifier is being evaluated with the use of an appropriate measures. This method is applied on the TCGA DNA methylation data for 28 various cancer types. DNA methylation are from site https://portal.gdc.cancer.gov/repository. Figure 1 highlights the Workflow chart of data generation and analysis.

2.1   Pre-processing

Data. The TCGA project publishes DNA methylation profile data for 28 cancer types. Raw data ($0 \leq x \leq 1$) is available online and mapped to a specific data location or range via the TCGA Data Portal (e.g. chr17: 17033575 indicates position 17033575 on chromosome 17). Since the raw data from TCGA was inherently complex, data preprocessing by Broad Institute's FireBrowse was used.

Unlike TCGA data, FireBrowse [4] maps values to specific human genes based on HGNC nomenclature notes [5]. Mapping makes feature selection easier, as being discussed in the next section. Each sample file is annotated with the TCGA identifier value [6], which indicates whether the sample is tumor or normal (e.g. TCGA -2F -A9KW-01:

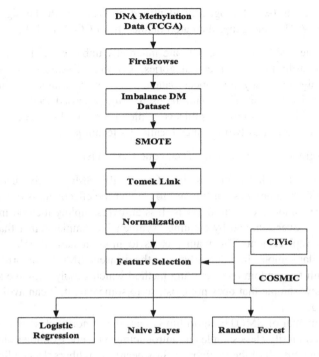

Fig. 1. Workflow chart of data generation and analysis

Table 1. Statistics of TCGA DM Data used in the paper

Tumor type	Abbrev.	# Patients	Tumor - 1	Normal - 0
Breast Invasive Carcinoma	BRCA	886	789	97
Lung Adenocarcinoma	LUAD	493	461	32
Urothelial Bladder Carcinoma	BLCA	433	412	21
Prostate Adenocarcinoma	PRAD	548	498	50
Lung Squamous Cell Carcinoma	LUSC	414	372	42
⋮	⋮	⋮	⋮	⋮
Thyroid Cancer	THCA	563	507	56
Head-Neck Squamous Cell Carcinoma	HNSC	578	528	50

tumor type 01 - 09 (Class 1), normal type 10 - 19 (Class 0). The statistical data for the 28 tumor types considered on this study are shown in Table 1.

Feature Selection. The DNA methylation data contained in various types of cancer in TCGA has a characteristic variable of more than 20,000 protein-coding genes. On this case, feature selection is found to be important, however the study limits only to those information that has been biologically identified as Genes with the significance of cancer mutations. The said genes data are obtained from COSMIC [7] and CIViC [8].

Sampling. Table 1 shows that the data are inherently unbalanced. This is due to the non-uniform distribution of target categories. Current classification methods can achieve very high accuracy for cancer patients. However, results will indicate less sensitive to normal subjects. Thus, with the used of the proposed method, a model that maximizes sensitivity to normal subjects while achieving high accuracy can be achieved. The method uses two advanced sampling techniques.

1. Synthetic Minority Oversampling Technique (SMOTE)

The SMOTE algorithm is proposed by Chawla [9]. SMOTE (Synthetic Minority Oversampling Technique) was put forward to counter the effectiveness of having a few instances of the minority class in datasets. It is an oversampling method in which the minority class is over-sampled by creating "synthetic" examples rather than by over-sampling with replacement. The main idea is to insert a new sample into a small number of similar samples in order to balance the datasets. SMOTE algorithm is not in accordance with the random oversampling method which simply copy the sample, but which add a new sample that does not exist, so to some extent, it can avoid excessive classifier fitting.

The fundamental of SMOTE algorithm is that, the minority class is over-sampled by taking each minority class sample and introducing synthetic examples along the line segments joining any/all of the k minority class nearest neighbors. Depending upon the amount of over-sampling required, neighbors from the k nearest neighbors are randomly chosen. Synthetic samples are generated in the following way: Take the difference between the feature vector (sample) under consideration and its nearest neighbor. Multiply this difference by a random number between 0 and 1, and add it to the feature vector under consideration. This process causes the selection of a random point along the line segment between two specific features. This approach effectively forces the decision region of the minority class to become more general. The basic principle of SMOTE algorithm is shown in Table 2.

$$p_i = x + rand(0, 1) * (y_i - x), i = 1, 2, \ldots, N \tag{1}$$

Where p_i is new minority sample, x is sample, and y_i is randomly selected k nearest neighbors.

2. Tomek Link (T-Link)

Tomek link is a data cleaning technique proposed by Ivan Tomek [10]. A Tomek link is defined as a pair of minimally Euclidian distanced neighbors (x_i, x_j) with x_i belonging to the minority class and x_j to the majority class. Let $d(x_i, x_j)$ denote the Euclidian distance between x_i and x_j. If there is no sample x_k satisfies the following condition:$d(x_i, x_k) < d(x_i, x_j)$ or $d(x_j, x_k) < d(x_i, x_j)$ then the pair of (x_i, x_j), is a Tomek link.

Table 2. Algorithm SMOTE

Algorithm SMOTE:
For each point p in S:
1. Compute its k nearest neighbors in S.
2. Randomly choose r ⩽ k of the neighbors (with replacement)
3. Choose a random point along the lines joining p and each of the r selected neighbors.
4. Add these synthetic points to the dataset with class S.

2.2 Classification

The core aspect of the study is problem classification. For this case, three different supervised learning methods for comparison have been used.

1. Gaussian Naive Bayes (GNB)

   Since the data are continuous values as feature space, each feature space belongs to the category $k \in \{0, 1\}$, so they are assumed to be distributed according to the given Gaussian distribution [12],

$$p(x_j = x | y_j = k) = \frac{1}{\sqrt{2\pi \sigma_{jk}^2}} \exp\left[-\frac{1}{2}\left(\frac{x - \mu_{jk}}{\sigma_j k}\right)^2\right] \tag{2}$$

2. Logistic Regression (LR)

   LR [13] finds the best fitting model to describe the dichotomous characteristic of the classification problem and the set of feature variables. Formally, model logistic regression models are,

$$\log \frac{p(x)}{1 - p(x)} = \beta_0 + x.\beta \tag{3}$$

3. Random Forest (RF)

   Random Forest classifier [14] was trained with decision trees constructed by random sampling over the reduced feature space. The prediction on a new sample is done by majority voting over every decision tree constructed.

2.3 Comparison

The comparison phase involves the comparison of the model sensitivity/recall towards the minority class using different machine learning algorithms as indicated in the above section. The model performance was also evaluated using measures such as precision and F-1 score. Following is a description of each of the measure.

- Sensitivity/Recall - Commonly referred to as the True Positive Rate. It measures the proportion of positives that are correctly identified.

$$Sensitivity = \frac{True\ Positive}{True\ Postive\ +\ False\ Negative} \tag{4}$$

- Precision - Positive predictive value.

$$Precision = \frac{True\ Positive}{True\ Postive\ +\ False\ Positive} \tag{5}$$

- F-1 score - Measure of test's accuracy.

$$F_1 = 2 * \frac{precision\ *\ recall}{precision\ +\ recall} \tag{6}$$

3 Discussion

TCGA DM data for 28 cancer types were downloaded from FireBrowse and the molecular profiling data was parsed in order to retrieve the sample type $y \in \{0, 1\}$. To reduce the feature space of the data, only the genes present in the Census of Cancer gene and the Civic directory were used. Training set: test sets a ratio of 7: 3. Figure 2 shows a two-dimensional PCA map of pre-training data, from which it can be found that the data distribution in the target variable is not balanced.

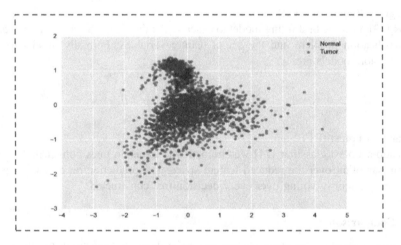

Fig. 2. Class distribution before sampling

To demonstrate the issues of imbalance data the classification methods were initially tested and the evaluation metrics calculation result was shown in Table 3.

Table 3. Evaluation metrics before sampling and normalization

Classification methods	Sensitivity/Recall		Precision		F-1 Score	
	0	1	0	1	0	1
GNB	**0.48**	0.83	0.35	0.99	0.40	0.90
LR	**0.78**	0.99	0.89	0.98	0.83	0.98
RF	**0.79**	1.00	0.87	0.98	0.83	0.99

Table 3 shows that all the three classification methods have a high overall accuracy, but the sensitivity of the Class 0 is worst primarily attributed to the imbalance that exists within the data.

In order to overcome the problem, the SMOTE and the T-Link algorithms were applied which resulted a balance target class distribution. This can be seen in the PCA 2D plots of the sampled data as shown in Fig. 3.

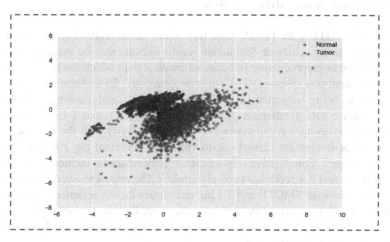

Fig. 3. Class distribution, after sampling and normalization

Table 4 depicts the classification results. The result represents the evaluation metrics obtained by using of only the SMOTE preprocessing method.

Table 4. Evaluation metrics only using SMOTE

Classification methods	Sensitivity/Recall		Precision		F-1 Score	
	0	1	0	1	0	1
GNB	0.80	0.83	0.45	0.99	0.58	0.90
LR	0.88	0.89	0.59	1.00	0.70	0.94
RF	**0.90**	**0.92**	0.84	0.99	0.87	0.95

Note, that data are first normalized, and then the classification methods were reapplied on the sampled data. The resulting metric values are presented in Table 5.

Table 5 clearly shows that feature selection with the used of SMOTE and T-Link sampling algorithms provides better prediction statistics. The results indicate that prior biomedical knowledge and sampling algorithms are essential to build better classifiers.

Table 5. Evaluation metrics after using SMOTE and T-Link

Classification methods	Sensitivity/Recall		Precision		F-1 Score	
	0	1	0	1	0	1
GNB	0.90	0.89	0.45	0.99	0.60	0.94
LR	0.93	0.98	0.59	1.00	0.72	0.99
RF	**0.93**	**0.99**	0.84	0.99	0.88	0.99

4 Conclusion and Future Work

The used of SMOTE with Tomek Link techniques in preprocessing the imbalanced medical data is being achieved. Simulation results indicate that the methods employed is very effective in preprocessing imbalanced medical data which leads to an effective prediction of common diseases. The study preprocessed 28 imbalanced type cancer data which are being subjected to SMOTE and T-Link analysis which results to a relatively balance data. Evaluation of the combined SMOTE and T-Link algorithm is being realized by using a comparative performance analysis with the used of 3 other classifiers. Experiments are carried out to show the effects of the combined preprocessing technique. Comparative classification performances including Sensitivity/recall, Precision and F-1 score has been undertaken. The results showed that evaluation metrics with combined SMOTE and T-Link techniques improves better as compared to other classifier without any preprocessing procedure. Further studies using advanced classification techniques of neural and deep learning methods which will farther improve the system are suggested.

References

1. Hao, X., Luo, H., Krawczyk, M., et al.: DNA methylation markers for diagnosis and prognosis of common cancers. PNAS **114**(28), 7414–7419 (2017)
2. Kursa, M.B.: Robustness of Random Forest-based gene selection methods. BMC Bioinformatics **15**(1), 8 (2014)
3. Zeng, H., Gifford, D.K.: Predicting the impact of non-coding variants on DNA methylation. Nucleic Acids Res. **45**(11), e99 (2017)
4. Elhassan, T., Aljurf, M., et al.: Classification of imbalance data using Tomek Link (T-Link) combined with random under-sampling (RUS) as a data reduction method. J. Inform. Data Min. **1**(2), 1–12 (2016)
5. HGNC Database of Human Gene Names, HUGO Gene Nomenclature Committee. https://www.genenames.org. Accessed 6 Mar 2018
6. Home, NCI Genomic Data Commons. https://portal.gdc.cancer.gov/. Accessed 7 Mar 2018

7. Forbes, S.A., Beare, D., et al.: COSMIC: exploring the world's knowledge of somatic mutations in human cancer. Nucleic Acids Res. **43**, D805–D811 (2015)
8. Griffith, M., Spies, N.C., et al.: CIViC: a knowledge base for expert-crowdsourcing the clinical interpretation of variants in cancer. bioRxiv (2016)
9. Blagus, R., Lusa, L.: SMOTE for high-dimensional class-imbalanced data. BMC Bioinform. **14**(1), 1–16 (2013)
10. Angermueller, C., Lee, H.J., Reik, W., et al.: DeepCpG: accurate prediction of single-cell DNA methylation states using deep learning. Genome Biol. **18**, 67 (2017)
11. Xu, R.H., Wei, W., Krawczyk, M., et al.: Circulating tumour DNA methylation markers for diagnosis and prognosis of hepatocellular carcinoma. Nat. Mater. **16**(11), 1155–1161 (2017)
12. Li, Y., Luo, Z.G., Guan, N.Y., et al.: Applications of deep learning in biological and medical data analysis. Prog. Biochem. Biophys. **43**(5), 472–483 (2016)
13. Hoadley, K.A., Yau, C., Wolf, D.M., et al.: Multiplatform analysis of 12 cancer types reveals molecular classification within and across tissues of origin. Cell **158**(4), 929–944 (2014)
14. Zhu, M., Xia, J., Jin, X.Q., et al.: Class weights random forest algorithm for processing class imbalanced medical data. IEEE J. Mag. **6**, 4641–4652 (2018)

Auxiliary Disease and Treatment System of Aortic Disease Based on Mixed Reality

Zishan Qiu[1], Jian Zhang[2], and Hui Gao[3(✉)]

[1] Harbin Normal University High School, Harbin, China
249600398@qq.com
[2] Northeast Forestry University, Harbin, China
107099694@qq.com
[3] Harbin Huade University, Harbin, China
44117252@qq.com

Abstract. With the development of science, intelligent medical technology is constantly improving. In the past, doctors were unable to perform an operation that could not correspond to the actual anatomy of the 3D and the exact anatomy of the patient. Based on the 3D reconstruction process from lesions, we combine VR and actual scene to discuss the application of mixed reality technology in the process of diagnosis and treatment of aortic disease according to the case of the auxiliary diagnosis and treatment of aortic disease [1]. Practice has proved that doctors who used our technology involved in the cross space real-time interaction with remote operation. This technology can provide a simple optimal solution for China's medical conjoined and grading treatment system with a high efficient and low cost [2].

Keywords: Mixed reality · 3D reconstruction · Aortic disease
Auxiliary disease and treatment

1 Introduction

In addition to coronary and peripheral artery diseases, aortic diseases contribute to the wide spectrum of arterial diseases: aortic aneurysms, acute aortic syndromes (AAS) including aortic dissection (AD), intramural haematoma (IMH), penetrating atherosclerotic ulcer (PAU) and traumatic aortic injury (TAI), pseudoaneurysm, aortic rupture, atherosclerotic and inflammatory affections, as well as genetic diseases (e.g. Marfan syndrome) and congenital abnormalities including the coarctation of the aorta (CoA). Similarly to other arterial diseases, aortic diseases may be diagnosed after a long period of subclinical development or they may have an acute presentation. Acute aortic syndrome is often the first sign of the disease, which needs rapid diagnosis and decision making to reduce the extremely poor prognosis [3] (Fig. 1).

Abdominal ultrasound remains the mainstay imaging modality for abdominal aortic diseases because of its ability to accurately measure the aortic size, to detect wall lesions such as mural thrombus or plaques, and because of its wide availability, painlessness, and low cost. Duplex ultrasound provides additional information on aortic flow. In the acute setting, MRI is limited because it is less accessible, it is more difficult

© Springer Nature Singapore Pte Ltd. 2018
Q. Zhou et al. (Eds.): ICPCSEE 2018, CCIS 902, pp. 10–16, 2018.
https://doi.org/10.1007/978-981-13-2206-8_2

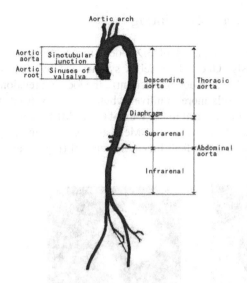

Fig. 1. Segments of the ascending and descending aorta

to monitor unstable patients during imaging, and it has longer acquisition times than CT.79,88 Magnetic resonance imaging does not require ionizing radiation or iodinated contrast and is therefore highly suitable for serial follow-up studies (younger) patients with known aortic disease. Standardized measurements will help to better assess changes in aortic size over time and avoid erroneous findings of arterial growth. Meticulous side-by-side comparisons and measurements of serial examinations (preferably using the same imaging technique and method) are crucial, to exclude random error [4].

The main treatment methods are endovascular treatment and surgical treatment. Thoracic Aorta repair objective is to eliminate the aortic lesion in circulation, such as the false lumen produced after an aneurysm or aortic clip. To prevent further enlargement and ultimately rupture of the aorta by implanting a mulch stent that spans the lesion. The main principle of surgical treatment for ascending aortic aneurysms is to reduce the risk of dissection and rupture by restoring the normal internal diameter of the ascending aorta.

If the proximal end of the aneurysm enlargement exceeds the sinus duct joint and there is a plurality of aortic sinus dilatation, surgery is required according to the aortic valve ring and valve involvement [5]. In the endovascular treatment and surgical treatment, we need to reconstruct 2D image accurate data to 3D model. Through 3D visualization technology, 3D printing technology and holographic imaging technology can help doctors to customize the operation of personalized surgery. It still shorten the operation time and reduce the risk of surgery [6].

2   3D Modeling and 3D Printing

In recent years, the development of 3D model reconstruction technology has been improved continuously. One 3D model is synthesized by multi-layer images. It is placed in the tablet PC to help doctors recognize the location relationship of lesions and tissues, organs and vessels more intuitively, but it is still viewed in two-dimensional display. In the study, the medical image data of CT or MRI are obtained and we get the 3D model by using the "the software of Medical diagnostic image processing", which is developed by TuoMeng Medical Image System [7] (Figs. 2 and 3).

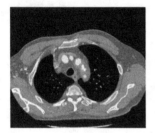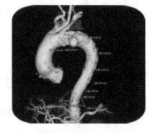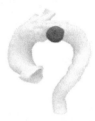

Fig. 2. Aortic aneurysm

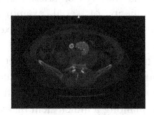

Fig. 3. Aortic dissection

Full color 3D printing not only can help the doctor to observe the situation of the focus and surrounding tissue more intuitively, but help the doctor to design the operation plan. Ultimately, we regret that it costs at least a few hours. When some patients with aortic diseases occurs, they need surgery immediately but 3D printing is not enough. Besides, full color 3D printing cost is stay at a high cost. The auxiliary diagnosis and treatment system of aortic disease based on mixed reality technology developed by TuoMeng Medical Image System can solve this problem well [8, 9]. Technical flow chart is shown in Fig. 4.

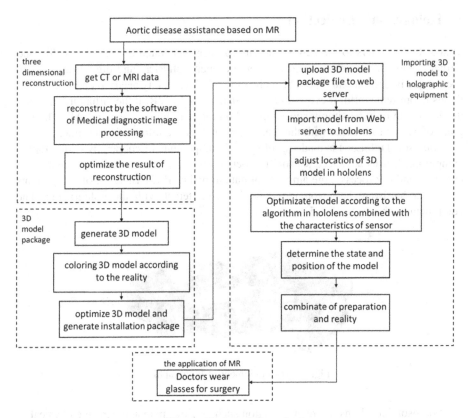

Fig. 4. Technical flow chart

The 3D reconstruction technique is mainly based on the theory of curve and surface, which usually uses the curves represented by parametric equations, such as parametric curves. The parametric curve contains only one independent variable, the coordinate of the point on the curve is the parameter of the curve. If the parameter is T, the equation of the curve as 2-1 follow:

$$\begin{cases} x = x(t) \\ y = y(t), t \in [a, b] \\ z = z(t) \end{cases} \tag{1}$$

Thus, we give the equation a value of T, the coordinates of a point on the curve can be listed. If T is in [a,b], the entire curve can be obtained. A complete curve is composed of a number of sub curves, it may be assumed that the parameter interval of a curve segment is [a,b], the interval [a,b] is normalized to [0,1], the required parameter is converted to $\bar{t} = t - a/b - a$. In general, $P = [x,y,z]^T$, $P(t) = [x(t), y(t), z(t)]^T$, we can get $P = P(t)$ which is the vector form of the curves parameter. In $t \in [0, 1]$, t may represent any amount and it does not have a clear meaning.

3 Holographic Projection

Mixed Reality refers to the new visual environment generated by merging reality and virtual world. In the new visualization environment, physical and digital objects can coexist and interact at real time. The realization of mixed reality needs to interact with all things in our real world. If everything is virtual, we call it VR. If the virtual information displayed can only be simply superimposed on the real things, we call it AR. The key point of MR is to interact with the real world and get the information in time.

Microsoft HoloLens is made up of special components of cooperative holographic computing. It uses advanced sensors to operate optical system in lock step, and make every second to process large amounts of data to make HPU easier. All of components and other hardware help you move freely and interact with the holographic. The equipment is shown in Fig. 5.

Fig. 5. The equipment of holographic

Microsoft HoloLens has many function such as view, map and perceive physical of locations, spaces and objects around you. HD perspective holographic lens use a high-level optical projection system which can generate multi-dimensional full color images with a low latency. It allows you to see holograms in real time. Holographic Processing Unit(HPU) is a custom silicon chip that can handle a lot of data from the sensor per second. Microsoft HoloLens can also understand gestures and human sight, and map the environment around you. All of things are done in real time.

Mixed reality resolves this problem completely. It the 3D model is completely holographic to the real world. This is the real 3D object. You can see different effect from various angles. Figure 6 shows the projection of the aorta by holographic.

The Second Affiliated Hospital of Harbin Medical University used HoloLens in the operation of the aorta. TuoMeng Medical Image System has built a holographic navigation platform for HoloLens. It can cover the 3D model on the patient (HoloLens Superposition Technology) during operation and assist doctors in operation. During the operation, the surgeon used a custom version of the Microsoft HoloLens and checked the 3D model data at any time [10].

In the course of the clinical treatment of the aortic disease, combined with the technique of mixed reality, the 3D reconstruction technique can display the various branches of the blood vessels in a 3D manner. We can see the lesion structure in the lumen. It also makes the pathological structure of the vessel can be seen further clearly and accurately.

Fig. 6. Projection of the aorta by holographic

In the diagnosis and treatment of the aortic disease, with the significant application of this technology, the doctor and patient can understand the 3D state of the arterial system in a multifaceted, multi angle and 3D manner. It provides a morphological basis for the teaching of clinical surgery and basic medicine as well as the construction of a virtual platform for operation. What is more, it is able to provide reliable and objective basis for interventional surgical treatment.

4 Summarize

Not only it achieves the participation of doctors in cross space real-time interaction and remote operation, but also provides an efficient, low-cost and simple optimal solution for our medical association and grading treatment system. The software system can present the structure of the human body by 3D Holographic projection. Surgeons and radiologists can cut virtual structures at any angle from the patient's body [11]. Our study also can help the doctor diagnose disease more quickly and accurately and design a better operation plan.

References

1. Tian, J.-H., Jiang, D.: The reform of medical education in the future based on the development of virtual reality. China Manag. Informationiz. (06), 209–210 (2017)
2. Luo, H.-Y.: Application of VR virtual reality technology in Medical college education. Electron. Technol. Softw. Eng. (04), 10 (2017)
3. Xiao, X.-G.: Application of multi-slice CT scan and volume rendering reconstruction technique in lower abdominal aortic aneurysm. Chin. J. CT MRI (04), 117–119+153
4. Xie, C.-X., Long, T.-H., Zhao, H.-B., Deng, Y.-Y., Liao, M.-Z.: Diagnostic value of multi-slice spiral CT for acute aortic syndrome. Chin. Med. Equip. J. (07), 85–87 (2015)

5. Shun, Q.-L., Wang, Y., Zhao, B.-Y.: Value of MSCTA and CPR in the diagnosis of abdominal aortic aneurysm. Chin. J. Med. Guide (03), 264–265 (2015)
6. Ji, L.-Z., Liu, X.-P., Li, H.-T., Liu, Y.-K., Deng, Q.-C., Yang, J.-L.: The value of multislice spiral CT angiography in the diagnosis of angiogenic acute abdomen. Chin. J. Gen. Pract. (03), 443–445+505 (2015)
7. Yin, L.-L., Pan, Y.-X., Chen, J.-Y., Xie, H., Chen, X.-Y., Li, Y.-C.: Clinical value of dual-source CT angiography in diagnosis and following-up postoperative endovascular stent graft exclusion of abdominal aortic aneurysm. Sichuan Med. J. (02), 162–166 (2015)
8. Zheng, Z.-Y., Ye, Z., Huang, Y.-X., Ye, J.-L., Liu, J.-H., Huang, Y., Wang, K.-K., Zhan, H.: Factors affecting the prognosis of ruptured abdominal aortic aneurysm. Chin. J. Emerg. Med. (11), 1253–1258 (2014)
9. Xia, X.-L., Chen, Y., Qiu, Y.-Y., Yang, X.-J., Zhang, W.-B., Yang, Z.-Y.: Experimental study on application of 3D printing technology to print personalized vertebral body. Chin. J. Bone Joint Inj. (03), 247–250 (2016)
10. Huang, J.-Y., Huang, W., Hunag, F.: Effects of 64 slice spiral CT reconstruction techniques (VR, MIP) and doppler ultrasound in the diagnosis of the degree of internal carotid artery stenosis. Chin. J. CT MRI (12), 19–22 (2017)
11. Cai, F.-W., Hong, W.: Aortic segmentation and three-dimensional reconstruction based on CT images. J. Dongguan Univ. Technol. (05), 40–44 (2017)

An Algorithm for Describing the Convex and Concave Shape of Protein Surface

Wei Wang[1,2(✉)], Keliang Li[1], Hehe Lv[1], Lin Sun[1], Hongjun Zhang[3], Jinling Shi[4], Shiguang Zhang[1], Yun Zhou[1], Yuan Zhao[1], and Jingjing Xv[1]

[1] Department of Computer Science and Technology, College of Computer and Information Engineering, Henan Normal University, Xinxiang 453007, Henan Province, China
weiwang@htu.edu.cn
[2] Laboratory of Computation Intelligence and Information Processing, Engineering Technology Research Center for Computing Intelligence and Data Mining, Xinxiang 453007, Henan Province, China
[3] School of Aviation Engineering, Anyang University, Anyang 455000, Henan Province, China
[4] School of International Education, Xuchang University, Xuchang 461000, Henan Province, China

Abstract. Protein surface plays a key role in many biological processes. Most proteins participate in the life activities of cells via binding to other proteins or ligand molecules. It is an important work to study protein structure and function by analyzing the protein surface shape. Based on the CX algorithm and the 2D fngerprint-base method, we proposed a FCX method to identify the morphology of bulges and depressions on the protein surface. The experimental results show that the FCX algorithm has a more desirable outcome than CX algorithm. The FCX algorithm has a higher correlation with the convex and concave features than CX values with solvent accessibility, solvent accessibility, and B-factor's Pearson correlation coefficient. This result shows that the FCX algorithm can describe the shape of the protein surface residues more accurately than the CX algorithm.

Keywords: Protein three-dimensional structure
Relative solvent accessibility · Solvent accessibility · B-factor

1 Introduction

With the advent of post-genomic era, proteomics has attracted more and more attention of scholars. The study of proteomics includes two aspects, one is the study of the structure and composition of the proteome, and the other is the study of the functional patterns of the proteome [1–3]. With the continuous development of structural determination techniques and high-throughput sequencing technologies, a large amount of protein structure data has been generated, which lays the foundation for the study of proteins from a structural

© Springer Nature Singapore Pte Ltd. 2018
Q. Zhou et al. (Eds.): ICPCSEE 2018, CCIS 902, pp. 17–26, 2018.
https://doi.org/10.1007/978-981-13-2206-8_3

perspective [4,5]. The surface structure of proteins shows different forms, which is very helpful for us to understand the principle of interaction between proteins and ligands. The probability of ligands binding to the groove area on the surface of the protein is greater [6–13]. Analyzing and calculating groove and gap areas on the protein surface is a very important method for identifying the binding sites of small molecules of the ligand.

Shazman and Coleman proposed an algorithm based on the curvature for the description of the region shape of the protein interface [14,15]. While Curvature-based methods are not simple and fast, which is not conducive to rapid modeling during work. Pintar's CX algorithm based on three-dimensional structure data of proteins, it can calculate the numerical measurement of the convex or concave surface of each protein surface [16]. The CX algorithm is very simple and efficient, and is used by many experiments [17–21]. But the amino acid residues on the protein surface are usually very randomly distributed. In our work, we found that the algorithm does not accurately describe the morphology of all surface amino acid residues.

Based on the 2D fingerprint-base method and the CX algorithm, we proposed the FCX algorithm to describe the protein surface bulges and depressions [22,23]. By continuously expanding the sampling range, and averaging the results of the CX algorithm over multiple levels, the results are used as descriptors of protein surface bulges and depressions. We used the CX and FCX calculations for the same dataset separately, and do a correlation analysis with solvent accessibility, relative solvent accessibility, and b-factor. Finally, it was found that FCX showed a higher correlation with these properties, indicating that FCX can better describe the concave and convex morphology of protein surface residues. At the same time, we also extracted the binding sites of the dataset proteins and ligands, and obtained the same results using the two algorithms. This shows that FCX is also applicable to the morphological studies of the amino acid residues on the surface of protein and ligand binding domains.

2 Methods and Materials

2.1 Methods

The CX algorithm is a simple and rapid method to describe the concave and convex morphology of a protein surface. The algorithm uses the non-hydrogen atoms of protein surface residues as the center of the sphere, and then calculates the ratio of the molecular volume of the protein in the sphere range to the remaining volume. The ratio result is CX.

$$CX = V_{ext}/V_{int} \tag{1}$$

V_{int} is the volume of protein atoms inside the sphere, and V_{ext} is the remaining volume in the sphere. CX can characterize the geometry of the local concave and convex of the residue. The larger the CX value is, the more the local morphology of the protein surface tends to bulge, whereas the smaller the CX is, the more it tends to be concave.

The FCX algorithm combines the CX algorithm with the 2D fingerprint-base method. Fingerprint biometric is one of the most successful biometrics applied in both forensic law enforcement and security applications. The algorithm computes sampling for local feature information at multiple levels, which is also suitable for calculating the morphological features of projections and depressions on the protein surface. Firstly, the CX algorithm was used to calculate the cx value of the protein surface residues, and then the constant calculation of the radius R was continued. The shapes of the depressions and bulges around the residues in different ranges were calculated. Finally, an average value is calculated for these CX values. The final result can remove some interference and describe the local geometry of the residue more appropriately. We use the CA atom of the surface residue of the protein as the center of the sphere to calculate the surface morphology inside the multi-scale sphere. Finally, we found that when the FCX algorithm is set to a three-layer sphere and the sphere radius is 8Å, 10Å, and 12Å, respectively, FCX values can more accurately describe the concave and convex morphology of amino acid residues on the protein surface.

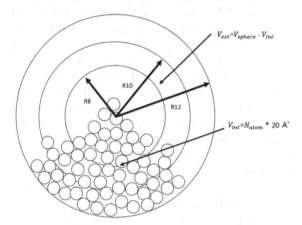

Fig. 1. Schematic diagram of the FCX algorithm. N_{atom} is the number of non-hydrogen atoms found within radius around the amino acid residue CA atom. The default radius of the sphere is set to 8Å, 10Å, 12Å in three cases. The default non-hydrogen atomic volume is 20Å^3. CX is the ratio of the atomic volume V_{int} of the protein to the empty volume V_{ext} in the sphere, and FCX is the average of the CX obtained from the three radius spheres.

FCX algorithm principle showed in Fig. 1. N_{a-i} is the number of non-hydrogen atoms of the protein inside the ith sphere, and the default non-hydrogen atomic volume is 20Å^3. V_{int-i} is the volume of the residue inside the ith sphere, and the remaining volume of the sphere V_{ext-i} is the difference between sphere volume and V_{int-i}. CX value is defined by V_{ext-i}/V_{int-i}. CX_i is the CX calculation result in the ith sphere:

$$V_{int} = N_a * V_a \tag{2}$$

$$CX_i = V_{ext-i}/V_{int-i} \tag{3}$$

$$FCX = (CX_8 + CX_{10} + CX_{12})/3 \tag{4}$$

FCX is the average of the CX values for the three cases with a radius of 8Å, 10Å, and 12Å. FCX value can be more evenly represented the local concave-convex shape of the surface residue.

2.2   Materials

For the surface amino acid residue showed in Fig. 1 that exhibits a convex morphology, the number of atoms around the sphere center is small, and the FCX value will be large. However, for the surface amino acid residues that show depressions, the number of surrounding atoms is large, and the value of FCX is very small. For the depression, the number of atoms around the center of the sphere is large, and the value of FCX is very small. In order to test the FCX algorithm, we downloaded two non-homologous protein datasets of DNA-417 and RNA-282 from the ccPDB database, with a resolution higher than 2.0Å [24]. ccPDB contains datasets that are collected and edited in many documents. It also supports users to create their own datasets. DNA-417 contains structural data for 417 DNA-binding proteins and RNA-282 contains 282 RNA-binding protein structure data. From the two datasets, 40 DNA-binding proteins and 28 RNA-binding proteins were randomly selected to form the BindingPorein-68 dataset. To compare the CX and FCX algorithms, we calculated the solvent accessibility, relative solvent accessibility and B-factor of protein surface amino acid residues. The solvent accessibility and relative solvent accessibility of residues can be calculated by the DSSP program (http://swift.cmbi.ru.nl/gv/dssp/) [25,26]. B-factor value can be obtained from the PDB file. Judging whether the residue is on the surface of the protein with a relative solvent accessibility value of more than 5%. Finally, we performed a correlation analysis of the calculation results and features. It is well known that relative solvent accessibility, solvent accessibility, and b-factor are morphologically related parameters. If the correlation coefficient between the feature and algorithm results is larger, the better the performance of the algorithm.

3   Result and Discussion

In the process of obtaining the optimal parameters of the FCX algorithm, a set of FCX values is obtained by calculating the number of different sphere layers and different radius values. Subsequently, the Pearson correlation coefficient of the FCX value and the solvent accessibility value was calculated, the Pearson

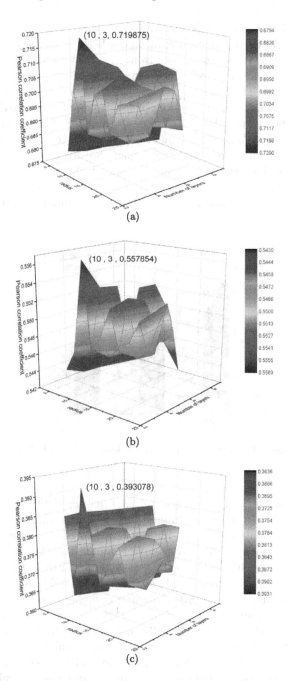

Fig. 2. (a) Number of spheres, average sphere radius, and relative solvent accessibility three-dimensional plot; (b) Number of spheres, average sphere radius, and solvent accessibility three-dimensional plot; (a) Number of spheres, average sphere radius, and B-factor three-dimensional plot. From the three graphs, it can be observed that when the FCX algorithm selects 3 layers and the average radius is 10Å, the correlation coefficient between the calculation results and solvent accessibility, relative solvent accessibility, and B-factor is the largest.

correlation coefficient of the FCX value and the relative solvent accessibility value was calculated, and the Pearson correlation coefficient of the FCX value and the b-factor was calculated. It can be seen from Fig. 2 that with the increase of the sphere radius, the Pearson correlation coefficient also increases. When the radius is 10, the correlation coefficient is the largest, and then decreases as the radius increases. At the same time, the number of spheres and the correlation coefficient also shows a linear relationship. The larger the number of layers, the smaller the correlation coefficient. When the number of layers is 3, the correlation coefficient obtains the maximum value. Therefore, when the statistical layer number is 3 and the sphere radius are 8, 10, and 12 respectively, the FCX algorithm has the highest performance. In this case, the FCX algorithm can more accurately describe the local concave and convex morphology of amino acid residues on the protein surface.

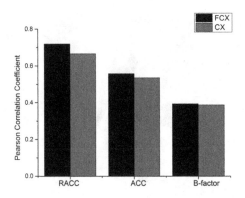

Fig. 3. Comparison of Pearson correlation coefficients between FCX algorithm and CX algorithm

Solvent accessibility and B-factor are two characteristics closely related to morphology [27–29]. In order to compare the FCX algorithm with the CX algorithm. First, we calculated the CX and FCX values of the amino acid residues exposed to the surface of all proteins. We compare the Pearson correlation of the results of the algorithm. The results are shown in Fig. 3. Since the datasets we selected are all nucleic acid-binding proteins, it is also very meaningful to study the surface morphology of the binding domains of proteins. We selected the binding site residues in all proteins and the distance between the residues and the nucleic acid molecule should be less than 3.5Å. For the binding site residues, we also compare the Pearson correlation coefficients of the algorithm results. The result is shown in Fig. 4. The correlation coefficient between the three characteristics of the binding site residues and the CX value is smaller than that of the FCX. From Fig. 5, we can find that the FCX algorithm has a higher correlation between the results of the protein binding domain and the three features. This shows that the FCX algorithm shows more excellent performance in calculating the amino acid depressions and bulges on the surface of

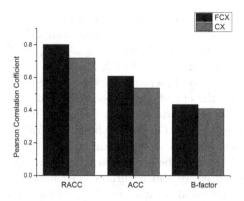

Fig. 4. Comparison of Pearson correlation coefficients of FCX and CX algorithms for binding site residues

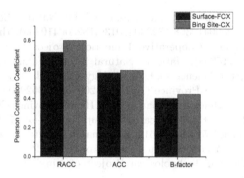

Fig. 5. Comparison of the Pearson correlation coefficients of FCX values of binding site residues and FCX values of all surface residues

the protein binding domain. The FCX algorithm is of great interest for the work focused on studying the morphological structure of the protein binding region. Researchers can use the FCX algorithm to wait for more significant eigenvalues and apply them to build a machine learning algorithm models.

4 Conclusion

In our experiment, we used the dataset of BindingPorein-68 nucleic acid binding protein. First, the relative solvent accessibility values, solvent accessibility values and B-factor values of amino acid residues on the protein surface were determined. The relative solvent accessibility values of amino acid residues on the protein surface, the solvent accessibility value and the B factor value are three characteristics closely related to the protein interface. The projections and depressions of the surface amino acid residues are linearly related to their relative solvent accessibility, and the B-factor of the residue also represents the degree of activity of amino acid residues. Then the correlation coefficients between the

FCX values and the features were analyzed to select the optimal parameters of the FCX. In the second step, CX and FCX values were calculated for all protein surface residue and binding site residues. By analyzing the Pearson correlation coefficient, we found that our proposed FCX algorithm has better performance. In recent years, more ideas and methods have been proposed for protein-ligand research. Among them, the classification model of predicting binding sites was extracted from the structural data to extract effective feature information and made a relatively large progress. On the one hand, the features can be extracted based on a three-dimensional space algorithm, and on the other hand, features can also be extracted through the structural properties of the protein. Finally, the FCX algorithm can easily and accurately calculate the concave and convex morphology of amino acid residues on the protein surface. The algorithm helps to quickly build models in research processes such as proteins and proteins, proteins and drugs, and proteins and nucleic acids.

Acknowledgements. This work was supported by Natural Science Foundation of Henan province (182300410368,182300410130,182300410306), the Production and Learning Cooperation and Cooperative Education Project of Ministry of Education of China (201702115008), National Natural Science Foundation of China (No. 61772176, 61402153), the Science and Technology Research Key Project of Educational Department of Henan Province (No. 16A520016, 17B520002, 17B520036), Key Project of Science and Technology Department of Henan Province (No. 1821022 10208, 142102210056, 17B520002), Ph.D. Research Startup Foundation of Henan Normal University (Nos. qd15130, qd15132, qd15129), China Postdoctoral Science Foundation (No. 2016M602247), the Young Scholar Program of Henan Province (2017GGJS041), and the Key Scientific and Technological Project of Xinxiang City Of China (No. CXGG17002).

References

1. Pandey, A., Mann, M.: Proteomics to study genes and genomes. Nature. **405**(6788), 837–846 (2000)
2. Fleming, K., Kelley, L.A., Islam, S.A., MacCallum, R.M., et al.: The proteome: structure, function and evolution. Philos. Trans. R. Soc. London. **361**(1467), 441–451 (2006)
3. Thul, P.J., Åkesson, L., Wiking, M., et al.: A subcellular map of the human proteome. Science. **356**(6340) (2017)
4. Burley, S.K., Berman, H.M., Kleywegt, G.J., et al.: Protein Data Bank (PDB): the single global macromolecular structure archive. Methods Mol. Biol. **1607**, 627 (2017)
5. Biasini, M., Bienert, S., Waterhouse, A.: SWISS-MODEL: modelling protein tertiary and quaternary structure using evolutionary information. Nucleic Acids Res. **42**, W252 (2014)
6. Jorgensen, W.L.: Rusting of the lock and key model for protein-ligand binding. Science **254**(5034), 954 (1991)
7. Morrison, J.L., Breitling, R., Higham, D.J., et al.: A lock-and-key model for protein-protein interactions. Bioinformatics **22**(16), 2012 (2006)

8. Mullard, A.: Protein-protein interaction inhibitors get into the groove. Nat. Rev. Drug Discov. **11**(3), 173–5 (2012)
9. Chen, L., Frankel, A.D.: A peptide interaction in the major groove of RNA resembles protein interactions in the minor groove of DNA. Proc. Nat. Acad. Sci. USA **92**(11), 5077 (1995)
10. Spola Jr., R.S.: Coupling of local folding to site-specific binding of proteins to DNA. Science **263**(5148), 777–84 (1994)
11. Wang, W., Liu, J., Sun, L.: Surface shapes and surrounding environment analysis of single- and double-stranded DNA-binding proteins in protein-DNA interface. Proteins **84**(7), 979–989 (2016)
12. Wang, W., Liu, J., Zhou, X.: Identification of single-stranded and double-stranded DNA binding proteins based on protein structure. BMC Bioinf. **15**(S12), S4 (2014)
13. Wang, W., Sun, L., Zhang, S., et al.: Analysis and prediction of single-stranded and double-stranded DNA binding proteins based on protein sequences. BMC Bioinf. **18**(1), 300 (2017)
14. Coleman, R.G., Burr, M.A., Souvaine, D.L., et al.: An intuitive approach to measuring protein surface curvature. Proteins **61**(4), 1068–1074 (2005)
15. Shazman, S., Elber, G., Mandel-Gutfreund, Y.: From face to interface recognition: a differential geometric approach to distinguish DNA from RNA binding surfaces. Nucleic Acids Res. **39**(17), 7390–9 (2011)
16. Iwakiri, J., Tateishi, H., Chakraborty, A., et al.: Dissecting the protein-RNA interface: the role of protein surface shapes and RNA secondary structures in protein-RNA recognition. Nucleic Acids Res. **40**(8), 3299 (2012)
17. Albou, L.P., Schwarz, B., Poch, O., et al.: Defining and characterizing protein surface using alpha shapes. Proteins **76**(1), 1–12 (2009)
18. Kraynov, V., Knudsen, N., Hays Putnam, A.M.A., et al.: Modified interferon beta polypeptides and their uses. US. 8329869 B2[P] (2012)
19. Xia, J.F., Zhao, X.M., Song, J., et al.: APIS: accurate prediction of hot spots in protein interfaces by combining protrusion index with solvent accessibility. BMC Bioinf. **11**(1), 174 (2010)
20. Sikić, M., Tomić, S., Vlahovicek, K.: Prediction of protein-protein interaction sites in sequences and 3D structures by random forests. PLoS Comput. Biol. **5**(1), e1000278 (2009)
21. Wang, W., Liu, J., Sun, L.: Surface shapes and surrounding environment analysis of single-and double-stranded DNA-binding proteins in protein-DNA interface. Proteins **84**(7), 979–989 (2016)
22. Pintar, A., Carugo, O., Pongor, S.: CX, an algorithm that identifies protruding atoms in proteins. Bioinformatics **18**(7), 980–984 (2002)
23. Duan, J., Dixon, S.L., Lowrie, J.F., et al.: Analysis and comparison of 2D fingerprints: Insights into database screening performance using eight fingerprint methods. J. Mol. Graph. Model. **29**(2), 157–170 (2010)
24. Harinder, S., Singh, C.J., Michael, G.M., et al.: ccPDB: compilation and creation of data sets from Protein Data Bank. Nucleic Acids Res. **40**, 486–9 (2012)
25. Carter, P., Andersen, C.A., Rost, B.: DSSPcont: continuous secondary structure assignments for proteins. Nucleic Acids Res. **31**(13), 3293–3295 (2003)
26. Touw, W.G., Baakman, C., Black, J., et al.: A series of PDB-related databanks for everyday needs. Nucleic Acids Res. **43**, 364–8 (2015)
27. Kim, H., Park, H.: Prediction of protein relative solvent accessibility with support vector machines and long-range interaction 3D local descriptor. Proteins **54**(3), 557 (2004)

28. Janin, J., Bahadur, R.P.: Relating macromolecular function and association: the structural basis of protein-DNA and RNA recognition. Cell. Mol. Bioeng. **1**(4), 327–338 (2008)
29. Yang, J., Wang, Y., Zhang, Y.: ResQ: an approach to unified estimation of B-factor and residue-specific error in protein structure prediction. J. Mol. Biol. **428**(4), 693–701 (2016)

Establish Evidence Chain Model on Chinese Criminal Judgment Documents Using Text Similarity Measure

Yixuan Dong[1,2], Yemao Zhou[1,2], Chuanyi Li[1,2], Jidong Ge[1,2(✉)],
Yali Han[1,2], Mengting He[1,2], Dekuan Liu[1,2], Xiaoyu Zhou[1,2],
and Bin Luo[1,2]

[1] State Key Laboratory for Novel Software Technology,
Nanjing University, Nanjing 210093, China
gjdnju@163.com
[2] Software Institute, Nanjing University, Nanjing 210093, China

Abstract. One of the most prominent issues in criminal judgment documents is the insufficient evidence. In order to raise the level of judgment documents reasoning, we need to evaluate the quality of evidence in judgment documents. In the recent informatization of Chinese courts, the huge amount of law cases made it necessary to automate the evaluation of evidence. Constructing the model of evidence chain is the basis for assessing the quality of the judgment documents as evidence chain model can describe the relationship between evidence and fact as well as the relationship between evidence more intuitively. In trying to achieve all above mentioned, we propose a model of evidence chain based on Chinese criminal judgment documents. Automated text preprocessing of Chinese criminal documents creates semi-structured XML documents and in XML file, we can get evidence set and fact set. Key element extraction based on syntactic parsing is used to get keywords of each evidence and fact. Text similarity measure based on Word2vec and keyword overlap ratio calculation is used to get the connection point of evidence chain. Predefined weight of different kinds of evidence can be used to measure the importance of each evidence chain. Table format and graphical format make it possible for us to see the structure of evidence chain.

Keywords: Criminal judgment documents · Judgment documents reasoning
Big data · Evidence chain · Text similarity measure · Word2vec
Weight of evidence chain

1 Introduction

In the process of hearing a case, the Chinese People's Court writes judgment document according to law and details of the case, which records the process and result of each case. The judgment document is one of the most important instrument for resolving procedural issues and substantive issues. The reasoning of judgment documents refers to the explanation of the reasons in legal domain.

© Springer Nature Singapore Pte Ltd. 2018
Q. Zhou et al. (Eds.): ICPCSEE 2018, CCIS 902, pp. 27–40, 2018.
https://doi.org/10.1007/978-981-13-2206-8_4

At present, many courts appear to pay little attention to the reasoning of judgment documents. In recent years, most of the injustices that have strong social influence are due to the error of fact instead of applying wrong law. The error of fact is most likely to happen when we do not have enough evidence or the quality of evidence is low [1]. At present, it is generally believed by all walks of life in the community that the reasoning of judgment documents is an important mechanism for promoting judicial fairness, implementing judicial openness and enhancing the credibility of the judiciary [2]. In order to achieve the goal of raising the level of reasoning of judgment documents, it is necessary to study the automated assessment techniques of judgment documents.

In order to extract evidence and construct evidence chain model of each judgment document, we have some challenges as follows:

(1) Although Chinese judgment document has a general fixed format, it is still written in natural language. So, how can we get fact and evidence information from the judgment document?
(2) There is so much information in an evidence record and a fact record, which kinds of key elements should be selected? How to extract these key elements?
(3) As every fact is inferred by a set of evidence, how can we build the relationship between fact and evidence?
(4) In criminal case, different evidence may combine together to prove one fact, how can we find the link point of different evidence?
(5) A criminal judgment document usually has much evidence, so we may generate many evidence chain units for one document, in this case, how can we rank these evidence chain units?
(6) How to store the results of evidence chain model and how to display the relationship clearly?

In this paper, we propose an approach to establish evidence chain model based on Chinese Criminal judgment documents. In order to extract the content from judgment document, we use Chinese natural language processing including Chinese words segmentation, syntactic parsing and regex match based on keywords. Also, in order to calculate the link point in evidence chain model, we use text similarity measure based on Word2vec.

2 Related Work

It is generally believed that the reasoning of judgment documents is an important criterion for judging the quality of judgment documents. Therefore, improving the level of reasoning in judgment documents is an important breakthrough to enhance the level of judicial services. For a long time, Chinese judges have a bad reputation for being unreasonable in judgment documents. This situation has been highly criticized by the academic circles [3]. In order to improve the level of reasoning in judgment documents, we can start from the following three aspects: First, the reasoning about evidence, second, the reasoning about facts and the third is about the reasoning of law [4]. There is an internal connection between evidence and fact, that is, evidence is the basis for proving the facts of a case [5]. One of the essential attributes of evidence is relevance.

Evidence unrelated to fact has no value in judgment documents. The vast majority of evidence cannot be directly related to the facts, and need to be arranged with each other. The combination of the two evidences means that the two have reached a steady and solid link and formed the most basic evidence chain model [6].

The Study on Criminal Evidence Chain is the representative work of the research on evidence chain in China. This thesis defines the basic model of chain unit, main part of unit (chain unit body), key of unit (chain unit head) and connection point in evidence chain, and proposes different kinds of link, for example simple link, multiple link, net link and so on [7]. This thesis points out that each evidence chain has a weight according to its evidence content and the number of evidence in evidence chain.

In order to establish evidence chain model from criminal judgment documents, we need to extract evidence and fact content from criminal judgment document using Chinese Information Extraction. The main function of Information Extraction is to extract specific factual information from the text, which can be structured, semi-structured or unstructured text. Overall, the method of information extraction is divided into two categories: one is based on KDD (Knowledge Discovery in Databases) and data mining methods, mainly from structured and semi-structured data extraction information and the other one is using NLP and text mining methods. The goal is to discover new knowledge from unstructured, open texts and turn them into understandable and useful information [8]. News reports are based on events, in order to clearly illustrate the environment of an event, news reports usually contain the organic combination of six elements: Who, Where, What, When, When, Why, How. If the content of a news article contains these elements of the news, then the news content is considered complete [9]. Although the application scenarios of these papers are not as the same as this one, they are all based on Chinese text.

In order to calculate the link point of different evidence, natural language processing including Chinese Word Segmentation, syntactic parsing, word vector model and text similarity measurement is applied. Chinese word segmentation theory can be attributed to: three main word segmentation algorithm, combinatorial algorithm research, Chinese word segmentation disambiguation, unrecognized word recognition and research on Word Segmentation and POS Marking Evaluation [10]. In general, syntactic parsing methods can be divided into rule-based analysis methods, methods based on statistical analysis and methods of combining statistics with rules [11].

3   Approach

In this section, we describe our approach to establish evidence chain on Chinese criminal judgment documents in detail as follows. Section 3.1 presents the model of evidence chain and introduces an overview of the workflow in our approach. Section 3.2 introduces preprocessing of Chinese criminal judgment documents and how to get fact set and corresponding evidence set of the documents. Section 3.3 introduces the means of extracting keywords of each evidence item using natural language processing. Section 3.4 introduces how to get link points of each evidence chain using text similarity measure. Section 3.5 introduces the approach to calculate the weight of the evidence chain. Section 3.6 introduces different ways to display evidence model.

3.1   Overview

Figure 1 presents the form of evidence chain model. An evidence chain model is a three-layer hierarchical structure. First layer: Each evidence node is composed of evidence unit body and evidence head. Second layer: Link point is composed of evidence heads of two or more different evidence nodes if the heads can match each other. Third layer: evidence node set is connected to fact by link point.

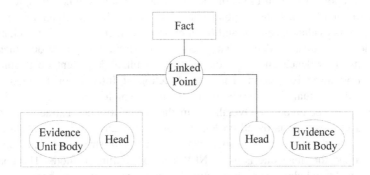

Fig. 1. Form of evidence chain model

Figure 2 presents an overview of workflow of our approach. Because of the characteristics of Chinese natural language, the process is based on Chinese word segmentation. The steps of our workflow are as follows:

(1) Preprocess the text of Chinese criminal judgment documents and extract fact content and corresponding evidence set of each fact.
(2) Extract five different kinds of key elements of each fact and evidence using natural language processing. In this method, we use 4W1H (What, Where, When, Who and How much) as five different kinds of key elements.
(3) Calculate the link points of each evidence head using Word2vec as the text similarity measure methods.
(4) Calculate the weight of each evidence chain.
(5) Display evidence chain model in both table format and graphical format.

3.2   Text Preprocessing

Text preprocessing step can be divided into three sub steps as follows: (1) Extract useful paragraph; (2) Get fact and corresponding evidence set from the paragraph; (3) Label the attributes of evidence.

3.2.1   Extract Useful Paragraph

Chinese judgment documents are usually written in a fixed format. According to the position of a paragraph, the keywords from the first sentence in each paragraph and the writing regulation of each part, we can split an article into seven different parts: headline, litigant participant, and litigant record, basic information of case, trail

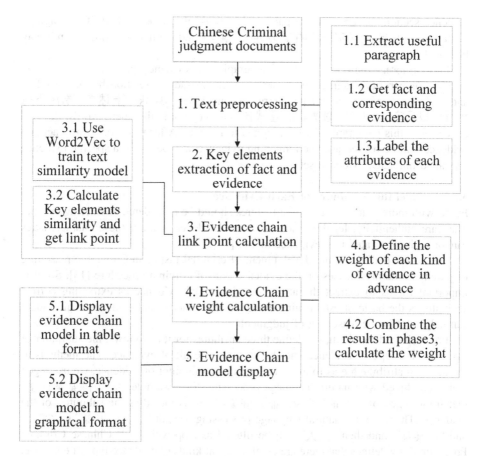

Fig. 2. Overview of the workflow of establish evidence chain model

process, judgment result and end. If the first sentence in the paragraph has keywords like "法院", "书" and "号", then the paragraph belongs to headline. If the first sentence has words like "纠纷" or "起诉书", then it may not be the litigant participant part. The litigant record appears after the litigant participant part and the basic information of case comes after the litigant record. If the first sentence has keywords like "本院认为" or "审查认为", then it belongs to the trial process part. If the first sentence has keywords like "裁定如下" or "达成如下结论", then it belongs to the judgment result part. If the first sentence has keywords like "速录" and "书记", then it belongs to the end part.

3.2.2 Extract Fact and Corresponding Evidence

In order to get fact and evidence information, basic information is the most important parent paragraph. For example, if a paragraph of basic information contains words like "证据如下", "书证", "物证", "证人证言" and similar keywords, it may be an evidence section. In terms of fact paragraph, we use the same method, if a paragraph contains words like "经审理查明", "认定XX事实" and similar keywords, then it may be a fact section.

Furthermore, when we get fact paragraph and evidence paragraph from basic information part of a judgment documents, we can extract the relationship between fact and evidence. For instance, if a paragraph contains keywords like "上述事实有经下列证据予以证实", "认定上述事实的证据有" and keywords like these words, the evidence set in this paragraph will be assigned to the fact. After this processing, fact and evidence will be related and the remaining evidence will be put in a specific evidence set called unrelated evidence set.

3.2.3   Label the Attributes of Each Evidence

Faced with more evidence, the three features should be considered as filtering criteria. They are authenticity, legality and relevance. Evidence without these characteristics cannot be incorporated into evidence model [12].

Due to the perfection and development of criminal justice in China, the reasoning of judgment mainly focuses on the evidence part of criminal procedure [13]. So, it is important for us to extract the attributes of criminal evidence as extracting correct attributes is the basic of establishing evidence chain model and evidence chain model can help a lot in the reasoning of judgment.

For each evidence item, we define three attributes for it as follows: the submitter of evidence, the type of evidence and the reasoning result of evidence. The methods to extract each attribute are as follows: (1) to extract submitter of evidence, we use regex match combined with natural language processing. For example, "证人陈某某证实" match the regex of "(.*)证实", so "证人陈某某" is extracted as the submitter of the evidence. Then, we use natural language processing to split "证人陈某某" to "证人" and "陈某某", and then "证人" will be filtered as stop words. (2) Chinese Criminal Procedure Law defines that there are eight different kinds of evidence type of evidence, so we use keywords matching to extract evidence type. (3) Reasoning result of evidence have two different types: accepted and not accepted, so we use keywords matching to extract reasoning result. For example, if the evidence contains words like "不采信", "不采纳", "不予采信" and "不予采纳", then the reasoning result is labeled as not accepted.

3.3   Keywords Extraction

As is mentioned in Sect. 3.1, we use 4W1H as the type of keywords. Because 4W1H have different characteristics, we need to apply different methods to extract each kind of keywords. In this section, we will introduce the methods used to extract keywords.

3.3.1   Extract What

What means the object in evidence and fact. For instance, What usually refers to the murder weapon in a murder case and refers to the stolen things in a theft case. What elements are all nouns and most of them belong to subject or object in a sentence.

In order to extract What, we combine regex match with syntactic parsing as follows:

(1) Regex match: for things that have fixed structure, we can use regex match to extract this element. For example, the title of a book between " 《 " and " 》 ".

(2) Syntactic parsing: we use HanLP as Chinese syntactic parsing tool. HanLP is based on Chinese dependent syntax analysis based on CRF (Condition Random Field) sequence annotation algorithm and maximum entropy dependency syntax analysis algorithm. This step can be divided into four sub steps as follows: Firstly, we use HanLP to analyze the structure of a sentence. Secondly, we get the description of each word using the sentence structure. For example, "黄色" is used to describe "上衣" and "塑料" is used to describe "热水瓶" in the sentence structure. Thirdly, we can get all subjects and objects of a sentence using syntactic parsing. Fourthly, we combine descriptions and subjects or objects extracted from the sentence to get What element. Finally, we filter stop words from the result.

3.3.2 Extract Where

Where refers to the site information in evidence and fact, for example, the crime spot or witness spot. Where elements are all nouns or location words and most of them appear after prepositions like "在", "于", "至" or similar prepositions.

Two methods are used to extract Where element as follows:

(1) Participle the sentence and extract the words that are labeled as "S" which means site.

(2) Regex match: define a list of prepositions and the words after the prepositions will be extracted as Where elements.

3.3.3 Extract Who

Who refers to the participants mentioned in evidence and fact. POS (Part of Speech) is used to extract Who elements from a sentence, we participle the sentence and get the words that are labeled as "nr" which means name of people or organizations. These words will be extracted as Who element.

3.3.4 Extract When

When refers to time information mentioned in evidence and fact. Time information in evidence or fact is usually in a fixed format so we use regex match to get time information. Here are two examples of the regex:"\d年\d月\d日" and "\d月\d日".

3.3.5 Extract How Much

How much refers to quantitative phrases in evidence and fact and amount information. They are usually in a fixed format: numbers always appear before quantifiers. So we use regex match to get amount information as well.

3.3.6 Generate Stop-Word List Based on Judgment Documents

Stop-word means word that have high frequency of occurrence but do not have important meaning in text such as "is", "of" or "the". The processing of stop words can greatly accelerate the speed of word segmentation and subsequent parsing [14]. In terms of judgment documents, we have more words that need to be filtered as stop-word. For example, "原告", "被告", "证言" or similar words, these words have high frequency in judgment documents but have little meaning in law related information extraction. To construct a stop-word list that meets requirements talked above, we choose 100000 criminal judgment documents as our corpus and calculate IDF of each unique word. IDF (Inverse Document Frequency) refers to the inverse fraction of documents that contain a specific word. Words with low IDF can be regarded as potential stop words. After scanning all judgment documents and filtering common stop words, we calculate IDF of rest words and sort words in ascending order. Top N words have been chosen as stop words. Stop words we choose are added into the set of common stop-word list to construct the stop words list we use to filter candidate set of chain unit heads.

3.4   Link Point Calculation

As we have introduced in Sect. 3.1, link point is the fundamental component in evidence chain, so link point calculation is of great importance in establishing evidence chain. Link point calculation can be divided into two sub steps: (1) Calculate the Link point between evidence. (2) Build relationship between evidence and fact.

3.4.1   Calculate the Link Point Between Different Evidence

In terms of calculating the head of evidence we mainly use two methods:

(1) Calculate link point using equality of keywords extracted in Sect. 3.3.

In this method, we compare keywords of an evidence with keywords of another evidence in a specific keyword type such as "What". If these two keywords are equal to each other, then these two keywords will be extracted as the head of each evidence and the head will be the link point of these two evidence.

After the loop, we can get the link points between evidence.

(2) Calculate link point using text similarity measure of keywords extracted in Sect. 3.3 based on Word2vec.

In this method, we compare keywords of an evidence with keywords of another evidence in a specific keyword type such as "What". If these two keywords are similar to each other semantically, then these two keywords will be extracted as head of each evidence.

In calculating the similarity between words, we convert words into word vectors and measure the similarity between words by calculating the cosine distance of word vectors. We use Word2vec model to convert words into word vectors. The specific steps are as follows:

Firstly, train Word2vec model on all Chinese criminal judgment documents in the corpus. In this step, we choose all criminal documents and get the evidence part and fact part of each article. Then we participle the evidence part and fact part and get all the words except stop-word and use these words as the training corpus and train the Word2vec model on this training corpus. After this step, we get a Word2vec model that contains the word vector of each word.

Secondly, determine the threshold for similarity calculation. As there is no uniform standard to determine in which case two words vectors are similar. In this step, we choose 100 judgment documents from the training corpus randomly and participle evidence and fact part of each document. Then we calculate the cosine distance of every two words that lies in different evidence and respectively output the word pairs that have the cosine distance larger than 0.95, 0.9, 0.85 and 0.8. Combine the precision and recall of the similar words extraction, we choose 0.9 as the threshold value in similar words calculation.

Thirdly, participate the keywords that need to be compared and get the word list of it and then filter the stop-word from it.

Fourthly, compare the word similarity of each word that we get from the two keywords respectively by using Word2vec model we trained in step1 and get highest word similarity as the similarity between the two keywords. If the similarity of two keywords are higher than the threshold that we set in step two, these two keywords will be extracted as the head of each evidence.

As we can see from the Table 1, the first method is more efficient but may leave out some information and the second method is somehow less efficient but can extract more useful information. So we apply the first method on the comparison of keywords that belong to "When", "Who" and "How much" as these three types have a fixed format. We apply the second method on the comparison of keywords that belong to "What" and "Where" as these two types have the characteristic that the same thing may have different expression.

Table 1. Comparison of two methods in link point calculation

Method type	Advantage	Disadvantage
Equality of keywords	Simple calculation and efficient	Leave out keywords that have different expression but have same meaning
Text similarity of keywords	Extract more keywords that have different expression but have similar meaning	More complex calculation

3.4.2 Build Relationship Between Evidence and Fact

Section 3.2.2 shows that we extract fact and corresponding evidence from every judgment documents, and the remaining evidence will be put in unrelated evidence set.

So for the evidence which has not be related to a fact, we compare its similarity with each fact by calculating its overlap of keywords and fact's keywords and the evidence will be related to the fact which has the highest overlap with this evidence.

3.5 Weight of Evidence Chain Model

In this section, we mainly focus on how to calculate the weight of evidence chain model. As we may generate a lot of evidence chain models in a judgment document and some of them are of great importance while some are less important in helping judges analyze the evidence information. So it is essential for us to calculate the weight of each evidence chain model.

In calculating weight of evidence chain model, we ask some experts in legal domain to decide how to calculate the weight of each evidence and get a conclusion that the weight should be decided by the evidence type and reasoning result. In terms of evidence type, authoritative evidence such as documentary evidence or physical evidence provided by police have higher weight than non-authoritative evidence such as testimony of witness. We define that the weight of authoritative evidence is 1.0 and the weight of non-authoritative evidence is 0.8. In terms of reasoning result of evidence, evidence that is confirmed by the judges has higher weight than unconfirmed evidence. So, each evidence item has a weight according to evidence type and reasoning result of evidence in Sect. 3.2.3. Then we add the weight of each evidence item in an evidence chain model and get the weight of an evidence chain model.

As for now, we can get the weight of each evidence chain model in a criminal judgment document automatically; in the future we will apply it by sorting the weight and give the top n evidence chain models of the judgment document as a reference for judges.

3.6   Display Evidence Chain Model

In this section, we will introduce two main ways that we apply to display evidence chain model: the excel format and the graphical format. Excel sheets can clearly present the content and attributes of each fact and evidence and graphical format can directly shows the relationship between fact and evidence. Evidence chain model information is now saved in JSON format. JSON means JavaScript Object Notation, it is a syntax for storing and exchanging data and it is in text format written with JavaScript object notation. Experimental data show that JSON is obviously superior to other data transmission formats in terms of data transmission efficiency, which provides an optimized reference scheme for the selection of data transmission formats in light-weight applications [15]. We can easily convert Json file to object and it is easy for person to read or write.

3.6.1   Generate Excel File of Evidence Chain Model

Evidence chain model of each Chinese criminal judgment documents is saved as an excel file with two sheets:

(1) Evidence sheet focuses on evidence part of a criminal judgment document. In this sheet, we display the evidence and its attributes including evidence content, evidence type, reasoning result and the head list of each evidence that we calculated in Sect. 3.4.1.

(2) Fact sheet focuses on the fact content and relationship between fact and evidence. Identifier, name and content of each fact are shown in the sheet in turn. At the same time, the corresponding evidence and their chain unit heads are displayed in the sheet.

3.6.2 Graphical Display of Evidence Chain Model

Graphical display of evidence chain model focuses on demonstrating the relationship of fact and evidence. As Fig. 3 shows, we can use the system with evidence chain model information in the excel file as input and extracts the evidence content from excel file. Then the evidence chain model we build can be displayed on the interface clearly.

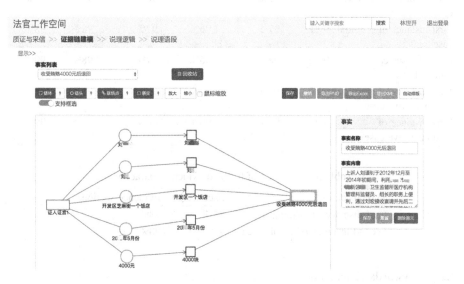

Fig. 3. Form of evidence chain model

4 Experiment and Result

In this chapter, we design some experiments to evaluate the effectiveness of each step in the approach of establishing evidence chain and give the result of each experiment.

4.1 Text Preprocessing

To evaluate the results of paragraph partition, we compare the results of our approach with the information of each case in the trial system database. As Table 2 shows, the accuracy of splitting basic information of a case if higher than 90% and for criminal case, the accuracy is even up to 97%.

As criminal case has more detailed evidence record, so we choose criminal judgment documents to extract fact and evidence. To evaluate the results of extracting fact and evidence set, we compare our results with the evidence and fact information in the trial system database. Table 3 shows the accuracy of extracting fact section and evidence section from basic information of case. It is approximately equal to 90% which means that our approach in text preprocessing is effective.

Table 2. Accuracy of paragraph partition

Type of case	Headline	Litigant participant	Litigant record	Basic information	Trial process	Result
First-instance of civil case	99%	93%	88%	94%	89%	91%
Second-instance of civil case	99%	92%	90%	90%	91%	97%
First-instance of criminal case	99%	97%	96%	97%	98%	98%
Second-instance of criminal case	99%	98%	96%	97%	98%	99%
First-instance of administrative case	99%	94%	88%	92%	91%	98%
First-instance of administrative case	99%	97%	96%	96%	95%	99%

Table 3. Accuracy of fact and evidence extraction

Type of case	Fact section	Evidence section
First-instance of criminal case	88%	88%
Second-instance of criminal case	90%	90%

4.2 Keywords Extraction

As there are no existing tools to extract keywords in terms of 4W1H, to evaluate the results of our approach, we randomly choose 100 criminal judgment documents and label the correct key elements by skilled legal practitioners. Then we compare our results with it and get the results shown in Table 4. As the table shows, recall and precision are almost more than 70%. The evaluation results prove that our methods to extract key elements can lay a solid foundation for finding out link points and establishing evidence chain model in later steps.

Table 4. Evaluation of key elements extraction

Key elements	Precision	Recall	F1
What	68.3%	79.1%	73.3%
Where	81.2%	85.4%	83.2%
Who	90.6%	87.9%	89.2%
When	100%	92.7%	96.2%
How much	98.7%	92.5%	96.9%

4.3 Link Point Calculation

In order to verify the effectiveness of link point calculation, we randomly choose 100 criminal judgment documents and ask some skilled experts in legal domain to label the correct keywords of each evidence and label the head of each evidence chain model in each judgment document. Then we did two experiments, in the first experiment, we calculate the head simply using equality calculation based on the correct keywords that are labeled by skilled experts and in the second experiment, we calculate the head using text similarity measure combined with equality calculation and the accuracy of the two types are shown in Table 5. As we can see from Table 5, the F1 of link point calculation of What and Where elements rise from 81.6% to 90.5% when we use text similarity calculation instead of equality calculation.

Table 5. Accuracy of two methods in link point calculation

Method type	Precision	Recall	F1
Equality of keywords	97.7%	70.1%	81.6%
Text similarity of keywords	92.4%	89.8%	90.5%

5   Conclusion and Future Work

In this thesis, we propose an approach to establish evidence chain model from criminal judgment documents which is based on Chinese natural language processing. In the experiments, we choose 4W1H as key elements of each evidence and fact and combine equality and text similarity measure based on Word2vec to calculate the relationship of different evidence.

However, establishing evidence chain model from judgment documents is the very first step of assessing the reasoning of judgment documents. In the future, more work need to be done to evaluate the reliability of the fact according to the evidence content in judgment documents. Also, we need to tell which part of fact in judgment document do not have enough evidence to support it. This results will help judges write more reasonable judgment documents, thus promoting judicial fairness and implementing judicial openness.

Acknowledgements. This work was supported by the National Key R&D Program of China (2016YFC0800803).

References

1. Zhang, D.: The optimization of China criminal evidence system. Soc. Sci. Chin. **07**, 125–148 (2015)
2. Lei, X.: Predicament and way out of criminal judgment documents - taking death penalty cases as analysis sample. J. Grad. Sch. Chin. Acad. Soc. Sci. **15**(6), 85–89 (2015)
3. Ling, B.: How judges reason: Chinese experience and universal principles. Chin. Leg. Sci. **15**(5), 99–117 (2015)
4. Yang, Y.: A study of paragraphs of evidence reasoning in judgment documents. Appl. Linguist. **05**(3), 111–115 (2005)

5. Liu, Z., Xiong, Z.: Factual information of evidence. J. Southwest Chin. Norm. Univ. **31**(5), 87–91 (2005)
6. Li, Z.: Evidence chain and structuralism. Chin. Legal Sci. **2017**(02), 173–193 (2017)
7. Chen, W.: The study on the evidence chain. J. Natl. Prosec. Coll. **15**(4), 128–136 (2007)
8. Guo, X., He, T.: Survey about research on information extraction. Comput. Sci. **42**(02), 14–17 (2015)
9. Wang, W., Zhao, D.: Chinese news event ontology construction and auto-population. Comput. Eng. Sci. **34**(4), 171–176 (2012)
10. Feng, G., Zheng, W.: Review of Chinese automatic word segmentation. Libr. Inf. Serv. **55**(2), 41–45 (2011)
11. Shi, C.: Dependency parsing research. Intell. Comput. Appl. **3**(6), 47–49 (2013)
12. Cai, Z.: Criteria of evidence chain integrity and approach to examine. Lawyer World **3**, 11–13 (2003)
13. Chen, J.: On fact-finding in the court and evidence. Southwest University of Political Science and Law (2009)
14. Hua, B.: Stop-word processing technique in knowledge extraction. New Technol. Libr. Inf. Serv. **2**(8), 48–51 (2007)
15. Gao, J., Duan, H.: Research on JSON data transmission efficiency. Comput. Eng. Des. **32**(7), 2267–2270 (2011)

Text Sentiment Analysis Based on Emotion Adjustment

Mengjiao Song[(✉)], Yepei Wang, Yong Liu, and Zhihong Zhao

Software Institute, Nanjing University, Nanjing 210093, China
selene_smj@163.com

Abstract. Text sentiment analysis is used to find out how much the public's appreciation and preferences for specific events or objects. In order to effectively extract the deep emotional features of words, this paper proposes two sentiment analysis methods, which are emotion adjustment method based on semantic similarity and skip-gram model. In these two methods, word vectors containing semantic information obtained from Word2vec and emotional seeds are used to adjust the sentiment orientation of the words so that word vectors can trained both the semantic information and the sentiment contents. And the TF-IDF method is used to calculate the word's weight in the text, the vector of the whole text is represented by adding the weighted word vectors. Experiments show that the emotion-adjusted word vector improves the accuracy of the text sentiment analysis more effectively than the traditional method, and proves the validity of these two methods in the sentiment analysis task. At the same time, the emotion adjustment method based on skip-gram model is more effective than the method based on semantic similarity.

Keywords: Sentiment analysis · Skip-gram · Word2vec · Emotion adjustment

1 Introduction

In the Web 2.0 era, people express their opinions through the Internet and post comments with personal feelings about social hot issues, products and services. In China, with its quick and easy features, Weibo becomes the platform for most people to express their ideas, disseminate content and exchange learning. According to the statistics of CNNIC, as of December 2017, the number of netizens in China reached 772 million, the utilization rate of Weibo reached 40.9%, an increase of 16.4% over 2016 [1]. At the same time, Weibo texts has attracted a large number of scholars to carry out various studies [2]. Analyzing the emotions of Weibo texts can be used to find out the public's preference of specific events or objects, and is of great significance for applications such as market research, network public opinion discovery and early warning. Therefore, sentiment analysis of microblogging has become one of the hot research directions.

Machine learning method is the mainstream of current research on text sentiment analysis. The primary task of machine learning is to use the words contained in the text to construct the representation text, usually use the word vector to represent the words. This idea was first proposed by Hinton in 1986, by mapping words to K-dimensional

© Springer Nature Singapore Pte Ltd. 2018
Q. Zhou et al. (Eds.): ICPCSEE 2018, CCIS 902, pp. 41–54, 2018.
https://doi.org/10.1007/978-981-13-2206-8_5

real vectors and judging the semantic similarity between them based on the distance between word vectors [3]. Mnih and Kavukcuoglu proposed a simpler, faster and more efficient representation of word vector using the method of noise estimation comparison [4]. Google introduced Word2vec in 2013 which use the idea of deep learning to train a great deal of predictive word vector [5]. However, these methods mainly focus on the grammar of words and do not consider the emotional features of words. Words with similar contextual information but different emotional polarities may be mapped into similar vectors. Therefore, in this paper, word vectors are used to project texts into low-dimensional dense semantic spaces. At the same time, large-scale unlabeled corpora can be used to mine lexical richness semantic information, and the semantic similarity of words is introduced into the classification so as to improve the performance of sentiment analysis. Considering the emotional features of words and selecting the emotional seeds as the reference, two methods of emotion adjustment of word vectors are proposed. One is based on the semantic similarity of word vectors for the calculation of affective features and the other is the emotional transformation of the Skip-gram neural network structure. By these two methods, the sentiment words and sentiment vectors with semantic features and sentiment features are obtained. The TF-IDF method is used to weighted sum the sentiment word vectors according to the word weights to obtain the vector representation of the texts. The validity of these two kinds of emotional features in semantic representation and emotional representation is verified by experiments, which improves the accuracy of text sentiment analysis.

The remainder of this paper is laid out as follows. Section 2 introduces related work. Section 3 introduces our methods in detail. Section 4 shows the implementation of our methods and the experiments and Sect. 5 makes conclusion and discusses the future work.

2 Related Work

2.1 Sentiment Analysis

Sentiment analysis, also known as propensity analysis and opinion mining, is a process of analyzing, processing, inducing and reasoning subjective texts with emotions [6]. The methods used in sentiment analysis mainly divided into methods based on sentiment lexicon and machine learning.

The method based on sentiment lexicon mainly through the development of a series of sentiment lexicons and rules, paragraph disassembly and syntactic analysis of the text, analyzing the number of positive sentiment words and negative sentiment words in the text, calculating the emotional differences and the text's sentiment orientation is judged by this value. Turney proposed to judge the sentiment orientation of the whole text according to the relevance between the words in the test text and the words in the seed dictionary [7]. Pan proposed six kinds of sentiment, and the application of emoticons significantly improved the accuracy and coverage of sentiment analysis [8].

Usually, Methods based on machine learning transform this problem to a classification problem. For the judgment of sentiment orientation, the target text is divided into two categories as positive and negative emotions. By manual annotation of training corpus and test corpus, sentiment orientation is carried out using classifier such as Support Vector Machine (SVM), Maximum Entropy Model and K-nearest neighbor (KNN). For example, Pang first proposed using machine learning methods to classify movie review data into positive and negative categories [9]. Mullen used support vector machine to analyze the syntactic relationships and features of texts to achieve sentiment partitioning [10]. Wang [11], Li [12] and others improved the accuracy by combining different classifiers, and proposed a semi-supervised propensity analysis model to solve the imbalance corpus problem under certain circumstances. Sun and others proposed to combine the sentiment lexicon with machine learning method [13], using the method of location weight calculation based on the characteristic polarity and support vector machine as the machine learning model to classify the Weibo text into three sentiment categories: positive, negative and neutral, which effectively improved the accuracy of sentiment analysis. Jiang and others combined methods based on sentiment lexicon and machine learning to improve the accuracy of sentiment analysis [14].

2.2 Word Vector

The basic idea of word vector is to map each word into k-dimensional real number vector to obtain the feature representation of text data. This method is superior to the traditional One-hot representation method, which not only avoids the dimensionality disaster but also includes the semantic relationship between words. Google's Word2vec is the most commonly used word vector tool currently. Word2vec is based on two models, CBOW (Continuous Bags-of-Words) model and Skip-gram model, and its structure is shown in the Figs. 1 and 2. The CBOW model uses the context of the current word to predict the probability of the current word, while the Skip-gram model uses the current word to predict the probability of its context. This chapter mainly introduces Skip-gram model.

The Skip-gram neural network model can be constructed by two frameworks, including Hierarchical Softmax [15] framework and Negative Sampling (NEG) [16] framework. In this paper, the negative sampling method is used. This model includes the input layer, the projection layer and the output layer. A piece of text

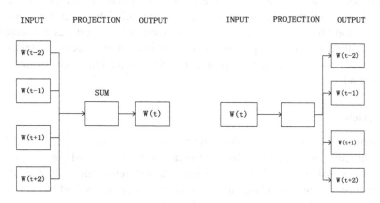

Fig. 1. Cbow model   **Fig. 2.** Skip-gram model

$w_{t-n}, \ldots w_{t-1}, w_t, w_{t+1} \ldots w_{t+n}$ in corpus C is recorded as a sample $(w, \text{context}(w))$. Negative sampling method considers the word in the sample as positive and negative, the center word w as a negative sample, and the rest as positive samples. In the output layer using Negative Sampling, for any $z \in \{u\} \cup NEG(u)$, $NEG(u)$ means the negative sample subset that when processing the word u generated, the formula of the posterior probability is as follows:

$$p(z|w) = \begin{cases} \sigma(v(w)^T \theta^z), & L^u(z) = 1 \\ 1 - \sigma(v(w)^T \theta^z), & L^u(z) = 0 \end{cases} \tag{1}$$

The Skip-gram model requires maximizing the probability of words appearing in the context. The objective function of optimization is as follows:

$$G = \prod_{w \in C} \prod_{u \in Context(w)} g(u), \quad g(u) = \prod_{z \in \{u\} \cup NEG(u)} p(z|w) \tag{2}$$

Take the logarithm of G, the final objective function is:

$$\mathcal{L} = \sum_{w \in C} \sum_{u \in C(w)} \sum_{z \in \{u\} \cup NEG(u)} \begin{Bmatrix} L^u(z) \cdot \log[\sigma(v(w)^T \theta^z)] + \\ [1 - L^u(z)] \cdot \log[\sigma(v(w)^T \theta^z)] \end{Bmatrix} \tag{3}$$

3 Approach

Using Word2vec to learn word vectors for texts and by random negative sampling methods can improve training speed and improve the quality of word vectors. Word vectors can reflect the similarity of words in the semantic level, but they do not consider the sentiment orientation information. Words with similar contexts may have opposite sentiment tendencies. Therefore, this paper proposes two sentiment adjustment methods based on semantic similarity and Skip-gram model.

In this section, we describe in detail how we judge the sentiment tendencies of the text, as follows. Section 3.1 introduces the sentiment analysis model and presents an overview of the workflow in our approach. Section 3.2 introduces the sentiment adjustment method based on semantic similarity. Section 3.3 introduces the sentiment adjustment method based on skip-gram model. Section 3.4 introduces the text representation method considering the weight of words. Section 3.5 introduces the sentiment analysis method.

3.1 Overview

Figure 3 shows the model for emotion adjustment of word vectors. The model is a three-layer structure, the first layer: the word vector of sentiment seeds and corpus after segmentation by using Word2vec. The second layer: sentiment feature extraction method based on semantic similarity and sentiment adjustment method based on skip-gram model. The third layer: Emotionally adjusted word vectors.

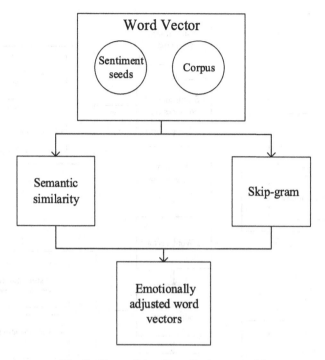

Fig. 3. Form of emotion adjustment model

To perform sentiment analysis on Weibo text, in addition to the emotional adjustment model proposed in this paper, there are many other steps. Figure 4 presents an overview of our approach. The steps are as follows:

(1) Emotional seeds preprocessing, including picking the emotional seeds with strong emotions and assigning sentiment orientation.
(2) Corpus preprocessing, including word segmentation, removal of stop words.
(3) The Word2vec method is used to compute the processed corpus to get the word vector.
(4) Use the method of emotional adjustment to get emotion vector of training set.
(5) Get the text vector with TF-IDF method and weighted sum.
(6) Use machine learning method to obtain the sentiment orientation of the test set.

3.2 Emotion Adjustment Method Based on Semantic Similarity

In order to consider the affective propensity information of words in the unknown sentiment polarity Corpus $= \{c_1, c_2 \ldots c_m\}$, the affective distribution is calculated by the similarity between the words c_i in the Corpus and the sentiment seeds. For any word $c_i \in$ Corpus, calculate the cosine similarity between it and the emotional seed $s_i \in$ *Sen*, the formula is as follows, where w_{ci} is the word vector of word c_i and w_{sj} is the word vector of word s_i.

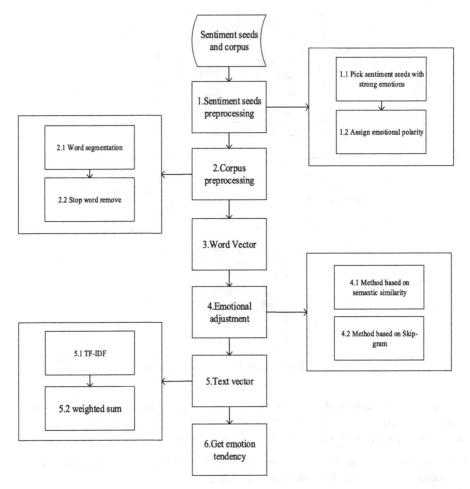

Fig. 4. Overview of the workflow of establish emotion analysis model

$$Sim_{ci,sj} = f(ci, sj) = \frac{w_{ci} \cdot w_{sj}}{|w_{ci}| * |w_{sj}|} \tag{4}$$

According to the cosine similarity $Sim_{ci} = (Sim_{s1}, Sim_{s2}...Sim_{sn})$ of word c_i and emotional seeds, we multiply them by the word vectors of the emotional seeds respectively, and get an n-dimensional vector, which is the emotion word vector w_{ci}^* of the word, the formula is as follows.

$$w_{ci}^* = (Sim_{ci,s1} \odot w_{s1}, Sim_{ci,s2} \odot w_{s2}...Sim_{ci,sn} \odot w_{sn}) \tag{5}$$

The following Table 1 gives the pseudocode of this method.

Table 1. Pseudocode of the emotion adjustment method based on semantic similarity.

[1] for $c \in Corpus$ DO:
[2]   for $s \in Sentiment$ DO:
[3] $q = w_c \cdot w_s$
[4] $g = \|w_{ci}\| * \|w_{sj}\|$
[5] $sim = q/g$
[6] $w_c^* := (w_c^*, sim \odot w_s)$

3.3 Emotion Adjustment Method Based on Skip-gram Model

Based on the Skip-gram neural network model, this paper proposes a method of emotion adjustment. The input layer and the projection layer unchanged, and the output layer becomes sentiment orientation. For the sample $(w, context(w))$, $|context(w)|$ times negative sampling are taken for the words, the sentiment orientation of word w is a positive sample, and the rests are negative samples. Each sentiment orientation corresponds to a corresponding auxiliary vector θ^{Si}, which is the parameter to be trained. Maximize the probability of predicting the sentiment orientation of the word, that is, to maximize the formula as follows:

$$g(w) = \prod_{w \in Content(w)} \prod_{u \in \{w\} \cup NEG^w(w)} p(u|\widetilde{w}) \tag{6}$$

$$p(u|\widetilde{w}) = [\sigma(v(\widetilde{w})^T \theta^s)]^{L_{su}^w} \cdot [1 - \sigma(v(\widetilde{w})^T \theta^s)]^{1 - L_{su}^w} \tag{7}$$

$$L_{su}^w = \begin{cases} 1, & S_u = S_w \\ 0, & others \end{cases} \tag{8}$$

Where $NEG^w(w)$ is the subset of negative samples generated when the word $\widetilde{w}$ is processed. For a given corpus C, $G = \prod_{w \in C} g(w)$ is the overall optimization goal, taking the logarithm of G, which is the final objective function:

$$\mathcal{L} = \sum_{w \in C} \sum_{\widetilde{w} \in C(w)} \sum_{u \in \{w\} \cup NEG^w(w)} \begin{Bmatrix} L^w(u) \cdot \log[\sigma(v(\widetilde{w})^T \theta^{su})] + \\ [1 - L^w(u)] \cdot \log[\sigma(v(\widetilde{w})^T \theta^{su})] \end{Bmatrix} \tag{9}$$

Take the contents of triple sum denoted as $\mathcal{L}(w, \widetilde{w}, u)$, gradient calculation on θ^{su}, the formula is as follows:

$$\frac{\partial \mathcal{L}(w, \widetilde{w}, u)}{\partial \theta^{su}} = [L_{su}^w - \sigma(v(\widetilde{w})^T \theta^{su})]v(\widetilde{w}) \tag{10}$$

From the above formula, θ^{su} can be updated as:

$$\theta^{su} := \theta^{su} + \eta[L_{su}^w - \sigma(v(\widetilde{w})^T\theta^{su})]v(\widetilde{w}) \tag{11}$$

In the same way, gradient calculation of $v(\widetilde{w})$, the formula is as follows:

$$\frac{\partial\mathcal{L}(w, \widetilde{w}, u)}{\partial v(\widetilde{w})} = [L_{su}^w - \sigma(v(\widetilde{w})^T\theta^{su})]\theta^{su} \tag{12}$$

From all the above formulas, word vector $v(\widetilde{w})$ can be updated as:

$$v(\widetilde{w}) := v(\widetilde{w}) + \eta \sum_{u\in\{w\}\cup NEG^w(w)} \frac{\delta\mathcal{L}(w, \widetilde{w}, u)}{\delta v(\widetilde{w})} \tag{13}$$

The training sample is built on the emotional seeds and their contexts $(w,\text{context}(w))$ in the corpus.

The following Table 2 gives the pseudocode of this method.

Table 2. Pseudocode of the emotion adjustment method based on Skip-gram model.

[1]for $\widetilde{w} \in Context(w)$ DO:
[2] e=0
[3] for $u \in \{w\} \cup NEG^{\widetilde{w}}(w)$ DO:
[4] $q = \sigma(v(\widetilde{w})^T\theta^{su})$
[5] $g = \eta[L_{su}^w - q]$
[6] $e := e + g\theta^{su}$
[7] $\theta^{su} := \theta^{su} + gv(\widetilde{w})$
[8] $v(\widetilde{w}) := v(\widetilde{w}) + e$

3.4   Text Vector Representation

The text vector is the input to the machine learning method and is calculated from all the word vectors of the words in this text. Each word has its importance in the text, if simply adding the emotion-adjusted word vectors, the word's importance information will be lost. Therefore, this paper adopts the method of term frequency–inverse document frequency(TF-IDF) to consider the weight of words in the text. This method considers the importance of words in the text, and also considers the importance of words in all corpus. The text vectors are superimposed by the weighted word vectors. Through this method, the words with distinguishing ability play its most important role.

For a text i, the weight of the word j in the text is calculated as follows:

$$W_{ij} = \frac{tf_{ij} \cdot \log(N/n_j + a)}{\sqrt{\sum_j (tf_{ij} \cdot \log(N/n_j + a)^2)}} \tag{14}$$

Among them, tf_{ij} represents the number of words j appears in the text i; N represents the number of texts in the corpus; n_j represents the number of word j appearing in the corpus; and a is the adjustment parameter.

The vector of text i is obtained by the following formula. The dimension of the text vector is consistent with the dimension of the word vector, V_j indicates the word vector of the word j.

$$V_i = \sum_{j \in i} W_{ij} \cdot V_j \tag{15}$$

3.5 Sentiment Analysis Method

In this section, we mainly introduce machine learning algorithms used for sentiment analysis. Text sentiment analysis can also be seen as a classification problem, this article defines the classification of emotions as positive emotions and negative emotions two categories. Support Vector Machine (SVM) algorithm was chosen for text classification research in this paper.

The basic idea of SVM algorithm is to convert the input of input space into feature space, and find an optimal classification hyperplane in feature space, then use this hyperplane for classification. The criterion for finding the best classification hyperplane is that it separates two sets of points and distances as large as possible. The hyperplane formula is $f(x) = Wx + b = 0$, when $f(x)$ is greater than 0, the text is positive emotion, otherwise the text is negative emotion.

4 Experiment and Analysis

4.1 Experimental Data

In this paper, the initial word vector is the word vectors trained by word2vec, and emotion adjustment is performed according to the two algorithms proposed in this paper, resulting in two sets of emotional seeds vectors that take into account both semantic and emotional features. The verification data of this paper is obtained by crawling Weibo data and indicating that 3375 positive emotions and 3625 negative emotions by using manual annotation, with a positive and negative balance. The total verification data is expected to be 7,000 microblogging articles. 70% of the corpus is used as a training set and 30% as a test set. A total of 650,000 Wiktionary and Weibo materials are used as training data sets for initial word vector by Word2vec. The training corpus consists of 250,000 Chinese Wikipedia and 400,000 Sina Weibo texts. Skip-gram model is used as training model of Word2vec to obtain model Model_m2v.

Emotional seeds set Sen $= \{(s_1, w_{s1}, S_{w1}), (s_2, w_{s2}, S_{w2})\ldots(s_n, w_{sn}, S_{wn})\}$, where $s_i(i = 1, 2\ldots n)$ is the emotional seeds, w_{si} is the word vector corresponding to the emotional word s_i, and S_{wi} is the sentiment polarity.

In this paper, sentiment polarity is divided into positive emotion and negative emotion. Emotional seeds selected with explicit sentiment polarity, positive emotional seeds such as "happy", "like", negative emotional seeds such as "sad" "disgust" and so on. The Dictionary of Affective Sentences of the HowNet [17] and the Dictionary of Tsinghua Chinese Poetic Depreciation [18], taking the intersection of them as the source of emotional seeds used for experiments.

4.2   Evaluation Standard

In this paper, Accuracy and Recall are the criteria for evaluating the quality of sentiment analysis, the accuracy of the overall performance using Accuracy as an indicator. The following is the description of each evaluation index.

(1) Precision, which indicates the proportion of the texts correctly classified in this category to the texts categorized in that category, as shown in the formula 16.

$$\text{Precision}_p = \frac{TP}{TP + FP}, \text{Precision}_n = \frac{TN}{TN + FN} \tag{16}$$

(2) Recall, indicating that the proportion of the texts correctly classified into such texts in the texts is as shown in the formula 17.

$$\text{Recall}_p = \frac{TP}{TP + FN}, \text{Recall}_n = \frac{TN}{FP + TN} \tag{17}$$

(3) For the overall performance, using the Accuracy as an indicator, the calculation method is shown in the formula 18.

$$\text{Accuracy} = \frac{TP + TN}{TP + TN + FP + FN} \tag{18}$$

4.3   Experimental Design and Result Analysis

In the last section, we introduced two methods of emotion adjustment of word vectors: semantic similarity and Skip-gram model based method. In order to verify the validity of the emotion adjustment methods proposed in this paper, we combine the two methods with the SVM classification algorithm to construct the sentiment analysis model and compare with the traditional word vector model.

In order to verify the effectiveness of the two methods, we compared the methods of designing different features to design experiments, including the following methods.

(1) Word vectors based on one-hot
(2) Feature representation based on the initial word vector
(3) Feature representation based on the initial word vector and weighted sum

(4) Emotion adjustment method based on semantic similarity
(5) Emotion adjustment method based on semantic similarity and weighted sum
(6) Skip-gram based emotion adjustment method
(7) Skip-gram based emotion adjustment method and weighted sum

The experimental results shown in Table 3, the results are as follows:

Table 3. Table of the experimental results.

Method	Positive		Negative		Acc
	Pre	Recall	Pre	Recall	
Method 1	68.16%	83.98%	77.78%	58.84%	71.71%
Method 2	83.44%	73.60%	76.73%	85.64%	79.67%
Method 3	87.87%	88.59%	88.24%	87.50%	88.05%
Method 4	88.29%	88.03%	88.62%	88.86%	88.46%
Method 5	91.71%	89.73%	89.63%	91.40%	90.54%
Method 6	89.68%	89.17%	89.73%	90.22%	89.71%
Method 7	92.78%	90.27%	89.97%	92.55%	91.38%

(1) The accuracy of the tagged "positive emotions" is higher than the accuracy of "negative emotions". This may be due to the continuous updating of words on the Internet, in which the negative emotional words are updated more quickly, resulting in the inability to learn their characteristics in time, and the accuracy is lower.
(2) Compared with the method which use word vector based on one-hot representation, all methods which use Word2vec to generated word vectors have better results on text sentiment analysis.
(3) Compared with the traditional method of word vector, the two methods of emotion adjustment proposed in this paper have improved the accuracy and recall rate, which shows that the method based on word-vector emotion adjustment is helpful to improve the accuracy of text sentiment analysis.
(4) The accuracy and the recall rate of the method in this paper are all above 88%, which shows that the method in this paper have good performances in the task of sentence sentiment analysis.
(5) Compared with the emotion adjustment method based on semantic similarity, the evaluation performance of the emotion adjustment method based on Skip-gram model has been significantly improved, indicating that emotion adjustment method based on Skip-gram can achieve better results.
(6) Fig. 5 presents the comparison of whether to use weighted sum under different methods, it is obvious that using word vector constructed by weighted sum method to construct text feature space is better than non-weighted representation, all indicators are improved, which can effectively improve the accuracy of text sentiment analysis.

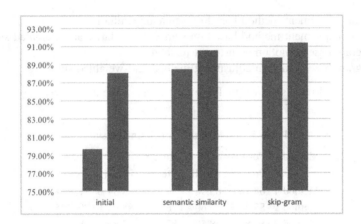

Fig. 5. Comparison of whether to use weighted sum under different methods

This paper also finds an example to show how the ten words closest to the same word "高兴" change after the emotional adjustment method is used, we choose the method based on Skip-gram to get the result, as shown in the Table 4.

Table 4. The ten closest words

Before	After
欣慰	开心
懊恼	惊讶
荣幸	欣喜
惊讶	欢乐
欣喜	懊恼
恼火	喜欢
吃惊	积极
恼怒	惊喜
难过	难过
着急	轻松

From the Table 4, we can clearly find that before emotion adjust, the ten closest words of the positive emotion word "高兴" contain some negative words like "懊恼"、"恼怒", the reason of this situation is that they have same context when use. After use our emotion adjustment method, eight of the ten closest words are positive words, although not converted all words into positive words, this result has made great progress compared with the previous ones which proves that our method is useful but has a lot of room for improvement.

5 Conclusion and Future Work

In this paper, a learning method of emotion adjustment word vector based on semantic similarity and skip-gram model is proposed, which solves the problem that the traditional method based on neural network is only based on the context of words, ignoring the sentiment polarity of words, which can not be effectively used in the task of sentiment analysis. The original semantic word vector is regarded as the initial word vector. Through the two methods proposed in this paper, we can get the word vector which takes into account the semantic and sentimental tendency at the same time, which is used to sentiment classification of the text. The method of this paper avoids the defects of semantic approximation of the original word vector but the big difference of emotion, and can learn the semantic information of the word from a large amount of unlabeled corpus, which achieves good affective classification effect. Among them, the skip-gram model based emotion adjustment method is more effective than the method based on semantic similarity.

There is still much space for improvement in this article. At present, the model proposed in this paper is limited to a single text that contains only one emotion. In reality, a single text may contain many opinions or emotions, whether these methods are suitable remains to be further explored.

References

1. CNNIC. http://www.cnnic.net.cn/hlwfzyj. Accessed 31 Jan 2018
2. Kwak, H., Lee, C., Park, H., et al.: What is Twitter, a social network or news media? In: International Conference on World Wide Web, pp. 591–600 (2010)
3. Paccanaro, A., Hinton, G.E.: Learning distributed representations of concepts using linear relational embedding. IEEE Trans. Knowl. Data Eng. **13**(2), 232–244 (2002)
4. Pennington, J., Socher, R., Manning, C.: Glove.: global vectors for word representation. In: Conference on Empirical Methods in Natural Language Processing, pp. 1532–1543 (2014)
5. Mikolov, T., Corrado, G., Chen, K., et al.: Efficient estimation of word representations in vector space. In: International Conference on Learning Representations, pp. 1–12 (2013)
6. Hatzivassiloglou, V., Mckeown, K.R.: Predicting the semantic orientation of adjectives. In: Eighth Conference on European Chapter of the Association for Computational Linguistics, pp. 174–181. Association for Computational Linguistics (1997)
7. Turney, P.D.: Thumbs up or thumbs down?: Semantic orientation applied to unsupervised classification of reviews. In: Meeting on Association for Computational Linguistics, pp. 417–424 (2002)
8. Niu, Y., Pan, M., et al.: Emotion analysis of Chinese microblogs using lexicon-based approach. Comput. Sci. **41**(9), 253–258, 289 (2014)
9. Pang, B., Lee, L., Vaithyanathan, S.: Thumbs up? Sentiment classification using machine learning techniques. In: ACL-02 Conference on Empirical Methods in Natural Language Processing, pp. 79–86. Association for Computational Linguistics (2002)

10. Mullen, T., Collier, N.: Sentiment analysis using support vector machines with diverse information sources. In: Conference on Empirical Methods in Natural Language Processing, EMNLP 2004, A Meeting of Sigdat, A Special Interest Group of the ACL, Held in Conjunction with ACL 2004, 25–26 July 2004, Barcelona, Spain, pp. 412–418. DBLP (2004)
11. Wang, S., Manning, C.D.: Baselines and bigrams: simple, good sentiment and topic classification. In: Meeting of the Association for Computational Linguistics: Short Papers, pp. 90–94. Association for Computational Linguistics (2012)
12. Li, S., Huang, L., Wang, J., et al.: Semi-stacking for semi-supervised sentiment classification. In: Meeting of the Association for Computational Linguistics and the, International Joint Conference on Natural Language Processing, pp. 27–31(2015)
13. Sun, J., Xueqiang, L., Zhang, L.: On sentiment analysis of Chinese microblogging based on lexicon and machine learning. Comput. Appl. Softw. **31**(7), 177–181 (2014)
14. Jiang, J., Xia, R.: Microblog sentiment classification via combining rule-based and machine learning methods. In: NLPCC (2016)
15. Morin, F., Bengio, Y.: Hierarchical probabilistic neural network language model. In: AISTATS (2005)
16. Mikolov, T., Sutskever, I., Chen, K., et al.: Distributed representations of words and phrases and their compositionality, vol. 26, pp. 3111–3119 (2013)
17. HowNet. http://www.kenage.com
18. tsinghua.:http://nlp.csai.tsinghua.edu.cn/site2/index.php/zh/resources/13-v10

Text Sentiment Analysis Based on Convolutional Neural Network and Bidirectional LSTM Model

Mengjiao Song[✉], Xingyu Zhao, Yong Liu, and Zhihong Zhao

Software Institute, Nanjing University, Nanjing 210093, China
selene_smj@163.com

Abstract. Text sentiment analysis is used to discover the public's appreciation and preferences for specific events. In order to effectively extract the deep semantic features of sentences and reduce the dependence of long distance information dependency, two models based on convolutional neural network and bidirectional long short-term memory model, CNN-BLSTM and BLSTM-CNN are proposed. Convolutional neural networks can get text features better. Bidirectional long-short time memory model can not only capture long-range information and solve gradient attenuation problem, but also represent future contextual information semantics of word sequence better. These two network architectures are explained in detail in this paper, and performed comparisons against some normal methods, such as methods based emotion lexicon, machine learning methods, LSTM and other neural network models. Experiments show that these two proposed models have achieved better results in text sentiment analysis. The best model CNN-BLSTM is better than the normal neural network models in accuracy.

Keywords: Sentiment analysis · Long short-term memory
Convolutional neural network · Bidirectional LSTM

1 Introduction

In the Web 2.0 era, people express their opinions through the Internet and post comments with personal feelings about social hot issues, products and services. Social media services like Facebook, Twitter and LinkedIn grew exponentially. In China, Weibo becomes the platform for most people to express their ideas, disseminate content and exchange knowledge with its quick and easy features. According to the statistics of CNNIC, the number of netizens in China reached 772 million, the utilization rate of Weibo reached 40.9%, an increase of 16.4% over 2016 [1] by the end of December 2017. The number of Weibo texts has also witnessed large-scale growth, which attracting a large number of scholars to do various studies based on weibo data [2]. The data from these platforms is a valuable source of information and can be used to understand customers more by analyzing it.

This paper focuses on the sentiment analysis of Weibo texts, analyzing the emotions of Weibo texts can be used to find out the public's evaluation and preference of specific events or objects, and is of great significance for applications such as market

© Springer Nature Singapore Pte Ltd. 2018
Q. Zhou et al. (Eds.): ICPCSEE 2018, CCIS 902, pp. 55–68, 2018.
https://doi.org/10.1007/978-981-13-2206-8_6

research, network public opinion discovery and early warning. Therefore, the analysis of the emotional bias of Weibo texts has become one of the hot researches.

Traditional research on sentiment analysis can be divided into method based on sentiment lexicon and machine learning. Method based on sentiment lexicon usually require manual construction of special sentiment lexicons for different fields. The quality of sentiment analysis is closely related to the quality and coverage of sentimental lexicon. The construction and maintenance of sentiment lexicon also requires a lot of manpower. With the emergence of new words on the Internet, approach based on sentiment lexicon can no longer meet the needs. The method based on machine learning relies on the manual selection of features. Different feature selections can cause differences in the results of sentiment analysis. There is a certain degree of difficulty in model generalization.

In recent years, the method of sentiment analysis based on deep learning has become a hot topic of research. Compared with traditional machine learning methods, deep learning methods have stronger expressive capabilities and do not require manual feature selection and construction. They have a good effect on sentiment analysis tasks.

Based on the above discussion, this paper chooses deep learning method to do sentiment analysis. Feature selection and sentence sequence contextual dependency information plays important roles in the accuracy of the sentiment analysis. Therefore, this paper proposes to combine convolutional neural network and bidirectional long-short time memory model, and propose two different neural network models: CNN-BLSTM model and BLSTM-CNN model. Based on the convolutional neural network can extract hidden features from the text and feature combination; Bidirectional long-short time memory model can better use the text sequence relationship to learn sentence semantics, store contextual information while taking into account future contextual information, which can have a better performance in text sentiment analysis. Through experiments, the accuracy of lexicon-based methods, traditional machine learning method, LSTM method, bidirectional long-short time memory model method, CNN-BLSTM method and BLSTM-CNN method in sentiment analysis are compared. The results show that the two models proposed in this paper are more effective for the sentiment analysis bias of Weibo texts.

The remainder of this paper is laid out as follows. Section 2 introduces related work. Section 3 introduces our approach in detail. Section 4 shows the implementation of our approach and the experiments and Sect. 5 makes conclusion and discusses the future work.

2 Related Work

2.1 Sentiment Analysis

Sentiment analysis, also known as propensity analysis and opinion mining, is a process of analyzing, processing, inducing and reasoning subjective texts with emotions [3]. The methods used in sentiment analysis at this stage are mainly divided into methods based on sentiment lexicon and machine learning.

The method based on sentiment lexicon mainly through the development of a series of sentiment lexicons and rules, paragraph disassembly and syntactic analysis of the text, analyzing the number of positive sentiment words and negative sentiment words in the text, calculating the emotional differences and the text's sentiment orientation is judged by this value. Turney proposed to judge the sentiment orientation of the whole text according to the relevance between the words in the test text and the words in the seed lexicon [4]. Pan proposed six kinds of sentiment and the application of emoticons significantly improved the accuracy and coverage of sentiment analysis [5].

Methods based on machine learning most transform this problem to a classification problem to treat. For the judgment of sentiment orientation, the target text is divided into two categories as positive and negative emotions. By manual annotation of training corpus and test corpus, sentiment analysis is carried out using classifier such as Support Vector Machine(SVM), Maximum Entropy Model and K-nearest neighbor(KNN). For example, Pang used machine learning methods to classify movie review data into positive and negative categories [6]. Mullen used support vector machines to analyze the syntactic relationships and features of texts to achieve sentiment partitioning [7]. Wang [8], Li [9] and others improved the accuracy by combining different classifiers, and proposed a semi-supervised propensity analysis model to solve the imbalance corpus problem under certain circumstances. Sun and others proposed to combine the sentiment lexicon with machine learning method. Based on the method of location weight calculation based on the characteristic polarity, SVM was used as the machine learning model to classify the Weibo text into three categories: positive, negative and neutral, which effectively improved the accuracy of sentiment analysis [10]. Jiang and others combined rules based on sentiment lexicon and machine learning methods to improve the accuracy of sentiment analysis [11].

Recently, many people use deep learning method to do sentiment analysis. Sunderm and others proposed using a long-short memory neural network to construct a language model and achieved good results [12]. Kim first proposed using convolutional neural networks to classify texts and achieved excellent results [13]. Tang and others used convolutional neural networks or long-term and short-term memory neural networks to learn sentence expressions, and used a gated recurrent unit (GRU) to learn document representations and achieved a text sentiment classification task [14].

2.2 Convolutional Neural Network

Convolutional Neural Network (CNN) is an improvement on Backpropagation Neural Networks. In 1962, Hubel and Wiesel [15] found that the transmission of visual information from the retina to the brain was accomplished through multiple levels of sensory field excitation. This kind of structure can effectively reduce the complexity of traditional Backpropagation neural network calculation, and thus put forward the concept of CNN. CNN usually contains three parts, including a Convolutional layer, a Pooling layer and a Fully-Connected layer. The following describes the function of each component.

Convolutional layer: Each convolutional layer in convolutional neural network is composed of certain number of convolution units, each convolution unit parameters are optimized through backpropagation algorithm. The purpose of the convolution operation is to extract the different features of the input. The first convolution layer may only extract some low-level features, but more layers of networks can iteratively extract more complex features.

Pooling layer: Usually after the convolution layer, it will get a large dimension features, the pooling layer cuts the features into several regions, get the maximum or average and deserve a new and smaller dimensions.

Fully-Connected layer: Combine all the local features into a global feature, which is used to calculate the score for each of the last categories, and the classification results are obtained at the end.

2.3 Long Short-Term Memory

The LSTM model was first proposed by Hochreiter and Schmidhuber in 1997 [19]. Felix Gers et al. improved in 2000 to solve the problem that the Recurrent Neural Network (RNN) has a gradient explosion or disappearance caused by long-term dependencies and overlong sequence. LSTM is an extension of RNN that records long-term information through a special structure of memory cell and removes or increases the information to the cell state through the structure of "gate".

The LSTM neural network model used in this paper is shown in Fig. 1. It mainly consists of four parts: an input gate, a neuron with a self-recurrent connection (a connection to itself), a forget gate and an output gate. The input gate can allow the input signal to change the state of the memory cell or block it. The self-recurrent connection has a weight of 1.0 and ensures that, barring any outside interference, the state of a memory cell can remain constant from one time step to another. The output gate may allow the state of the memory cell to affect or prevent it from other neurons. Finally, forget gate can modulate the memory cell's self- recurrent connection, allowing the cell to remember or forget its previous state as needed. The concrete calculation is as shown as follows.

$$i_t = \sigma(Wx_i x_t + U_i h_{t-1} + b_i) \tag{1}$$

$$\widetilde{C}_t = tanh(W_c x_t + U_c h_{t-1} + b_c) \tag{2}$$

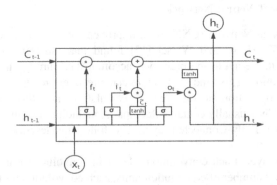

Fig. 1. LSTM neural network structure

$$f_t = \sigma\left(W_f x_t + U_f h_{t-1} + b_f\right) \tag{3}$$

$$C_t = i_t * \widetilde{C}_t + f_t * C_{t-1} \tag{4}$$

$$o_t = \sigma(W_o x_t + U_o h_{t-1} + b_o) \tag{5}$$

$$h_t = o_t * tanh(C_t) \tag{6}$$

Where x_t is the input to the memory cell layer and h_t is the representation sequence corresponding to x_t at time t; and f_t is the activation of the memory cells' forget gate; i_t is the activation of the input gate; o_t is the input of the output gate; C_t is the memory cell's new state; σ represents the sigmoid activation function; W_*, U_* and V_* are weight matrixes, b_* is the bias vector, their subscripts indicate their classification. Through the control of these gates, LSTM has the ability to save, read and update long-distance information, which effectively reduces the gradient attenuation problem of RNN.

3 Approach

In this section, we describe in detail how we analysis the sentiment of the text as follows. Section 3.1 presents an overview of the workflow in our approach. Section 3.2 introduces text preprocessing. Section 3.3 introduces how to get word vector. Section 3.4 introduces the structure of CNN-BLSTM model and Sect. 3.5 introduces the structure of BLSTM-CNN model.

3.1 Overview

Figure 2 presents an overview of our approach. The steps are as follows:

(1) Data preparation, crawl data online and annotate sentiment orientation manually.
(2) Corpus preprocessing, including data cleaning, word segmentation, removal of stop words and so on.
(3) The word2vec method is used to compute the processed corpus to get the word vector.
(4) Using different neural network model to train and get text sentiment analysis results.

3.2 Corpus Preprocessing

As the Chinese microblogging text grammar is not standardized, including spoken language, Internet words, expressions and links, etc., which need to preprocess the text, including the following steps:

1. Data Cleaning: Since the microblog text contains information such as links, for-wards, @username and so on, the presence or absence of such information will interfere with the results of the sentiment analysis task, but the information itself is

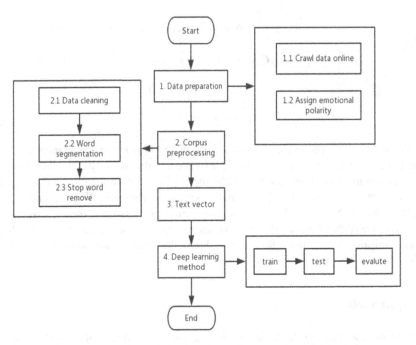

Fig. 2. Overview of the workflow of text sentiment analysis

irrelevant to the sentiment analysis task. Data cleaning is similar to the process of removing a stop word. However, if data cleaning is not performed in advance, the information may be improperly segmented under the subsequent word segmentation operation, resulting in a decline in the quality of the word vector and a failure to better represent the meaning of the word. Therefore, such information needs to be replaced with a uniform mark in advance through a regular expression.

In dealing with the information like @username, there are three situations need to be considered separately and take different measures. First of all, "context1 @username: context2". In this case, since context2 is written by other user, the sentiment orientation of the context may be the same as or different from the current user. If may affect the current user's sentiment orientation analysis when keeping context2, so we need to replace it. The second case is "@username context", where this context is written by the current user, so simply replace @username. The last case is "reply@ username: context". In this case, the speech is also written by the current user. Therefore, it is also sufficient to replace "reply@username"l.

The specific regular expression and replacement mark are shown in Table 1.

2. Chinese Word Segmentation: The segmentation of the processed text is used as the input for the next stage. The common Chinese word segmentation tools including Jieba [20], NLPIR [21], LTP [22] etc., this paper uses LTP tool to do word segment.

3. Remove Stop Words: In order to effectively save storage space and running time, this paper adopts the method of filtering stop words. The use of stop words can reduce the dimension of feature space, reduce the amount of data and exclude

Table 1. Regular expression and replacement mark.

Context	Regular expressions	Mark
HTML	</?\w +[^>]*≫	Html
URL	((https|http|ftp|rtsp|mms)?:VV)[^\s]+	Url
#topic#	#[^#]+#	Tag
context1//@username:context2	@.+MYM	Forward
@username context	@[^\s]+	At
reply@username:context	reply@[^:]+	Reply

interference characteristics. In Chinese, words expressing emotions include nouns, verbs, adjectives and adverbs, as well as emotional words such as diacritics, interjection, onomatopoeia, pronouns, idioms, and abbreviations. The stop words selected in this model are words that do not have emotional colors, including meaningless mood words, prepositions, conjunctions, and numbers, Chinese numbers, English letters and special symbols. There are emoticons in the Weibo text, since the emoticons in the crawled texts have been replaced by the corresponding Chinese characters, they are not processed. However, there are some expressions made by stitching punctuation marks and special symbols, which are temporarily unrecognized and deleted by using stop words.

3.3 Word Vector

Text vectorization means that the word segmentation result is expressed by the word vector, which was first proposed by Hinton [23] in 1986. By mapping words to K-dimensional real vectors and judging the distance between word vectors Semantic similarity. In this paper, we use word2vector [24] proposed by Google to train word vectors on large unlabeled corpus, and the word vector obtained by training is used as the word vector of this model. A total of 650,000 Wiktionary and Weibo materials are used as training data sets for initial word vector by Word2vec. The training corpus consists of 250,000 Chinese Wikipedia and 400,000 Sina Weibo texts.

3.4 CNN-BLSTM

The CNN-LSTM model (Fig. 3) proposed in this paper consists of an initial convolution layer, a max pooling layer, a bidirectional LSTM layer and a fully-connected layer. The convolution layer will receive word vectors as input and extract local features. Its output will then be pooled to a smaller dimension which is then fed into the bidirectional LSTM layer. The bidirectional LSTM layer will then use the ordering of features to learn about the input's text ordering. The fully-connected layer combines all the local features into a global feature, which is used to calculate the score for each of the last categories and the classification results are obtained at the end.

Convolution Layer. For a given text sample x, which can be represented as $x = (x_1, x_2, \ldots, x_n), x_i \in R^d$, where d is the dimension of the word vector n represents the number of words in the text. To ensure that words are the smallest granularity in the

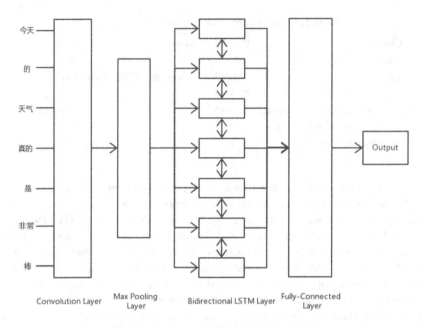

今天
的
天气
真的
是
非常
棒

Convolution Layer   Max Pooling Layer   Bidirectional LSTM Layer   Fully-Connected Layer   Output

Fig. 3. Structure of CNN-BLSTM model

language, filter only slides in height and the width is same as the dimension of the word vector. This guarantees that every time the window is scrolled, it is a complete word vector, and it does not convolve the partial word vectors of several words. The number of words contained in the filter is k, and its vector $m \in R^{k \times d}$. For each position in the text sample x, there is a window vector containing a continuous word vector, expressed as follows:

$$w_i = [x_i, x_{i+1}, \ldots, x_{i+k-1}] \tag{7}$$

The feature map $c \in R^{n-k+1}, c = [c_1, c_2, \ldots, c_{n-k+1}]$ is calculated by sliding the window one by one. The calculation is as follows, Where $.*$ means element-wise multiplication, b is bias and f is the activation function.

$$c_i = f(w_i.*m + b) \tag{8}$$

Pooling Layer. After the Convolutional Layer, the max pooling layer is used to reduce the computational complexity. By extracting the local dependencies of different convolution kernels, the most significant information is retained.

$$\tilde{c}_i = \max(c_i) \tag{9}$$

Bidirectional LSTM Layer. LSTM can capture longer-distance information, but as the text vector input is from left to right, which leads to the later words in the text which have greater influence on the sentiment analysis results. This is not appropriate, so this paper chooses bidirectional LSTM to improve this problem. The basic idea is that each training sequence contains two LSTM units, which are connected forward and backward respectively. The final output is obtained by combining the forward output vector $\vec{h}$ and the reverse output vector $\overleftarrow{h}$.

$$\vec{h}_{f_t} = \sigma\left(W_{f_t} x_t + U_{f_t} \vec{h}_{f_{t-1}} + b_{f_t} \right) \tag{10}$$

$$\overleftarrow{h}_{b_t} = \sigma\left(W_{b_t} x_t + U_{b_t} \overleftarrow{h}_{b_{t-1}} + b_{b_t} \right) \tag{11}$$

$$o'_t = [\vec{h}_{f_t}, \overleftarrow{h}_{b_t}] \tag{12}$$

Full-Connected Layer and Classifier. After the sentence is expressed by the bidirectional LSTM layer, all features are connected through the full connection layer, and the output value is used as the input of the softmax classifier to obtain the final classification result.

$$p(i) = \frac{exp\left(\theta_i^T x\right)}{\sum_{k=1}^{K} exp\left(\theta_k^T x\right)} \tag{13}$$

3.5   BLSTM-CNN

Our BLSTM-CNN model (Fig. 4) consists of an initial bidirectional LSTM layer, a convolution layer, a max pooling layer and a fully-connected layer.

The contents of all the layers are similar to those in Sect. 3.4. and are only described briefly here. The bidirectional LSTM layer will receive word embedding for each token in the Weibo text as inputs. Its output tokens will store information not only of the initial token, but also all the other tokens in the original input. Which is better than normal LSTM layer, cause LSTM layer can only store information of the tokens in front of the initial token. In other words, the bidirectional LSTM layer is generating a new encoding for the original input. The output of the bidirectional LSTM layer is then fed into a convolution layer which will extract local features. The output of the convolution layer will be pooled to a smaller dimension and ultimately outputted as either a positive or negative label by the fully-connected layer.

4   Experiment and Analysis

4.1   Data Preparation

In order to further analyze the influence of two neural network models proposed in this paper on sentiment analysis in Weibo texts, several specific topics of Weibo texts were

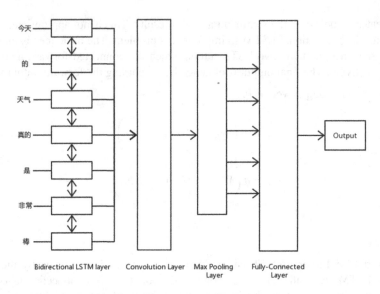

今天

的

天气

真的

是

非常

棒

Bidirectional LSTM layer   Convolution Layer   Max Pooling Layer   Fully-Connected Layer

Output

Fig. 4. Structure of BLSTM- CNN model

crawled from Sina Weibo, and the texts were emotionally classified by manual annotation. In this paper, we divide the simple emotion into two categories: positive emotion and negative emotion. The total verification data is expected to be 7,000 microblogging articles, 3375 positive emotions and 3625 negative emotions with a positive and negative balance. After labeling, 70% of all texts were extracted as training data and 30% as test data. Training data sets are used as training models and test data sets are used for performance evaluation.

4.2 Evaluation Standard

In this paper, the Precision and Recall as the index to evaluate the sentiment analysis of good and bad, the overall performance using Accuracy as an indicator. The following is the description of each evaluation index.

(1) Precision, which indicates the proportion of the texts correctly classified in this category to the texts categorized in that category, as shown in the formula.

$$Precision_p = \frac{TP}{TP + FP}, Precision_n = \frac{TN}{TN + FN} \tag{14}$$

(2) Recall, indicating that the proportion of the texts correctly classified into such texts in the texts is as shown in the formula.

$$Recall_P = \frac{TP}{TP + FN}, Recall_n = \frac{TN}{FP + TN} \tag{15}$$

(3) For the overall performance, using the Accuracy as an indicator, the calculation method is shown in the formula.

$$Accuracy = \frac{TP + TN}{TP + TN + FP + FN} \qquad (16)$$

4.3 Parameters

The best parameters for our models were initially selected based on previous implementations of LSTMs, and further fine-tunned through manual testing (Table 2).

Table 2. Parameters selected for our CNN-BLSTM and BLSTM-CNN models

Embedding dimension	256
Epoch	5
Batch size	16
Filter	128
Kernel size	3
Pool size	3
Dropout	0.5

4.4 Experimental Design and Result Analysis

In the last section, we introduced two neural network models: CNN-BLSTM and BLSTM-CNN. In order to verify the validity of the neural network models proposed in this paper, comparing with different methods including the following methods.

(1) Lexicon-based method
(2) Support Vector Machine (SVM) method
(3) LSTM model
(4) Bidirectional LSTM model
(5) CNN-LSTM model
(6) CNN-BLSTM model (Proposed by this paper)
(7) LSTM-CNN model
(8) BLSTM-CNN model (Proposed by this paper)

The experimental results shown in Table 3, the results are as follows:

(1) The method of using neural network model compared with the traditional method based on the lexicon and machine learning, the evaluation performance has significantly improved, indicating that the use of deep learning method can achieve better results on text sentiment analysis.
(2) As can be clearly seen from the Table 4, the method using the bidirectional LSTM model is obviously improved compared with the method using the traditional LSTM model, indicating that the bidirectional LSTM can fully capture the text context information.

Table 3. Experimental results.

Method	Positive		Negative		Accuracy
	Pre	Recall	Pre	Recall	
Lexicon	68.16%	83.98%	77.78%	58.84%	71.71%
SVM	85.71%	53.93%	67.46%	91.40%	73.08%
LSTM	90.16%	84.81%	83.33%	89.13%	86.80%
BLSTM	91.02%	86.30%	84.84%	90.00%	88.00%
CNN-LSTM	89.62%	86.30%	84.58%	88.26%	87.20%
CNN-BLSTM	89.43%	87.78%	85.96%	87.83%	87.80%
LSTM-CNN	91.60%	88.89%	87.39%	90.43%	89.60%
BLSTM-CNN	92.02%	89.63%	88.19%	90.87%	90.20%

(3) The CNN-BLSTM model does worse than a regular bidirectional LSTM model, and the reason may be that the initial convolutional layer of our CNN-BLSTM is loosing some of the text's sequence information. Which makes the bidirectional LSTM layer act as just a fully connected layer.

(4) The BLSTM-CNN model is the best model which achieved an accuracy of 3.9% better than LSTM model, 2.5% better than the bidirectional LSTM model and 1% better than LSTM-CNN model. The initial bidirectional LSTM layer act as an encoder, for every token in the input there is an output token that contains information not only of the original token, but all other tokens. And the output token in LSTM-CNN model can only contains all other precious tokens. Then, the CNN layer find local patterns using this richer representation of the original input. The BLSTM-CNN model can effectively improve the accuracy of text sentiment analysis.

5 Conclusion and Future Work

In this paper, aiming at the problem of text sentiment analysis, this paper presents two methods based on convolutional neural network (CNN) and bidirectional long-short memory network (bidirectional LSTM) to discriminate the sentiment orientation of Weibo texts, including CNN-BLSEM model and BLSTM-CNN model. Experiments show that the BLSTM-CNN model can learn deep features and order features better, with an accuracy rate of 90.2%, which is 2.5% better than the bidirectional LSTM model and 1% better than LSTM-CNN model. It proves that the model has good expansibility and application value in the task of text sentiment analysis.

There is still much space for improvement in this article. At present, the model proposed in this paper may limited to a single text that contains only one emotion. In reality, a single text may contain many opinions or emotions, whether these methods are suitable remains to be further explored.

References

1. CNNIC. http://www.cnnic.net.cn/hlwfzyj. Accessed 31 Jan 2018
2. Kwak, H., Lee, C., Park, H., et al.: What is Twitter, a social network or news media? In: International Conference on World Wide Web, pp. 591–600 (2010)
3. Hatzivassiloglou, V., Mckeown, K.R.: Predicting the semantic orientation of adjectives. In: Eighth Conference on European Chapter of the Association for Computational Linguistics, pp. 174–181. Association for Computational Linguistics (1997)
4. Turney, P.D., Pantel, P., et al.: From frequency to meaning: vector space models of semantics. J. Artif. Intell. Res. **37**(1), 141–188 (2010)
5. Niu, Y., Pan, M., et al.: Emotion analysis of chinese microblogs using lexicon-based approach. Comput. Sci. **41**(9), 253–258, 289 (2014)
6. Pang, B., Lee, L., Vaithyanathan, S.: Thumbs up? Sentiment classification using machine learning techniques. In: ACL-02 Conference on Empirical Methods in Natural Language Processing, pp. 79–86. Association for Computational Linguistics (2002)
7. Mullen, T., Collier, N.: Sentiment analysis using support vector machines with diverse information sources. In: Conference on Empirical Methods in Natural Language Processing, EMNLP 2004, A Meeting of SIGDAT, A Special Interest Group of the ACL, Held in Conjunction with ACL 2004, 25–26 July 2004, Barcelona, Spain, pp. 412–418. DBLP (2004)
8. Wang, S., Manning, C.D.: Baselines and bigrams: simple, good sentiment and topic classification. In: Meeting of the Association for Computational Linguistics: Short Papers, pp. 90–94. Association for Computational Linguistics (2012)
9. Li, S., Huang, L., Wang, J., et al.: Semi-stacking for semi-supervised sentiment classification. In: Meeting of the Association for Computational Linguistics and the, International Joint Conference on Natural Language Processing, pp. 27–31 (2015)
10. Sun, J., Xueqiang, L., Zhang, L.: On sentiment analysis of Chinese microblogging based on lexicon and machine learning. Comput. Appl. Softw. **31**(7), 177–181 (2014)
11. Jiang, J., Xia, R.: Microblog sentiment classification via combining rule-based and machine learning methods. In: NLPCC (2016)
12. Sundermeyer, M., Schlüter, R., Ney, H.: LSTM neural networks for language modeling. In: Interspeech, pp. 601–608 (2012)
13. Kim, Y.: Convolutional Neural Networks for Sentence Classification. Eprint Arxiv (2014)
14. Tang, D., Qin, B., Liu, T.: Document modeling with gated recurrent neural network for sentiment classification. In: Conference on Empirical Methods in Natural Language Processing, pp. 1422–1432 (2015)
15. Hubel, D.H., Wiesel, T.N.: Receptive fields, binocular interaction and functional architecture in the cat's visual cortex. J. Physiol. **160**(1), 106–154 (1962)
16. Schuster, M., Paliwal, K.K.: Bidirectional recurrent neural networks. IEEE Press (1997)
17. Ma, X., Hovy, E.: End-to-end sequence labeling via bi-directional lstm-cnns-erf. arXiv preprint arXiv (2016)
18. Wang, C., Yang, H., Bartz, C., et al.: Image captioning with deep bidirectional LSTMs, pp. 988–997 (2016)
19. Hochreiter, S., Schmidhuber, J.: Long short-term memory. Neural Comput. **9**(8), 1735–1780 (1997)
20. Jieba. https://github.com/fxsjy/jieba
21. NLPIR. http://ictclas.nlpir.org

22. LTP. https://www.ltp-cloud.com
23. Paccanaro, A., Hinton, G.E.: Learning distributed representations of concepts using linear relational embedding. IEEE Trans. Knowl. Data Eng. **13**(2), 232–244 (2002)
24. Mikolov, T., Corrado, G., Chen, K., et al.: Efficient estimation of word representations in vector space. In: International Conference on Learning Representations, pp. 1–12 (2013)

Research on Dynamic Discovery Model of User Interest Based on Time and Space Vector

Jinxiu Lin, Zhaoxin Zhang, Lejun Chi$^{(\boxtimes)}$, and Yang Wang

School of Computer Science and Technology, Harbin Institute of Technology,
Harbin 150001, People's Republic of China
qdclj@163.com

Abstract. It is difficult for forum administrators to grasp the timeliness and accuracy of forum hotspot distribution and changes, which is detrimental to the management of forum hotspot trends, and the timeliness and accuracy of user groups with the same concerns, which is not conducive to the forum user's mutual communication. Therefore, a dynamic interest discovery model based on space-time vector is proposed. In this paper, based on the traditional vector space model (VSM), the TF-IDF algorithm is used to calculate the keyword weights of user interests and construct the user interest vector matrix. The dynamic discovery of the user's focus through two dimensions of time and space, improving the timeliness and accuracy of the administrator's grasp of the forum hotspots, and quickly and accurately finding the user groups with the same focus, is conducive to the realization of knowledge sharing. The case study shows that the dynamic discovery model of user interest based on space-time vectors can effectively find out the dynamic changes of the attention points of users and users that meet the concerns.

Keywords: User interest model · VSM · Time and space vector

1 Introduction

In today's web2.0 era, the breadth and depth of the network are constantly expanding, and the network of people in the society are getting closer and closer. Social networking sites play an increasingly important role in people's work and life. Forum is a large number of users online social networking platform, users can freely comment on this platform [1]. Such a huge amount of data, both forum managers and forum users are both wealth and trouble. It is difficult for the forum administrator to timely and accurately grasp the hot spots of the forum, which is not conducive to the daily management of the forum. It is difficult for the users to quickly and accurately search for the posts of interest and their users in the forum [2]. Therefore, it is very necessary to excavate and dynamically discover forum hot spots and user's concerns. In this paper, a Vector Space Model (VSM) is introduced and a time mechanism is introduced to build a user interest matrix. The forum hot spots and user's concerns are dynamically discovered in two dimensions of time and space. The case study shows that the dynamic interest discovery model based on spatio-temporal vectors can discover the

© Springer Nature Singapore Pte Ltd. 2018
Q. Zhou et al. (Eds.): ICPCSEE 2018, CCIS 902, pp. 69–80, 2018.
https://doi.org/10.1007/978-981-13-2206-8_7

dynamic changes of the forum hot spots and the user's concerns timely and accurately in two dimensions of space and time

2 Extraction of User Interest Features

The accurate extraction of user interest features is the key to construct the user interest model [3].

According to the degree of user's participation, user interest feature extraction can be roughly divided into two kinds. The first is the need for user's participation, based on their information to build a user interest model. The second is to construct the user interest model by mining the text content and tracking the user's behavior without user's participation [4]. However, in comparison, the second method is superior to the first method, because in most practical situations, the users may be unwilling to provide the information voluntarily or the information provided is not comprehensive and accurate [5].

Therefore, this article will take a second user-independent approach to extract user interest features. This article will develop the deployment of distributed crawler crawling the forum user's information, such as user's name, url, the post title and content, time of publication and other data.

After collecting the relevant information data of the forum users, the data needs to be pre-processed, including data cleaning, Chinese word segmentation, and feature word extraction.

2.1 Data Cleaning

There are some special symbols, foreign languages and other dirty data in the forum data crawled by the crawler to remove the abnormal data by matching regular expressions and store the cleaned data into a database.

2.2 Chinese Word Segmentation

Word segmentation refers to the continuous sequence of words in accordance with a certain standard re-segmentation and combination of the formation of the sequence process. It is the foundation of all Chinese natural language processing [6]. This article uses the Chinese word segmentation module developed by Python Jieba participle.

The basic principle of Jieba segmentation is to realize efficient wordmap scanning based on the Trie tree structure and generate a directed acyclic graph of all the possible words formed in the sentence in the sentence and to find the maximum probability path by using dynamic programming to find the maximum segmentation based on word frequency combination [7].

After word segmentation of the text content, the corresponding part of speech, the noun, the verb and the noun are reserved.

2.3 Feature Word Extraction

Text using Jieba participle contains a large number of words, the use of simple word frequency information to extract the content keyword effect is acceptable. Because the most meaningful words for the difference document should be those that appear more frequently in the document but less frequently in other documents in the entire document collection, if the feature space coordinate system takes the TF word frequency as a measure, Textual characteristics [8]. The TF-IDF algorithm considers the smaller the frequency of a word appears, the greater its ability to distinguish different types of text. Therefore, the concept of inverted text frequency IDF is introduced, the product of TF and IDF is taken as the value measure of the feature space coordinate system and used to complete the adjustment of weight TF. The purpose of weight adjustment is to highlight important words, wanted words. Therefore, we use the TF-IDF method to extract the feature words of user's content [9], and extract the important words that can express the text content better. The TF-IDF calculation formula is shown as (1) below.

$$tfidf_{i,j} = tf_{i,j} \times idf_i \tag{1}$$

Which, $tfidf_{i,j}$ that the entry t_i in the document d_i weight, tf_{ij} for the word frequency, idf_i is the inverse document frequency.

$$tf_{i,j} = \frac{n_{i,j}}{\sum_k n_{k,j}} \tag{2}$$

Where $n_{i,j}$ is the number of occurrences of the word t_i in the document d_i and the denominator is the sum of the occurrences of all the words in the document d_i.

$$idf_i = \log \frac{|D|}{|\{j : t_i \in d_j\}|} \tag{3}$$

Where $|D|$ is the total number of documents in the corpus, $|\{j : t_i \in d_j\}|$ number of documents containing the term t_i.

3 User Interest Model Representation

The user interest model was originally proposed in the information filtering system and is a description of the user's information needs. The user interest model represents the user's relatively stable interest needs for specific topic information, including dynamic updates to accurately reflect the particular interests of the user. Therefore, the user interest model has a certain period of stability and tendencies [10].

User interest representation can be divided into the following categories: thematic representation, keyword list vector method, bookmark representation, ontology-based representation, interest granularity representation, neural network representation model, vector space model representation [11].

Among them, the Vector Space Model (VSM) representation is a vector representation of user models in vector space of key words, which is the most commonly used interest representation model [12].

Vector Space Model is an n-dimensional eigenvector to represent user interest, as shown in (4):

$$\{(t_1, w_1), (t_2, w_2), \ldots, (t_n, w_n)\} \tag{4}$$

Each dimension is composed of the keyword t_n and its corresponding weight w_n. The weight can be a Boolean value or a real number, which indicates the user's interest in the content of the keyword [13]. For example, the interest feature vector $\{(literary, 0.3), (mathematics, 0.6), \ldots, (music, 0.4)\}$ means that the degree of interest in literature is 0.3, the degree of interest in mathematics is 0.6, the degree of interest in music 0.4.

This paper introduces a time mechanism based on the traditional Vector Space Model to represent the user's concern in the time dimension, as shown in Eq. 5. Among them, $time_1$ said the interest of the post where the publication of the time.

$$\{(t_1, w_1), (t_2, w_2), \ldots, (t_n, w_n), time_1\} \tag{5}$$

4 User Interest Model Construction

User interest modeling is the foundation and core of interest discovery [14]. User interest modeling involves the following three steps [15]:

1. Collect and process user-related information data;
2. Build a user interest model;
3. Update the user interest model.

Hao and others proposed a user interest model based on the hierarchical Vector Space Model (VSM) representation and updated processing mechanism [12]. Diao Zulong uses domain ontology to build user interest model [13]. Qiu uses the Vector Space Model to establish the interest space of microblog users [16]. Li and others put forward a method of constructing user interest model based on domain ontology [17]. The research results of Ren et al. showed that the Vector Space Model is usually used to judge the relativity between the data in the network and the determined topics. But when the elements of the theme reappear, their order will not be considered by the Vector Space Model. Adding a new element provided an evolutionary vector space model [18]. Zhao et al. developed an indicative opinion generation model using BM25 to identify important texts and use syntactic parsing to obtain a brief opinion representation [19]. Dong et al. used the Vector Space Model to generate alternative microblogging heat groups in a centralized manner, and optimized hotspot microblogging discovery algorithms so that the sorting results were more in line with information propagation rules [20]. Based on the original Vector Space Model, Cheng et al. proposed the time-consciousness and gray incidence rate based on the theory of user interest model to improve the literature recommendation [21].

The main ideas of this paper are:

1. Take the distributed crawler to collect user's information in the forum page and perform pretreatment operations such as data cleaning and word segmentation;
2. TF-IDF algorithm is used to calculate the weight of the key words, and the key words of the top5 weight value are extracted as the characteristic items to represent the contents of the post.
3. Mark and classify the feature items and introduce the time mechanism to form the interest feature vector, and then construct the user interest feature matrix;
4. Analyze the user's concerns in the spatial dimension, and get the different types of hot spots as well as the persistent hot spots and intermittent hot spots in the forum. Analyze the user's concerns in the time dimension to get the dynamic changes of the users' concerns and the continuous concerns (Table 1).

Table 1. User interest model building process algorithm

Input: Forum
Output: User Interest Model

Begin:

1. crawl forum user's information data(username, url, title, content, posttime)
2. clean data and structured storage to the database
3. data preprocessing like Chinese word segmentation and marked part of speech
4. use TF-IDF algorithm to calculate the keywords weight
5. label keywords categories and add time
6. build user interest vector, vector $= \{(t_1, w_1), (t_2, w_2), ..., (t_n, w_n), time_1\}$
7. build user interest matrix, Matrix $= [vector_1, vector_2, ..., vector_n]$
8. complete building user interest model

END

This paper builds the user's content feature and its weight into the interest vector. Each interest vector represents information of a post content, t_i represents an item of interest, and w_i represents a weight corresponding to the feature, as shown in Eq. 6.

$$\{(t_1, w_1), (t_2, w_2), ..., (t_n, w_n)\} \tag{6}$$

A user interest matrix is composed of all the interest vectors of the same user. User A's interest matrix is shown in Eq. 7 below.

$$\begin{bmatrix} (t_{11}, w_{11}) & \cdots & (t_{1i}, w_{1i}) \\ \vdots & \ddots & \vdots \\ (t_{n1}, w_{n1}) & \cdots & (t_{ni}, w_{ni}) \end{bmatrix} \qquad (7)$$

Where each row represents the interest vector for each post content published by user A. Constructing user interest vector of all the contents of user A into interest matrix.

Based on the traditional Vector Space Model, this paper introduces a time mechanism to analyze the dynamic changes of users' concerns from both the space and the time. At this point, the user interest model is shown in Eq. 8. Which, $time_i$ said the content of the publication time.

$$\begin{bmatrix} (t_{11}, w_{11}) & \cdots & (t_{1i}, w_{1i}), time_1 \\ \vdots & \ddots & \vdots \\ (t_{n1}, w_{n1}) & \cdots & (t_{ni}, w_{ni}), time_i \end{bmatrix} \qquad (8)$$

5   User Interest Dynamic Discovery

Data preprocessing is performed on the contents of posts of all the users in the database, TF-IDF algorithm is used to calculate key words' weights, key words with high representative content are extracted, the feature items are marked and classified, and the time mechanism is introduced to complete the user interest model building [13]. In the spatial dimension, the user's interest model can be used to mine the number and the distribution of forum hot spots, as well as to discover the distribution of the focus of a single user. In the time dimension, user interest models can be used to mine the number of forum hot spots over time changes and distribution of changes in the situation, and can also dig out a single user's concerns over time changes.

The traditional user interest model only excavated the user interest from the space dimension, can not reflect the change of the user interest distribution in the time dimension, and thus can not realize the dynamic change of the user interest. In this paper, the user interest model is analyzed through two dimensions of space and time, which can get the hot spot distribution of the forum, the dynamic change process of the forum, the dynamic change of the user's focus, and the continuous focus, so that the dynamic discovery can be more intuitively and effectively Forum hot spots and user's concerns.

6   Example Analysis

This article uses distributed crawler technology to crawl the forum "bbs.kafan.cn" user-related content information, after cleaning the data into the database. The following table is part of the user's message content as shown in Table 2.

Table 2. Some user post content

Username	Content
hochanmoke	7-data recovery suite是由香港sharpnight公司出品的一款数据恢复软件。7-data recovery suite自带简体中文，使用方便。7-data recovery suite采用向导式提醒，可以从硬盘、内存卡、闪存驱动器、usb设备和手机等设备中恢复已丢失/已删除的照片、文件、分区等等。
wangyunxi80	lz是什么系统？本机装什么别的安软了没有？系统32位win7旗舰版装套装的时候都没有其他安软，装单独墙的时候有红伞。
BHHZDQL	安全卫士20.0概念版昨天晚上的库kill44个其中自学习引擎kill21个ave引擎干掉10个启发引擎干掉15个。

Post text content using Jiaba segmentation, the part of speech tagging, retaining all the names, verbs and gerunds. Then, the TF-IDF algorithm is used to calculate the weight of the key words, and the key words of the weight value top5 are extracted as the feature item of the text content. As shown in Table 3.

Table 3. Some user interest features and their weight

Username	Keywords/Weights
hochanmoke	活动0.7 /软件0.5 /赠送0.4 /公司出品0.3 /闪存0.2
wangyunxi80	本帖1.6 /机装1.0 /旗舰版0.9 /套装0.9 /系统0.8
BHHZDQL	引擎2.7 /干掉2.2 /安全卫士1.5 /启发0.9 /概念0.6

And then use the key words' tag classification algorithm to mark the content of the post classification, divided into four categories: "Firewall Discussion", "HIPS Discussion", "Virus Sample Discussion", "Auxiliary Tools Discussion". The classification result is shown in Table 4.

This article will use this forum data as an example to analyze the hotspots of the forum and the distribution and changes of the user's focus from the spatial and temporal dimensions, in order to prove the advantages of this model in interest discovery.

The traditional Vector Space Model only analyzes forum hotspots from the spatial dimension. From Fig. 1, the forum's hot spots are "HIPS Discussion", "Virus Sample Discussion", "Auxiliary Tools Discussion" and "Firewall Discussion", but they cannot analyze the changes in the hot spots of the forum during a certain period of time.

Table 4. Classification marking results table

Keywords	Label
活动/软件/赠送/公司出品/闪存	辅助工具讨论
本帖/机装/旗舰版/套装/系统	防火墙讨论
引擎/干掉/安全卫士/启发/概念	病毒样本讨论
破解版/去除/破解/提示/发表	HIPS讨论

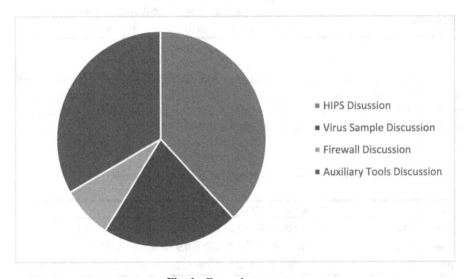

Fig. 1. Forum hot spot map

Figures 2 and 3 show the hotspots distribution of the forum in 2017. In the traditional Vector Space Model, a time mechanism was introduced to analyze the hot trend of the forum.

In the first six months of 2017, 45% of the forums were hot in "HIPS Discussion", "Virus Sample Discussion" accounted for 16%, "Auxiliary Tools Discussion" accounted for 29% and "Firewall Discussion" accounted for 11%.

In July-December 2017, the hot spot "HIPS Discussion" accounted for 32%, "Virus Sample Discussion" accounted for 26%, "Auxiliary Tools Discussion" accounted for 37%, and "Firewall Discussion" accounted for 6%.

From the time dimension, the share of hotspot "HIPS Discussion" and "Firewall Discussion" in the second half decreased compared to the first half of the year, while the share of hotspot "Auxiliary Tools Discussion" and "Virus Sample Discussion" in the second half of the year compared with the first half, the ratio had risen. Therefore, it could be concluded that there was a slight shift in the center of gravity of the forum.

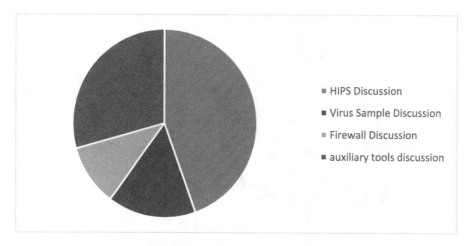

Fig. 2. January-June 2017 Forum hot spots map

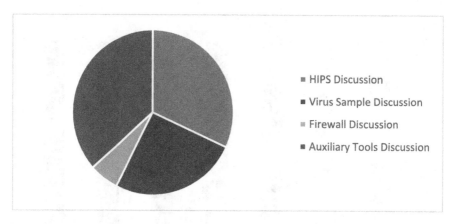

Fig. 3. July-December 2017 Forum hot spots map

In this paper, the user interest matrix whose username is "1654637359" will be taken as an example to analyze the changes of single user's concerns in the time dimension and the spatial dimension (Fig. 4).

In terms of spatial analysis, there are three points of interest for user "1654637359": "HIPS Discussion", "Auxiliary Tools Discussion" and "Firewall Discussion" (Fig. 5).

In the time dimension, the user "1654637359" focused on the "Auxiliary Tools Discussion" in January-February 2017 and shifted the focus to "HIPS Discussions" in June-August. In the first six months, he focused on three aspects Information content, July-August concerns decreased.

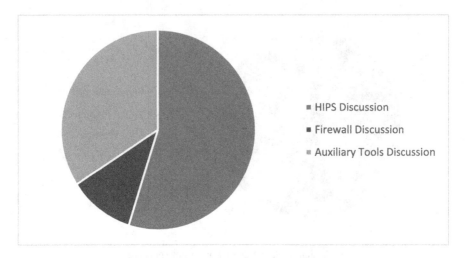

Fig. 4. User "1654637359" focus point map

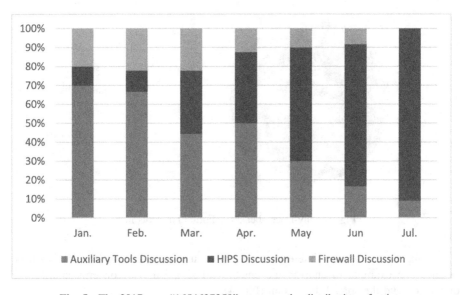

Fig. 5. The 2017 user "1654637359" concerns the distribution of points

Therefore, by dynamically discovering the model based on the user interest of spatio-temporal vectors, the forum's hot spot distribution and changes in spatial and temporal dimensions as well as the user's change of focus and continuous focus can be drawn. The next step is how to optimize the user interest model, improve the accuracy of user interest features, to ensure that user interests are found more valuable, and to apply the research results to the practice system.

7 Conclusion

This paper designs a dynamic discovery model of user interest based on space-time vector for the forum, and dynamically discovers the forum hot spots and the user's concerns in the space and time by introducing the time change mechanism, which could be help the forum administrator to timely and accurately grasp the forum hot spot distribution and change Situation, to help forum users find more quickly and accurately the same concerns the user community. Through case analysis, the dynamic discovery model of user interest based on space-time vector has better interest discovery ability in time and space dimension.

Acknowledgement. The thesis is supported by the National Key Research & Development Program of China under Grant NO. 2017YFB0803001, National Natural Science Foundation of China under Grant NO. 61370215, 61370211; and National Information Security Program of China under Grant NO. 2017A065, 2017A111.

First and foremost, I would like to show my deepest gratitude to my two teachers Prof. Chi and Prof. Zhang, who had provided me with valuable guidance in every stage of the writing of this thesis. Without their enlightening instruction, impressive kindness and patience, I could not have completed my thesis.

Then, I'd like to thank all my friends for their encouragement and support.

Last but not least, I shall extend my thanks the experts for taking the time to review my thesis and making valuable suggestions.

References

1. Ren, B., Liang, Y., Zhao, J., Lian, W., Li, Y.: A user interest model based on multi-dimensional weights dynamic update. Comput. Eng. **40**(9), 42–45 (2014)
2. Hao, S.: User interest model representation and update research for personalized services. Hefei University of Technology master's degree thesis, Anhui (2012)
3. Song, Y., Chen, Z.: Research on user interest model in personalized retrieval system. Comput. Math. Eng. **41**(2), 271–274 (2013)
4. Zhu, Q., Shyu, M., Wang, H.: Video topic: modeling user interests for content-based video recommendation. Int. J. Multimed. Data Eng. Manag. **5**, 1–21 (2014)
5. Xian, X., Li, Y.: Proposed a Related PageRank algorithm based on the user's interest. Adv. Mater. Res. **718-720**, 2040–2044 (2013)
6. Qiu, Y., Wang, L., Shao, L.: User interest modeling method based on weibo short text. Comput. Eng. **40**(2), 275–279 (2014)
7. Liang, X., Gu, L.: Chinese analysis and part of speech tagging. Comput. Technol. Dev. **25**(2), 175–181 (2015)
8. Lin, H., Yang, Y.: User interest model representation and update mechanism. Comput. Res. Dev. **39**(7), 843–847 (2002)
9. Teng, X.: Based on Struts2 and Hibernate community website system design and implementation. Huazhong University of Science and Technology master's degree thesis, Wuhan (2011)
10. Diao, Z.: Research on personalized recommendation system based on ontology user interest model. Master's degree thesis of Taiyuan University of Technology, Taiyuan (2013)

11. Chen, W., Zhang, X., Li, Z.: Analysis of constructing topic model of Weibo user interest model. Comput. Sci. **40**(4), 127–130 (2013)
12. Hao, S., Wu, G., Hu, X.: User interest expression and updating based on hierarchical vector space model. J. Nanjing Univ. (Natural Science) **48**(2), 190–197 (2012)
13. Tian, J.: Research on User Interest Model Based on Social Network. Master's Degree in Electronic and Science University, Chengdu (2010)
14. Tao, W.: Realization of Chinese word segmentation algorithm based on bidirectional maximum matching method in policing applications. Electron. Technol. Softw., 153–155 (2016)
15. Liu, B.: Research on personalized user interest modeling in mobile environment. Master's thesis, Beijing University of Posts and Telecommunications, Beijing (2009)
16. Qiu, J.: Research on user interest model based on Weibo social network. Shanghai Jiaotong University master's degree thesis, Shanghai (2013)
17. Li, Z., Wang, D., Guan, Z.: Research on construction of user interest model based on domain ontology. J. Inf. Sci. **33**(11), 69–73 (2015)
18. Ren, Y.: An improved method of judging the theme relativity based on vector space model in vertical search engine. Appl. Mech. Mater. **411-414**, 106–109 (2013)
19. Zhao, Q., Niu, J., Chen, H.: An indicative opinion generation model for short texts on social networks. Future Gener. Comput. Syst. **86**, 1471–1480 (2017)
20. Dong, Y., Liu, Y., Luo, J.: The optimization algorithm based on vector space model to determine the influence of microblogging text. Appl. Mech. Mater. **571-572**, 278–281 (2014)
21. Cheng, S., Liu, Y.: Time-aware and grey incidence theory based user interest modeling for document recommendation. Cybern. Inf. Technol. **15**, 36–52 (2015)

Automatic Generation of Multiple-Choice Items for Prepositions Based on Word2vec

Wenyan Xiao[1,2] ⓘ, Mingwen Wang[1(✉)] ⓘ, Chenlin Zhang[1],
Yiming Tan[1], and Zhiming Chen[1]

[1] School of Computer Information Engineering,
Jiangxi Normal University, Nanchang 330022, China
mwwang@jxnu.edu.com
[2] Jiangxi University of Science and Technology, Nanchang 330013, China

Abstract. The automatic generation of multiple-choice item (MI) has attracted amounts of attention. However, only a limited number of existing research address automatic MI generation for prepositions, and even fewer consider learners' need in the generation process. In this paper, we propose an approach to generate preposition MIs suitable for non-native English learners of different language proficiency. First we select sentences with similar difficulty level to that of a given textbook as stems by using the sentence difficulty model we constructed. Then, we use the Word2vec model to retrieve a preposition list of distractor candidates where three of them are chosen as distractors. To validate the effectiveness of our approach, we produce four tests of preposition MIs at different difficulty levels and conduct a series of experiments regarding evaluations of stem difficulty, distractor plausibility and reliability. The experimental results show that our approach can generate preposition MIs targeting learners at different levels. The results of distractor plausibility and reliability also point to the validity of our approach.

Keywords: Multiple-choice item · Automatic generation · Word2vec

1 Introduction

Preposition is one of the most frequently used lexical category in English language. According to the word frequency list of British National Corpus (BNC), there are 13 prepositions in the top 100 frequently used words. Nevertheless, preposition usage is one of the most difficult aspects of English grammar for non-native speakers to master due to its complexity and variability [1]. Incorrect usage of prepositions constitutes the most common error category in various learners' corpora, accounting for a significant proportion of all English as a Second Language (ESL) grammar errors [2–4]. Thus, preposition teaching and learning for non-native English learners is of great importance.

As a type of objective items, MIs have a high reliability and efficiency of scoring. Therefore, they are extensively used not only in classroom achievement tests, but also in large-scale, high-stake proficiency tests. Figure 1 shows a sample MI for preposition learning and testing, where one word is removed from a sentence, forming a blank, the

© Springer Nature Singapore Pte Ltd. 2018
Q. Zhou et al. (Eds.): ICPCSEE 2018, CCIS 902, pp. 81–95, 2018.
https://doi.org/10.1007/978-981-13-2206-8_8

sentence with the gap is called the stem or carrier sentence, and the removed word is referred to as the key while the three alternatives in the item are called as distractors.

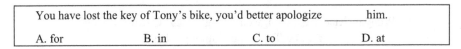

You have lost the key of Tony's bike, you'd better apologize _____him.			
A. for	B. in	C. to	D. at

Fig. 1. An example of preposition MI

However, it is costly and time-consuming to manually produce those items as writing incorrect choices that distract only the non-proficient test-taker is a highly skilled business [5]. To address this problem, we propose a novel approach to automatic generation of preposition MIs based on the Word2vec model. In addition, MIs should be at a level appropriate to the learners' proficiency level [6]. Thus, we also pay special attention to the difficulty level of stem sentences.

To our knowledge, little research has been done on the automatic generation of MIs for prepositions based on the Word2vec model, and even fewer related publications report the stem sentence difficulty in the generation process. To automatically generate preposition MIs suitable for learners of different language proficiency, first, we select sentences with similar difficulty level to that of a given textbook as stems by using the sentence difficulty model we constructed. Next, we use the Word2vec model to retrieve a preposition list of distractor candidates where three of them are chosen as distractors. Finally, we show that our approach can generate useful preposition MIs with reliable and plausible distractors. Moreover, the generated MIs can be tailored to suit learners of different language proficiency.

The rest of the paper is organized as follows. Section 2 provides an overview of related work; Sect. 3 provides details of the approach to generating preposition MIs, including stems and distractors; Sect. 4 describes the evaluation procedures and analyzes the results; and Sect. 5 provides some concluding remarks and future work.

2 Related Work

2.1 Automatic MI Generation

So far, some methodologies of automatically generating MI have been proposed. In general, the MI generation includes two key procedures: stem generation and distractors generation. Most previous works on stem generation selected sentences from random text [7], a large corpus [8–10], or authentic documents from the web [11].

Compared with stem generation, more effort has been put on the distractor generation, as the quality of distractors, to a large extent, affects the quality of the whole MI. Some attempts were made using WordNet [8, 12–14] to obtain semantically alternative terms. Other work [15–18] selected distractors by using frequency and language learner's corpus.

Note that most previous work has focused on the automatic MI generation for English content words, like verbs, nouns, adjectives, etc., while less attention has been paid to items for function words, such as modals, articles and prepositions. It is possibly caused by people's neglecting their importance. Moreover, the above-mentioned approaches are not quite applicable to generating distractors for function words. Although the papers [10, 17] tackled preposition MIs, their work lack enough attention on stem difficulty so that the generated items cannot meet different learners' need.

Different from the previous approaches, we address the automatic MI generation for prepositions based on the Word2vec model, while paying special attention to sentence difficulty. The aim is to generate MIs for prepositions, and more importantly to suit learners of different language proficiency.

2.2 Sentence Difficulty

Substantially, less research has focused directly on sentence difficulty, however, it is often studied in the field of text readability. Researchers have used various factors to measure readability, such as speed of perception, perceptibility at a distance, reading speed, eye movement, etc. Some traditional approaches focus on measuring the familiarity of semantic units such as words or phrases, and the complexity of syntax. The most widely-used traditional measure is the Flesch-Kincaid score, and some other formulas include the Gunning-Fog Score, Coleman-Liau Index, Automated Readability Index (ARI), SMOG Index, FORCAST, etc. Other traditional approaches estimate the semantic difficulty of words in a text by assigning individual words a familiarity or difficulty level based on their occurrence in a pre-specified vocabulary resource. Notable works are the Revised Dale-Chall formula [19], the Lexile measure [20]. They adopt simply two or three variables to investigate readability. The limitations of the traditional measures have recently inspired researchers to explore how richer linguistic features combined with machine learning techniques could lead to a new generation of more robust and flexible readability assessment algorithms. Studies, such as those by Pitler and Nenkova [21], Tanaka-Ishii et al. [22], Heilman et al. [23], François and Miltsakaki [24], have mostly used a data-driven machine learning approach to computationally measure readability.

In the above-mentioned research on readability, sentence difficulty is necessarily examined and considered for measuring text readability. Less direct attention has been paid to measuring the single sentence difficulty. Perera [25] and Troria [26] indicated that the average sentence length influences the complexity of the sentences. Lin [27] proposed a definition of structural complexity as the dependency relationships between words in a sentence. His assumption was extended by Liu [28] and Lu et al. [29]. Liu [28] proposed dependency distance, in the framework of dependency grammar, as an insightful metric of complexity. Mishra et al. [30] investigated three properties of the input sentence, namely, length, degree of polysemy and structural complexity when predicting sentence translation difficulty. Specia et al. [31] proved language model (LM) probability an efficient feature in quantifying the translation complexity.

Inspired by those previous studies, we investigate the sentence difficulty by combining the six following features: sentence length, syntactic depth, amount of polysemy, length of dependency links, word frequency, and LM probability of sentence.

3 Item Generation

In this section, we first describe our dataset and the corpora used as the stem source. We then present our approach to item generation in terms of stem extraction and distractor extraction.

3.1 Dataset

We use the BNC[1] as the source of stems. It consists of a 100-million-word collection of samples of written and spoken language from a wide range of resources. We choose it over the other corpora because the language register of 90% of the collected texts is written and formal, which is more consistent with that of language learned in school setting. In order to reduce computational cost and improve representativeness of the sentences selected as stems, we apply K-means clustering method to classify the BNC data into 2000 clusters. Then, we randomly extract one sentence from each cluster. By regex and manual check, we delete the sentences which contain only a single name (e.g. Chris), a date (e.g. Fri, 14 Jan 1994), a title (e.g. Obituary: Ian Scott-Kilvert), a phrase (e.g. By Derek Pain), etc. Finally, the BNC sub-corpus including 1765 sentences is developed. This sub-corpus is used as the more specific and representative source of stem extraction.

Longman New Concept English series is a typical English language textbook series teaching British rules of English. Four books are compiled for English learners of different language proficiency. Compared with many other English textbooks, this set is more obviously different in difficulty as the difficulty level of each textbook increasing apparently. This set of textbooks is used as a reference for stem difficulty, i.e. in our proposed approach, the difficulty of stem should match the average sentence difficulty of a given textbook.

3.2 Stem Generation

The first stage of the MI generation is to extract an appropriate sentence as stem. We use the BNC sub-corpus we produced as the stem source. The sentence extracted as stem should meet two requirements. First, the sentence difficulty should be similar to the average sentence difficulty of the book under consideration. Second, the sentence should have at least one preposition.

In order to extract appropriate sentences approximating the average sentence difficulty level, we first extract six sentence difficulty features, then validate the effectiveness of the features, and finally generate stems by similarity computation represented by the Euclidean distance.

Extraction of Sentence Difficulty Features. In this step six features are extracted as factors mainly affecting sentence difficulty. Table 1 gives a summary of the features used.

---

[1] http://www.natcorp.ox.ac.uk.

Table 1. Feature set for sentence difficulty measuring

Feature symbol	Description
Len_{sen}	Number of tokens in sentence
Fre_{low}	The lowest word frequency in a sentence
Syn_{dep}	The depth of syntactic tree of a sentence
Sen_{sum}	The sum of word senses in a sentence
Dep_{len}	The Length of dependency links in a sentence
LM_{pro}	Language model probability

Sentence length: Len_{sen} is the count of words in a sentence after tokenization. The tool we use is the tokenization script of Moses which is an open source statistical machine translation system [32].

Frequency: Feature Fre_{low} is the frequency of the word that has the lowest word frequency based on the BNC. The word frequency represents the number of times a word occurs in the corpus. Generally, the low frequency words are more difficult than the high frequency words, so we consider the smaller value of Fre_{low} is, the harder the sentence is.

Syntactic depth: Syn_{dep} refers to the number of layers of a sentence syntax tree. Syntactic parsing is a key technique in natural language processing, often used to analyze the sentence syntactic structure or the word dependency. We parsed the BNC, as well as the New Concept English textbooks with the Stanford parser[2]. Figure 2 shows the syntax tree for an example sentence. Excluding the ROOT node, there are 6 layers of the syntax tree, thus the Syn_{dep} is 6.

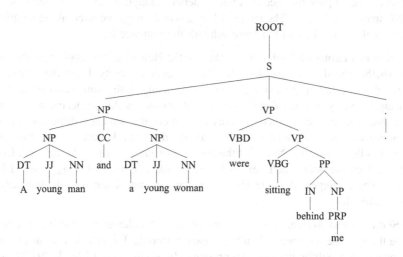

Fig. 2. Syntax tree for the example sentence: a young man and a young woman were sitting behind me.

---

[2] https://nlp.stanford.edu/software/lex-parser.shtml.

Word Senses: Sen_{sum} is the sum of senses possessed by each word in sentence. We retrieve the polysemic words from WordNet[3] and then compute the total number of senses in the sentence.

Length of dependency links: Dep_{len} is the length of the dependency links in the dependency structure of the sentence [27]. It captures the complex relationship among words in a sentence. Figure 3 shows the graphical representation of the Stanford dependencies for the above example sentence. Excluding the ROOT node, the length of the dependency links can be computed by adding all the distances between a governor and a dependent. For the example sentence, the Dep_{len} is 20 through continuously adding the distance.

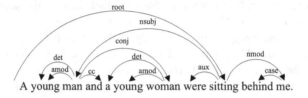

Fig. 3. Graphical representation of the Stanford dependencies for the example sentence

Sentence language model probability: A language model is a probability distribution over sequences of words. It can reflect how frequently a string occurs as a sentence and the probability of a sentence being grammatical. In a sense, the smaller LM_{pro} indicates the sentence is more complex, or less frequently used in daily language. In the experiment, we employ the recurrent neural network language model trained based on the BNC to capture LM_{pro}[4]. The value of LM_{pro} is often negative after taking logarithm, the smaller the value is, and the more difficult the sentence is.

Validation of Features Effectiveness. We use the New Concept English textbooks to verify whether the six features can reflect the sentence difficulty. From the perspective of textbook compilation, the average sentence difficulty of the four books presents the increasing tendency. If the value of the six features shows this tendency, we consider them valid, namely the selected features can reflect the sentence difficulty. Table 3 summarizes the average value of each sentence difficulty feature for the four New Concept English textbooks. Note that the averages of Len_{sen}, Syn_{dep}, Sen_{sum}, and Dep_{len} calculated increase with each textbook, while the averages of Fre_{low} and LM_{pro} decrease. These suggest that the six features we extracted are reliable indicator of sentence difficulty.

Stem Sentence Extraction. We use the values of each column in Table 2 as vectors to indicate the average sentence difficulty for each textbook. Take Book 1 as an example, its average sentence difficulty can be represented by vector $X_1 = [7.16, 16291.27, 6.04,$

---

[3] https://wordnet.princeton.edu.

[4] http://www.fit.vutbr.cz/~imikolov/rnnlm/.

49.90, 9.89, −15.10], and X_2, X_3, X_4 for Book 2, Book 3 and Book 4 respectively. To select the sentences with similar sentence difficulty to each textbook, first, we extract the six sentence difficulty features of each sentence from the BNC sub-corpus as feature vector Y. Then, we normalize each feature value by z-scores, and compute the similarity by the Euclidean distance between X_i and Y using the following formula:

$$Dist = \sqrt{\sum_{j=1}^{N} (X_i(j) - Y(j))^2} \quad i = 1,..,4 \tag{1}$$

where N is the number of features, which is 6 in this experiment. The smallest Euclidean distance value determines which textbook the sentence selected resembles concerning sentence difficulty. If it is Book 1, it indicates that the difficulty of the extracted sentence is similar to that of Book 1.

After extracting the sentences with similar difficulty to each textbook, we use Stanford POS tagger[5] to identify prepositions and then select the sentences with prepositions. The sentences filtered are stems for the MIs. The preposition contained in the sentence is thus left as a blank in the stem, which also serves as the key to the item. For the sentences with more than one preposition, we randomly select one as the blank word. Below is a generated stem with similar difficulty to that of Book 1.

Example: This places too much burden ____ (on) others.

3.3 Distractor Generation

Mitkov and Ha [14] proposed a principle according to which distractors were chosen: semantic similarity. They pointed out that the semantically closer were the distractors to the correct answer, the more plausible they were supposed to be. As an open source tool of calculating similarity distance, Word2vec transforms every word in the sentence into low-dimension dense real number vectors and outputs the ranked word lists according to similarity. Compared with the previous distractor generation approaches such as using the WordNet, the Word2vec-based approach can output words that share common contexts in the corpus, thus it is more capable of generating plausible distractors. Note that retrieving the most relevant words to the key preposition by the Word2vec is not very ideal and meaningful. Besides, many preposition usage errors are actually caused by the similar words to the word before and after the certain preposition. Therefore, we utilize the Word2vec to generate distractors by considering the words occurring before and after the key preposition. The procedures are as follows.

First, the previous word A and the next word B to the stem's preposition are identified, then their most relevant words C and D which are also of the same POS with words A and B are selected by the Word2vec trained on the BNC. We then use regex to search out in the BNC all the prepositions occurred before or after C, as well as prepositions occurred before or after D. Those prepositions are then ranked by frequency. For each item, we finally retrieved a set of ten prepositions excluding the

---

[5] http://nlp.stanford.edu/software/tagger.shtml.

Table 2. The average value of each sentence difficulty feature for New Concept English.

Feature	Book 1	Book 2	Book 3	Book 4
Len_{sen}	7.16	15.23	19.34	26.80
Fre_{low}	16291.27	1659.98	940.86	390.90
Syn_{dep}	6.04	9.60	11.31	13.34
Sen_{sum}	49.90	121.76	150.08	190.34
Dep_{len}	9.89	30.29	41.95	68.49
LM_{pro}	−15.10	−36.09	−48.14	−70.43

correct answer as distractor candidates. For example, for the blank of the above stem example, the most relevant words to "burden" and "others" are respectively "debt" and "individual". The retrieved list of distractor candidates consists of the following prepositions: *towards, into, by, along, among, over, of, between, through, during.* Three prepositions are then randomly selected from the list as distractors. Therefore, a MI is finally generated, as shown in Fig. 4.

This places too much burden ___ (on) others.
A. on   B. by   C. into   D. towards

Fig. 4. An example of the generated item

4 Evaluation and Results

A series of experiments were conducted to evaluate the multiple-choice preposition items in terms of stem difficulty, distractor plausibility and reliability.

4.1 Stem Difficulty Evaluation

To evaluate the stem difficulty, we compared the difficulty level with humans' judgment. First, based on the proposed stem generation approach, we randomly extracted 25 sentences with similar difficulty to each New Concept textbook's average sentence level from the BNC, forming a list of 100 sentences in total. Each sentence has labelled with a number from 1 to 4 indicating difficulty level. Then they were mixed in random order for fair comparison.

After removing the difficulty label of sentences, two professional college teachers with over a decade ESL teaching experience were invited to fulfill the human judgment by annotating each sentence's difficulty with number 1 to 4 based on their experience. In this evaluation, we assumed that the sentence difficulty level is in accordance with the textbook volume number. Namely, if the sentence is annotated with 1, its difficulty level is similar to the average sentence difficulty of New Concept textbook 1.

The two teachers' judgments had a Cohen's kappa agreement of 0.91. We therefore considered the agreement between them to be "almost perfect" according to Landis and Koch [33] 's label of strength of agreement associated with kappa statistics. Then, we compared their respective judgment with the actual difficulty level by computing the accuracy rate. The accuracy rates are 0.58 and 0.76 respectively. This result indicated that the difficulty of sentences selected as stems approximates each textbooks' average sentence level to a great extent. It apparently showed that our stem generation approach based on similarity computation can generate stems with similar difficulty level to a given English textbook. Therefore, the aim of creating MIs suitable for ESL learners of different language proficiency can be achieved.

4.2   Distractor Evaluation

We evaluated the distractors' effectiveness in terms of their plausibility and reliability.

Learner Evaluation on Plausibility. For the evaluation, we generated a preposition test for each New Concept English textbook based on our approach. Each contains 25 multiple choice items with the full score of 25. For a better comparison with the distractor generated randomly, two of the three distractors in several items were selected by Word2vec, while the other was selected randomly from the list of prepositions. We then gave the tests to students in four different grades and asked the students to finish in the testing setting. According to the students for whom the New Concept English textbooks are suitable, we selected second-year junior high school students to take test paper 1, senior high school sophomores to take test paper 2, firs-year non-English major students to take test paper 3 and first-year English major students together with some graduates to take test paper 4.

As shown in Table 3, test paper 1 and 2 are taken in a classroom testing mode where students independently finish the paper in 30 min without review or preparation before taking the test. We collected 88 and 100 valid papers respectively after deleting similar answer sheets caused by test-taking fraud and incomplete ones. Different from the classroom testing for test paper1 and 2, test paper 3 and 4 were distributed on the professional questionnaire website[6] through crowdsourcing. The participants are required to provide information on major and education level so that we can guarantee the subjects we need. In total 136 and 167 valid results for each were obtained after deleting abnormal data such as the collected papers with identical answers and answer time.

It can be observed that the average score of each test paper is relatively low, with 10.90, 10.96, 10.75, and 10.21 respectively. This result again confirms the previous finding that preposition usage remains one of the most difficult aspects of English grammar for ESL learners to master.

We then analyzed the distractors plausibility by calculating the ratio of the times of a certain distractor chosen by students to the total times of the three distractors chosen, as shown in the following formula:

$$R(i) = \frac{Error(i)}{\sum_{j=1}^{3} Error(j)} \tag{2}$$

---

[6] http://www.wjx.cn.

Table 3. Experiment settings for distractor plausibility evaluation

Test	Subjects	Testing mode	Number of valid tests	Average score
1	Second-year junior high school students	Classroom test	88	10.90
2	Senior high school sophomores	Classroom test	100	10.96
3	First-year non-English major students	Online test	136	10.75
4	First-year English major students and graduates	Online test	167	10.21

Table 4. Distractor plausibility of some items

$R(i)$	Test paper 1			Test paper 2			Test paper 3			Test paper 4		
	Q1	Q2	Q3	Q1	Q2	Q3	Q1	Q2	Q3	Q1	Q2	Q3
$R(A)$	0	0.18	0	0.32	0	0.26	0.73	0	0.31	0.25	0.19	0.24
$R(B)$	0.24	0.74	0.48	0.41	0.45	0.35	0	0.38	0.33	0.48	0.15	0.26
$R(C)$	0.38	0	0.31	0.27	0.37	0	0.06	0.33	0.36	0	0.66	0.5
$R(D)$	0.38	0.08	0.21	0	0.18	0.39	0.21	0.29	0	0.27	0	0

where i denotes a certain choice of the item, such as A, B, C or D, j is the three distractors. If $R(i)$ is 0, then it indicates that choice i is the correct answer.

From the results, we observed that every distractor of each item has a certain plausibility for all the $R(i)s$ of distractors are greater than 0. Therefore, our Word2vec based approach to distractor generation can produce plausible distractors which perform efficiently in distracting students from choosing the correct answer. Table 4 shows the distractor plausibility of some items which are listed in the appendix. The results are analyzed in details as follows.

(1) For some items, three distractors possess similar degree of plausibility. Take Q3 in test paper 3 for an example, the three distractors generated are *in, to, above*, while the key is *on*. The plausibility of each distractor is 0.31, 0.33, and 0.36 respectively. Almost the same amount of students chose each distractor. This indicates that for the students who failed in this item haven't master the usage of the verbal phrase followed with preposition, namely *based on*. Thus distractors *in, to, above* successfully distracted many of them choosing correct answer. By examining the three preposition distractors in the distractor candidates list we extracted, we found that they are adjacent, which to a certain degree led to the similar plausibility.

(2) On contrary, for some items, three distractors' plausibility are rather different. For instance, the plausibility of Q2 distractors (*with, by, inside*) in test paper 1 is 0.18, 0.74, 0.08 respectively. Among the three distractors, *with* and *by* were randomly selected from the distractor candidates, while *inside* was a distractor we extracted randomly from the list of prepositions. This exactly shows that the distractor generated beyond the distractor candidates list play a much smaller role in distracting students from choosing the correct answer.

(3) By comparing the distractor plausibility rank and its rank in the candidates list based on the Spearman's correlation coefficient (ρ), we observed 65% of the rank pairs are positively correlated with ρ as 0.5 or 1. The distractors ranked high on the candidates list tend to be more plausible than those ranked low on the list. Correspondingly, the items become less difficult if the words ranked low on the list of distractor candidates are selected as distractors, since they are less plausible. Therefore, we can update the distractor selecting model so as to flexibly decrease the item difficulty.

Evaluation by Comparison with Other Approaches. Given the distractor plausibility is largely determined by its semantic similarity to the key, we compared the distractors of the 100 items generated by our approach, with those generated by corpus-based approach and the random approach on the distractors' semantic similarity to the key. As for the corpus-based approach, the corpus we used is the Chinese Learner English Corpus (CLEC) with error taggers. First, we extracted the sentences with prepositional use errors. Because the CLEC is not big enough to generate distractors for our stems by considering the prepositional object or the NP/VP head used in Lee et al. [10], we instead extracted 186 patterns of the incorrect preposition use formed by the part of speech (POS) of the word before and after the preposition. Then according to the preposition patterns of the generated stems and the CLEC sentences with preposition errors, we extracted the prepositions as distractors. If less than three distractors are generated, we randomly selected from the preposition list as supplement(s). The random approach extracts three prepositions randomly from the preposition list as the distractors. Finally, the three approaches were compared in terms of the distractors with the minimum and maximum similarity to the key based on Word2vec. The results of the comparisons are presented in Table 5.

Table 5. Comparisons results

Comparison	The number of more plausible distractors by comparing the distractors with the minimum semantic similarity to the key (min)	The number of more plausible distractors by comparing the distractors with the maximum semantic similarity to the key (max)
1 (Ours vs. Random)	58	54
2 (CLEC vs. Random)	69	56
3 (Ours vs. CLEC)	43	56

From the comparison 1 and 2, we found that both our approach and the corpus-based are effective in generating distractors, performing better than the random approach. The results of the comparison 3 revealed that our approach is more capable of generating distractors with higher plausibility than corpus-based approach, as shown by the max, while the latter has the advantage of ensuring the lower limit of the distractor plausibility.

Human Evaluation on Reliability. Another consideration of MIs' effectiveness is the reliability of their distractors. A distractor is considered "reliable" if it yields an incorrect sentence, whereas "unreliable" if it might yield correct sentences. To investigate the distractor reliability, we asked the two ESL teachers to judge whether the item has only one correct answer. In these 100 items, 93 of them are considered to have only one correct answer by the first ESL teacher, while 95 by the second. It means that 97.67% and 98.33% of the distractors are viewed as reliable. By analyzing the items with an unreliable distractor, we found that the existence of more than one correct answer is mainly caused by too little context and the semantic similarity between preposition phrases. As Heaton [6] pointed out that too little context is insufficient to establish any meaningful situation, thus leading to ambiguity. Besides, short decontextualized sentences are usually open to several interpretations when used as stems for MIs. Take a look at the following item sample.

I said, "Harvey thinks I am employed by Midwinter to check _____ on him?"

A. of B. with C. in D. up

The word removed as key to the item in the stem is *up* according to the stem source. However, due to the limited context, the preposition *in* can also yield correct sentence, yet expressing a different meaning. Besides, the subtle difference between prepositional phrases make both prepositions acceptable as correct answer, such as *in the end of* and *at the end of*, *on the internet* and *in the internet* occurred in the MIs we generated. In this case, manual adjustments can be made to ensure the distractor reliability, such as replacing the unreliable distractor with a reliable one.

On the whole, the evaluation results show that our Word2vec based approach can generate plausible and reliable distractors. Compared with the corpus-based approach, our approach performs better in generating more plausible distractors. It is worthy of note that we can decide the item difficulty by adjusting the distractor selecting model based on the degree of plausibility. The overall effectiveness of the automatically generated MIs can be improved by minor distractors revision and human supervision.

5   Conclusions and Future Work

Automatic generation of MI for prepositions could save teachers time and energy from doing it manually, and bring many benefits to non-native English learners. The MIs generated should be at an appropriate difficulty level suitable for learners of a certain language proficiency. However, so far few approaches have been proposed which would fully consider sentence difficulty level during the MI generation process.

In this paper, we have described a novel approach to automatically generating MIs for prepositions based on the Word2vec model. When selecting sentences from the BNC as stems, we extracted six features of sentence difficulty where LM_{pro} was first used. In this way, the generated items can be tailored to match learners' English proficiency. The evaluation results showed that stem difficulty approximates each textbooks' average sentence level to a great extent, and the generated MIs are effective in terms of distractor plausibility and reliability. As a feasible and effective aid, the proposed approach to the automatic generation of preposition MIs will greatly enhance language learning and teaching.

In the future, we would like to devote to three aspects. One is to further measure the sentence difficulty by considering more features and data, and conduct a more in-depth evaluation on the quality of the generated MIs. Another is to extend the automatic MI generation approach to other question types and learner errors. In addition, we will strive to build a preposition learning platform.

Acknowledgments. This paper is supported by the National Science Foundation of China (No.61462045).

Appendix: Sample Items of Preposition Test Paper for Each Textbook

Textbook	Sample items
I	1. So you're happy ____ your work?
	A. with B. inside C. of D. on
	2. She has a low opinion ____ herself.
	A. with B. by C. of D. inside
	3. I'm shocked ____ my husband's behavior.
	A. by B. of C. in D. for
II	1. ____ addition, as we have just seen, there was a growing control of family size.
	A. For B. On C. Beside D. In
	2. Both stared at it almost ____ horror for a moment, listened to the somehow insistent, angry ringing.
	A. in B. round C. inside D. along
	3. The life of a subsistence farmer simple does not accord____ our notion of labor.
	A. of B. on C. with D. in
III	1. Section 61 also covers the case where A holds a deposit ____ trust for B and C jointly.
	A. of B. on C. in D. with
	2. A thesaurus module that is integrated ____ the software used for database creation, maintenance and searching is particularly valuable.
	A. into B. of C. in D. for
	3. Most theories agree that a concept is based ____ objects or events that have features in common.
	A. in B. to C. above D. on
IV	1. Opera is the communal act of listening to this silent voice from inside: it is not surprising that so many people are hungry ____ the experience it awakens.
	A. of B. in C. for D. at
	2. His attempts to link reductions in working hours ____ more flexible work practices, for example, have run into powerful union resistance.
	A. in B. of C. for D. to
	3. His colleagues had expressed anxiety ____ pension rights but saw possible benefits in attracting and retaining staff if pay constraints could be freed.
	A. in B. upon C. with D. over

References

1. Chodorow, M., Tetreault, J.R., Han, N.R.: Detection of grammatical errors involving prepositions. In: Proceedings of the Fourth ACL-SIGSEM Workshop on Prepositions, pp. 25–30. Association for Computational Linguistics (2007)
2. Izumi, E., Uchimoto, K., Saiga, T., Supnithi, T., Isahara, H.: Automatic error detection in the Japanese learners' English spoken data. In: Proceedings of the 41st Annual Meeting on Association for Computational Linguistics, vol. 2, pp. 145–148. Association for Computational Linguistics (2003)
3. Dahlmeier, D., Ng, H.T., Schultz, T.: Joint learning of preposition senses and semantic roles of prepositional phrases. In: Proceedings of the 2009 Conference on Empirical Methods in Natural Language Processing, vol. 1, pp. 450–458. Association for Computational Linguistics (2009)
4. Lee, J., Yeung, C.Y., Zeldes, A., Reznicek, M., Lüdeling, A., Webster, J.: Cityu corpus of essay drafts of English language learners: a corpus of textual revision in second language writing. Lang. Resour. Eval. **49**(3), 659–683 (2015)
5. Alderson, J.C.: Do corpora have a role in language assessment? In: Using Corpora for Language Research, pp. 248–259 (1996)
6. Heaton, J.B.: Writing English Language Tests: A Practical Guide for Teachers of English as a Second or Foreign Language. Longman Publishing Group, Harlow (1975)
7. Hoshino, A., Nakagawa, H.: A real-time multiple-choice question generation for language testing: a preliminary study. In: Proceedings of the Second Workshop on Building Educational Applications Using NLP, pp. 17–20. Association for Computational Linguistics (2005)
8. Smith, S., Avinesh, P.V.S., Kilgarriff, A.: Gap-fill tests for language learners: corpus-driven item generation. In: Proceedings of ICON-2010: 8th International Conference on Natural Language Processing, pp. 1–6. Macmillan Publishers (2010)
9. Sumita, E., Sugaya, F., Yamamoto, S.: Measuring non-native speakers' proficiency of English by using a test with automatically-generated fill-in-the-blank questions. In: Proceedings of the Second Workshop on Building Educational Applications Using NLP, pp. 61–68. Association for Computational Linguistics (2005)
10. Lee, J., Sturgeon, D., Luo, M.: A CALL system for learning preposition usage. In: Proceedings of the 54th Annual Meeting of the Association for Computational Linguistics, pp. 984–993 (2016)
11. Pino, J., Eskenazi, M.: Semi-automatic generation of cloze question distractors effect of students' L1. In: International Workshop on Speech and Language Technology in Education (2009)
12. Brown, J.C., Frishkoff, G.A., Eskenazi, M.: Automatic question generation for vocabulary assessment. In: Proceedings of the Conference on Human Language Technology and Empirical Methods in Natural Language Processing, pp. 819–826. Association for Computational Linguistics (2005)
13. Lin, Y.C., Sung, L.C., Chen, M.C.: An automatic multiple-choice question generation scheme for English adjective understanding. In: Workshop on Modeling, Management and Generation of Problems/Questions in eLearning, the 15th International Conference on Computers in Education (ICCE 2007), pp. 137–142 (2007)
14. Mitkov, R., Ha, L.A.: Computer-aided generation of multiple-choice tests. In: Proceedings of the HLT-NAACL 03 Workshop on Building Educational Applications Using Natural Language Processing, vol. 2, pp. 17–22. Association for Computational Linguistics (2003)

15. Coniam, D.: A preliminary inquiry into using corpus word frequency data in the automatic generation of English language cloze tests. Calico J. **14**(2–4), 15–33 (1997)
16. Agarwal, M., Mannem, P.: Automatic gap-fill question generation from text books. In: The Workshop on Innovative Use of NLP for Building Educational Applications, pp. 56–64. Association for Computational Linguistics (2012)
17. Lee, J., Seneff, S.: Automatic generation of cloze items for prepositions. In: Conference of the International Speech Communication Association, INTERSPEECH 2007, Antwerp, Belgium, pp. 2173–2176 (2007)
18. Sakaguchi, K., Arase, Y., Komachi, M.: Discriminative approach to fill-in-the-blank quiz generation for language learners. In: Proceedings of the 51st Annual Meeting of the Association for Computational Linguistics, pp. 238–242 (2013)
19. Chall, J.S., Dale, E.: Readability Revisited: The New Dale-Chall Readability Formula. Brookline Books, Northampton (1995)
20. Lennon, C., Burdick, H.: The lexile framework as an approach for reading measurement and success (2004). Electronic Publication on http://www.lexile.com
21. Pitler, E., Nenkova, A.: Revisiting readability: a unified framework for predicting text quality. In: Proceedings of the Conference on Empirical Methods in Natural Language Processing, pp. 186–195. Association for Computational Linguistics (2008)
22. Tanaka-Ishii, K., Tezuka, S., Terada, H.: Sorting by readability. Comput. Linguist. **36**(2), 203–227 (2010)
23. Heilman, M., Collins-Thompson, K., Eskenazi, M.: An analysis of statistical models and features for reading difficulty prediction. In: Proceedings of the Third Workshop on Innovative Use of NLP for Building Educational Applications, pp. 71–79. Association for Computational Linguistics (2008)
24. François, T., Miltsakaki, E.: Do NLP and machine learning improve traditional readability formulas? In: The Workshop on Predicting and Improving Text Readability for Target Reader Populations, pp. 49–57. Association for Computational Linguistics (2012)
25. Perera, K.: The assessment of linguistic difficulty in reading material. Educ. Rev. **32**(2), 151–161 (1980)
26. Troia, G.A. (ed.): Instruction and Assessment for Struggling Writers: Evidence-Based Practices. Guilford Press, New York (2011)
27. Lin, D.: On the structural complexity of natural language sentences. In: Proceedings of the 16th Conference on Computational Linguistics, vol. 2, pp. 729–733. Association for Computational Linguistics (1996)
28. Liu, H.: Dependency distance as a metric of language comprehension difficulty. J. Cogn. Sci. **9**(2), 159–191 (2008)
29. Lu, Q., Xu, C., Liu, H.: Can chunking reduce syntactic complexity of natural lan-guages? Complexity **21**(S2), 33–41 (2016)
30. Mishra, A., Bhattacharyya, P., Carl, M.: Automatically predicting sentence translation difficulty. In: Proceedings of the 51st Annual Meeting of the Association for Computational Linguistics, vol. 2, pp. 346–351 (2013)
31. Specia, L., Shah, K., Souza, J.G., Cohn, T.: QuEst-A translation quality estimation framework. In: Proceedings of the 51st Annual Meeting of the Association for Computational Linguistics: System Demonstrations, pp. 79–84 (2013)
32. Koehn, P., et al.: Open source toolkit for statistical machine translation: factored translation models and confusion network decoding. In: Final Report of the 2006 JHU Summer Workshop (2006)
33. Landis, J.R., Koch, G.G.: The measurement of observer agreement for categorical data. Biometrics **33**(1), 159–174 (1977)

ABPR– A New Way of Point-of-Interest Recommendation via Geographical and Category Influence

Jingyuan Gao and Yan Yang[(⊠)]

Key Laboratory of Database, Heilongjiang University, Harbin, China
yangyan@hlju.edu.cn

Abstract. Point-of-Interest (POI) recommendation has been an important topic on Location-Based Social Networks (LBSN). It could recommend the POI point for users that they have never been. During the latest research, when adding geographical influence, recent research always pick up all the POI points to learn the influence it makes to the users. However, this may reduce the precision of experiment, for it does not take into consideration the reason that influences users in their frequent check-in activity region. To solve this problem, we propose a new POI recommending approach with the activity region, named Activity region Bayesian Personalized Ranking (ABPR), which adds geographical influence into the basket of BPR. This paper outlines the experiments done with Gowalla and Foursquare datasets to demonstrate the effectiveness and advantage of our approach.

Keywords: Location-Based Social Networks (LBSN)
Point-of-Interest (POI) recommendation · Geographical influence

1 Introduction

In recent years, as the rapid developing of social networks, people could share their experiences on Location-Based Social Network services, like Foursquare, Facebook and so on. People could also share their attitude to the POI points they have been on network services. To help people find the place they prefer much more, POI recommending has been proposed and plays an significant role in the stage of LBSN research.

POI Recommending: The goal of POI recommendation is learning the preference of users and recommend the place to users which they are fond of. By now, to divide the types of users, Li [1] proposed the approach of recommend next POI for users by learning the intrinsic and extrinsic interests, which is named as IEMF approach. It argues that users sign outside the activity region only because of the extrinsic reason. But the problems with the IEMF approach are that users may not check in outside the activity region only because of extrinsic reasons, and the creation of the activity region is poorly defined. Our work is different from IEMF in building activity region and the

© Springer Nature Singapore Pte Ltd. 2018
Q. Zhou et al. (Eds.): ICPCSEE 2018, CCIS 902, pp. 96–107, 2018.
https://doi.org/10.1007/978-981-13-2206-8_9

recommend approach. Another current approach is the PG-rank [2] approach. It combines the geographical influence and the category influence. However, not all the users are affected by the geographical influence, for the POI points inside the activity region produces few influence.

We suggest that POI recommending should add in geographical influence appropriately for it is a significant factor when users are making decisions. For example, user may check-in in a restaurant famous in social networks, or he may give up that thought because of the long distance. On the other hand, we suggest that, we couldn't add geographical influence to all the POI recommendation, for it doesn't play a significant role when it comes to the situation inside their activity region. For example, you may not change your attitude of whether or not check-in in the nearer restaurant because of the distance.

To solve the aforementioned problem, we propose a new Bayesian recommendation approach via location and category influence named ABPR (Activity region Bayesian Personalized Ranking), the purpose of this approach is to analyze the record of POI point of users, then build the activity region. For the POI point inside the region, we are doing the recommending via category approach; for the outside ones, we perform the recommending by using the geographical influence and the category influence, we also updated the parameter with SGD algorithm. Try to imagine that you want to have a meal, there is a restaurant you always check-in, will you change your decision to have a meal in other POI point much closer to where you live?

We think that the related research has the following insufficient parts:

(1) When adding the geographical influence, most of the research add the geographical influence into all the POI points, but when it comes to the POI points that are close the user, geographical influence is not the most significant reason influencing the users' decision;
(2) The research involving activity region always builds it with circles, and it may not accurately catch the users' frequently activity region;
(3) Some users like to explore new POI points, others may prefer the similar one. This has not been well analyzed by recently study.

To solve these problems, we propose the solution of the above– our ABPR approach. The advantage of our approach is:

(1) We do recommendation via category inside the activity region without the geographical influence;
(2) We calculate the activity region by using the approach of loss function to promote the precision;
(3) We add a parameter to measure the users' preference to explore new places.

2   Related Work

POI recommending has received a lot of attention recently, and it could be classified into the following parts:

Geographical Influence Based Recommending
There are tremendous amount of research concerning the geographical influence based recommendation. Some scholars argue users may be affected by the geographical influence to change their mind in check-in behavior in the LBSN. By this mean, [3] first added the geographical influence into the POI recommendation. Then, [4] prominent the influence of geographical influence by using the Gaussian Mixture Model (GMM) and the Matrix Factorization approach (MF), but the reason of check-in is complex and it isn't being analyzed concretely. [5, 6] proved that geographical influence could increase the precision of algorithms, however, both of them didn't take in the geographical influence appropriately. [7] proposed the Potential Dirichlet Distribution to integrate the location-based model. This research indicated that adding the geographical influence may promote the performance of algorithm.

Category Influence Based Recommending
Category based recommending is the earliest and the most widely used approach. [8] proposed the Bayesian Personalized Ranking approach (BPR) which add the thought of Bayesian into POI recommending research in order to analyze the intrinsic information concealed in the check-in data. By this means, [9] proposed the approach of LBPR via listwise Bayesian Personalized Ranking to solve the problem among the successive recommending. [10] also analyzed the last patterns of users to recommend next POI point that user may be most interest in; however, the disadvantage of their research is that they didn't separate the different types of users.

Time-Aware Based Recommending
The newly developed approach aimed to catch the temporal feature of activities, which is well studied recently. Because of the regularity of human daily activity, temporal influenced recommendation helps researchers to catch the preference of POI points much more precisely. [15] firstly add the temporal influence into POI recommendation, [11] divide the time slot into days, months and years, but the time slot is too long to catch the users' activity pattern.

Social Influence Based Recommending
Social influence based recommend is inspired by the intuition that there are some common interests of friends in the LBSN. [12] showed the social relations exert more influence on long distanced travel. [13] proposed the model of PMRE-GTS to show the connect of social relation and temporal influence.

Because of the sparsity and time complexity of the dataset, most of literature only build their models with two influences. And for the long-distance POI points, most of them are not ideal.

3 The ABPR Model

The recommendation task is defined as following:

Given the check-in behaviors of users and the locations of the POI points, we aim at recommending the Top-K locations users might be interested in but never visited before or not often check-in.

3.1 Our Framework

Activity Regions. People's activity has some type of scope [14], and the recent research mostly added the geographical influence into all the POI points. But the geographical influence effects are much lower than we considered inside the region.

Based on such considerations, we propose the concept of activity region. Users usually check-in around one or more regions, such as school, work place and home. We consider that if adding the geographical influence into the recommendation, it will reduce its precision. So, we use the SGD method to create the activity region. We performed category based recommending inside the region, performed the category and geographical based recommending outside the region. For the geographical influence may tremendously influence the check-in decision. Figure 1 shows the idea of activity region.

Fig. 1. The activity region (Color figure online)

The blue circle is the activity region, different color of points shows the different reason of check in.

For the outside points, we add in the geographical influence, here is the lost function of building the activity region:

$$J(\theta_a) = \frac{1}{2m} \sum_{i=1}^{m} (t_{\theta_a}(x^{(i)}) - y^{(i)})^2 \tag{1}$$

$x^{(i)}$ is the coordinate of check-in location, $y^{(i)}$ is activity region, t_{θ_a} is the matrix of POI points, θ_a is the parameter matrix of t, $(t_{\theta_a}(x^i) - y^i)^2$ is the distance from outside POI points to the activity. We minimize the sum of the distance:

$$A(\theta_a) = \arg\min \frac{1}{2m} \sum_{i=1}^{m} (t_{\theta_a}(x^{(i)}) - y^{(i)})^2 \tag{2}$$

To minimize this formula, we use the SGD method to calculate the parameter metrix θ_a and get the activity region, the process of calculating θ_a is:

$$\frac{\partial A(\theta_\alpha)}{\partial \theta_a} = \frac{1}{m} \sum_{i=1}^{m} \left(t_{\theta_a}(x^i) - y^i \right) x^i \tag{3}$$

If we want to get the Minimum value of $A(\theta)$, then $\frac{\partial A(\theta_\alpha)}{\partial \theta_a}$ should be 0. During the iteration, we use α_1 as the step of iteration, the process of iteration is as the following:

$$\theta_a = \theta_a - \alpha_1 \left(\frac{1}{m} \sum_{i=1}^{m} \left(t_{\theta_a}(x^i) - y^i \right) x^i \right) \tag{4}$$

until it convergences.

The Process of Naive Bayesian. posteriori probability could be define as:

$$P(l|l_u) \propto P(l)P(l_u|l) \tag{5}$$

And it equals to:

$$P(l|l_u) = P(l) \prod_{l' \in l_u} P(l'|l) \tag{6}$$

Where $P(l)$ is the posteriori probability of check-in in POI point l, l_u is the set of history check-in points, l' is the point of history check-in.

The geographical Influence. Users may be influenced by the geographical influence and category influence to check-in outside the activity region. By this means, we add the geographical influence into the process of POI recommendation.

We use $U \in R^{k*1}$ matrix as the set of users, $l \in L^{n*1}$ as the set of POI points, θ_l to as the parameter of matrix. Here we use the Euclid distance $D_{l_i l_j}$ between the users and the POI points.

Users may be influenced by the geographical influence, in other words, the distance between the users and the POI points may influence the check-in decision. For example, because of the distance, we may go to check-in in the shop 3 km from where you live, and we may not check-in in the 6 km one. We define $|X(l_i) - X(l_c)|^2$ as the distance from POI points to activity region. If it is much more near to the user, we define it as follow:

$$|X(l_i) - X(l_c)|^2 - |X(l_j) - X(l_c)|^2 \tag{7}$$

We use this formula to calculate l_i or l_j is closer to the activity region. If $D_{l_c l_i} - D_{l_c l_j} < 0$, To make it easier to array, we negate it as:

$$-\left(D_{l_c l_i} - D_{l_c l_j} \right) > 0 \tag{8}$$

The Desire to Explore the New POI Point. In our daily lives, some people desire to explore the new POI points, while some people prefers to check-in to the similar ones. Until now, few research involves the desire of individuals to explore. We would like to introduce the desire parameter χ to define the desire to explore the new POI point. If the users may desire to explore the similar ones, we may recommend the POI points based on the category factor.

We define the check-in set as U, ϑ_u as the number of the check-in category, η as the frequency of check-in, we divide ϑ_u by U as follow:

$$\chi = \frac{\vartheta_u}{\eta} \tag{9}$$

Clearly, it is a constant, if it is above the value θ_q, we may consider that the user prefers to check-in in the similar one.

3.2 Parameter Estimation

According to the Bayesian personalized ranking thought, we use the maximum likelihood posteriori probability to build the target formula, and then doing iteration to calculate the parameter and combine the activity region, geographical influence and desire parameter. To optimize the parameter, its derivation process is given by the following formula:

$$\gamma = \mathrm{argmax} \prod_{u \in U} \prod_{l \in L^{n*1}} \prod_{l_i \in L^{n*1}} \prod_{l_j \in L^{n*1}} \chi \mathrm{P}(>_{u,l \in a} |\theta) \mathrm{P}(\theta) \tag{10}$$

We define γ as the probability of check-in, The parameter $\theta = \{X(L), X(U), X(a), X(q)\}$ is the parameter that needs to be optimized, we change it by using the sigmoid formula $\sigma = \frac{1}{1+e^{-x}}$, we define the probability as:

$$\mathrm{P}(>_{u,l \in a} |\theta) = \sigma(x_{u,i,j}) \tag{11}$$

So the optimize formula could be changed as:

$$\gamma = \mathrm{argmax} \log \prod_{u \in U} \prod_{l \in L^{n*1}} \prod_{a \in R^+} \prod_{q \in R^+} \chi \mathrm{P}(>_{u,l \in a} |\theta) \mathrm{P}(\theta)$$
$$= \mathrm{argmin} \sum_{u \in U} \sum_{l \in L^{n*L}} \sum_{l_i \in L^{n*1}} \sum_{l_j \in L^{n*1}} - \log \sigma(x_{u,i,j}) + \log \mathrm{P}(\theta) + \theta_q \chi \tag{12}$$

The above formula is inside the activity region, when it comes to the outside the region, it can be defined as:

$$= \operatorname{argmin} \sum_{u \in U} \sum_{l \in L^{n*L}} \sum_{l_i \in L^{n*1}} \sum_{l_j \in L^{n*1}} [-\log \sigma(x_{u,i,j}) + \lambda_\theta \|\theta^2\| + \theta_q \chi] + A(\theta_a) \qquad (13)$$

For the outside ones, we define the geographical influence as the most significant factor, based on this idea, we change the above formula as:

$$\gamma = \operatorname{argmax} \prod_{u \in U} \prod_{l \in L^{n*1}} \prod_{l_i \in L^{n*1}} \prod_{l_j \in L^{n*1}} P(>_{u, l \in a} | \theta) P(\theta) \qquad (14)$$

Well the $P(>_{u, l \in a} | \theta) = P(-(D_{l_c l_i} - D_{l_c l_j}) > 0 | \theta)$

$$= -\sigma(D_{l_c l_i} - D_{l_c l_j}) \qquad (15)$$

So it could be changed as:

$$\gamma = \operatorname{argmax} \log \prod_{u \in U} \prod_{l \in L^{n*1}} \prod_{a \in R^+} \prod_{q \in R^+} P(>_{u, l \in a} | \theta) P(\theta)$$
$$= \operatorname{argmin} \sum_{u \in U} \sum_{l \in L^{n*L}} \sum_{a \in R^+} \sum_{q \in R^+} -\log \sigma(D_{l_c l_i} - D_{l_c l_j}) + \lambda_\theta \|\theta^2\| \qquad (16)$$

3.3   Algorithmic Description

We firstly extract all the train set, then optimize the parameter, we could get the check-in dataset ψ, it contains $\{u, l^c, \vartheta_u, \eta\}$, the four parameter defines users; the users' location; the location of POI points and its category; the frequency of check-in. The approach of update the parameter is:

$$\theta \leftarrow \theta - \alpha \frac{\partial \gamma}{\partial \theta} \qquad (17)$$

The α defines the learning step of SGD. To express the idea of the algorithm more Intuitively, we list the process of the algorithm as follow:

The algorithm of ABPR

Input:check-in dataset,learning step

α, regularization parameter λ_θ

Bulid the activity region;

Repeat

  For every $\{u,l^c, \vartheta_u, \eta\}$

    If $A(\theta_a)^t = 1$ (inside the activity region)

      Update X(L),X(U),X(a),X(q);

    Else $A(\theta_a)^t = 0$

      Update X(L),X(U),X(a);

  Until convergence

  Return the parameter$\{X(L),X(U),X(a),X(q)\}$;

---

4 Experiments

4.1 Dataset

We use the Gowalla and Foursquare datasets to evaluate the model performance. each check-in record contains the user's ID and the POI point location, each location has the latitude and the longitude information. The data statistics are shown in Table 1.

Table 1. The data statistics

Dataset	No. Check in	No. Users	No. POI
Foursquare	18492	1228	3360
Gowalla	34120	1838	2952

4.2 Evaluate Parameter

We quantitatively evaluate the model performance by using top-K recommending performance, i.e, Precision@K and Recall@K. They are defined as following:

$$\text{Precision@K} = \frac{1}{n}\sum_{i=1}^{n}\frac{S_i(K)\cap\tau_i}{K} \tag{18}$$

$$\text{Recall@K} = \frac{1}{n}\sum_{i=1}^{n}\frac{S_i(K)\cap\tau_i}{|\tau_i|} \tag{19}$$

Where $S_i(K)$ is a set of top-K unvisited locations recommend to user i excluding those locations in the training, τ_i is the set of locations that are visited by users i.

4.3 Performance Comparison

We compare the proposed model with the following methods:

BPR [8]: optimizes the ordering relationship of users' preferences for the observed and unobserved location;

USG [3]: incorporates the geographical influence,social network and user interest into collaborative filtering in an additive manner.

IEMF [1]: learning users' intrinsic and extrinsic interests for POI recommendation.

We experiment these methods and our ABPR with the real world datasets, and here is the comparison result (Fig. 2):

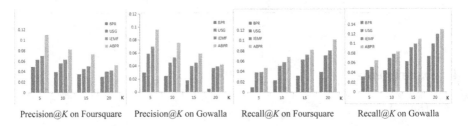

Precision@K on Foursquare   Precision@K on Gowalla   Recall@K on Foursquare   Recall@K on Gowalla

Fig. 2. Performance comparison

We evaluated the model performance by using precision@K and recall@K, our model gains about 15.4% and 17.3% improvement in precision@5 and recall@20 in Gowalla data; 12.6% and 13.8% in Foursquare data. Through analyzing the experimental result, we found the following observations.

First of all, our method outperforms all other methods. This superior result is for modeling the users' preference. By this means, we've done it better.

Secondly, the methods with geographical influence (USG, IEMF, ABPR) performs much better than the one without geographical influence (BPR). This further illustrates the benefits of doing POI recommending with geographical influence.

Lastly, we'd like to take about the influence of the number of iterations of building activity region, and the outperform of the method of building the activity region. The influence of it will be illustrated in Fig. 3. The red line of the figure shows our precision of our activity region, the green one shows the method of STELLAR [16], it adds the geographical influence to separate the POI points only with distance.

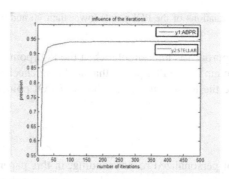

Fig. 3. Influence of the number of iterations (Color figure online)

Clearly, our method of building the activity region of ABPR performs much more better than the STELLAR [16] of adding geographical influence. STELLAR abandons the POI points over 10 km far. To illustrate that, we compared with the two methods to show the advantage of our method.

We can also conclude from the figure that when the iteration comes to more than 100, it convergences. So we suggest that the iteration may be getting its best situation when more than100 iteration is performed.

We also analyzed how does POI points effect the performance of activity region. The Fig. 4 shows the influence:

Precision of creating activity region   Recall of creating activity region   F1 score of creating activity region

Fig. 4. The connection between activity and the POI number

We found that our method does well when there are more than 30 POI points. The precision is not affected as much when the number of POI points changes, it remains at 80%. However, it affects much on the percentage of recall. The F1 score shows when POI point drops down to lower than 30, the score disparity of precision and recall gets higher. If the POI points are lower than 30, it is undesirable in building the activity region. Here are four examples of the situation of it (Fig. 5):

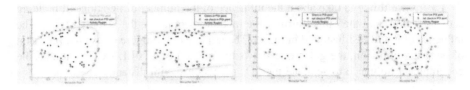

Fig. 5. Four situations of the POI points lower than 40 and more than 40

It seems that our approach is undesirable when the POI points is lower than 30 for the algorithm try to contain all the POI points that are being checked-in, but it does well in other situations, like the forth figure which has 137 POI points.

5   Conclusion

To further the study of personalized recommending, in this paper, we propose a new method to catch the preference of users based on the geographical influence and the category influence. We use the approach of SGD to estimate the parameter, and performance evaluation in the real world LBSN datasets shows our approach of POI recommending outperforms other methods. More specifically, our proposed method is more capable in distinguishing different types of users. For future works, we could consider the social relations and temporal influences on personalized recommending.

References

1. Li, H.Y., Ge, Y., Lian, D.F., Liu, H.: Learning user's intrinsic and extrinsic interests for point-of-interest recommendation: a unified approach. In: University of North Carolina at Charlotte, America (2017)
2. Feng, S., Li, X., Zeng, Y., Cong, G., Chee, Y.M., Yuan, Q.: Personalized ranking metric embedding for next new POI recommendation. In: IJCAI, pp. 2069–2075 (2015)
3. Ye, M., Yin, P., Lee, W.-C., Lee, D.-L.: Exploiting geographical influence for collaborative point-of-interest recommendation. In: SIGIR, pp. 325–334 (2011)
4. Cheng, C., Yang, H., Lyu, M.R., King, I.: Where you like to go next: successive point-of-interest recommendation. In: IJCAI 2013 (2013)
5. He, J., Li, X., Liao, L.: Category-aware next point-of-interest recommendation via listwise Bayesian personalized ranking. In: Proceedings of the Twenty-Sixth International Joint Conference on Artificial Intelligence (IJCAI 2017). Beijing Institute of Technology (2017)
6. He, J., Li, X., Liao, L.: Inferring a personalized next point-of-interest recommendation model with latent behavior patterns. In: AAAI 2016. Beijing Institute of Technology (2016)
7. Kurashima, T., Iwata, T., Hoshide, T., Takaya, N., Fujimura, K.: Geo topic model: joint modeling of user's activity area and interests for location recommendation. In: WSDM, pp. 375–384 (2013)
8. Rendle, S., Christoph, F., Zeno, G.: BPR: Bayesian Personalized Ranking from Implicit Feedback. University of Hildesheim, Germany (2009)

9. Cho, E., Mysers, S.A., Leskovec, J.: Friendship and mobility: user movement in location-based social network. In: Proceedings of the 17th ACM SIGKDD International Conference on Knowledge Discovery and Data Mining, USA, pp. 1082–1090 (2011)
10. Han, Z., Liu, Y.: PMRE-GTS: a new type of continuous POI recommending model via temporal influence and social influence. In: CCF of 2017 (2017)
11. Liu, Y., Wei, W., Sun, A., Miao, C.: Exploiting geographical neighborhood characteristics for location recommendation. In: Proceedings of CIKM, pp. 739–748 (2014)
12. Yao, L., Sheng, Q.Z., Qin, Y., Wang, X., Shemshadi, A., He, Q.: Context-aware point-of-interest recommendation using tensor factorization with social regularization. In: SIGIR, pp. 1007–1010 (2015)
13. Li, X., Cong, G., Li, X., Pham, T.N., Krishnaswamy, S.: Rank-GeoFM: a ranking based geographical factorization method for point of interest recommendation. In: SIGIR, pp. 433–442 (2015)
14. Hu, B., Ester, M.: Social topic modeling for point-of-interest recommendation in location-based social networks. In: ICDM, pp. 845–850 (2014)
15. Yuan, Q., Cong, G., Ma, Z., Sun, A., Thalmann, N.M.: Time-aware point-of-interest recommendation. In: SIGIR 2013, 28 July–1 August 2013
16. Zhao, S., Zhao, T., Yang, H., Micheal, R., King, I.: STELLAR: spatial-temporal latent ranking for successive point-of interest recommendation. In: AAAI 2017, pp. 315–321 (2017)
17. Jinghua, Z., Xuming, Y., Yake, W., et al.: Structural holes theory-based influence maximization in social network. In: The 12th International Conference on Wireless Algorithms, Systems, and Applications, WASA 2017, Guilin, China, pp. 860–864 (2017)

A Study on Corpus Content Display and IP Protection

Jingyi Ma[1], Muyun Yang[1(✉)], Haoyong Wang[2], Conghui Zhu[1],
and Bing Xu[1]

[1] Computer Science and Technology, Harbin Institute of Technology,
92 West Dazhi Street, Harbin 150001, China
mr_lovegreen@163.com, {yangmuyun,conghui}@hit.edu.cn,
xubing@hit-mtlab.net
[2] Institute of Foreign Languages, Agricultural University of Hebei,
289, Raining Temple Street, Baoding 071001, Hebei, China
hy701001@126.com

Abstract. Corpus has played an important role in most of research fields, especially in natural language processing. Some research demos provided detailed corpus content to highlight the contribution they have made, while overlook the security of corpus. In this paper, we explore content leakage resulted from the content display through a crawler. A website for displaying corpus is selected to be crawled by a simply crawler algorithm with some strategies we present. It is estimated that over 85% of the corpus can be downloaded, which means a substantial threaten to its IP right. Finally, we discuss the protection measures for content display, and give some valid suggestions for information content protection in technology and law.

Keywords: Corpus content display · Corpus security
Information content protection

1 Introduction

Traditional corpora are built by corpus linguists for linguistic research. As the development of deep learning, a large number of corpus content is used as research data in natural language processing, which has greatly promoted the development of corpus. Renouf has reviewed the evolution of text corpora over the period 1980 to the present day, focusing on three milestones. The first milestone is the 20-million-word Birmingham Corpus (1980–1986); the second is the 'dynamic' corpus (1990–2004); the third is the 'Web as corpus' (1998–2004) [1, 2]. The growth of corpora has expanded very fast with the world entering the age of the Internet just as the 'Web as corpus'. Especially in recent years, deep learning becomes more and more popular, facilitating widely use of corpus. For example, Google's Machine Translation has made great success on tens of millions of sentences corpus last year [3].

Most of research achievements would be displayed for people to prove how great contribution they have made in research area, and corpus is no exception. Especially in the age of Internet, the way of research achievements displayed become more various.

Q. Zhou et al. (Eds.): ICPCSEE 2018, CCIS 902, pp. 108–119, 2018.
https://doi.org/10.1007/978-981-13-2206-8_10

And the content of achievements would be more persuasive with more detailed display. However, because the public can get easy access to the research achievements via Internet, the security of the achievements may be overlooked by the owner. During the data-driven world, the corpora become more and more significant in research field, especially in Nature Language Processing (NLP) research [4]. Since development of the Internet, data acquisition from the web becomes increasingly common. So the technology of crawling data from web developed faster and faster recently [5]. Stealing the corpus in the display website encounters quite few safe measures, which is much easier than before. However, the security of corpus has not caught people's attention on most cases, even if their corpora will be purloined.

In this paper, we select a corpus content display website as the simulated crawled data. The purpose of website selection is that displaying data on website is the most represented way in various achievements exhibition. Then, we perform an experiment on crawling the corpus content from its display website by means of some sample strategies presented in this paper. The goal of this experiment is to simulate how many pairs corpus we can get by a simple crawler program. To fully illustrate that displaying corpus content on website is unsafe, we present an evaluation to estimate the seriousness of corpus leakage. Additionally, we discuss methods for content display protection from text corpus to video, and summarize technical protection measures and legal protection measures.

The rest of paper is organized as follows: Sect. 2 gives the corpus display website features and some crawling strategies for them; Sect. 3 shows our experiment result through strategies designed, and evaluation method with its result; Sect. 4 discusses some protection methods from text corpus to video, and gives protection measure from two aspects of technology and law; The conclusions and future work were arranged in Sect. 5.

2 Corpus Website: A Case Study

Considering that there is no evidence to prove how unsafe the corpus display is, we design an experiment on crawling corpus. We utilize web crawler technology to acquire data from a website displayed corpus content and chosen particularly. To get the corpus from its display website, we adopt strategies for crawling data, designed according to the selected website. And we draw up an evaluation to estimate the percentage of crawled corpus which can reflect on the danger in corpus content exhibition.

2.1 Website Content Display

In natural language processing research, especially in machine translation, corpus is the footstone in experiments. Displaying content on the web becomes the basic way in corpus display field. While most of corpora is displayed on website, there are still some differences in different website. Following are some main represented ways:

(1) A download link is put on the display website, through which the corpus could be downloaded simply. This kind of corpora displayed in this way, is usually shared for everyone by its owner. Owning to this corpus's sharing, it does not have any security.

(2) A website could be visited only by logged-on user, displayed a small part of corpus's download links. Corpus displayed on such website is more safe, because of its authorization visit and incompleteness data. On account of above method to limit to display, the purpose of display cannot be reached.

(3) The displayed corpus content is showed on webpage, which is searched through an online retrieval system. On such website, all corpus data cannot be downloaded directly, but can be displayed through searching different query keywords input by user. This displayed way seems to be more secure than the first way, and more public than the second way. Thus, displaying corpus on such website becomes the mainstream way in practice.

Although the third way seems much safe and more public, is it really safe to display corpus content in such way? A website for bilingual corpus exhibition representing the third way, was selected for our experiment [6]. There are three main features of this website: Firstly the website offers a query interface used to search for the corpus, whose content includes query keywords input by users. The search result consists of the parallel corpus pairs and their order for returned web page. Then, for security concerns, the only first 15 pairs content are shown in the webpage, while the total number of result corpus is far more than 15. The real number of result for each query is shown in the bottom of return page. Finally, there are no limits to prevent users to visit site on high frequency, like restricting the frequency of user's IP visit, providing verification code, etc. Just owning to above features, this website is chosen for our experiment.

2.2 Strategy Design

Aimed at the bilingual corpus website, a simple crawler algorithm is devised for this experiment. Although the algorithm is easy to be implemented by programming, the result corpus cannot be obtained entirely. Because there are only top 15 pairs parallel corpus on the result page through searching query keywords, crawler could not get the corpus locating at the back from this return page. That a pair corpus consists of two sentences including many keywords, which makes it possible. Because the corpus at the end searched from one query keyword return, may be in the front of result returned by another keyword query in its bilingual sentences. Thus, lacking of appropriate query keywords dictionary becomes the core problem. How to build a suitable query dictionary is the next necessary thing to do. Based on the features in the selected website, we present three main strategies for establishing and appending this dictionary in this paper.

Strategy a: selecting English and Chinese dictionary to form a query keywords dictionary. In consideration of the query Application Program Interface (API) and unknowing what keywords are in need, common dictionary is a good choice [7]. Because the content of exhibition website is English-Chinese Parallel Corpus, an English dictionary including 45093 words and a Chinese dictionary with 3000 characters are used. Algorithm 1 provides the detailed process in pseudo code.

Algorithm 1. CrawlingData(Q)

      input : query keywords Q, size of query keywords dictionary m

      output: all crawled corpus content saved in Data Base(DB)

1     $D \leftarrow (Q_1, Q_2, ..., Q_m)$

2   **for each** $q \in D$ **do**

3   *result ←query from website(q)*

4   **if** *result isn't empty* **then**

5   *result_total ← result.num*

6   *result_content ← result.corpus*

7   **for each** $p \in result\_total$ **do**

8   *DB.append(p.english, p.chinese)*

9   **end**

10   *DB.append(result_total)*

11   **end**

12   **end**

13   return *DB*

Strategy b: supplying keywords dictionary from the crawled corpus like self-learning. We have got 32494 pairs parallel corpus through the above keywords dictionary. However, so huge dictionary cannot cover all the keywords in this real corpus whose distribution of keywords is unknown. A good idea occurring to us is that the keywords in crawled corpus could be used to expand our query dictionary. As a result, some words in corpus sentences are not in our query dictionary. Thus, the crawled bilingual corpus is transformed into a word dictionary through a word segmentation system. After filtering words existing in query dictionary, this word dictionary could be the new query dictionary used to the continuation of crawling. Repeat the above steps until the new query dictionary cannot be generated. This strategy for supplying keywords dictionary automatically is defined as corpus self-learning, showing in Algorithm 2 in detail.

Strategy c: generating query string from the existing corpus. After the building of words dictionary to acquire corpus content, there are still some corpus in the end of result from querying keywords, which cannot be crawled by keywords query. What we find is that the query API could use string consisting of words to search like searching keywords. In theory, searching by string comprised by words is more accurate than only by word. Thus, building a string dictionary generated from our existing corpus is beneficial to obtaining more corpus content. The query string is made up by the keyword and words around it in parallel sentences of crawled corpus. The length of words gotten window (denoted by h) in query string used in experiment, are two and three. The pseudo algorithm of crawling corpus by generated query string is shown in Algorithm 3.

Algorithm 2. Self-SupplyingCorpus(C)

---

      input : Crawled corpus C, query keywords dictionary D
      output: Crawled corpus C through strategies.

1   *Sentences ← GetSentencesfromCorpus(C)*
2   **while** True **do**
3   *SuppliedKeywordsDic ←* {}, *SuppliedStringDic ←* {}
4   **for each** *s in Sentences* **do**
5   *en_sen = s["en_data"]*
6   *cn_sen = s["cn_data"]*
7   **for each** *en_w in en_sen* **do**
8   **if** *en_w* **not** *in D* **do**
9   *SuppliedKeywordsDic.append(en_w)*
10   **end if**
11   **end for**
12   **for each** *cn_w in cn_sen* **do**
13   **if** *cn_w* **not** *in D* **do**
14   *SuppliedStringDic.append(cn_w)*
15   **end if**
16   **end for**
17   **end for**
18   **if** *isEmpty(SuppliedKeywordsDic) and isEmpty(SuppliedStringDic)* **do**
19   **break**
20   **end if**
21   *C ←CrawlingData(SuppliedKeywordsDic)*
22   *C ←CrawlingData(SuppliedStringDic)*
23   *Sentences ←GetSentencesfromCorpus(C)*
24   **end while**
25   return C

Algorithm 3. CrawledByString(C, D)

	input : query keywords dictionary D, crawled corpus C
	output: all crawled corpus content saved in DataBase(DB)
1	$StringDic \leftarrow \{\}$
2	$h \leftarrow 2$
3	$Sentences \leftarrow GetSentencesfromCorpus(C)$
4	**for each** s **in** $Sentences$ **do**
5	**for** i **in** range(len(s)) **do**
6	**if** s[i] **in** D **do** \\find the keywords in sentences
7	**if** $i+h$ <len(s) **do**
8	StringDic.append(s[i:i+h])
9	**end if**
10	**if** $i-h$ > 0 **do**
11	StringDic.append(s[i-h:i])
12	**end if**
13	**end if**
14	**end for**
15	**end for**
16	$C \leftarrow CrawlingData(StringDic)$
18	return C

3 Experiment

3.1 Corpus Download

After designing above crawling strategies which aimed at selecting website peculiarity, a practical experiment is implemented. Table 1 gives an example result content of query keywords 'density' in crawling data, while the amount of real result is far more than these. In Table 1, the 'Order' is the top4 in the original order in return pages, 'Corpus pair' gives the English sentences and the corresponding Chinese sentences.

Strategy-a in previous is adopted to simulate the relationship between the number of acquiring data and the crawling time. According the experiment records, the following mathematical formula can express their relationship.

$$\log(Y) = a + b \bullet \log(X) \tag{1}$$

where Y and X are the pairs of crawled corpus and the time for crawling these data respectively, a and b are parameters for fitting the data by functions. After simulation, we got that $a = 3.2247, b = 0.7315, R^2 = 0.9941$, which R is the correlation coefficient of X and Y. Figure 1 shows the change situation of the crawled corpus pairs with the increase of hours. At the beginning, the amount of crawled data increases fast,

Table 1. A part of results corpus of query "density"

Order	Corpus pair
1	X-ray density with central radiolucency and fluid level. X线密度增加伴有中心部透光及液平面。
2	Intravenous injection of iodinated contrast media followed by repeat scanning is capable of enhancing the density of lesions with an increased blood pool (e.g., meningioma, arteriovenous malformation) or alteration in the blood brain barrier (e.g., metastatic tumor and inflammatory lesions). 静脉注射碘对比剂，然后重复扫描描能够增强伴有血池增加病变(如脑膜瘤、动静脉畸形)的密度或血脑屏障的改变(如转移灶和炎症)。
3	Localized air-fluid levels, localized ileus, or increased soft tissue density in the right lower quadrant is present in $50\sim$ of patients with early acute appendicitis. 50%的早期急性阑尾炎病人中可看到局限性气液面、局限性肠梗阻或右下腹软组织密度增深。
4	Density of lesions may be important, especially in lesions less than 3 cm in diameter, these lesions are often malignant. 病变的密度可能重要,特别是病变直径小于3 cm者,常为恶性。

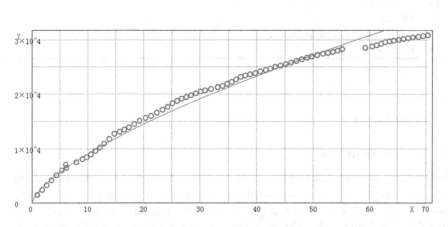

Fig. 1. The relationship between the pairs of crawled data and cost hours

while its speed represented by gradient reduces gradually. The cause of this appearance is that the corpus sentences queried by the keywords at the end of dictionary, are likely to have appeared in the previous keywords' result, so that these sentences wouldn't be saved.

After crawling by strategy a, 32494 pairs corpus is acquired in 80 h. Then the number of crawled corpus increases to 42958 by employing strategy-b. Finally, using query string extracted from corpus sentences in strategy-c helps us get 46594 pairs corpus. Table 2 gives the crawled result data in detail.

Table 2. Amounts of corpus crawled after adopting strategies gradually

Strategy	Corpus	Hours
a	32494	80
+b	42958	20
+c	46594	60

3.2   Content Leakage Estimation

Obtaining the above experiment result is enough for us, if its usefulness just stays in as the base data for common NLP research. However, the purpose of our experiment is to illustrate how harmful for corpus owner in corpora exhibition. Thus, an evaluation to estimate its perniciousness is necessary to be designed. The most straight-forward method is the ratio of the crawled corpus' content amount to the real corpus's. Nevertheless, what is we cannot get by crawler program is the number of the online corpus content.

According to features of the website selected in experiment, the number of bilingual corpus for each query can be obtained easily from its result page. Therefore, we assume that the proportion of crawled corpus amount including some query words to online corpus amount including the same query words, roughly equals to the ratio of obtained corpus amount to the real corpus's. Table 3 provides some example query words with their query return amount in crawled corpus and online corpus.

$$v = \frac{\sum_{i=1}^{n} c_i}{\sum_{j=1}^{n} c_j'} \tag{2}$$

$$\tilde{v} = \frac{\sum_{i=1}^{5} v_i}{5} \tag{3}$$

To formalize this measure process, we defined some parameters and some mathematical formula. We defined that n is size of the test query keywords dictionary which formulated $D = \{q_1, q_2, \ldots, q_n\}$, that q_i means the i-th word in dictionary. The amount of each query words result in crawled corpus is represented by $C_1 = \{c_1, c_2, .., c_n\}$,

Table 3. Amount of query result between crawled corpus and online corpus

Query words	Crawled corpus	Original corpus
Many	967	971
ectopic	50	50
X-Ray	105	137
表明	374	380
子	2537	2583

where c_i means the number of query words q_i result. $C_2 = \{c_1', c_2', .., c_n'\}$ indicates amounts and c_i' is the same result number but in online corpus. The ratio v representing the proportion of crawled corpus to original corpus, can be calculated by Eq. 2. To reduce the measurement error value, the average $\tilde{v}$ is gotten from Eq. 3.

3.3 Results

First, we set that a number of query words for evaluation are 100, 200, 500, 1000, etc. To get these different number test data, 10000 English words are selected from the original English dictionary in random. And to embody the change of result with query words expending, 5000 words are randomly selected from the previous 10000 words. To reduce the measurement error in calculation, five test query dictionaries in same number are generated at the same time and the final proportion is the average of the five results. And the 2000 words dictionary is originated from 5000 words dictionary. The rest of number dictionary is generated in the same way, and the method of computing ratio is the same. As shown in Table 4, the average ratio of different test keywords number, reaches 64% only by strategy-a, and could be up to 84% finally.

Table 4. The result ratio $\tilde{v}$ of crawled corpus by strategies

Words number	strategy a	+strategy b	+strategy c
100	0.6799	0.8052	0.8524
200	0.6666	0.8200	0.8771
500	0.6131	0.7656	0.8168
1000	0.6132	0.7683	0.8182
2000	0.6319	0.7909	0.8413
5000	0.6478	0.8103	0.8600
Average	0.6421	0.7934	0.8443

From Fig. 2, we can discover that the ratio could exceed 60% only by so simple strategy-a. What is more noteworthy is that the final ratio can reach 87% superlatively through our strategies, which means that most of the corpus in website could easily be stolen. Another thing which must not be ignored, is the promotion after adopting strategy-b and strategy-c, indicating that supplying query keywords and string is very powerful. There are other ways to generate query keywords and string, such as selecting keywords from sentence, changing the window in making up query string. No hard to guess that the whole corpus may be acquired by user through more appropriate strategies.

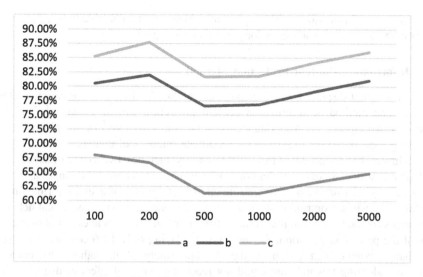

Fig. 2. The ratio vary after crawling by strategies

4 Content Protection: Measures and Practice

4.1 Text Protection

Corpus content display in public website is insecure indicted by above result of experiment. However, corpus is only one type of numerous text data. Other types of text data displaying in website could be crawled in the same way as corpus.

Text content on website is even more unsafe, one of the major reasons is that everyone can use it through copying and pasting. Prohibition of copying and pasting is one way for protection. In practice, the technology of prohibiting user to copy and paste in website has been fully developed. Since picture cannot be utilized directly after acquisition, converting text into picture for display is another good idea. Although someone wants to use the picture straightforward, digital watermark could be another protective screen [8].

4.2 Video Protection

Video content display is generally present in playing browser or others terminal software. Because text display protocol is Hyper Text Transfer Protocol (HTTP), which is different from video protocol Real Time Streaming Protocol (RTSP) [9]. Thanks to this difference, video content is not easy to get as text by web crawler. Acquiring video is not like getting text through copying and pasting, but by caching technology which not all browsers have. In the light of this feature of video display, providing no cache function in player browser and hiding local cache files can easily prevent video from being acquired illegally. While user cannot obtain the video from its cache files, screen recorder technology could acquire video directly during playing video [10]. Confronted with this problem, most of these video companies adopt a strategy that they charge for some more significant videos. Although some videos are unavoidable to be acquired, it has become a legal transaction as long as users have paid.

4.3   Information Content Protection Method

Whether it is text display or video display, all can be classified as information content display. The method of information content protection can be explained from the following two aspects. (1) Adopting technical measures to protect information content. (2) Using legal measures to protect the legitimate interests of owners.

Technical Protection Measures (TPM) mainly includes two types of techniques [11]. The first one is control technology, which means to control the range of IP that has access to information content website and to set the part of actual information for display in websites. Controlling the range of IP like only allowing educational network IP to access is more safe, but it also limits the content display openness. It is securer to control the part of information, for instance, displaying only 10% content in the real information content can guarantee the absolute safety of the other 90% content. However, adopting this measure could not reach the expected effect of display if the amount of overall data is small. Another is anti-crawler technology, which is to solve the problems of web crawler. Anti-crawler technology mainly relies on following strategies: checking the preprocessing request header in user's network request, setting graphic verification code during user first visit for some time, adopting Asynchronous Loading technology to present data on webpage, controlling the visit frequency of the same IP, etc. [12].

What should not be overlooked is the Legal Protection Measures (LPM) for information content protection. Technology gaps can never be eliminated with development and progress of technology. Thus, it is essential to protect rights by using the weapon of law. The article fifty of Copyright Law of People's Republic of China illustrates that "A copyright owner or copyright-related right owner who has evidence to establish that another person is committing or will commit an act of infringing his right, which could cause a remediless loss to his legitimate rights and interests if the act is not prevented immediately, may apply to a people's court for adopting such measures as order to stop the relevant act and property preservation before he initiates an action". According to above regulations, when we cannot stop content acquired by TPM and have enough evidence to show someone's infringement, we should seek help from the law immediately like applying to a people's court. We call on people to pay attention to the security of information content display.

5   Conclusion and Future Work

In this paper, we choose a corpus content display website to simulate a crawling experiment by a simple crawler algorithm with some strategies. While it seems that showing only top 15 pairs corpus content could prevent the online corpus from danger of being stolen, some other easy strategies can be employed to obtain them. Aimed at these features of the selected website, three main strategies are presented to assist to improve data acquiring in this work. To evaluate the seriousness of corpus content leakage, we designed how to estimate the total number of original corpus. The experiment shows that 87% of corpus could be obtained easily by these simple strategies finally, indicating that it is not impossible to copy the whole corpus secretly by website visitors. We discuss some protection measures from text corpus to video and give some suggestions in two aspects of technology and law. An appeal for attaching importance to corpus protection is made for those corpus owners who have not realized the seriousness.

For the future work, many more problems need to be solved. The existing protection measure cannot resolve the contradictions between product display and content protection. Exploring safer methods in content display is a both significant and challenging. Owning to the fact that few people have fully realized the insecurity in achievement displays, we call on more people to attach attention on this issue.

Acknowledgments. The work of this paper is funded by the project of National Natural Science Foundation of China (No. 2017YFB1002102) and the project of National key research and development program of China (No. 91520204).

References

1. Renouf, A.: Corpus development 25 years on: from super-corpus to cybercorpus. Lang. Comput. Stud. Pract. Linguist. **62**(1), 27–49 (2007)
2. Kennedy, G., Ooi, V.B.Y.: An Introduction to Corpus Linguistics. Studies in Language and Linguistics (1998)
3. Vaswani, A., Shazeer, N., Parmar, N., et al.: Attention is all you need. In: Advances in Neural Information Processing Systems, pp. 6000–6010 (2017)
4. Cohen, K.B., Ogren, P.V., Fox, L., et al.: Corpus design for biomedical natural language processing. In: ACL-ISMB Workshop on Linking Biological Literature, Ontologies and Databases: Mining Biological Semantics, pp. 38–45. Association for Computational Linguistics (2005)
5. Heydon, A., Najork, M.: Mercator: A scalable, extensible Web crawler. World Wide Web-Internet Web Inf. Syst. **2**(4), 219–229 (1999)
6. Koehn, P.: A parallel corpus for statistical machine translation. In: Proceedings of the Third Workshop on Statistical Machine Translation, pp. 79–86 (2005)
7. Bergler, F.: Application program interface: US, US 5572675 A[P] (1996)
8. Mehrabi, H.: Digital watermark. In: Constantopoulos, P., Sølvberg, I.T. (eds.) ECDL 2001. LNCS, vol. 2163, pp. 49–58. Springer, Heidelberg (2001). https://doi.org/10.1007/3-540-44796-2_5
9. Adji, F.R., Saputra, H.M.: Perbandingan Hyper Text Transfer Protocol (HTTP) dengan Real Time Streaming Protcol (RTSP) menggunakan Video Streaming. In: Prosiding Seminar Nasional Rekayasa & Desain Itenas (2016)
10. Sun, H., Tang, Y., Liang, C., et al.: High speed computer screen recorder system based on FPGA+ARM. Application of Electronic Technique (2011)
11. Dong, A.: Question inquiry on the copyright protection of foreign language corpus. J. Beijing Inst. Graph. **25**, 68–70 (2017)
12. Liu, SL.: The strategy of coping with anti-crawler website. Comput. Knowl. Technol. **13**, 19–21 (2017)

Construction and Application of Diversified Knowledge Model for Paper Reviewers Recommendation

Hua Zhao$^{(\boxtimes)}$ ⓘ, Wei Tao ⓘ, Ruofei Zou ⓘ, and Chunming Xu ⓘ

College of Computer Science and Engineering, Shandong University of Science and Technology, Qingdao 266590, China
huamolin@163.com

Abstract. Reviewer assignment problem is a time-consuming work, so an automatic reviewer recommendation method is important and necessary, where how to represent the reviewers' research interest was the key. In order to mine the research interests, the published research papers are firstly collected from Internet; secondly a diversified knowledge modeling method based on the cluster algorithm is proposed; finally carry out research on the similarity computing between the pending paper content and the experts' research interests, and then recommend experts based on the similarity. Experimental results show that the proposed diversified knowledge model can find the appropriate paper reviewers effectively, which can reduce the workload of the editors greatly.

Keywords: Diversified knowledge model · Reviewer recommendation
Research interest

1 Introduction

More and more research areas have appeared with the development of the scientific research, which made the Reviewer Assignment Problem (RAP) be a time-consuming work [4, 8]. On the one hand, RAP is usually needed to be done in a limited time by only one or several editors, especially for the papers submitted to a conference [5]; on the other hand, there are many problems needed to be considered in RAP, for example whether the reviewer's research interests are consistent with the content of the pending paper? How many research interests does a reviewer have? Whether the reviewer's research interests will change? With the refinement of the branch of the research subjects, it is more and more difficult to get the research interest of every expert. So it is necessary to design an automatic method to recommend the paper reviewers, which has become a hot topic [9].

In this paper, a diversified knowledge model to represent the expert's research interests based on the clustering is firstly proposed. The similarity computation method between the paper and the expert's research interests is explored. And finally the paper reviewers are recommended based on the similarity.

© Springer Nature Singapore Pte Ltd. 2018
Q. Zhou et al. (Eds.): ICPCSEE 2018, CCIS 902, pp. 120–127, 2018.
https://doi.org/10.1007/978-981-13-2206-8_11

2   Related Works

The common methods used to mine the expert's knowledge model were to produce the topic distribution based on some topic model [1, 3, 7, 10], such as LDA. The produced distributions represented the expert's knowledge model. Some researchers considered the semantic information when constructing the knowledge model [2].

For the reviewer recommendation, [5] used the concept tree to compute the similarity between the paper content and the expert's knowledge model, and then recommended the reviewers based on the similarity. The thought the recommendation problem as the classification problem, and then used the classification algorithm to recommend. Firstly used the terminology extraction method to solve the new words appeared in the paper, and then recommended the reviewers based on the vector space retrieval.

From the above analysis, we can see that many works had been done for RAP, and had gotten many achievements. But there are still some things that can be improved: firstly, there are little work being done on how to construct the knowledge model accurately; secondly, although the knowledge model produced by the topic model can represent the research interest to some extent, but every expert may have several different research interests, and this kind of knowledge model cannot distinguish between these different research interests. So we propose to create the diversified knowledge model. Experimental results confirm the good performance of the proposed model.

3   Frame of the Paper Reviewer Recommendation System

This section will introduce the frame of the RAP system, which includes three modules and is shown in Fig. 1. But only several parts are mainly discussed in this paper, which include "Research Interest" and "Similarity Computation".

(1) Data collection module (M1). This module firstly automatically and periodically (10 days) downloads the papers from the online paper platform, which is Wan-Fang (http://www.wanfangdata.com.cn/) in our system. Secondly, the metadata will be extracted, which include the title, abstract, keyword, author name, author workplace, journal name, and publish date, the papers number the authors have published, and the citation number of the paper. The former three are used to mine the research interests of the experts and the other metadata are used to compute the authority and the activity, and to mine the relationships between the experts. Not all the authors could be the experts. Price theory [6] thinks that the authors who have more than N papers are outstanding experts, where N is computed by the formula (1). In our experiments, we finally choose the authors who have published more than 22 (that is N = 22) papers as the initial experts, and as a result, we collect 3,603 initial experts.

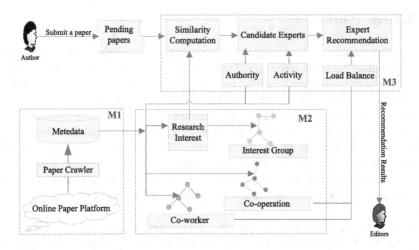

Fig. 1. Frame of the reviewer recommendation system

$$N = 0.749(\sigma_{max})^{1/2} \tag{1}$$

(2) Data analysis module (M2). This module firstly detects the research interests of every author, which will be introduced in detail in the next section. Secondly, the system recognizes the co-operation relationship and the co-worker relationship based on the authors' information in the research papers.

(3) Data computation module (M3). This module firstly computes the similarity between the pending paper and the experts' research interest, which is the basis of the recommendation. Secondly, detect the paper reviewers, which are based on the multidimensional features, which include similarity, authority, activity, expert credit evaluation, and so on. And finally, recommend the paper reviewer. Expert recommendation is different from the product recommendation. So we need to take many things into account, for example, the reviewer should not be the co-workers with the authors of the pending paper. The system do some combination optimal works to the recommendation after the paper reviewer detection.

4   Diversified Experts Knowledge Model Based on Clustering

Constructing the exact expert knowledge model is the first and the key part in the automatic detection and recommendation of the paper reviewers. In order to mine more elaborative research interests, we propose to create the diversified experts knowledge model based on the clustering.

The published papers are the representative research achievements of the experts, which can represent their research interests to a great extent. So, we propose to use these papers to mine the knowledge model of the experts. In order to distinguish between the several research interests of one expert, we propose to create the diversified knowledge model, and the algorithm of which is as follows:

Algorithm1: Diversified Experts Knowledge Model Construction Algorithm

---

Input: All the papers in the database

Output: $KnowModel(expert)$, the diversified knowledge model of $expert$

1. Extract all the experts' papers from the database, marked as expertpaper = $\{paper_1, paper_2, \dots paper_n\}$;

2. Extract the keywords, marked as $PaperKeyword = \{keyword_i | keyword_i \in paper_j, paper_j \in expertpaper\}$;

3. Create the co-occurrence matrix, marked as $Co - occur = \begin{bmatrix} co_{11} & \cdots & co_{1m} \\ \cdots & \cdots & \cdots \\ co_{m1} & \cdots & co_{mm} \end{bmatrix}$,
    where co_{ij} is the times of the co-occurrence frequency between $keyword_i$ and $keyword_j$;

4. Adopt GN algorithm, the classical community detection algorithm, to cluster the co-occurrence matrix;

5. Each community produced by GN represents a research interest of the expert;

6. Output the diversified knowledge model of expert.

Suppose there are n communities produced by GN, we then consider that this expert has n research interests. We will create the diversified knowledge model: $KnowModel(expert) = \{C_1, C_2, \dots, C_n\}$, where $C_i(1 \leq i \leq n)$ is the ith research interest of $expert$, $C_i = \{c_ik_1, c_iw_1; c_ik_1, c_iw_1; \dots; c_ik_t, c_iw_t\}$ t is the number of the keywords of the ith research interest (different interest maybe has the different number of the keywords), $c_ik_j(1 \leq j \leq t)$ is the jth keyword of the ith research interest, $c_iw_j(1 \leq j \leq t)$ is the interest degree which is computed by formula (2):

$$c_iw_j = \frac{tf(c_ik_j)}{KeywordNum} \times \log \frac{PaperN}{PNum(c_ik_j)} \tag{2}$$

Where $tf(c_ik_j)$ is the frequency of the keyword c_ik_j which is counted in the $PaperKeyword$, $PaperKeyNum$ is the number of the keywords in $PaperKeyword$, $PaperN$ is the number of the papers in the database, $PNum(c_ik_j)$ is the number of the papers which include the keyword c_ik_j.

We adopt two different measures for the GN, which are *Betweenness* centrality and *Degree* centrality, respectively. The *Betweenness* of a node s is the number of the shortest path between the every two nodes which is across the node s. The counting method of the *Betweenness* centrality of the node s is as follows:

$$C_B(s) = \sum_{u \neq v \neq s \in V} \frac{\sigma_{uv}(s)}{\sigma_{uv}} \tag{3}$$

Where $C_B(s)$ is the *Betweenness* centrality of the node s, V is the set of all the nodes within the network, $\sigma_{uv}(s)$ is the shortest path between the node u and the node v across the node s. Intuitively, $C_B(s)$ reflects the importance of the node s which acts as bridge. The larger the $C_B(s)$, the more the shortest path across the node s, the larger the probability of the node connecting two different communities.

The *Degree* centrality adopted in our experiment is computed as follows. $C_D(s)$ is the degree centrality of the node s. $\pi_{us} \in \{0, 1\}$ represents the co-occurrence score of the node u and the node s: if the node u and the node s co-occur in the keywords of the papers, then $\pi_{us} = 1$; or $\pi_{us} = 0$.

$$C_D(s) = \sum_{u \neq s \in V} \pi_{us} \tag{4}$$

5 Auto-Detection Method for the Paper Reviewers

Vector space model (VSM) is adopted to model the pending paper. Suppose the pending paper has ω keywords. The model of the paper can be represented by $paperVec = \{keyword_1, w_1; keyword_2, w_2; \ldots; keyword_\omega, w_\omega\}$, where $keyword_i$ is the keyword appeared in the pending paper, w_i is the weight of the keyword, which is computed by TF-IDF.

We then define the similarity between the expert's research interest and the pending paper as the maximum value of all the similarity between the pending paper's VSM and expert's each research interest, where is shown in the following formulas:

$$Sim(paper, expert) = Max_{1 \leq i \leq n}(\cos(paperVec, C_i)) \tag{5}$$

$$\cos(paperVec, C_i) = \frac{\sum_{k=1}^{m} w_k \times c_i w_k}{\sqrt{\left(\sum_{k=1}^{m} (w_k)^2\right)\left(\sum_{k=1}^{m} (c_i w_k)^2\right)}} \tag{6}$$

Where $Sim(Paper, expert)$ is the similarity between the *paper* content and the research interest of *expert*. *paperVec* is the vector space model of the *paper*. C_i is the ith research interest.

The following process is used to detect the paper reviewer.

(1) Extract the authors' names and workplace from the pending paper. Extract the experts (which have no cooperation with the authors in the pending paper) from the initial expert database, and use these experts to create the candidate experts set.
(2) Compute the similarity between the pending paper and the research interest of each expert in the candidate experts set. Extract these experts whose similarities are larger than the threshold θ, and use these experts to create the relative experts set.
(3) Sort (in descending order) these experts in the relative experts set based on the similarities and h index, respectively. Select the top-n expert as the recommended paper reviewers.

6   Experiments and Results

Our recommendation system crawls 195,440 papers from WanFang, with the publish date from 1998 to 2016. These papers include 151,858 authors and 128,257 keywords.

6.1   Visualization Results of the Construction of the Diversified Knowledge Model

Figures 2 and 3 show the diversified knowledge model construction visualization result based on the *Betweenness* centrality and *Degree* centrality, respectively, where different colors represent different clusters. From the results shown in the Figs. 2 and 3, we can see that the two diversified knowledge models on the basis of the two different centralities are consistent, from which we can find that the knowledge model construction method based on the clustering are feasible and effective.

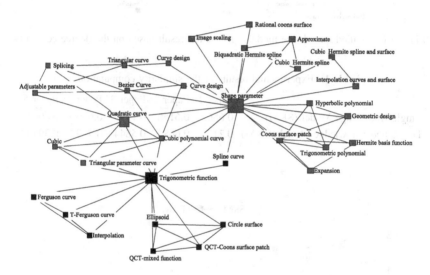

Fig. 2. Diversified knowledge model based on the betweenness centrality

6.2   Results of the Similarity Between the Pending Paper and the Expert Research Interests

Our recommendation method is mainly based on the similarity, so now we evaluate the similarity method to confirm the performance of the diversified knowledge model. In order to evaluate the effectiveness of the proposed knowledge model, we design the following experiment: extract an expert randomly from the expert database and his/her five papers, which are marked as Paper 1, Paper 2, Paper 3, Paper 4, and Paper 5. We can evaluate the effectiveness by computing the similarities between the experts' research interests and the five papers, respectively. The experiments results are shown in Table 1, which are compared in Fig. 4 in detail. Single knowledge model means that the research interest is represented by a single vector space model.

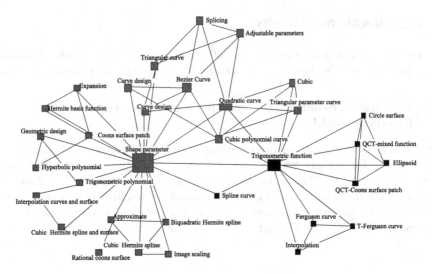

Fig. 3. Diversified knowledge model construction result based on the degree centrality

Table 1. Experimental results of the similarity computation

	Paper 1	Paper 2	Paper 3	Paper 4	Paper 5
Single Knowledge Model (SKM)	0.6562	0.6156	0.1599	0.0561	0.4017
Diversified Knowledge Model (DKM)	0.7121	0.6680	0.5844	0.4207	0.3949

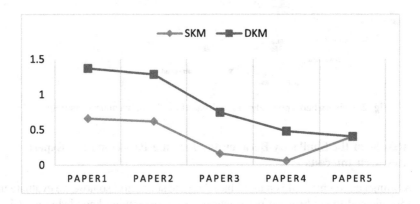

Fig. 4. The similarity comparison between different experiments

From the above results showed in Table 1 and Fig. 4, we can see that the proposed diversified knowledge model is effective. The proposed model performs better for every paper, especially for the Paper 3 and Paper 4. For these two papers, the similarities are relatively lower under the single knowledge model, which are 0.1599 and 0.0561, respectively. But the similarities are greatly improved under the diversified knowledge model, which are 0. 5844 and 0. 4207, respectively. The higher similarities are more reasonable; after all, all these papers are belonging to the expert.

7   Conclusions and Future Work

In order to mine the experts' research interests and then to recommend paper reviewers, we propose to create a kind of diversified knowledge model. The model uses several separate vectors to represent the research interest of a certain expert based on the clustering. The experimental results show that the proposed model is very effective.

During the experiments, we find that there is a common "different words having the same meaning" problem. So the future work will include how to adopt the semantics into the diversified knowledge model.

Acknowledgements. This work is supported by the China NSFC program (No. 61602278, 31671588, 61702306, 71704096); Sci. & Tech. Development Fund of Shandong Province of China (2016ZDJS02A11 and ZR2017MF027); the social science and humanity on Young Fund of the ministry of Education (16YJCZH154, 16YJCZH041, 16YJCZH012); the SDUST Research Fund (No. 2015TDJH102); CAS Key Lab of Network Data Science and Technology Open Fund Project (No. CASNDST201706); International cooperation and training program for outstanding young teachers in Shandong colleges and Universities.

References

1. Ali, D.: Using time topic model for semantics-based dynamic research interest finding. Knowl. Based Syst. **26**(2), 154–163 (2016)
2. Ba, Z.C., Li, G., Zhu, S.W.: Similarity measurement of research interests in semantic network. New Technol. Libr. Inf. Serv. **32**(4), 81–90 (2016)
3. Huang, S., Zhang, J., Dan, S., Wang, L., Hua, X.S.: Two-stage friend recommendation based on network alignment and series expansion of probabilistic topic model. IEEE Trans. Multimed. **19**(6), 1314–1326 (2017)
4. Daş, G.S., Göçken, T.: A fuzzy approach for the reviewer assignment problem. Comput. Ind. Eng. **72**(1), 50–57 (2014)
5. Li, X.L., Watanabe, T.: Automatic paper-to-reviewer assignment, based on the matching degree of the reviewers. Procedia Comput. Sci. **22**, 633–642 (2013)
6. Price, D.D.S.: Little Science, Big Science. Columbia University Press, New York (1965)
7. Shi, Q.W., Li, Y.N., Guo, P.L.: Dynamic finding of Authors' research interests in scientific literature. J. Comput. Appl. **33**(11), 3080–3083 (2013)
8. Wang, F., Shi, N., Chen, B.: A comprehensive survey of the reviewer assignment problem. Int. J. Inf. Technol. Decis. Making **9**(4), 645–668 (2010)
9. Wang, J., Yue, F., Wang, G., Xu, Y.H., Yang, C.: Expert recommendation in scientific social network based on link prediction. J. Intell. **34**(6), 151–156 (2015)
10. Yu, J., Yu, Z.T., Yang, J.F., Guo, J.Y., Yan, X.: Expert recommendation method for project evaluation based on topic information. Comput. Eng. **40**(6), 201–205 (2014)

Study on Chinese Term Extraction Method Based on Machine Learning

Wen Zeng[1(✉)], Xiang Li[2], and Hui Li[3]

[1] Institute of Scientific and Technical Information of China,
Beijing 100038, China
zengw@istic.ac.cn
[2] High Technology Industry Development Center, Beijing 100045, China
[3] Beijing Institute of Science and Technology Information,
Beijing 100044, China

Abstract. Term is the linguistic expression of the concepts in professional knowledge, which are accumulated through incremental exploration and research in specific fields. In the study of intelligence analysis and knowledge organization, term extraction is an important research subject. Deep neural network is an algorithm based on machine learning. It aims to obtain high-level features that can better represent raw data through learning by multilayer structure. Though machine learning has been widely used in studies in many fields, it is rarely mentioned in term extraction. The paper combines traditional method of extraction term with the new method of machine learning that is deep neural network. And it uses the method to extract terms from the real and effective corpus for experiments. Compared with methods based solely on language rules, language rules & statistical calculation, this method can improve the accuracy rate by about 47% and 8% respectively. This method gets some new terms that are not contained in the thesaurus. It verifies the effectiveness of machine learning in term extraction.

Keywords: Machine learning · Deep neural network · Term extraction

1 Introduction

Term is defined as word reference to concept representation in specific professional fields [1]. Term extraction is also called term identification, which is the process of obtaining terminologies representing professional concepts from texts in respective special fields. Term extraction is an important research subject in both the study of intelligence analysis and knowledge organization. Manual extraction of from corpus is very inefficient and insufficient to meet the needs of intelligence analysis, especially in the era of big data. Therefore some methods are required to fulfill the task of term extraction. The concept of deep neural network is derived from the study of artificial neural network. It is an emerging machine learning method based on multi-layer neural network, which aims to obtain high-level features that can better represent raw data through learning by multilayer structure. At present, study of deep neural network has become trendy and its application has been employed in some fields, but there is little

© Springer Nature Singapore Pte Ltd. 2018
Q. Zhou et al. (Eds.): ICPCSEE 2018, CCIS 902, pp. 128–135, 2018.
https://doi.org/10.1007/978-981-13-2206-8_12

research on its application in term extraction. Therefore, this study in the paper attempts to apply deep neural network to term extraction and develops an extraction method based on it, which hopefully can provide more accurate identification of terms from corpus at certain scale so as to help people engaged in intelligence analysis to improve their work and lay foundation for the future research.

2 Related Research

2.1 Research on Term Extraction Method

There are mainly three methods of term extraction at present: the method based on language rules, the method based on statistical calculation and the method based on both language rules & statistical calculation. Besides, there are also methods based on machine learning and related methods of various extensions. In the case of the method based on language rules, generally speaking, commonly formats of language rules are formed to identify terms of corpora. This method is effective for term extraction in specific fields or types. But as it is impossible to cover all the morphological rules of terms, there exists restrictions to its application. The method based on language rules is rarely used alone. The method based on statistical calculation identifies terms based on the distribution and related statistical calculation of the terms in the corpora. This type of methods mainly includes TD-IDF (Term Frequency-Inverse Document Frequency), mutual information and information entropy and so on. The idea of TF-IDF is that the more word in the texts appears frequently, the lower the value of the word, meaning that the word is more likely to be a common word. On the other hand, if the high frequency is coupled with appearance in only few texts, the value of the word is high, implying that the word is more likely to be a term in certain field [2]. Mutual information method computes the binding strength of word strings to decide if two strings can form a terminology. Information entropy method mainly calculates the uncertainty of boundaries between word strings. The more uncertain the boundary, the higher the information entropy, the more likely the two word strings are one complete word. The calculation of left and right information entropies of word strings leads to determining the uncertainty of left and right boundaries of word strings. When left or right information entropy exceeds certain threshold value, the word can be identified as a term. Methods based on both language rules and statistical calculation use language rules to collect candidate term sets at first, and then filter the word sets with statistical information to get the final terms. This sort of methods include C-Value, NC-Value, LIDF-Value and L-Value. This method first extracts candidate terms through lexical rules model, and then calculate the frequency of candidate terms appear in the corpora, the frequency of candidate terms as a part of other long terms, the number of long terms and the length of candidate terms. With the above statistical calculation, the value of a candidate term can be calculated. NC-Value takes into consideration the contexts of terms in addition to the method used by C-Value. LIDF-Value combines the methods of grammatical rules, IDF and C-Value, and measures grammatical rules by probabilities, thus obtaining related probabilities of candidate terms. L-Value is a metamorphosis of LIDF-Value, which does not take into account the value of IDF, but emphasizes more on the property of terminology of a candidate term in a single document or a corpus [3–7]. Extension method is mainly to extend a term based on such core parts of a term as seeded term, central word string and term components, and make judgement based on parts of speech and length of words during extension, and terminate when the word-building conditions are met. One of the research fields of term extraction is one based on machine learning. The main steps are: first, construct a training corpus, and then use some machine learning method to learn from the training corpus according to some artificially structured features, and get related model parameters, and finally extract terms from the test corpus bused on the trained model. The paper will combine traditional method of extraction term with the new method of machine learning that is deep neural network.

2.2 Research on Deep Neural Network

Deep neural network is a machine learning method of emerging multi-layer neural network, with multi-layer and non-linear mapping deep structure. The essence of deep neural network is a kind of feature extraction. It constructs a deep neural network model to learn features of the inputted model. It realizes to abstract expression of features by non-linear mapping of features through some activation function and transmits deviation between abstract features and inputs through back propagation to instruct the model to optimize its parameters for better abstract representation of the connotation of inputs. The difference between methods of traditional machine learning and deep neural network lies in the way they capture the features of data. Traditional machine learning method relies on manual work to construct some representative features of data for learning whereas deep neural network attempts to structure features of data automatically and complete the extraction of data features. Besides as its model has several layers, it is more capable of enhancing the explanatory ability of raw data and the generalization ability of models. The typical models of deep neural network include the following [8]: convolutional neural network (CNN), recurrent neural network (RNN), autoencoder (AE) and so on. The CNN is a feed-forward multilayer neural network mainly consisting of the convolution part and the fully connected part. The convolution part is made of the convolution layer that conducts convolutional computation of original features and the pooling layer that extract features after convolutional computation. The fully connected part drags the features obtained from the convolutional part to a fully connected layer for related computation and get the ultimate output. The convolutional part trains data in a supervised way, that is, it compares the output with the labeled original data, then feeds back the errors through back propagation and makes improvement on model parameters with related optimized method. Besides the input, output and hidden layers, RNN has self-connections and inter-connections in the hidden layer. It can structure delay signals that can reproduce the input data before transmitting in the matrix, and the following node can retain information of the precedent node, so that the model obtains memory properties. Auto-Encoder (AE) is a model composed of encoder and decoder to reproduce the original data as much as possible. AE usually trains with unlabelled data. The encoder recodes the raw data and the decoder decodes the recoded data, and then optimizes based on the deviation between the decoded data and the raw data. At present, deep neural network has achieved certain advance in some fields [9, 10].

From the above mentioned research or development status, deep neural network is rarely employed in term extraction. Compared to traditional machine learning method, deep neural network makes improvement on eigen-structure and conducts pre-training of unlabeled data with AE, which better fits the facts that unlabeled data is much more common than labeled data. Therefore, this paper attempts to make research use the core model of deep neural network for term extraction.

3   Term Extraction Method Based on Deep Neural Network

3.1   Constructing Candidate Term Dictionary

As the tool of word analyzers usually divide words into one-character word, whereas the length of terminologies varies a lot, the results of term extraction would be under direct influence of original word division if we use the original corpus to train word vector and model of deep neural network, e.g. AE. Therefore, it is necessary to construct a candidate term dictionary for the tool of word analyzers during divide words. By analysis and summary of word formation properties about terms, the paper find that most terms are binary or ternary terms. Terms are usually certain parts of speech such as nouns, verbs, adjectives, conjectives, prepositions and interjections are hardly used as terms. Therefore, the research considers the part of speech and its combination to construct part of speech rules template of terms. Besides, even such parts of speech as nouns, verbs and adjectives that can become terms, there are some words that cannot become terms. Before matching terms using rules template, the paper need to filter words whose part of speech cannot be terms. The main method is to count word frequency of texts that have done word division and sum up high-frequency words whose part of speech are included in the language rules template but obviously cannot be terms, and finally make preliminary selection of candidate terms according to language rules template. The candidate terms are only preliminary ones, and many of them are not terms, for instance, some common word collations and meaningless collations. Therefore, the paper need some method to select terms from candidate terms. C-Value is a widely used method that combines rules with statistics.

3.2   Word Representation with Word Vectors

To enable computers to process texts and extract terms based on model of deep neural network, the work need to represent words in the corpus for term extraction with word vectors, and use the word vectors as feature to input the model. Traditionally one-hot representation method is used for word vector representation, which represents every word as a long word vector with values of either 0 or 1, mostly 0, so that the dimension of word vector is high and sparse [11]. The distributed representation method of word vector is an improvement to the traditional one-hot representation method, the main idea of which is to obtain the distributed representation feature of a word by training the neural network language model, i.e., representing a word in the text by a continuous real number vector, and measure the semantic similarity between words by the distance between the words [12]. The word vectors acquired by distributed representation can abstractly express the semantic meaning of words, that is, semantically similar words are similar in vector space [13]. Therefore, word representation through word vector distributed representation method can express the features of words as much as possible, so that the model can better learn the advanced features of the represented words. Word2vec is an efficient tool for characterizing words as real number valued vector, it uses distributed representation method to represent word vectors [14]. Therefore, this paper uses Word2vec to represent the words with word vectors.

3.3   Extraction Features with Model of Deep Neural Network

Auto-Encoder model is a kind of neural network. We use the trained word vectors as the original input to AE model for training, we can extract high-level features in low dimensional. The deep AE is stacked by multi-layer AEs, which has one input layer, one output layer and at least 3 hidden layers. Theoretically speaking, the number of nodes in the input layer is same as the number of dimensions of the original input data, the number of nodes in the hidden layer should be less than the number of nodes in the input layer, the number of nodes in the latter hidden layer should be less than that in the previous hidden layer, and the number of nodes in the final output layer should be the smallest. The model can realize dimensionality reduction and feature reconstruction of the original input data, and each layer is the reconstruction expression of the previous layer, and the output model of final output layer expresses the characteristics of original input data in low dimensional and abstract manner. The principle of dimensionality reduction, feature expression and reconstruction of original input data by the model.

3.4   Construction Classifiers for Feature Classification

Generally speaking, high-level features in low dimensional extracted by AE can not be directly used to judge whether the related words are terms or not, thus a classifier is needed to attach to AE. The training of a classifier calls for marking of some candidate terms with 0 or 1 in reference to thesaurus that is used for evaluation standards. And input the features and marks of candidate terms obtained by AE into the classifier for training. The term that serves as evaluation standards includes terms concerning new energy vehicles and their related fields obtained from ISTIC and Chinese thesaurus.

Classifiers include logistic classifiers, SVM (Support Vector Machine) classifiers, K-Nearest Neighbors (KNN) classifiers and so on. Logistic classifiers put weights on features for classification and use related classification function to map weighted value of features to between 0 and 1, and making classification by probability. SVM classifiers try to classify samples by finding the maximum margin super plane between binary variables. KNN classifiers adopt a method based on actual examples. They calculate distances between samples for classification and all other samples with known classification and take the nearest k points in distance to identify the samples with unknown classification as the classification type that turn out to be the largest number in the neighboring k samples.

4   Experiment and Analysis

The experimental data set is from the National science and Technology Library (NSTL). The experimental data are the patents and journal paper in the field of new energy vehicle. The patent data includes 90000 pieces of patent, and the journal paper data includes 30000 pieces of paper abstracts. The thesaurus that serves as evaluation standards includes the following: terms concerning new energy vehicles and their related fields obtained from ISTIC and Chinese thesaurus.

Experiment model is 3 hidden layers, activation function is ReLU, optimization method is SGD and classifier is SVM. The experiment compares with the term extraction method based on Language rules, Language rules & statistical calculations. Extract terms based on language rules, language rules & statistical calculations. We choose the first 100, 200, 500 and 1000 terms to measure the accuracy rate. Choose the first 100, 200, 500 and 1000 word features extracted by AE as the testing sets for trained classifiers, and measure the accuracy rate of terms. The experiment results are shown in Table 1.

Table 1. Comparative results of different term extraction methods.

	P@100	P@200	P@500	P@1000
Based on language rules	0.192	0.209	0.192	0.169
Based on language rules & statistical calculations	0.670	0.670	0.602	0.517
Based on the model of deep neural network	0.7145	0.7045	0.695	0.67

From the experiment results, the paper find that the extraction method based on language rules has the lowest accuracy rate, which explains why language rule based method is rarely used alone in term extraction now. The extraction method based on language rules & statistical calculations and AE models turned out relatively good results in the respective term selection. When the paper uses this method for term extraction, we identified some words that are judged by classifiers as terms but do not exist in the Chinese thesaurus. These terms include: (1) Some terms are nested short terms of the long terms in Chines thesaurus. (2) Some terms are compound words made up of multiple parts, and the compound word is different in meaning from any component part. (3) Some terms are synonyms but are not included in the thesaurus. (4) Terms different from the above three types with a definite representation of professional concept and meaning.

5   Summary and Outlook

This paper proposes the idea of applying machine learning to extract term by going over the research status of current term extraction method. The paper constructs a set of term extraction methods based on real data and makes experiments on the real corpus. By comparing with extraction methods based on language rules, language rules &

statistical calculations and the model of deep neural network, the paper finds that the model of deep neural network can improve accuracy rate of term extraction by about 47% and 8%. And, the paper discovers some terms that are not included in the thesaurus, but have a definite representation of professional concepts and meaning. It proves to the effectiveness of term extraction method based on deep neural network. Nonetheless, this method has certain limitations. Firstly, this method calls for a massive scale of corpus to work, which takes a relatively long time for corpus pre-treatment and word vector training. Secondly, the relatively low dimensional features obtained by deep neural network are quite abstract with poor interpretability and it is hard to clarify the exact content represented by each dimensional feature. Lastly, though compared to other supervised learning model, AE reduces the demand on labeled data and the load of artificial data marking, it nonetheless requires some labeled data in the construction of classifier for training, and cannot be entirely free of supervision. Therefore, it is important to explore some more suitable learning model of neural network to apply to term extraction in the next work, further improve related theoretical knowledge and make the low dimensional features interpretable as much as possible.

Acknowledgment. This research is supported by the National Social Science Fund Project: Research on Information Analysis Method and Integrated Platform Based on Fact-type Scientific and Technical Big Data. [grant number 14BTQ038].

References

1. Jinsong, Y.: A Survey of terminology automatic extraction methods. Comput. Sci. **42**(8), 7–12 (2015)
2. Heylen, K., Dehertog, D.: Automatic Term Extraction. John Benjamins Publishing Company, Amsterdam (2014)
3. Frantzi, K., Ananiadou, S., Mima, H.: Automatic recognition of multi-word terms: the C-value/NC-value method. Int. J. Digit. Libr. **3**(2), 115–130 (2000)
4. Astrakhantsev, N.A., Fedorenko, D.G., Turdakov, D.Y.: Methods for automatic term recognition in domain-specific text collections: a survey. Program. Comput. Softw. **41**(6), 336–349 (2015)
5. Lossio-Ventura, J.A., Jonquet, C., Roche, M., Teisseire, M.: Yet another ranking function for automatic multiword term extraction. In: Przepiórkowski, A., Ogrodniczuk, M. (eds.) NLP 2014. LNCS (LNAI), vol. 8686, pp. 52–64. Springer, Cham (2014). https://doi.org/10. 1007/978-3-319-10888-9_6
6. LossioVentura, J.A., Jonquet, C., Roche, M.: Biomedical term extraction: overview and a new methodology. Inf. Retr. J. **19**(1), 59–99 (2016)
7. Wen, Z.: The exploration of information extraction and analysis about science and technology policy in China. Electron. Libr. **35**(4), 709–723 (2017)
8. Lecun, Y., Bengio, Y., Hinton, G.: Deep learning. Nature **521**(7553), 436–444 (2015)
9. Lecun, Y., Kavukcuoglu, K., Farabet, C.: Convolutional networks and applications in vision. In: IEEE International Symposium on Circuits and Systems, pp. 253–256. IEEE, New York (2010)
10. Leng, J., Jiang, P.: A deep learning approach for relationship extraction from interaction context in social manufacturing paradigm. Knowl. Based Syst. **100**, 188–199 (2016)

11. Glorot, X., Bordes, A., Bengio, Y.: Domain adaptation for large-scale sentiment classification: a deep learning approach. In: International Conference on Machine Learning, pp. 513–520. Omni Press, Bellevue (2011)
12. Xue, F., Guodong, Z.: Deep learning for natural language processing. J. Autom. **42**(10), 1445–1465 (2016)
13. Mikolov, T., Chen, K., Corrado, G.: Efficient estimation of word representations in vector space. Comput. Sci., 1–13 (2013)
14. Shui, L., Xi, C., Wenyan, G.: New progress in deep learning methods. J. Intell. Syst. **11**(5), 567–577 (2016)

Topic Detection for Post Bar Based on LDA Model

Muzhen Sun[1(✉)] and Haonan Zheng[2]

[1] College of Public Administration, Huazhong University of Science and Technology, Wuhan 430074, Hubei, China
sunmuzhen0515@163.com
[2] College of Information Engineering, Zhengzhou University, Zhengzhou 450001, Henan, China

Abstract. Since most college students use social networks in their daily lives, analysing them would help teachers grasp their students' thoughts and be consulted for ideological education. This paper aims to extract valuable information and analyse the topic distribution from a large number of posts on the Baidu Post Bar (BPB) using topic detection technologies. We first crawled the post's data from ten colleges' post bars and carried out topic detection based on the LDA model. We defined the word weight according to the tf-idf, length, cover and whether it was in the title. We also defined the topic heat ranking model according to the support documents, reply and time coverage. Two label words were automatically selected to represent the topics' meaning. Then, we analysed the topic distributions. The results of the empirical research showed that college students focused on topics of graduate, work, examinations, learning, campus life and consultation. They paid little attention in politics and society. Finally, we proposed suggestions to the colleges.

Keywords: Topic detection · Hot topic ranking · LDA model
Baidu Post Bar · Ideological education

1 Background

With the emergence of networks, increasing numbers of people like to post their opinions or emotions on the internet, such as on BBS, WeChat and microblogs. As a group with active thoughts, most college students use social networks in their daily lives to express their emotions and opinions. The enormous growth of social networks usage has led to an increasing amount of big data. Analysing big data from social networks would be beneficial for teachers to grasp their students' thoughts and be consulted for ideological education.

Supported by "Innovation and entrepreneurship project of college students in Huazhong University of Science and Technology: The study of Chinese undergraduates' ideological trend on the social platform".

Q. Zhou et al. (Eds.): ICPCSEE 2018, CCIS 902, pp. 136–149, 2018.
https://doi.org/10.1007/978-981-13-2206-8_13

Baidu Post Bar (BPB) is the largest Chinese network community and brings together people who are interested in the same topic. It provides a free network space for people to express and exchange ideas. BPB covers all aspects of society, life, education, entertainment stars, games, sports, enterprises and others. Currently, BPB includes 35 categories and 27,900 colleges' post bars,[1] thus providing online communication platforms for college students. Based on the data analysis of the BPB, we can understand college students' situations in their daily lives as well as their learning and ideological trends. Suggestions for college students' ideological education can be given based on the analytical results.

Although there is much research on opinions and topic detection for BPB or other BBS, most of them only focused on specific universities [1–4]. The data are thus limited, and the results can only represent a certain range. Because of the limited data, researchers can analyse the data manually. However, these methods are not suitable for huge amounts of data. To extract valuable information from big data, NLP and other content mining technologies have been used [5–8]. These works focused on microblog data or a special industry BBS, such as tourism or economics [9, 10].

This paper focuses on extracting valuable information from large amounts of posts and analyses the topic distribution on the BPB with using Chinese Natural Language Processing (NLP) and topic detection technologies. We use the LDA model to implement topic extraction. We consider the difference between words in post titles and content important and define a new word weight calculation method. We propose a topic heat ranking method based on the document support, reply number and time coverage. We also propose the automatic extraction of two label words to describe the topic. Our method is successfully used to analyse data from college post bars in this paper.

The rest of this paper is divided into four parts. Section 2 briefly reviews the related works on the topic analysis for the BBS. Section 3 introduces our method for analysing college posts. Section 4 introduces the experiment and results of the topic analysing based on our method. Some advice for universities is also proposed according to the results. Section 5 gives the conclusions.

2 Related Work

Extracting valuable information and analysing the topic distribution from large amount of posts on the BPB would help teachers grasp their students' thoughts and be consulted for ideological education.

Zhao [1] analysed the influence of BPB on college students and proposed a routine of value education. Yang [2] analysed colleges' post bars in Dalian and proposed new ideas for stability maintenance at the colleges. Lin [3] used the BBS forum of Wuhan University to analyse the hot spots of public opinion and determined the concerns and ideological trends of college students' public opinion. Teng [4] selected the Tsinghua Shuimu BBS and Beida Weiming BBS as research objects and found some of the

---

[1] http://tieba.baidu.com/f/fdir fd = University &i.e. = utf-8&sd = Henan College &i.e. = utf-8.

features of network public opinions in the colleges. Based on an in-depth analysis of the current situation of college students' public opinion in Xinjiang, Yuan [11] proposed countermeasures and suggestions to handle the crisis of public opinion and utilize the information about college students' public opinion. Li [12] studied the distribution of different types of posts on Baidu Post Bar from 10 colleges in Liaoning Province and provided a baseline for the guidance of network public opinion in colleges.

The aforementioned papers analysed the posts of special colleges on the BBS forum. The data are limited, and the researchers can analyse the data manually. However, these methods are not suitable for huge amounts of data. When we deal with massive data, NLP and topic detection technologies should be used.

Topic detection, also known as topic discovery and topic recognition, refers to the technology that divides information from news streams or news reports into different topics and builds new topics when necessary. Currently, this technology is widely used in analysing news, micro-blogs, WeChat and forum data. It can help people select from massive amounts of information, choose topics of interest, understand people's concerns, and detect public concerns. Hao [9] took the network economy forum as the research object and proposed a comprehensive model of hot topic mining, which included the hot topic discovery module based on iterative clustering, the topic heat assessment module and the opinion mining module of the topic. They took data from a given period of time from the economic BBS of the Tianya forum as samples for the experimental research. They uncovered hot topics on the economic forum and further provided a basis to identify the current economic hot topics. Wang [10] extracted Ctrip online travel text content and used a content analysis method to carry out the hot word analysis, co-occurrence word analysis and long tail analysis of the samples. They obtained a series of behaviour rules of tourists regarding their online travel intelligence decisions. Lan [13] implemented a hot topic discovery and monitoring system based on real-time BBS data. The system used a single-pass clustering algorithm to identify the topic. Based on topic recognition, the topics were ranked by the number of posts, ID number, number of replies and browsing number. Wang [14] proposed a topic detection model based on the topic extension and implemented a demo system to find hot topics on the BBS. The experimental results showed that the methods and systems proposed in this paper can effectively monitor the hot topics on the campus forum. Li [15] proposed a hot topic detection method based on an improved BP network. The experiment on the review posts of the CCTV Revival Forum showed that the hot topic classification result was in accordance with the reality of public opinion. Chen [16] presented a general detection framework and developed a variety of content and structure features to find high-quality threads. The experiment that focused on the Tencent Message Boards demonstrated the feasibility of modelling and detecting high-quality threads. The proposed feature extraction methods, feature selection algorithms, and detection frameworks can be useful for a variety of domains, such as blogs and social network platforms.

Different topic detection methods have also been used and studied. Ibrahim [17] categorized topic detection techniques into five categories: clustering, frequent pattern mining, Exemplar-based, matrix factorization, and probabilistic models. The most popular method for topic detection is the LDA model [18].

The LDA model has excellent abilities of dimension reduction, modelling and extension. It can mine the hidden semantics of massive texts and can be applied to text mining tasks such as text classification, topic discovery, automatic summarization and others. LDA also has great potential in short text topic mining. Fang [5] used the LDA model to identify the topic of a complaint. Weng [20] used all the micro-blog text from the same micro-blog user, which was placed into a long document. Then, they made use of LDA model for topic mining. Hong [21] proposed several schemes to train a standard topic model in micro-blog environments to solve the problem of the restricted length of messages.

In recent years, to improve the efficiency and accuracy of LDA models, many attempts have been made to improve LDA models. Zhao [22] proposed Twitter-LDA models to extract topics from the representative text in the entire Twitter platform. Ramage [23] proposed Labeled-LDA, which is a label-based topic model. Zhang [24] put forward the micro-blog generation model MB-LDA, which considered a micro-blog's text relation to help with topic mining. Yu [25] proposed an improved OLDA model that can more clearly explain the meaning of a topic.

The LDA model was used to analyse news, micro-blogs, WeChat and forum data for different platforms. However, no research was found that automatically detected college students' choice topics from the BPB. This paper uses the LDA model to detect topics from college post bars.

3 Methods

LDA is the most popular topic model used in topic detection, and it has advantages in short text topic detection. This paper uses the LDA model to find topics in college posts on BPB and then analyses the topic distributions. Figure 1 describes the procedure of our method, and the details of each step are described in the following text.

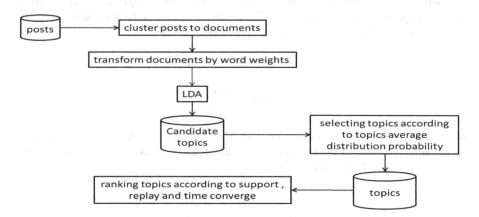

Fig. 1. Procedure for BBS topic detection based on the LDA model

Generally, BPB posts are short and informal and contain noisy symbols. Before extracting the topics, the posts should be reprocessed, such as by performing post clustering, cleaning, segmenting and transforming.

There are approximately 100 words in a post, including the post's title and content. It is difficult to infer the total semantics from the single short text. Currently, the most effective technology to achieve this is short text merging. The merging strategy includes two methods: according to the author [26, 27] and by clustering. The clustering method combines short texts with those of similar semantics. Similar posts contain commonness, and the clustering method can merge them together as a single document. The generated documents have rich content and are more suitable for the LDA model.

This paper uses spectral clustering [28] to cluster the posts. Spectral clustering is one of the most popular modern clustering algorithms and is simple to implement. The post text is represented by document embedding[2]. The document embedding is used to construct a similarity matrix based on the cosine distance. The clustered texts are saved as a document that will be used by the LDA model.

3.1 Document Transforming

The LDA model is based on a word bag. Thus, before using this model, the documents should be transformed to a word bag representation.

A word bag is a representation of a document that includes a sequence of words in the document with different word weights. Formally, the word bag is a vector whose length equals the size of the vocabulary that includes all words in the document set and whose item value is the word weight. The word weight can be represented as the word frequency, tf-idf and others.

The BPB post text includes the title and content. The title includes more topic information than content. Therefore, the words in topic are more important than are those in the content. However, the title is short and is not suitable for the LDA model by itself. Therefore, we merge the title and content into a single text and assign a greater weight to the words in the title. The weight of a word in the documents depends on its tf-idf and whether it is in title, as shown as Eqs. (1, 2):

$$weight = f1 \cdot tfidf + f2 \cdot intitle \tag{1}$$

$$intitle = \frac{title}{total} \tag{2}$$

where tf-idf is the value of the word's tfidf, title is the number of times that the word occurs in title and total is the number of times that the word occurs in the document set. The factors f1 and f2 are super parameters.

---

[2] https://github.com/RaRe-Technologies/gensim/blob/develop/docs/notebooks/doc2vec-lee.ipynb.

3.2 LDA

This paper uses the LDA [18] model as the topic model. The LDA model defines the topic as a sequence of semantic-related words and their distributions on the topic, as shown as Eqs. (3, 4):

$$z = \{(w_i, p(w_i|z, \beta)) | 0 < i < N\} \tag{3}$$

$$p(w_i|z, \beta) \sim Multinomial(\beta) \tag{4}$$

where z is a topic; w_i is the ith word in the vocabulary; N is the size of the vocabulary; and $p(w_i|z, \beta)$ is the probability of the ith word in topic z. The larger $p(w_i|z, \beta)$ is, the more relevant w_i and z are. The word distribution is multinomial with the parameter β.

The LDA model-defined document is a sequence of topics and their distributions in the document, shown as Eqs. (5, 6):

$$d = \{(z_j, p(z_j|d)) | 0 < j < M\} \tag{5}$$

$$z_j \sim Multinomial(\theta) \tag{6}$$

where d is the document; z_j is the jth topic in the topic space; M is the topic number; and $p(z_j|d)$ is the probability of the jth topic in a document d. The larger $p(z_j|d)$ is, the more relevant the topic j and document d are. The topic distribution is also a multinomial distribution with a Dirichlet random variable, as shown in Eq. (7):

$$\theta \sim Dir(\alpha) \tag{7}$$

Given the parameters α and β, the topic distribution for the ith document is θ_i:

$$\theta_i = (\theta_{i1}, \theta_{i2}, \ldots, \theta_{iM}) \tag{8}$$

The word distribution for jth topic is φj:

$$\varphi_j = (\varphi_{j1}, \varphi_{j2}, \ldots, \varphi_{jN})$$

Gibbs Sampling [19] can be used to solve the parameters θ and φ:

$$\begin{cases} \theta_{im} = \dfrac{count_i^m + \alpha_m}{\sum_{m=1}^{M} count_i^m + \alpha_m} \\ \varphi_{mn} = \dfrac{count_m^n + \beta_n}{\sum_{n=1}^{N} count_m^n + \beta_n} \end{cases} \tag{9}$$

where $count_i^m$ is the number of words that are included in the topic m of the ith document and $count_m^n$ is the number of times that word n occurs in topic m.

Before the inference parameters are determined, we should set the topic number M.

This paper uses perplexity [18] to estimate M:

$$perplexity = \exp\left(\frac{\sum_{i=1}^{D} \log(p(d_i))}{\sum_{i=1}^{D} L_i}\right) \tag{10}$$

where D is the document number, L_i is the length of the ith document and $p(d_i)$ is the probability of generating the ith document. The smaller the perplexity is, the better the results are.

The super parameters α and β can be defined by the topic number and vocabulary size as Eqs. (11, 12):

$$\alpha = \frac{50}{M} \tag{11}$$

$$\beta = \frac{200}{N} \tag{12}$$

3.3   Topic Label Words and Topic Selecting

Using the Gibbs Sampling will provide the topic-term and document-topic distributions. The more topics there are, the closer the semantics between different themes are. Therefore, we need to select topics that have significant differences.

Calculate the average probability of each topic in documents as follows:

$$\text{AVG}(z) = \frac{\sum_{j=1}^{D} p(z|d_j)}{m} \tag{13}$$

The topics' average distribution probability is ranked, and those topics whose average value is lower than a threshold are deleted. The remaining topics are selected to extract the label words.

The label word of each topic is extracted according to the word's weight. The label word is normally the word with the maximum weight in the topic [5]. However, only one label word is not sufficient to express the topic's actual meaning. Further, if each topic only has one label word, the situation in which different topics have the same label word will arise more frequently. One label word cannot distinguish the topics' subtle semantic differences; therefore, this paper uses two label words for each topic. The word weight is related to its distributed probability, frequency, length, cover and whether it is in the title, as shown as Eqs. (14–18):

$$lableword_z = large(weight_l, 2) \tag{14}$$

$$weight_l = probablity + count + l + cover + intitle \tag{15}$$

$$count = \frac{count(w)}{count(w) + 1} \tag{16}$$

$$l = \frac{length(w)}{\max(lenth(D))} \tag{17}$$

$$cover = \frac{last(w) - first(w)}{ctotal} \tag{18}$$

where $lableword_z$ is the label word of topic z; $count(w)$ is the occurrence number of word w in the documents with topic t; $length$ is the length of the word; $length(D)$ is the max word length in the document D; $ctotal$ is the position of the last word in the document; $first$ and $last$ are, respectively, the first and last positions of the word in document; $intitle$ represents whether the word in the title; and $probability$ gives the word's distributed probability in topic z. We assume that a word with a greater $weight_l$ is more likely to be the label word. The function $large()$ obtains the two words with the greatest $weight_l$.

Different topics may have the same or similar labels. It is assumed that the topics with the same or similar label words have similar semantics. When different topics have the same or similar label words, the topic with the largest average probability is chosen. The word similarity is calculated according to the method of [29]. The remaining topics are ranked according to the method described in Sect. 3.4.

3.4 Topic Ranking

After detecting topics with the LDA, we rank topics according to their heat. Most hot ranking methods consider the support documents and number of replies. According to our observations, the longer time a post covers, the more people are interested in it. Therefore, we assume that the number of posts that support a topic, the number of replies and the time coverage of posts reflect the heat of a topic, as shown in Eq. (19):

$$heat = w1 \cdot support + w2 \cdot reply + w3 \cdot tm \tag{19}$$

$$w1 + w2 + w3 = 1 \tag{20}$$

where support represents the number of posts that support the topic; reply represents the sum of the replies in all posts that support the topic; tm represents the time coverage between the post time and the last reply time; and w is the weight that reflects the contribution of each factor. The factor calculation equations are shown as Eqs. (21–23):

$$support = \frac{D_z}{D} \tag{21}$$

$$reply = \frac{R_z}{R} = \frac{\sum_{D_z} r_{d_i}}{R} \tag{22}$$

$$tm = \frac{T_z}{D_z} \tag{23}$$

where D_z is the number of posts that support the topic z; D is the number of all posts; R_z is the number of replies for all posts that belong to the topic z; R is the number of replies for all posts; and T_z is the sum of the time coverage of all posts that belong to the topic z. The document d is defined to support topic z when the probability of generating d with z is maximal.

4 Experiment and Analysis

4.1 Data

This paper focuses on extracting information from college posts on BPB. As of May 31, 2017, there were 2,631 colleges and universities in China, as obtained from the "Ministry of education sunshine entrance open platform"[3]. Most colleges have a post bar that their students visit frequently. The data set in the post bar is very large. For example, there are approximately eight hundred thousand themes with twenty million posts in the Zhengzhou University Bar. Because of our limited computational resources, it is impossible to access post data of all two thousand colleges. Therefore, we choose only the top 10 colleges with the largest number of members on BPB, as shown in Table 1.

According to the colleges listed in Table 1, we can find their post bar websites by using a crawler. We crawled the post bar data from every obtained post bar website. Each post includes the title, content, post time, poster, number of replies and reply time. The total number of posts is approximately 90,000 for these colleges.

4.2 Experiment

We extract the title and content from each post and merged them into a text string. The average length of each text is approximately 100 words, which is not suitable for the LDA model. We first cluster the texts with similar semantics using the spectrum cluster method. The cluster was implemented based on sklearn[4]. The clustered texts are merged together and saved as a document. The documents are cleaned to delete noisy characters and segmented using NLPIR[5]. Since the verbs and nouns represent the main semantics, we delete the other parts of speech. The LDA model is implemented using the Gensium package[6]. The LDA model parameters α and β are set to 50/M and 0.02, respectively. The word's tf-idf and whether it is in title are of the same importance. Based on experience, the word weight parameters f1 and f2 are both set to 1, and the ranking parameters w1, w2, and w3 are set to 0.5, 0.3, and 0.2, respectively. The topic number is selected according to the perplexity. Figure 2 shows the perplexity trends with the topic number. The topic number is set to 200 based on the perplexity trends.

---

[3] http://gaokao.chsi.com.cn/gkxx/zszcgd/dnzszc/201706/20170615/1611254988.html.

[4] http://scikit-learn.org/stable/modules/generated/sklearn.cluster.SpectralClustering.html.

[5] http://ictclas.nlpir.org.

[6] https://radimrehurek.com/gensim/models/ldamodel.html.

Table 1. Top 10 colleges in terms of number of BPB members

Colleges & Universities	Members number
Sichuan conservatory of music	137820
Zhengzhou University	90587
Sias International University	78549
Southwest Jiaotong University	60883
Henan University	60590
Sichuan University	53418
Wuhan University	51321
Henan University of Technology	45262
Xiamen University	45094
Central South University	44563

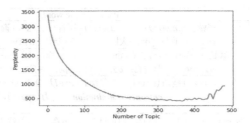

Fig. 2. The perplexity trends with topic number.

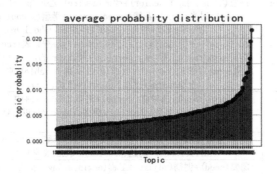

Fig. 3. The average probability distribution of each topic

The selected topics are ranked according to the method described in Sect. 3.4 (Fig. 3). Figure 4 and Table 2 show the topic words and top ten hot topics. In Table 2, the artificial labels are defined by people. The automatically selected label words are bold in the columns "Parts of topic words". The Italics bold is the automatically selected one-label word.

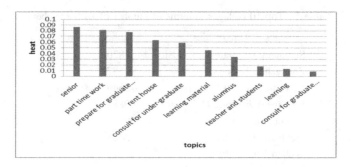

Fig. 4. Top ten hot topics

Table 2. Top 10 topics

Artificial label	Parts of topic words
大四(senior)	*毕业(graduate)*+论文(thesis)+课程(course)+询问(ask)+车站(station)+*专业(major)*+大四(senior)+到达(arrive)+评估(evaluate)
兼职(part time work)	家长(parents)+合作(cooperation)+加群(join group)+*创业(start business)*+*学校(school)*+开门(open)+打工(part-time work)
考研复习(prepare for graduate entrance examination)	免费(free)+技巧(skill)+*资料(material)*+分享(share)+*考研(graduate entrance examination)*+角力(wrestle)
租房(rent house)	空调(air condition)+垃圾(rubbish)+*出租(rent)*+洗衣机(wash machine)+房间(room)+单间(single)+家具(furniture)+体验(experience)+厨房(kitch)+*房子(house)*
本科咨询(consult for undergraduate)	高三(high school senior)+欣赏(admire)+心理(psychological)+来到(come)+时光(time)+花钱(spend money)+*学姐(school sister)*+学生(students)+咨询(consult)+*学长(school brother)*
学习资料(learning material)	数据(data)+分析(analyze)+工业(industry)+登记表(register form)+*同学(classmates)*+工程(engineering)+精通(proficient)+*资料(material)*
校友(alumnus)	回来(come)+想起(remind)+女孩(girl)+桌子(table)+味道(taste)+*毕业(graduate)*+感慨(lament)+班长(class leader)
师生(teacher and students)	教授(professor)+清华(Tsinghua)+洛阳(Luoyang)+中山大学(Zhongshan University)+提到(mention)+*学生(student)*+以为(think)+*老师(teacher)*
学习(learning)	*分数(grade)*+房租(rent)+提问(ask)+语言(language)+差别(difference)+问题(problem)+*专业*+文学(literature)
考研咨询(consult for graduate entrance examination)	需要(need)+调剂(adjust)+工程(engineering)+报考(apply for)+*学长(school brother)*+农大(university)+河南(Henan)+*求助(Help)*+总分(grade)

4.3   Analysis

Model Analysis

As Table 2 shows, the semantics of the automatically selected label words are similar to those of the artificial words. We believe that this occurs because we defined two label words that are not only related to the frequency, cover and length but also to the probability and whether they are in the title.

We selected two label words for each topic. This method elucidates the topic meaning compared with only one label. When only extracting one label word, the topics "大四(senior)" and "校友(alumnus)" have the same label "毕业(graduate)". One of these topics is deleted according to our model. However, the topic "大四(senior)" is related to the senior students, and the topic "校友(alumnus)" is related to the former students. This difference can be reflected by the second label. One label word for the topic "考研咨询(consult for graduate entrance examination)" is "求助(Help)". Although this word has a low frequency in the documents, it is selected as a label because it is often occurred in titles.

The topics that are found by our model are more detailed. "招生报考(Admission)" is only a topic on BPB. In our experiment, "考研咨询(consult for graduate entrance examination)" and "本科咨询(consult for undergraduate)" are found. "校友(alumnus)" and "大四(senior)" cannot be categorized by BPB but are also found by our model.

Topic Analysis and Advices

Figure 4 shows that most people on the university post bars pay more attention to the topics of graduation, work, examinations, learning, campus life and consultation. They want more information about the university, especially during the graduation season. High school students want to know more about university life, and undergraduate students want to know more about majors, teachers and admission information.

The topics also reflect that the university students not only care about leaning but also have more activities, such as part-time work and starting businesses. Their living circle is not limited to campus, as they sometimes rent houses and live off-campus.

The users on the post bar include current, future and former students. Different users care about different topics. The current students likely talk about campus life, learning and other activities. The future students who will enter or want to enter the university likely seek information regarding the university. The future undergraduate students are usually high school students who are curious about the university life and are full of admiration for the elder students. The future graduate students care about majors and admissions. The former students miss the time they spent at their alma mater.

According to the above analysis, we propose the following advice for universities and their administrators.

First, the universities should provide more information to introduce themselves. This information will not only attract more people to join them but also help fresh students become more familiar with the campus environment as quickly as possible.

Second, the topic analysis results showed that students are concerned about not only learning but also working and activities. The universities should provide guidance and opportunities for extra-curricular activities to avoid students going astray.

Third, alumni are valuable assets of the university, and the university should keep in touch with them. This will be beneficial for both the university and the students.

Finally, the current students pay little attention to politics and society. This indicates that students do not participate in society and do not care about political life. This phenomenon is not conducive to the future development of the country. Universities should enforce ideological education and encourage students to care about society.

5 Conclusion

Baidu Post Bar is the largest Chinese network community and brings together people who are interested in the same topic. Most college students like to express their opinions on their college or university post bars. This paper used the LDA model to analyse the posts and find hot topics. We considered the importance of the post title and defined word weights with the title information and tf-idf, coverage and length. The topic detection results showed that this design is effective. When ranking topics, we considered the reply time coverage. Two label words can clearly describe the topic's meaning. Our method can automatically obtain students' thoughts without more manual work.

Our result on the top ten universities post bars showed that college students' main concern is about graduation, work, examinations, learning, campus life and consultation. They pay little attention on politics and society. According to the analytical results, we suggest that universities and colleges provide more information about themselves, keep in touch with alumni, enforce ideological education and encourage students to care about society.

References

1. Zhao, H.W.: Research on the impacts of Baidu Post Bar to high technology students' value. J. Guilin Univ. Areospace Technol. **21**(2), 204–208 (2016)
2. Yang, X.Y., et al.: Analysis on the status of maintaining stability and network public opinion in Colleges and universities in Dalian – take WeChat, micro-blog, and Baidu on the bar as an example. Intelligence **17**, 206 (2017)
3. Lin, X.F., Xiao, Z.: Hot spot analysis of public opinion based on the BBS forum of college students– Taking Wuhan University LUOJIASHAN water forum as an example. Morden Bus. Trade Ind. **18**, 188–189 (2010)
4. Teng, Y., Chen, L.: An empirical study on the characteristics of network public opinion– text analysis based on BBS forum in Colleges and Universities. Shandong Soc. Sci. **03**, 181–186 (2014)
5. Fang, X.F., Huang, X.X., Wang, R.B.: Hot topic recognition of mobile complaint text based on LDA model. New Technol. Libr. Inf. Serv. **1**(2), 19–27 (2017)
6. Zhang, H.L.: Large text corpus analysis and hot topic discovery based on word co-occurrence network. Comput. Digit. Eng. **10**, 1729–1735 (2015)
7. Chen, X.G.: Research on fast self-clustering method of micro-blog public opinion based on large data technology. J. Intell. **36**(5), 113–117 (2017)
8. He, M.: Feature driven microblog topic detection. J. Chin. Inf. Process. **31**(3), 101–108 (2017)
9. Hao, X.L.: Research on hot topic mining based on economic forum data. Inf. Sci. **34**(5), 153–158 (2016)
10. Wang, X.T., Wang, L.: Analysis of hot word information of online tourism based on text mining – take ctrip as an example. Inf. Stud. Theor. Appl. **40**(11), 105–109 (2017)

11. Yuan, J., Huang, J.J., Tang, X.P.: The analysis of the public opinion of college students in Xinjiang and Countermeasures. Heihe J. **1**, 160–162 (2014)
12. Li, L.: Characteristics of network public opinion in different colleges and universities in the era of media—Take Baidu Post Bar as example. New Compus **9**, 30 (2016)
13. Lan, K.M.: The System of BBS Hot Topic Detection and Monitoring. Beijing Jiaotong University, Beijing (2011)
14. Wang, X.H.: Hot topic discovery method and system of campus forum based on topic extension. J. Tibet Univ. (Nat. Sci.) **31**(02), 110–116 (2016)
15. Li, Y.F., Chen, L.: Hot topic mining of network forum based on improved BP network. Comput. Syst. Appl. **25**(03), 113–118 (2016)
16. Chen, Y., Cheng, X.Q., Yang, S.: High quality topic discovery for network forums. J. Softw. **22**(08), 1785–1804 (2011)
17. Ibrahim, R., et al.: Tools and approaches for topic detection from Twitter streams: survey. Knowl. Inf. Syst. **6**, 1–29 (2017)
18. Blei, D.M., Ng, A.Y., Jordan, M.I.: Latent dirichlet allocation. J. Mach. Learn. Res. **3**, 993–1022 (2003)
19. Blei, D.M., Lafferty, J.D.: Dynamic topic models. In: Proceedings of the 23rd International Conference on Machine Learning, pp. 113–120. ACM (2006)
20. Weng, J., Lim, E.P., Jiang, J., et al.: TwitterRank: finding topic-sensitive influential Twitterers. In: Proceedings of the 3rd ACM International Conference on Web Search and Data Mining, pp. 261–270. ACM (2010)
21. Hong, L., Davison, B.D.: Empirical study of topic modeling in Twitter. In: Proceedings of the 1st Workshop on Social Media Analytic, pp. 80–88. ACM (2010)
22. Zhao, W.X., et al.: Comparing Twitter and traditional media using topic models. In: Clough, P., et al. (eds.) ECIR 2011. LNCS, vol. 6611, pp. 338–349. Springer, Heidelberg (2011). https://doi.org/10.1007/978-3-642-20161-5_34
23. Ramage, D., Hall, D., Nallapati, R., et al.: Labeled LDA: a supervised topic model for credit attribution in multi-labeled corpora. In: Proceedings of the 2009 Conference on Empirical Methods in Natural Language Processing, pp. 248–256 (2009)
24. Zhang, C.Y., Sun, J.L., Ding, Y.Q.: Topic mining for microblog based on MB-LDA model. J. Comput. Res. Develop. **48**(10), 1795–1802 (2011)
25. Yu, B.G., Zhang, W.C., Wang, F.L.: Topic detection and evolution analysis based on improved OLDA model. J. Intell. **36**(2), p. 102 (2017)
26. Hong, L., Davison, B.D.: Empirical study of topic modeling in Twitter. In: Proceedings of the SIGKDD Workshop on Social Media Analytics, pp. 80–88 (2010)
27. Rosen-Zvi, M., Griffiths, T., Steyvers, M., et al.: The author-topic model for authors and documents, pp. 487–494 (2012)
28. Ng, A.Y., Jordan, M.I., Weiss, Y.: On spectral clustering: analysis and an algorithm. In: Advances in Neural Information Processing Systems, pp. 849–856 (2002)
29. Li, H., Mu, L.L., Zan, H.Y.: Computation of word similarity based on the information content of sememes and pagerank algorithm. Chinese Lexical Semantics. LNCS (LNAI), vol. 10085, pp. 416–425. Springer, Cham (2016). https://doi.org/10.1007/978-3-319-49508-8_39

EventGraph Based Events Detection
in Social Media

Jianbiao He[1], Yongjiao Liu[2(✉)], and Yawei Jia[3]

[1] College of Information Science and Engineering, Central South University,
South Lushan Road 932, Changsha 410012, China
jbhe@mail.csu.edu.cn
[2] Central South University,
South Lushan Road 932, Changsha 410012, China
2518881390@qq.com
[3] University of Science and Technology of China,
Huangshan Road 443, Hefei 230027, China
ywjia@mail.ustc.edu.cn

Abstract. In the past few years, research about event detection has been devoted to a lot. In this paper, we propose an efficient method to detect hot events that spread within social media. Specifically, we build a directed weighted graph of words named EventGraph, in which events are embedded in the form of sub-graphs or communities. Lastly, we put forward a key node based event community detection method, which improve the efficiency of graph based event detection algorithms.

Keywords: Event detection · EventGraph · Co-occurrence · Key node

1 Introduction

With the fast growth of social media platform, such as micro-blog platform, which has attracted much more attention in the past years. Such micro-blog platform offers micro-blog service that enables its users to post and share short text messages of up to 140 words. Since people can post and share messages without constraint on locations and time, these systems produce a large amount of data each day, which contains large amount of valuable information as well as noises. In fact, the communication and interactions in social media reflect events and dynamics in real world. We can acquire the hot views in time from the micro-blog trends by text mining methods. And the vast majority of researches shows the fact that micro-blog truly reflects what's happening in real world. For example, Bollen *et al.* successfully predicted the stock market based on tweeter mood [1]. In addition, online social networks can be used as a sensor to detect frequent and diverse social and physical events in real time [2]. Besides, more and more people begin to focus researches about detecting events from social media.

© Springer Nature Singapore Pte Ltd. 2018
Q. Zhou et al. (Eds.): ICPCSEE 2018, CCIS 902, pp. 150–160, 2018.
https://doi.org/10.1007/978-981-13-2206-8_14

There still exits huge challenges in text mining, especially for detecting hot events from social media data, which contains a large volume of noises. The main challenges which we think are the following aspects: first, the volume of social media data is very huge and contains enormous amount of noises. Events detection method based on texts clustering has a very low efficiency when the dataset is very large. Therefore, the detection algorithms should keep efficient, even though the volume of data is very huge. Second, it is hard to determine the correlation of keywords related to event while it is the key step for events detection. We proposed an efficient events detection algorithm to attempt to solve the two challenges in this paper.

The events detection method that we prosed in this paper is a kind of graph based events detection method. Specifically, we build a weighted directed graph called EventGraph by the relationship of words co-occurrence in documents. The graph of nodes consist of words and there exists a directed edge between two nodes, if they co-occur in documents. The weights of edges reflect the correlation of words. The EventGraph exactly capture the semantics between words through this correlation. Words describing the common event tend to exist as one community in EventGraph. Sayyadi *et al.* proposed a graph analytical approach for topic detection [3]. By analysis and comparison, our approaches has two advantages: first, we quantize the correlation of words by a probabilistic feedback, which brings advantage to communities detection in the EventGraph. Second, we propose an efficient event communities detection algorithm which is based on key node in the graph. On the one hand, this algorithm attempts to avoid to regard communities consisting of noise words as events communities, since the algorithm first identify key node before extracting communities. On the other hand, it only needs to identify k key nodes that largely improves the efficiency of the algorithm.

The reminder of this paper is organised as follows. In Sect. 2 the related works are summarised. The basic ideas and model framework overview are described in Sect. 3. And we introduce how to build EventGraph in Sect. 4. In Sect. 5, we propose the key node based events detection algorithm and show experiment result and analysis in Sect. 6. Finally, the paper is concluded in Sect. 7.

2 Related Works

Recent years, enormous amounts of contents are being generated with the rise of social media, which brings new research questions as well as new challenges within the text data mining area. More and more researchers focus on the mining of social media. However, the technologies of detecting emergent events, more or less, are similar to Topic Detection and Tracking, which involves detecting the occurrence of a topic such as a plane crash in a stream of news stories from multiple source. Typically the event detection is an unsupervised clustering algorithm on documents that finds clusters of similar documents. Then the high frequency terms in documents within one cluster are regarded as an event. The related works include [4–6]. From another perspective, an event might be considered as a cluster of keywords or terms. So there are a lot of related works that applied text classification methods to events detection. Kumaran *et al.* showed how performance on new event detection can be improved, taking advantage of text classification technique [7]. And Yang *et al.* adopted several supervised text categorization methods specifically some variants of K Nearest Neighbour algorithm to track events [8]. In summary, all of the methods mentioned above are based on document-pivot clustering. In general, first, all documents are clustered into several groups. Then, it selects features or terms from the clusters of documents based on some feature selection approaches representing as an event.

As an alternative method, many related researches have attempted to skip documents clustering and directly to identify the bursty features or terms from a stream of texts. Afterwords, these bursty features are clustered based on certain approaches and each cluster represents as an event. [9] proposed a formal approach for modelling bursts and provided an organizational framework for analysing the underlying content based on modelling the stream using an infinite-state automaton, in which bursts appear naturally as state transitions. Based on the automaton model, He *et al.* proposed a bursty feature representation for clustering text streams [10], which in turn benefits the event detection based on documents-pivot method. Besides, Pui *et al.* developed a new novel parameter free probabilistic approach to identify burst features for a bursty event detection. They modeled the probability of the number of documents that contain the feature or term in the time window with a hyper-geometric distribution [11]. They identified the burst features according to the probability value. In addition to identifying burst keywords, graph based approaches detect keyword clusters based on their pairwise similarities. Sayyadi *et al.* built a term co-occurrence graph, whose nodes are clustered using a community detection algorithm [3]. [12] also proposed a graph analysis approach to detect events in social media. However, their methods don't keep efficiency when the data volume increase.

3   Model Framework Overview

Graph based method shows better result in events detection, especially in social media. However, the existing graph based method confront two problems: first, whether the built graph captures the words relationship in semantics. They are not good at dealing with short texts such as twitter or weibo documents, which contains very large noise. In such situation, they would take noise communities as event communities. Second, because the time complexity of community detection in graph is usually $O(n^2 logn)$, the efficiency of graph based events detection will decrease much as the volume of data increases. Attempting to solve such problems, as is shown in Fig. 1, we propose the EventGraph based events detection method. This model mainly consists of three modules: (1), build EventGraph. (2), identify key nodes. (3) events extraction based on key nodes. We will introduce each module in next sections.

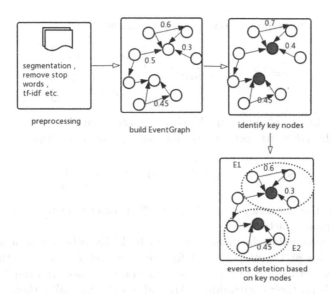

Fig. 1. EventGraph based model framework

4   EventGraph Build

In general, event is defined as a set of related words. In another word, the task of events detection is to extract these related words in semantics from a collection of documents. Intuitively, on the one hand, the related and collocated words will burst when the events spread within social media. On the other hand, the frequency of words co-occur in documents will increase rapidly. We build a directed, weighted graph based on words co-occurrence named EventGraph to capture words semantics relationship. Before dive into the event detection methods, let's give some definitions:

Definition 1. *Event: is defined as a set of descriptive, collocated keywords in this paper. Such group of keywords describing an event will be extracted from corpus through method proposed in the paper.*

Definition 2. *EventGraph: is a directed weighted network $G = (V, E, W)$. V is set of nodes representing word, E is the set of directed edge, representing semantics relationship of word pairs. W is the weight, which value measure how much words are related to each other.*

The basis of construct EventGraph is vector space model. We can build a document-words matrix for corpus, whose columns represent words in dictionary and each row represents a document, and element of matrix is the word's tf-idf value. Therefore, for a collection of documents D:

$$D = (d_1, d_2, ..., d_{|D|})^T = \begin{bmatrix} s_{11} & \cdots & s_{1|V|} \\ \vdots & \ddots & \vdots \\ s_{|D|1} & \cdots & s_{|D||V|} \end{bmatrix} \qquad (1)$$

where $|V|$ is the size of dictionary V, d_k is the kth document vector. while s_{kj} is value that the jth word occur in kth documents, which is computed as follows:

$$s_{kj} = (1 + tf_{kj}) \cdot idf_j = (1 + tf_{kj}) \cdot \frac{\lambda + |D|}{\lambda + df_j} \qquad (2)$$

where tf_{kj} is the frequency of word v_j and df_j is inverse document frequency. λ is Laplace smooth factor.

After preprocessing documents, we can build EventGraph as follows: take words as nodes of EventGraph and for any two nodes v_i and v_j, there is an directed edge between them if they co-occur in documents. And the direction of edge depends on their correlation $p(i|j)$ and $p(j|i)$. Specially, there is an edge from v_i to v_j if $p(i|j)$ is bigger than $p(j|i)$ and take $p(i|j)$ as the weight value. otherwise, the direction of edge is from v_j to v_i and the weight value is $p(j|i)$. the correlation is as follows:

$$w_{ij} = p(i|j) = \log \frac{n_{i,j}/(n_i - n_{i,j})}{(n_j - n_{i,j})/(N - n_i - n_j + n_{i,j})} \cdot \left| \frac{n_{i,j}}{n_i} - \frac{n_j - n_{i,j}}{N - n_i} \right| \qquad (3)$$

where:

$n_{i,j}$ the times of v_i and v_j co-occur in documents.
n_i the number of documents that v_i occur in.
n_j the number of documents that v_j occur in.
N the total number of documents.

As you see, the first item of formula 3 increase as $n_{i,j}$ increase and the second item of formula 3 decrease as n_j increase. In another word w_{ij} reflects how much a word occurring in document has an impact on that another word occurs in the same document. This can explain why collocated words tend to exist in the form of community in the EventGraph. As the Fig. 2 shows that the collocated, descriptive keywords tend to be within one group in EventGraph, which brings efficiency for community detection in the event detection algorithm.

5   Events Detection

Since the collocated words that describe an event tends to exist in the form of communities, the task of event detection is to extract subgraph or communities from EventGraph. However, the general community detection algorithms can not directly be applied to event detection from social media dataset. The existed algorithms is designed for general community detection, if directly

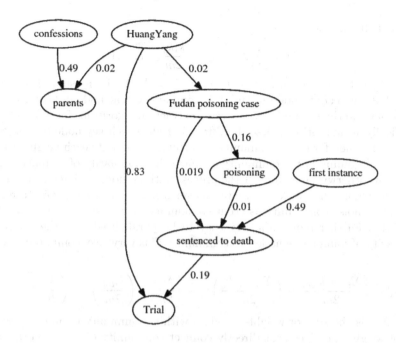

Fig. 2. EventGraph(part) example

applied to event detection, much noise words will be extracted as normal words that describe events. Another reason is that the efficiency of existed algorithms is very low as the scale of graph increase. Similar to [13] that detect events with burst information network, We propose the key node based events extract algorithm to avoid the problems.

Since we assume every event community contains at least a key node, we first identify key nodes of EventGraph. It is simple to identity key nodes with unsupervised methods for the EventGraph is an direct weighted network. Here we adapt PageRank [14] algorithm to identify key nodes. For node v_i in Event-Graph, its score is obtained as follow:

$$ws(v_i) = (1 - d) + d * \Sigma_{v_j \in In(v_i)} \frac{w_{ji}}{\Sigma_{v_k \in Out(v_j)}} ws(v_j) \qquad (4)$$

where $In(v_i)$ is the v_i In-Neighbor set and $Out(v_i)$ is Out-Neighbor set of v_i. $w_j i$ is weight value of edge $< v_i, v_j >$, $d \in (0, 1)$. The high score the node gets, the more import it is in EventGraph.

Communities detection in network structure is to divide the closely connected nodes into the same subgraph or community. Modularity is one measure of the structure of networks or graphs. It was designed to measure the strength of division of a network into modules or communities. Networks with high modularity have dense connections between the nodes within modules but sparse connections between nodes in different modules [15,16]. modularity is scalar between

−1 and 1. It's defined as follows:

$$Q = \frac{1}{2m} \Sigma_{i,j} \left[A_{ij} - \frac{k_i k_j}{2m} \right] \delta(c_i, c_j) \tag{5}$$

where A_{ij} is the weight of edge $< v_i, v_j >$. $k_i = \Sigma_j A_{ij}$ is the sum of weight of edges that connects to node v_i and c_i is community that node v_i belongs to.

We develop the key node based events extraction algorithms for EventGraph. Specifically, as alogrithm 1 describes, first, initialize each key node to be a community, and then for each community c_k, attempt to join its each neighbor node v_i into it. If joining v_i to community c_k make the modularity of c_k much bigger then do it, otherwise not. Repeat the process until no nodes is joined to community; The algorithm has two advantages over others: (1), It can avoid to regard much more noise community as event community for the algorithm is based on key nodes. Further more, it's very easy to be parallelized. (2), the change of modularity of joining one node i to community C is very easy computed in $O(1)$ time:

$$\Delta Q = \left[\frac{\Sigma_{in} + k_{i,in}}{2m} - \left(\frac{\Sigma_{tot} + k_i}{2m} \right)^2 \right] - \left[\frac{\Sigma_{in}}{2m} - \left(\frac{\Sigma_{tot}}{2m} \right)^2 - \left(\frac{k_i}{2m} \right)^2 \right] \tag{6}$$

where Σ_{in} is the sum of weights of edges within community C and Σ_{tot} is the sum of weights of edges that directly connect community C and the others. k_i is the sum of weights of edges that connect node i. $k_{i,in}$ is the sum of weight of edges that connect i to nodes within community C. the outputs of this algorithm is a list of events community and each community contains a set of descriptive, collocated words.

Data: EvenGraph:$G =< V, E, W >$, ordered key node list
$\qquad$:$keynodes = \{v_1, v_2, ..., v_k\}$
Result: Events communities list $C = \{C_1, C_2, ..., C_k\}$
//initialize K communites;
$C_1 = \{C_v\}$, $C_2 = \{v_2\}$, ..., $C_k = \{v_k\}$;
for *each community C_i, $i = 1, ..., k$* **do**
$\qquad$ N_i =set of C_i's neighbor nodes;
$\qquad$ **for** *each node $v_j \in N_i$* **do**
$\qquad\qquad$ compute ΔQ;
$\qquad\qquad$ **if** $\Delta Q \geq \delta$ **then**
$\qquad\qquad\qquad$ $C_i = C_i \cup v_j$;
$\qquad\qquad\qquad$ Add neighbors of v_j to N_i ;
$\qquad\qquad$ **end**
$\qquad$ **end**
end

Algorithm 1. Key Node Based Events Detection Algorithm

6   Experiment Analysis

To evaluate the algorithm we proposed, we conducted the experiments on four datasets: one Sina weibo dataset that we crawl through internet and other three

Table 1. Experiment result(part)

Key words	Events description
Sina weibo	
{Open, Champion, Final, Won, Women's Single, Dominika Cibulkova, Eugenie Bouchard, Li Na, Australian Open, Azarenka}	Li Na won Australian Open Women's Singles
FACup	
{lfc,facupfinal, facup, carroll, ynwa, game, goal, fans, good, andy}	Liverpool FC 1helsea 2 - FA Cup final defeat for LFC
{facupfinal, chelsea, liverpool, mikel, cards, yellow}	Liverpool was shown a yellow card for his challenge on Mikel
Super Tuesday	
{romney santorum idaho wins primary win supertuesday mitt news}	NBC call Idaho before polls were closed
{obama romney good president tonight back war iran people mitt}	Obama and mitt debate for president
US Election	
{obama president years America today vote congratulations good voted mr}	Obama win president
{election romney wins ohio votes polls win arizona}	Romney win votes from arizona

public twitter datasets, FACup, Super Tuesday and US Election [17]. Besides, compared to the algorithm described in paper [17], it concludes that our algorithm shows better result and performance.

6.1 Dataset

Our experiment data set is as follows:

(1) *Sina weibo*: The dataset is our owns that we collected through internet and we labeled 25 events according to the mainstream media reports and other news media.
(2) *FAcup*: The Football Association Challenge Cup or FA Cup, is the main knock-out competition in English football and is the oldest association football competition in the world. The ground truth comprise 13 events, including each of the three goals, some key bookings, and the start, middle and end of the match.
(3) *Super Tuesday*: In the US election system, each candidate goes through a series of primary elections. One of the most important moments is on the first Tuesday of March, during which most states will wait Elections vote. Therefore, this days election is an event that Americans are more concerned

about. This dataset contains 22 events and topics covering predictions of candidates, TV talks and other events.

(4) *US Election*: This dataset is also about the U.S. election. The dataset contains 64 events reported by the mainstream American media about the U.S. election.

6.2   Experiment Result and Analysis

We conduct experiments according to algorithms described in this paper. Specially, we first preprocess texts such as remove stop words etc. And then we build the EventGraph according to the methods that Sect. 4 described. Last, identify and extract events from EventGraph. Table 1 shows some results of four datasets.

In the evaluation of machine learning algorithm, recall is an important and often used indicator. Recall refers to how many positive examples in the sample are accurately predicted. In the algorithm presented in this paper, it refers to How many events are detected by the algorithm from dataset. To evaluate our algorithms, we compute the recalls of the four dataset under different K values and compared to the other methods. As is shown in Fig. 3, our methods performs better than the three others in four datasets except for the facup dataset.

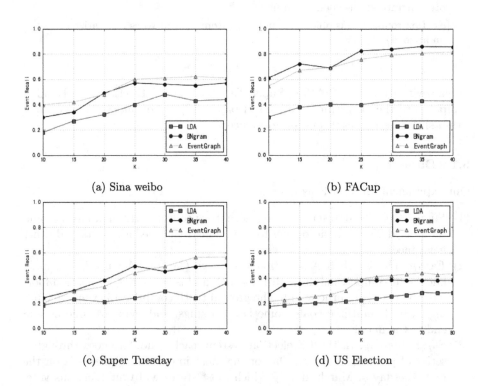

(a) Sina weibo   (b) FACup

(c) Super Tuesday   (d) US Election

Fig. 3. Events recall

This algorithm has two parameters: the number of events K and the threshold δ. The parameter K have an influence on the event recall. The parameter δ controls the number of words in the event community and how fast the algorithm converges. The higher the value of δ, the faster the algorithm converges and the fewer words in the event community. On the contrary, if the value of δ is small, it will reduce the convergence speed of the algorithm but it will increase the number of words of an event. We set a couple of values for K and δ and the results shows that the algorithm performs better on *FACup*, *Super Tuesday* and *US Election* when setting δ 0.25, 0.17 and 0.36.

7   Conclusions

We propose an efficient method for events detection in social media. We build a weighted directed graph, EventGraph whose nodes are words and exits an edge between nodes if the nodes co-occurrence in a document. The relationship between nodes in EventGraph is able to capture the semantics among them. The events are embedded in the EventGrahp in the form of subgraph or communities. We propose a key node based event community detection method, which improve the efficiency of graph based event detection algorithms. At last we conduct the experiment on four datasets and the results show that our algorithm performs better in four compared methods.

Acknowledgments. This work is partially supported by the National Natural Science Foundation of China (NO.61272147) and the Planned Science and Technology Project of Changsha City (No. KC17010266).

References

1. Bollen, J., Mao, H., Zeng, X.: Twitter mood predicts the stock market. J. Comput. Sci. **2**(1), 1–8 (2011)
2. Zhao, S., Zhong, L., Wickramasuriya, J., Vasudevan, V.: Human as real-time sensors of social and physical events: A case study of twitter and sports games. arXiv preprint arXiv:1106.4300 (2011)
3. Sayyadi, H., Raschid, L.: A graph analytical approach for topic detection. ACM Trans. Internet Technol. **13**(2), 4 (2013)
4. Allan, J., Carbonell, J.G., Doddington, G., Yamron, J., Yang, Y.: Topic detection and tracking pilot study final report (1998)
5. Allan, J., Papka, R., Lavrenko, V.: On-line new event detection and tracking. In: Proceedings of the 21st Annual International ACM SIGIR Conference on Research and Development in Information Retrieval, pp. 37–45. ACM (1998)
6. Brants, T., Chen, F., Farahat, A.: A system for new event detection. In: Proceedings of the 26th Annual International ACM SIGIR Conference on Research and Development in Information Retrieval, pp. 330–337. ACM (2003)
7. Kumaran, G., Allan, J.: Text classification and named entities for new event detection. In: Proceedings of the 27th Annual International ACM SIGIR Conference on Research and Development in Information Retrieval, pp. 297–304. ACM (2004)

8. Yang, Y., Ault, T., Pierce, T., Lattimer, C.W.: Improving text categorization methods for event tracking. In: Proceedings of the 23rd Annual International ACM SIGIR Conference on Research and Development in Information Retrieval, pp. 65–72. ACM (2000)
9. Kleinberg, J.: Bursty and hierarchical structure in streams. Data Mining Knowl. Discov. **7**(4), 373–397 (2003)
10. He, Q., Chang, K., Lim, E.-P., Zhang, J.: Bursty feature representation for clustering text streams. In: SDM Conference on SIAM, pp. 491–496 (2007)
11. Pui, G., Fung, C., Yu, J.X., Yu, P.S., Lu, H.: Parameter free bursty events detection in text streams. In: Proceedings of the 31st International Conference on Very Large Data Bases, pp. 181–192. VLDB Endowment (2005)
12. Jia, Y., Xu, J., Xu, Z., Xing, K.: Events Detection and Temporal Analysis in Social Media. In: Lin, C.-Y., Xue, N., Zhao, D., Huang, X., Feng, Y. (eds.) ICCPOL/NLPCC -2016. LNCS (LNAI), vol. 10102, pp. 401–412. Springer, Cham (2016). https://doi.org/10.1007/978-3-319-50496-4_33
13. Ge, T., Cui, L., Chang, B., Sui, Z., Zhou, M.: Event detection with burst information networks. In: Proceedings of COLING 2016, the 26th International Conference on Computational Linguistics: Technical Papers, pp. 3276–3286 (2016)
14. Page, L.: The pagerank citation ranking: bringing order to the web (1998). http://www-db.stanford.edu/backrub/pageranksub
15. Girvan, M., Newman, M.E.J.: Community structure in social and biological networks. Proc. Nat. Acad. Sci. **99**(12), 7821–7826 (2002)
16. Newman, M.E.J., Girvan, M.: Finding and evaluating community structure in networks. Phys. Rev. E **69**(2), 026113 (2004)
17. Aiello, L.M., Petkos, G., Martin, C., Corney, D., Papadopoulos, S., Skraba, R., Kompatsiaris, A.I., Jaimes, A.: Sensing trending topics in twitter. IEEE Trans. Multimed. **15**(6), 1268–1282 (2013)

A Method of Chinese Named Entity Recognition Based on CNN-BILSTM-CRF Model

Sun Long$^{(\boxtimes)}$, Rao Yuan, Lu Yi, and Li Xue

Lab of Social Intelligence & Complex Data Processing, School of Software,
Xi'an Jiaotong University, Xi'an 710049, China
491277866@qq.com

Abstract. The main task of naming entity recognition is to identify the person names, location names, organization names, meaningful time, dates and other quantitative phrases and also classifying them into different categories. Yet, there are lots of field terms, meaningful entities, complicated location names and complex name of organizations contained in all kinds of these field terms is the most important mission in digging out the text now. The extraction of Chinese is more difficult than English entities owing to the lack of definite boundary and size characteristics of Chinese words. Therefore, we would train and test the marked corpus which is based on the CNN-BILSTM-CRF neutral network model in order to make good use of CNN to get the presentation character of words and label the words by using BILSTM and CRF. It makes the extraction of entity in the common came true. The experimental results show that the accuracy rate, recall rate and F value that we got in an unaddressed artificial features condition are 98.81%, 90.70% and 91.57% respectively which is better than the results we got by using the method of BILSTM+CRF and conditional random field (CRF).

Keywords: Named entity recognition · CNN · BILSTM · CRF

1 Introduction

NER(named entity recognition) is a basic task of natural language processing, which aims to accurately identify the information from text for its important role in person names, location names, organization names, meaningful dates, currency and so on. So it can provide the practical information for natural language processing tasks such as information extraction, information retrieval, machine translation, entity co-reference resolution [1], question answering system, topic discovery, topic tracking, etc. [2]. The earliest was put forward by the Defense Advanced Research abroad planning committee (the Defense Advanced Research projects Agency, DARPA) funded news understand meeting (Message Understanding Conference, MUC), and to be held in September 1995 the 6th MUC meeting, the task of named entity evaluation was formally introduced. In meetings, named entities are defined as specific proper nouns and specific quantifiers that people are interested in [3]. Then, a new definition of named entity task was formulated in a conference in 1997 showed that it has three important

© Springer Nature Singapore Pte Ltd. 2018
Q. Zhou et al. (Eds.): ICPCSEE 2018, CCIS 902, pp. 161–175, 2018.
https://doi.org/10.1007/978-981-13-2206-8_15

tasks including quantitative expression, entity name and time representation, in which numerical representation include: currency phrase and comparison value; the names of entities includes: organization names, person names, location names; time identifiers include: date phrases and time phrases [4]. Based on this, a large number of researchers constantly expand the amount of named entity in terms of their different field demands and practical applications.

2 Relevant Works

2.1 A Subsection Sample

At home and abroad, the research on English named entity recognition started early, and has made great achievements. The accuracy rate and recall rate have reached about 97% of the data released by the MUC conference. The research on named entity in foreign countries mainly includes that Bikel D. et al. proposed a kind of named entity recognition method which is based on HMM. The results in MUC-6 text set are as follows: The recognition accuracy of English location names, organization names and person names reached 97%, 94% and 95% respectively, and the recall rate reached 95%, 94% and 94% respectively [5]. Borthwick proposed a named entity recognition method for English and Japanese based on maximum entropy model [6].

The Chinese named entity recognition is in the development stage which is very different from the English named entity recognition. Chinese named entity recognition is to recognize and extract the corresponding string of the named entities in Chinese text. The difficulty includes:

(1) The entity is upgraded constantly, unregistered words are constantly appearing, and it is difficult to enumerate by using dictionaries. For example, it is unrealistic to put the names of all people in the dictionary at present, as with the development of the times, new terms are constantly generated, so will cause in the process of identification recognition error, then word-based named entity recognition methods are proposed [7].

(2) At present, the Chinese named entity recognition focus on the research of recognition of organization name, person name and location name. These entity category doesn't have obvious feature. For example, entities in English mostly are all capitalized or capitalized in first letter while the Chinese named entity recognition will have to find the entity before determining the entity class. However the discovery entity must identify the proprietary entity by participle, but the spread of word segmentation errors and the lack of information will greatly reduce the accuracy of NER [8].

(3) Different nomenclature entities have ambiguity with each other, which can be divided into two types: classification ambiguity and boundary ambiguity. Classification ambiguity refers to a named entity with multiple entity types; Boundary ambiguity refers to the different recognition results could be different according to the different of the boundary of named entities [3].

(4) Acronyms appears in many articles. For example, the abbreviation form of "Xi'an Jiaotong University" is "Western Jiao University". The abbreviation from of "the central committee of the communist party of China" is "CPC central community". These abbreviations remove the key words that are used to indicate the type of entity words, making it difficult to identify entities. At present, the research methods are mainly divided into rule-based, statistical and neural network aspects [9].

Based on the rules and statistics: Feng used a few character groups and part of speech features to identify the named entity, and used the organization name and the end word of the location name as the small single character hints [3]. Hao tried to use rules to identify the named entity recognition in order to extract the needed expression from text and build a complete set of patterns [10]. All of the above research methods require researchers to manually extract relevant grammatical features and develop corresponding templates to identify entities, so the quality of feature templates directly affects the results of named entities.

Neutral network aspect: Li et al. achieved 88.6% F value on Biocreative II GM corpus and 72.76% F value on JNLPBA corpus by using bi-directional short and long term memory network (BILSTM) method [11]. Although there is no artificial extraction of the related features in these studies, it does not take into account the implicit semantic relationship between the vectors of words and the disadvantage that word vectors can not well represent character-level features.

In this paper, we pointed out a kind of neutral network model which is based on CNN+BILSTM+CRF according to the advantages and disadvantages of above works. It aims to identify effectively the person names, location names, and organization names by using neutral network in a situation with non-use of this model. First, we preprocessed the text data, using the self-defined unrecorded lexicon and word segmentation tools to segment the text, and then used word2vec to convert the words into word vectors. Through the sliding window of CNN to calculate the influence on the current words in before and after and the expression features of the words are generated. This feature is input into BILSTM, and the context information of each word is modeled to get the statement information feature. Then an optimal marker sequence is obtained with the CRF. The F value of the model can reach 91.57% without using any artificial marker on the SIGHAN 2006 Bakeoff-3 corpus.

3   Neural Network Structure

In this section we will describe the models used in this article: BLSTM+CRF and CNN+BLSTM+CRF.

3.1   BILSTM Network

The RNN has been used for various tasks, like the language modeling (Mikolov et al. [12]) and speech recognition (Graves et al. [13]). And it has achieved phased results. It should be noted that the context information that can be accessed using RNN is very limited, furthermore there are two problems, memory gradient explosion and memory gradient disappearing, one of these will exist on long sequences. So LSTM, which added some gates to the RNN, was proposed. It overcomes the above two problems memory gradient explosion or memory gradient vanishing which may exist long sequences of traditional RNN models. These special gates structures can effectively save, update, and eliminate context information. So LSTM has features that are suitable for named entity extraction. Figure 1 shows the structural unit of an LSTM:

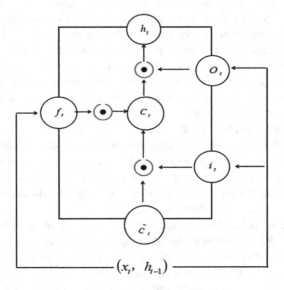

Fig. 1. The unit of LSTM

At some point, the structure of the LSTM network can be expressed as follows:

$$i_t = \sigma\left(x_t \cdot w_{xh}^i + h_{t-1} \cdot w_{hh'}^i + b_h^i\right) \tag{1}$$

$$f_t = \sigma\left(x_t \cdot w_{xh}^f + h_{t-1} \cdot w_{hh'}^f + b_h^f\right) \tag{2}$$

$$o_t = \sigma\left(x_t \cdot w_{xh}^o + h_{t-1} \cdot w_{hh'}^o + b_h^o\right) \tag{3}$$

$$\tilde{c}_t = \tanh\left(x_t \cdot w_{xh}^c + h_{t-1} \cdot w_{hh'}^c + b_h^c\right) \tag{4}$$

$$c_t = i_t \otimes \tilde{c}_t + f_t \otimes c_{t-1} \tag{5}$$

$$h_t = o_t \otimes \tanh(c_t) \tag{6}$$

In these, σ refers to the activation function named sigmiod; $\otimes$ represents the dot multiplication operation, tanh is the hyperbolic tangent activation function; i_t, f_t, o_t are input gates, forget gates and output gates at time t, respectively; c_t indicates the status of the t moment; h_t refers to the output at time t.

In our paper, we used the bidirectional LSTM network which proposed in (Graves et al. [14]), in order to get context information. It can make more efficient use of past features (through Forward state) and future features (via backward state) in specific time frame, by inputting each sentence forward (recursive from the first word to the last word) and reverse (recursive from the last word to the first word), we got two different hidden layers' representations, and then we spliced the forward and reversed hidden layers' representation vectors at each moment. (In Fig. 2, we express "I love China" in Chinese as "我爱中国").

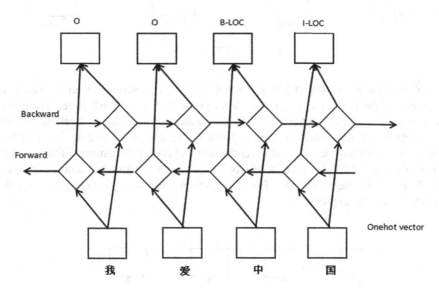

Fig. 2. A bidirectional LSTM network (BILSTM)

3.2   CRF Network

There are two different ways to use neighbor tag information when we predict the current tag. The first is to predict the label distribution for each time step and then use similar beam decoding to find the best tag sequence. The maximum entropy classifier (Ratnaparkhi [15]) and the maximum entropy Markov model (MEMMs) (McCallum et al. [16]) are belonging to this way. The second is to focus on the sentence level rather than the individual level, so such tasks can be handled by using the CRF model (Lafferty et al. [17]) (Fig. 3). However, statistical methods based on the CRF model still have many shortcomings, such as the need for features templates. (In Fig. 3, we express "Rockets beat the Los Angeles Lakers" in Chinese as "火箭队战胜洛杉矶湖人队").

3.3   BILSTM+CRF Network

This paper uses BILSTM and CRF together. Firstly, by combining the advantages and disadvantages of two models, the output of BILSTM is used as the input of CRF. Integrate the CRF into the BILSTM model (Fig. 4). (In Fig. 4, we express "I love

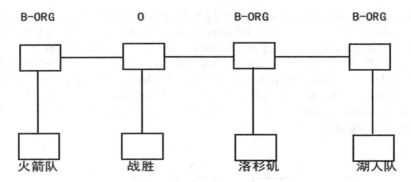

Fig. 3. A CRF network

China" in Chinese as "我爱中国"). The BILSTM layer is described as previously. The parameter of the CRF layer is a matrix of size (k + 2) * (k + 2), which represents the transfer score from the i-th tag to the j-th tag, so that when labeling a location, we can use labels which have been previously labeled. The reason for adding 2 is that the beginning and ending states are added at the start and end of the sentence. The matrix P is the output of BILSTM, where the size of P is N * K, N refers to the number of words, and K refers to the label type. For a prediction sequence $y = (y_1, y_2, \ldots, y_n)$, its probability can be expressed as:

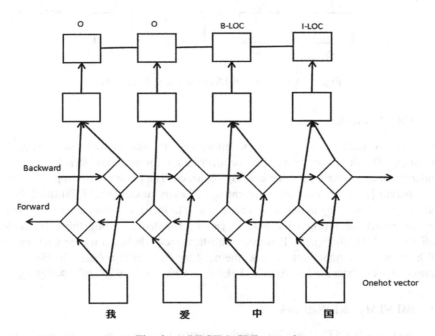

Fig. 4. A BILSTM+CRF network

$$score(x, y) = \sum_{i=1}^{n} P_{i,y_i} + \sum_{i=1}^{n+1} A_{y_{i-1},y_i} \tag{7}$$

Conclusion, the score of the whole sequence is the sum of the scores of each position. And the score of each position is obtained by two parts, one part is determined by the output matrix of the bidirectional LSTM, and the other part is determined by the transfer matrix of the CRF $(A_{i,j})$. Then the normalized probability through Softmax is:

$$p(y|x) = \frac{\exp(score(x, y))}{\sum_{y'} \exp(score(x, y'))} \tag{8}$$

The maximum logarithm likelihood function of the marker sequence in the training process is:

$$\log P(y^x|x) = score(x, y^x) - \log\left(\sum_{y'} \exp(score(x, y'))\right) \tag{9}$$

The optimal path is solved by using the dynamic programming Viterbi algorithm:

$$y^* = \arg\max_{y'} score(x, y') \tag{10}$$

3.4 CNN+BILSTM+CRF Network

BILSTM input can be a simple word embedding or a char embedding. You can also segment the sentence into some phrases or words and change the word vector initialization into the phrase vector in which the word is. It is also possible to use a CNN to get the representation feature of the words into the BILSTM by combining them. It can also be entered into the BILSTM by combining the expression features of words through a CNN. Some achievements have been achieved by using CNN for text processing, (such as Santos and Zadrozny, 2014/4 Chiu and Nichols [18]). Therefore, this paper adds a CNN layer to the model to deal with the problem of inter-word features. The experimental results show that it is feasible to optimize the input of BILSTM to improve the effect of named entity recognition.

3.4.1 The First Is Word Vector Processing

Symbol digitization is a way to turn natural language processing problem into machine learning problem. The idea of this paper is to use word segmentation tools and self-defined unrecorded lexicon to get each word. The quality of the named entity is determined by the quality of the participle. Therefore, this paper first uses Microsoft corpus to compare the four word segmentation tools: ICTCLAS, Paoding, Ansj and Jieba.

By comparing the experimental results (Fig. 5), this paper chooses ICTCLAS as the word segmentation tool.

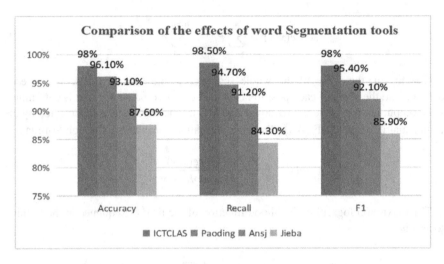

Fig. 5. Word segmentation tool effect contrast graph

Since there may be a large number of unregistered words in the text, the ability to accurately identify unregistered words directly affects the accuracy of the NER. Therefore, it is necessary to find unregistered words. The algorithm for finding unregistered words proposed in our paper is as follows: The first step is to count the number, select a fixed number n, count the number of Bigrams, Trigrams...and N-grams, calculate their internal solidification, and then only keep the j-grams (j = 2, 3, 4 ... n) higher than the threshold, finally these grams will form a set G. The threshold is defined by $\min\left\{\frac{P(abc)}{P(ab)P(c)}, \frac{P(abc)}{P(a)P(bc)}\right\}$. The second part is segmentation: use the above-mentioned grams to segment the corpus, and then count the frequency. If there is only one segment in the G set, this segment will be keep rather than to be segmented, like this "该项目", as long as the "该项" and the "项目" are in G, the "该项目" will not be segmented. The third step is to retrospectively check. If the length of a word is less than or equal to n, then check the word is in the G set or not. If it is not, remove the words; if a word is longer than n words, check whether each fragment of length n of the words is in the set G or not and if one of them is not in G, remove the words. For example, retrospectively checking that "该项目" is not in trigrams, it will not be removed. After a random sampling experiment on the generated thesaurus, the results are as follows:

Experimental data show that the algorithm proposed in this paper to find unrecorded words is feasible (Fig. 6).

3.4.2   CNN Layer

The use of CNN can be used to obtained the local features of words, sentence level features. Adding pool layer can maximize the local feature representation of words. Therefore, this paper extracts the context information of the words, the influence of the word before and after words on the current words, and generates the expression feature of the words as the input of BILSTM. The CNN structure is shown in Fig. 7.

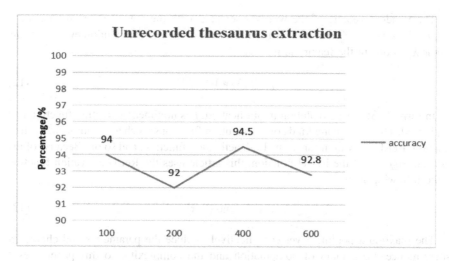

Fig. 6. Unrecorded thesaurus extraction

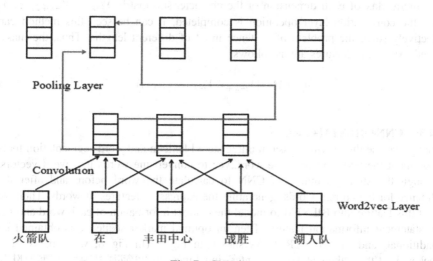

Fig. 7. CNN layer

(In Fig. 7, we express "The Rockets beat the Lakers at Toyota center" in Chinese as "火箭队在丰田中心战胜湖人队").

Set the word vector of word i is v_i, the dimension is d, the number of words in the input sentence is N, and the sliding window of CNN is K. the word vector in the j sliding window is $v_j, v_{j+1}, \ldots, v_{j+k-1}$ respectively, so the window vector is expressed as:

$$X_j = [v_j, v_{j+1}, \ldots, v_{j+k-1}] \tag{11}$$

The window vectors of the words associated with the j word v_j are respectively: $X_{j-k+1}, X_{j-k+2}, \ldots, X_j$. The convolution kernel w is used to convolution every window vector X_j to obtain the feature term.

$$Y_j = f(X_j \bullet W + b) \tag{12}$$

In formula 3: $\bullet$ is convolution multiplication, f is nonlinear activation function and b is biased. There are many kinds of convolution functions, such as: Sigmoid activation function, ReLU activation function, Tanh activation function, and so on. Because of the fast convergence of the ReLU function, this article uses the ReLU activation function in the following form:

$$g(x) = \max(0, x) \tag{13}$$

The maximum pooling layer can effectively reduce the parameters and characteristics, reduce the amount of computation and the complexity, so this paper uses a method called Max-over-time pooling. The principle of this method is to take out the maximum Max of each dimension of the characteristic matrix $Y_{j-k+1}, Y_{j-k+2}, \ldots, Y_j$ after the convolution layer operation is completed. It can be seen this method can effectively solve the problem of sentence input of different lengths. Thus, the maximized expression features are as follows:

$$a_j = Max(Y_{j-k+1}, Y_{j-k+2}, \ldots, Y_j). \tag{14}$$

3.4.3   CNN+BILSTM+CRF

Finally, we use the previous custom unregistered lexicon and word segmentation tools to segment the text, and then use word2vec to convert the words into word vectors. Through the sliding window of CNN to calculate the word before and after the influence for the current words, generating the expressive features of words. Then we input this feature into BILSTM to model the context information of each word and get the statement information feature. Then an optimal marker sequence is obtained by conditional random field (CRF). As shown in Fig. 8 (In Fig. 8, we express "The Rockets beat the Lakers at Toyota center" in Chinese as "火箭人在丰田中心战胜湖人队").

3.5   Training Parameters

In the model design, the optimizer adopts stochastic gradient descending SGD and moment; the learning rate is set to 0.001. At the same time, DROPOUT is added to the bidirectional LSTM model to reduce the over-fitting problem in the training process. The value of Dropout is set to 0. 5.

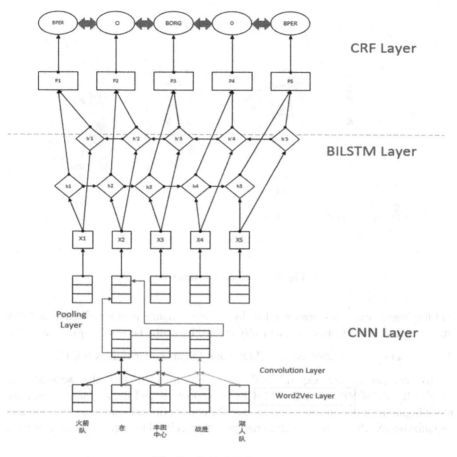

Fig. 8. CNN+BILSTM+CRF

4   Experimental Results

In order to be able to have a clear representation of named entities which is going to be identified. This article uses the BIO marking method to mark the entity. The BIO tag rule is that: B stands for the beginning of an entity, and I stands for the word in the middle of the entity and O represents the other non-entity words (Fig. 9).

The current use of corpus which is related to SIGHAN 2006 Bakeoff-3 is the main method to train this model in order to identify person names, location names and organizational names. Thus we have tried three schemes: CRF+, BILSTM+CRF, CNN+LSTM+CRF model.

(1) Person name entity (PER) identification extraction

The CRF+ model of selecting feature template is better than BILSTM+CRF in person name recognition, because the character of name is more flexible and shorter in length,

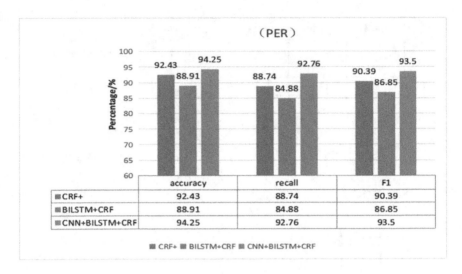

Fig. 9. Name entity extraction.

and the feature extracted from window by feature template is more effective and less interference than the feature automatically learned by neural network (Figs. 10 and 11).

(2) Extraction of Location entities (LOC) and Organization entities (ORG)

By comparing the experimental data, we founded that the accuracy of CNN+BILSTM+CRF and BILSTM+CRF in toponyms and organization names are much higher than that in CRF+, because of the complexity and length of Location and Organization(Such as: some location names with special abbreviation, national names,

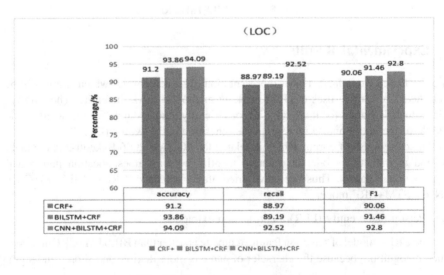

Fig. 10. Location entity extraction.

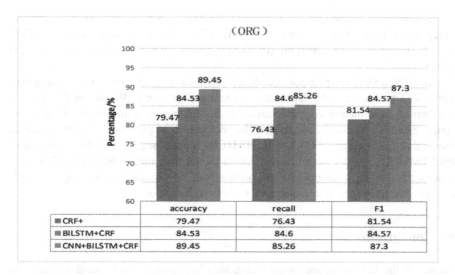

Fig. 11. Organization entity extraction.

etc.). However Two-way LSTM can make good use of sentence level semantic features, but feature template can only be extracted in a certain window and it cannot use the semantics of the whole sentence (Fig. 12).

(3) Contrastive Test of Corpus related to SIGHAN 2006 Bakeoff-3

At BILSTM model. By adding BILSTM into the model and comparing the results of the experimental CRF+ and BILSTM+CRF model. Both F value, accuracy rate and

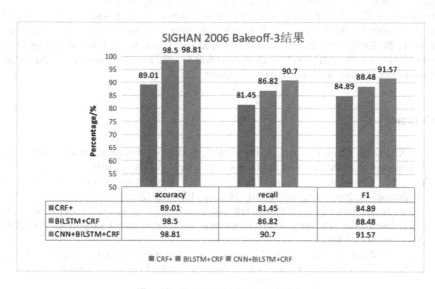

Fig. 12. SIGHAN 2006 Bakeoff-3.

recall rate were improved, and reached 88.48%, 98.50% and 86.82% respectively. From the numerical point of view, BILSTM+CRF is obviously superior to CRF model. The main reason is that the BILSTM+CRF model makes full use of the context information.

At CNN model. The purpose of adding CNN is to extract the context information of words and to generate the representation features of words. It can good solve the problem of combination characteristics and dependence to some extent. In order to verify the effectiveness of this method, the comparison of the above experimental data shows that the F value of CNN+BILSTM+CRF model is 3.09% higher than that of BILSTM+CRF, and the recall rate is 3.88% higher than that of BILSTM+CRF. It is concluded that the local context information can be obtained by adding the CNN module, and the expression features of words can be generated.

5   Conclusion

Aiming at the task of named entity recognition, this paper puts forward a method to calculate the influence of the words before and after by a CNN sliding window, and to generate the expression features of the words, so as to construct the CNN+BILSTM +CRF model. Without the use of artificial features, better results are obtained than a large number of machine learning methods with the usage of rich features. At the same time, we could supplement the lack of word vector input by adding the CNN network convolution layer. The output is decoded by CRF to obtain the optimal tag sequence. To sum up, the fusion of CNN, BILSTM and CRF can effectively improve the performance of recognition entities.

Acknowledgement. This article was supported jointly by the National Natural Science Foundation of China (No. F020807), Found Program of Ministry of Education of China for "Integration of Cloud Computing and Big Data, Innovation of Science and Education" (No. 2017B00030), Basic Scientific Research Operating Expenses of Central Universities (No. ZDYF2017006), Science and Technology Department Collaborative Innovation Program of Shaanxi Provincial (No. 2015XT-21) and Shaanxi Soft Science Key Program (No. 2013KRZ10). We would like to thank them for providing support.

References

1. Ma, X., Liu, Z., Hovy, E.: Unsupervised ranking model for entity coreference resolution. In: Proceedings of NAACL-2016, San Diego, California, USA, June 2016
2. Volk, M., Clematide, S.: Learn-filter-apply-forget. Mixed approaches to named entity recognition. In: Proceedings of NLDB 2001, pp. 153–163 (2001)
3. Grishman, R., Sundhenim, B.: Message understanding conference-6: a brief history. In: Proceeding of the 16th International Conference on Computational Linguistics (COLING 1996), vol. 8 (1996)
4. Bikel, D., Miller, S., Schwartz, R., Weischedel, R.: High-performance learning name-finder. In: Proceedings of the 5th Conference on Applied Natural Language Processing (1997)

5. Borthwick, J., Sterling, E., Agichtein, R., Grishman, N.Y.U.: Description of the MENE named entity system as used in MUC-7. MUC-7, Washington D.C, pp. 145–150 (1998)
6. Isozaki, H., Kazawa, H.: Efficient support vector classifiers for named entity recognition
7. Klein, D., Smarr, J., Nguyen, H., et al.: Named entity recognition with character-level models. In: Proceedings of the Seventh Conference on Natural Language Learning at HLT-NAACL 2003-Volume 4. Association for Computational Linguistics, pp. 180–183 (2003)
8. Wu, Y., Zhao, J., Xu, B., et al.: Chinese named entity recognition based on multiple features. In: Proceedings of the Conference on Human Language Technology and Empirical Methods in Natural Language Processing. Association for Computational Linguistics, pp. 427–434 (2005)
9. Feng, Y.: A rapid algorithm to chinese named entity recognition based on single character hints. J. Chin. Inf. Process. **22**(1), 104–110 (2008). (In Chinese)
10. Hao, W.: Named entity extraction model based on hierarchical pattern matching. New Technol. Libr. Inf. Serv. (5), pp. 62–68 (2007)
11. Li, L., Jin, L., Jiang, Y., Huang, D.: Recognizing biomedical named entities based on the sentence vector/twin word embeddings conditioned bidirectional LSTM. In: Sun, M., Huang, X., Lin, H., Liu, Z., Liu, Y. (eds.) CCL/NLP-NABD -2016. LNCS (LNAI), vol. 10035, pp. 165–176. Springer, Cham (2016). https://doi.org/10.1007/978-3-319-47674-2_15
12. Mikolov, T., Karafiat, M., Burget, L., Cernocky, J., Khudanpur, S.: Recurrent neural network based language model. In: INTERSPEECH (2010)
13. Graves, A., Schmidhuber, J.: Framewise phoneme classification with bidirectional LSTM and other neural network architectures. Neural Netw. **18**, 602–610 (2005)
14. Graves, A., Mohamed, A., Hinton, G.: Speech recognition with deep recurrent neural networks (2013)
15. Ratnaparkhi, A.: A maximum entropy model for part-of-speech tagging. In: Proceedings of EMNLP (1996)
16. McCallum, A., Freitag, D., Pereira, F.: Maximum entropy Markov models for information extraction and segmentation. In: Proceedings of ICML (2000)
17. Lafferty, J., McCallum, A., Pereira, F.: Conditional random fields: probabilistic models for segmenting and labeling sequence data. In: Proceedings of ICML (2001)
18. Santos, C.D., Zadrozny, B.: Learning character-level representations for part-of-speech tagging. In: Proceedings of ICML-2014, pp. 1818–1826 (2014)

Research Progress of Knowledge Graph Based on Knowledge Base Embedding

Tang Caifang[⊠], Rao Yuan, Yu Hualei, and Cheng Jiamin

Lab of Social Intelligence & Complex Data Processing, School of Software,
Xi'an Jiaotong University, Xi'an 710049, China
tang_c_fang@163.com

Abstract. The knowledge Graph (KGs) is a valuable tool and useful resource to describe the entities and their relationships in various natural language processing tasks. Especially, the insufficient semantic of entities and relationship in text limited the efficiency and accuracy of knowledge representation. With the increasing of knowledge base resources, many scholars began to study the knowledge graph's construction technology based on knowledge base embedding. The basic idea is that the knowledge graph will be treated as a recursive process. Through utilizing the knowledge base's resources and the semantic representation of text characteristic, we can extend the new features that improve learning performance and knowledge graph completeness. In this paper, we give a general overview of knowledge graph's construction research based on knowledge embedding, including knowledge representation, knowledge embedding and so on. Then we summarize the challenge for the knowledge graph and the future development trend.

Keywords: Knowledge graph · Knowledge representation
Knowledge embedding · Deep learning

1 Introduction

Since Google put forward the concept of knowledge graph in 2012 [1], it has become the focus of the concern of academic and corporate research field to build a < head entity (h), the relationship (r), tail entity (t) > triple set to try to solve and optimization based on the semantic content of knowledge representation and processing. On the one hand, enterprises put forward a large number of general or specific areas knowledge base such as Cyc [2], Freebase [3], YAGO [4], Wikidata [5], XLORE [6], NELL [7] and so on, to laid the important research foundation for natural language processing and the representation content of knowledge. On the other hand, Academia circle is also studying of knowledge graph in depth, there are two key problems in the application process: (1) In vast amounts of data, how to efficiently and accurately extract and get the semantic relationship between two different entities. (2) In practical application, how to use the limited textual features to express the high order semantics of information content and realize effective reasoning of knowledge. For example, the word "apple" may appear many times in a text, but it is difficult to accurately determine it is

© Springer Nature Singapore Pte Ltd. 2018
Q. Zhou et al. (Eds.): ICPCSEE 2018, CCIS 902, pp. 176–191, 2018.
https://doi.org/10.1007/978-981-13-2206-8_16

fruit, company, or products only by context features, which exactly requires related background knowledge in the knowledge base.

In order to solve the two problems, the researchers focus on the integration between the world knowledge and the target text. It has become the focus to embed the knowledge semantics of existing knowledge graph entities and relationship to realize the rich semantic knowledge representation at present. Figure 1 shows the knowledge graph construction diagram based on the embedding knowledge base. However, it is a very challenging problem how to realize automatic embedding, representation and alignment of text content and semanticist.

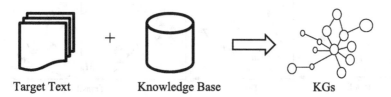

Target Text   Knowledge Base   KGs

Fig. 1. Knowledge graph construction schematic diagram based on knowledge base embedding.

On the basis of review of summarizing the classical models of knowledge embedding and the main methods of text representation, this paper reviews the construction method and key technology of knowledge graph based on knowledge embedding, and prospects the key challenges of technologies and future trends.

2 The Knowledge Base Embedding Model

Information about real-world entities and their relationships in a structured knowledge base (KBs) can implement various applications, such as structured search, fact-based Question-answering, and intelligent virtual assistants. However, one of the main challenges of using a discrete representation repository is the lack of ability to access the similarity between different entities and relationships. In recent years, knowledge base embedding technique is proposed in order to solve this problem. The main idea is to represent entities and relationships in dense, low-dimensional vector Spaces, and to learn the continuous representation of knowledge base in potential space by using machine learning techniques. For this research hotspot, the researchers propose a variety of models. The existing knowledge base embedding model is divided into two categories: (1) Model based on translation (2) Potential factor model.

2.1 Model Based on Translation

In the big data environment, the tensor dimension is very high and the computational complexity is large. Mikolov et al. found that there was a translate invariance in vector space by word2vec representation learning model. Inspired by this, Bordes et al. proposed a translation model-TransE [8], which regarded the relationship in the knowledge base as some translation vector between entities. For each of the triple (h, r, t) in the knowledge base, TransE expects they satisfy the following equation: $l_h + l_r \approx l_t$. The TransE model provides a new perspective for knowledge representation learning, and has achieved remarkable results in the field of knowledge graph. In addition to the simple 1-1 relationship, there is a complex relationship such as 1-n, n-1 and n-n in the knowledge base. Then a great deal of work has been done to extend and apply for TransE. As is shown in Fig. 2, entities and relationships in triples are represented in the same semantic space.

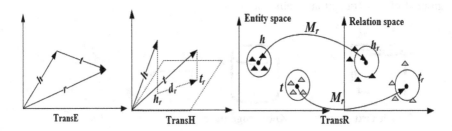

Fig. 2. TransE, TransH, TransR [8–10].

The TransH model proposed by literature [9] enables different entities to have different representations in different relationships, but it still assumes that entities and relationships are in the same semantic space, limiting TransH's expression ability. Literature [10] proposed the TransR model and CtransR to improve the above problems. However, TransR has the following two problems: (1) The head entity and the tail entity are mapped by the same projection matrix, which limits the expression ability of the model. (2) The projection matrix considers only different relationship types and ignores the interaction between entities and relationships. Therefore, the literature [11] proposed the TransD model, which respectively defined the projection matrix of the head entity and the tail entity in relation space. The heterogeneity and imbalance of entities and relationships in the knowledge base is the problem of learning, and in order to resolve the heterogeneity of the entity and the relationship, the literature [12] has proposed the TranSparse model, and the idea is that the use of the sparse matrix to replace the dense matrix in the TransR model. In the literature [13], DKRL model was proposed to consider the context description, and then the researchers proposed the TransG model [14], KG2E model [15]. The goal of training a translation model is usually to minimize the following marginal loss:

$$J_m = \sum_{(h,r,t) \in T} [\gamma + f(h, r, t) - f(h', r', t')]_+ \tag{1}$$

$[\cdot]_+ = \max(0, \cdot)$, γ is margin hyperparameter, usually set to 1, (h', r', t') is the set of incorrect triples generated by corrupting the correct triple triple (h, r, t). Table 1 summarizes the different scoring functions $f(h, r, t)$ and optimization algorithm (Opt) used to estimate model parameters.

Table 1. The scoring function and the optimization method of the translation model

Model	Score function $f(h, r, t)$	Opt.
TransE	$\| v_h + v_r - v_t \|_{l_{1/2}}; v_r \in \mathbb{R}^k$	SGD
TransH	$\| \left(\mathbf{I} - r_p r_p^T\right) v_h + v_r - \left(\mathbf{I} - r_p r_p^T\right) v_t \|_{l_{1/2}}$ $r_p, v_r \in \mathbb{R}^k; \mathbf{I}$: Identity matrix size $k \times k$	SGD
TransR	$\| \mathbf{W}_r v_h + v_r - \mathbf{W}_r v_t \|_{l_{1/2}}; \mathbf{W}_r \in \mathbb{R}^{n \times k}; v_r \in \mathbb{R}^n$	SGD
TransD	$\| \left(\mathbf{I} - r_p h_p^T\right) v_h + v_r - \left(\mathbf{I} - r_p t_p^T\right) v_t \|_{l_{1/2}}$ $r_p, v_r \in \mathbb{R}^k; h_p, t_p \in \mathbb{R}^k; \mathbf{I}$: Identity matrix size $k \times k$	SGD
TranSparse	$\| \mathbf{W}_r^h(\theta_r^h) v_h + v_r - \mathbf{W}_r^t(\theta_r^h) v_t \|_{l_{1/2}}; \mathbf{W}_r^h, \mathbf{W}_r^t \in \mathbb{R}^{n \times k}; \theta_r^h, \theta_r^t \in \mathbb{R}; v_r \in \mathbb{R}^n$	SGD
DKRL	$E_s + E_d$	SGD

2.2 Potential Factor Model

The LF model is given by the following logical link function to assume the probability of set of (h, r, t)

$$P((h, r, t)) = \sigma\left(X_{ht}^r\right) \tag{2}$$

$X^{(r)}$ represents the matrix of potential relationship scores. LF treats the relationship as a second-order relation between entities, it and defines the energy function:

$$f r(\mathbf{h}, \mathbf{t}) = \mathbf{h}^T X^{(r)} \mathbf{t} \tag{3}$$

This model makes a good interaction between entities and relationships, and the relationship between entities and relationships is fully reflected.

RESCAL [16] was the first embedded study based on the decomposition of potential matrix, using bilinear form as the scoring function. DistMult [17] simplified RESCAL by using diagonal matrices. However, the model loses a lot of expression ability because it is too simple and cannot accurately describe the asymmetric relation. To solve this problem, ComplEx [18] extends DistMult to Complex spaces. The HOLE [19] combines two entities with a cyclic correlation method. ConvE [20] uses the convolutional neural network as the scoring function. Table 2 summarizes the different scoring functions $f(h, r, t)$ and the optimization algorithm for estimating model parameters.

The model based on translation studies the validity of traditional relationship extraction model extensively. It can be seen that TransH, TransR, TransD and other models proposed after TransE solve the knowledge base embedding problem from different perspectives. However, the research on potential factor models is limited. In addition, the knowledge base embedding model has developed rapidly in recent years, but few efforts have been made to promote these advanced neural network models using the embedding of knowledge base [21–23].

Table 2. The scoring function and the optimization method of the potential factor model.

Model	Score function $f(h, r, t)$	Opt.
RESCAL	$e_h^T \mathbf{W}_r e_t$	SGD
DISTMULT	$v_r^h \mathbf{W}_r v_t; \mathbf{W}_r \in \mathbb{R}^{k \times k};$	AdaGrad
ComplEx	$\mathbf{R}_e(\langle e_h, \mathbf{w}_r, e_t \rangle)$	AdaGrad
HolE	$sigmoid\left(v_r^T (v_h \circ v_r)\right); \in \mathbb{R}; v_r \in \mathbb{R}^k, \circ$ denotes *circular correlation*	AdaGrad
ConvE	$v_t^T g(\text{vec}(g(\text{concat}(\bar{v}_h, \bar{v}_r) * \mathbf{\Omega}))\mathbf{W});$ g denotes a non $-$ linear function	Adam
ConvKB	$\mathbf{w}^T(\text{concat}(g([v_h, v_r, v_t])) * \mathbf{\Omega});$ $*$ denotes a convolution operator	Adam

3   Representation Learning of the Target Text

The representation learning of target text is to automatically obtain the semantic representation of target text through statistical learning. The German mathematician GottlobFrege proposed in 1892 that the semantics of a paragraph is determined by the semantics of its constituent parts and the combination of them [24]. The existing text semantic representation is also usually based on this idea and obtained through semantic combination. We commonly used combinational semantic composition functions, such as linear weighting, matrix multiplication, tensor multiplication, etc. [25]. From the structure of neural network, it can be divided into three types: recurrent neural networks, recurrent neural network and convolutional neural network.

3.1   Recursive Neural Network

The structure of Recursive neural network is shown in Fig. 3. The core of the Recursive neural network is to gradually synthesize the semantics of each phrase through a tree structure, and finally get the semantics of the whole sentence. The tree structure used by the recursive neural network is usually binary tree, and in some special cases (such as the dependency syntax analysis tree [26]), the multifork tree is also used. The text representation based on recursive neural network mainly includes the construction of syntax tree and the combination function of child node to parent node.

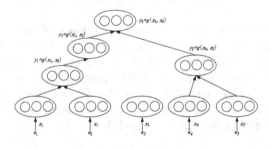

Fig. 3. Recursive neural network structure diagram.

Tree structure is generated in two ways: (1) construction of syntactic trees using syntactic analyzer [27, 28]; (2) use the greedy method to select the adjacent subtree with the smallest reconstruction error, and merge the layer by layer [29]. These two methods have their own advantages and disadvantages. The use of syntactic analyzer method can ensure that the generated tree structure is a syntax tree, each node in the tree is corresponding phrase in the sentence. The semantic representation of all nodes generated by network merging also corresponds to the semantics of each phrase. Building a tree structure by greedy method can automatically accomplish the process by automatically mining the rules in a large number of data, but each node of the tree has practical syntactic component cannot guarantee. The combination function of the child node to the parent node is y = f(a; B). There are three main types:

(1) Syntactic combination. In this way, the representation of the sub-node is vector a; b, the parent node can be obtained by matrix operation:

$$y = \phi(H[a; b])$$ (4)

where, the ϕ as nonlinear activation function, weight matrix H may be fixed, can also according to the subtree corresponding syntactic structure is different, and choose different matrix, this method is generally used in syntactic analysis.

(2) Matrix vector method [30]. In this representation, each node is composed of two parts, a matrix and a vector, for sub node A, a and B, b, and its combination function is.

$$y = \phi(H[Ba; Ab]) \quad Y = W_M \begin{bmatrix} A \\ B \end{bmatrix}$$ (5)

where, the $W_M \in R^{|a| \times |2a|}$, to ensure that the parent node corresponding semantic transformation matrix $Y \in R^{|a| \times |a|}$ consistent with A and B matrix dimensions of children. Using this method, every word has a semantic transformation matrix, the negative word has a similar effect on another part of syntactic structure such as words, common syntactic combinations can't very well to its modeling, while the matrix vector representation can solve this problem. Socher et al. used this method in relation classification.

(3) Tensor combination. The tensor combination method uses each matrix in the tensor to generate one dimension in the representation of the parent node.

$$y = \phi\left([a; b]^T W^{[1:d]}[a; b]\right) + H[a; b]$$ (6)

where, the $W^{[1:d]}$ denotes the first $\sim$ d slice matrix in tensor W. Different sections are used to generate different dimensions in the parent node y. This method is a generalization form of syntactic combination method and has stronger semantic composition ability.

The accuracy of recursive neural network depends on the accuracy of text tree when constructing text representation. No matter use which kinds of structured, which combination function, building the text tree is need time complexity of O(n2) at least, among them, n is the length of the sentence. When the model in the treatment of long sentences or document, the time spent is often unacceptable. Furthermore, the relationship between two sentences does not necessarily constitute a tree structure when making a document representation. Therefore, the recursive neural network performs well in a large number of sentence-level tasks, but may not be suitable for constructing long sentences or document-level semantics.

3.2 Recurrent Neural Network

The Recurrent neural network was first proposed by Elman in 1990 [31]. The core of the model is to enter each word in the text one by one and maintain a hidden layer, preserving all the above information. The cyclic neural network is a special case of recurrent neural network, which can be regarded as the corresponding tree of the right subtree of any non-leaf node. This special structure makes the circulating neural network have two characteristics: (1) because of the fixed network structure, the model can construct the semantics of the text only in order (n) time. This allows the cyclic neural network to model the semantics of the text more efficiently; (2) from the perspective of network structure, the number of layers of cyclic neural network is very deep. Each word in a sentence corresponds to a layer in the network. Therefore, the problem of gradient attenuation or gradient explosion is encountered when the traditional method is used to train the cyclic neural network. In order to optimize the problem, a long short-term memory model [32] is proposed. The model introduces the memory unit, which can save long distance information, and is a common optimization scheme of circulating neural network. The specific implementation formula of LSTM is as follows.

$$
\begin{aligned}
it &= \sigma(Wxixt + Whiht - 1 + Wcict - 1 + bi) \\
ft &= \sigma(Wxfxt + Whfht - 1 + Wcfct - 1 + bf) \\
ct &= ft \odot ct - 1 + it \odot \tanh(Wxcxt + Whcht - 1 + bc) \\
ot &= \sigma(Wxoxt + Whoht - 1 + Wcoct - 1 + bo) \\
ht &= ot fl \tanh(ct)
\end{aligned}
\tag{7}
$$

Considering the input text is chronological, and RNN is a sequential neural network, it can calculate the output of the current time node and the output of the previous node, and get the output result of the current time. Therefore, RNN cyclic neural network solves the correlation between the current node and the previous node, that is, the text feature vector will be well represented by this method. Yao et al. [33] used the BLSTM neural network to process the sequential data, and the method they adopted proved more effective than the LSTM through experiments. Chiu [34] designed a neural network model combining Bi - LSTM and CNN, which proved that the model obtained better results.

3.3   Convolutional Neural Network

The Convolutional neural network (CNN) was first proposed by Fukushima in 1980 [35], whose core is local perception and weight sharing. The convolutional neural network can model local information of each part of the text through its convolution kernel. Through the pooling layer, the full text semantics can be integrated from the local information, and the overall complexity of the model is O(n). In recent years, due to the successful application of the convolutional neural network (CNN) in computer vision [36], the researchers have put forward a number of study models [37–40] based on CNN. We mainly introduce Kim CNN [39].

Kim CNN [39] is a typical sentence that represents the CNN architecture, which uses the pre-trained word vectors to constrain the relationship through convolutional neural network (CNN) architecture to form an entity representation. Figure 4 illustrates the architecture of Kim CNN. Let $w1: n$ be the raw input of a sentence of length n, and $w1: n = [w1 \ w2 \ ... \ wn] \in \mathbb{R}^{d \times n}$ be the word embedding matrix of the input sentence, where $wi \in \mathbb{R}^{d \times 1}$ is the embedding of the i-th word in the sentence and d is the dimension of word embeddings. A convolution operation with filter $h \in \mathbb{R}^{d \times n}$ is then applied to the word embedding matrix $w1: n$, where l $(1 \leq n)$ is the window size of the filter. Specifically, a feature ci is generated from a sub-matrix $w \ i: i + l - 1$ by

$$c_i = f(h * wi : i + l - 1 + b), \qquad (7)$$

where f is a non-linear function, $*$ is the convolution operator, and $b \in R$ is a bias. After applying the filter to every possible position min the word embedding matrix, a feature map

$$c = [c1, c2, \ldots, cn - l + 1] \qquad (8)$$

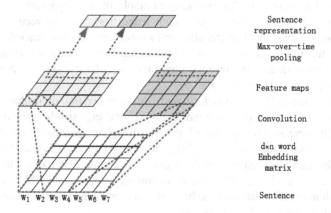

Sentence representation

Max-over-time pooling

Feature maps

Convolution

dxn word Embedding matrix

Sentence

$W_1 \ W_2 \ W_3 \ W_4 \ W_5 \ W_6 \ W_7$

Fig. 4. A typical architecture of CNN for sentence representation learning

is obtained, then a max-over-time pooling operation is used on feature map c to identify the most significant feature:

$$\tilde{c} = \max\{c\} = \max\{c1, c2, \ldots, cn - l + 1\} \tag{9}$$

One can use multiple filters (with varying window sizes) to obtain multiple features, and these features are concatenated together to form the final sentence representation.

4 Knowledge Graph Based on Knowledge Base Embedding

For the construction of knowledge graph, the traditional method only considers the characteristics of the target text, resulting in the semantic deletion of the knowledge map. Therefore, many researchers use existing knowledge base to enhance the text learning ability based on neural network. The core of these methods is to map the target text and knowledge base resources into the same low dimensional vector space. Combining the two types of knowledge, we can improve the effect of entity prediction and trituple classification, and construct the complete knowledge graph.

4.1 Relational Database Embedding

The traditional machine learning algorithm only combines the given characteristics to make the knowledge representation more suitable for learning algorithm. In many practical applications, simply combining existing features is not enough. Humans, however, can easily generalize and generate a new feature based on their previous background.

In view of the above problems, a new method of enhancing the existing learning algorithm based on the background knowledge of relational knowledge base is presented in the literature [41]. The working principle of this algorithm is to generate complex features by using the existing knowledge base, and the recursion learning problem based on existing characteristics and knowledge base is extracted. Then these problems is put into the induction algorithm, a classifier is the final output. The classifier is then converted to the original characteristics of the inductive problem. The result of this process, depicted in Fig. 5.

In the literature [41], an expander-based feature generation algorithm (Expander-FG) is proposed based on the extended feature. However, extending the algorithm in this way will increase the number of generated features exponentially. So, a new knowledge learning algorithm (FEAGURE) based on REcursive FEAture Generation is proposed. The algorithm can control the depth of recursion. This method takes a set of examples as input to the decision tree induction algorithm, applies FEAGURE algorithm on each node, and combines the background knowledge in the relational database to generate new features of these examples.

This method has universality, and the features generated by FEAGURE can be used by any inductive algorithm, which can inject external knowledge in a universal and independent way.

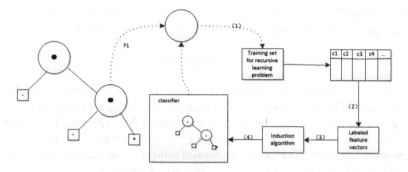

Fig. 5. Recursive construction of a learning problem. (1) Creating the objects for the new problem. (2) Creating features using the knowledge base. (3) Applying an induction algorithm. (4) The resulting feature.

4.2 Non Relational Database Embedding

Machine learning is the most typical solution to the AI problem, although some models can be merged with bayes' prior knowledge, but without the ability to access any organized world knowledge as needed. Therefore, literature [42] proposed to use the knowledge graph to enhance the neural network model, as shown in Fig. 6. The basic idea is that $\mathcal{X}$ is the feature input and $\mathcal{Y}$ is the predicted result. The relevant world knowledge for the task $\mathcal{X}_W$, is retrieved and augmented with the feature input before making the final prediction. Based on the above ideas, the literature [42] proposed two models: (1) Vanilla Model (2) Convolution-based Entity/Relationship Cluster Representation model.

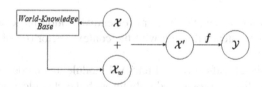

Fig. 6. Knowledge base embedding schematic diagram

(1) Vanilla model

The entities and relationships of the KG are encoded using DKRL [13], explained earlier. Let $e_i \in R_m$ stand for the encoding of the entity i and $r_j \in R_m$ stands for j^{th} relationship in the KG. The input text in the form of concatenated word vectors, $x = (x1, x2, \ldots, xT)$ is first encoded using an LSTM [32] module as follows,

$$h_t = f(x_t, h_t - 1)$$

$$\text{and } o = \frac{1}{T}\sum_{t=1}^{T} h_t, \tag{10}$$

$h_t \in \mathbb{R}^n$ is the hidden state of the LSTM at time t, f is a non-linear function and T is the sequence length. Then a context vector is formed from o as follows,

$$C = \text{ReLU}(o^T W), \tag{11}$$

where, $W \in \mathbb{R}^{n \times m}$ represent the weight parameters. The same procedure is duplicated with separate LSTMs to form two separate context vectors, one for entity retrieval (C_E) and one for relationship retrieval (C_R). As the number of fact triples in a KG is in order of millions in the vanilla model we resort to generating attention over the entity and relation space separately. The fact is then formed using the retrieved entity and relation. The attention for the entity, ei using entity context vector is given by

$$\alpha_{e_i} = \frac{\exp(C_E^T e_i)}{\sum_{j=0}^{|E|} \exp(C_E^T e_j)} \tag{12}$$

where, $|E|$ is the number of entities in the KG.
Similarly the attention for a relation vector r_i is computed as

$$\alpha_{r_i} = \frac{\exp(C_R^T r_i)}{\sum_{j=0}^{|R|} \exp(C_R^T r_j)} \tag{13}$$

where, $|R|$ is the number of relations in the KG. The final entity & relation vector retrieval is computed by the weighted sum with the attention values of individual retrieved entity/relation vectors.

$$e = \sum_{i=0}^{E} \alpha_{e_i} e_i \quad r = \sum_{i=0}^{R} \alpha_{r_i} r_i \tag{14}$$

Figure 7 shows the schematic diagram for entity/relation retrieval. This retrieved fact information is concatenated along with the context vector (C) of input x obtained using LSTM module.

When training the classification and retrieval module, the model often neglects the KG part, and the gradient is propagated only through the classification module. This is partly because the task from the training sample is the most relevant information, only the background information comes from KG. After several times of training, the fact that KG gets always converges to a fixed vector. To overcome this problem, we tried to retrieve the KG before training. The pre-trained KG model retrieves the fact and the classification module connection, and Fig. 8 describes the whole training process. During joint training, the program solves the gradient saturation problem of the KG retrieval part. But the big problem of entity/relational space to be resolved by the attention mechanism is still there.

(2) The convolution based entity and relational cluster representation.

In order to solve the above problems, a new mechanism is proposed to reduce the large number of entities/relationships that must be noticed in the knowledge graph. Reduce attention space by learning the representation of similar entity/relational

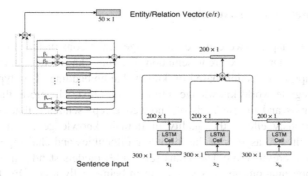

Fig. 7. Vanilla entity/relationship retrieval schematic diagram

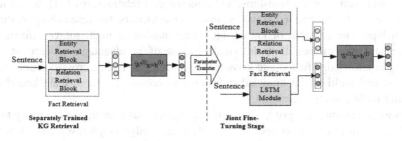

Fig. 8. The module diagram of whole training process.

vectors and pay attention to them. As shown in Fig. 9. Kmeans clustering is used to cluster similar entity/relational vectors and form the same number of entity/relational vector clusters in each cluster. Then, the convolution filter is used to encode each cluster, and finally the clustering outputs an entity/relation vector sequence.

The results of literature [41, 42] show that the performance of text semantic representation based on knowledge embedding is significantly improved. In recent years, with the increase of available knowledge base and the deepening of machine learning, it is believed that the performance of knowledge mapping based on world knowledge embedding will be promoted to the next level.

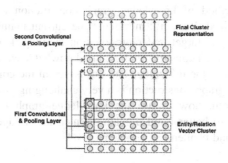

Fig. 9. Cluster representation in the convolution model.

5 Challenges and Prospects

From the above chapters, we can see knowledge graph construction mainly involves technologies like target text representation learning, knowledge embedding, and knowledge graph construction based on knowledge embedding. This type of method attempts to integrate world knowledge into the target text. Therefore, the integrity of knowledge graphs and the methods of knowledge representation that contain world knowledge seriously effects the quality of domain knowledge graph construction. There still remain two aspects of the following difficulties and challenges.

Knowledge representation: The Trans series [43] models stand out in the field of knowledge representation, but it is still far from being really used. Dividing complex relationships into only four types is too rough. The relationship of human knowledge also includes tree-like relationships, two-dimensional grid relationships, single-dimensional sequential relationships, and directed grid relationships [44], which need to combine with AI and cognitive science achievements for researching complex relationships. The research of knowledge representation in multi-source information fusion is still in its infancy. It needs to make full use of the advantages of representation learning in knowledge fusion and knowledge reasoning to realize cross-domain, cross-language and multi-modal knowledge fusion and construct a large-scale knowledge base and its fusion representation.

Knowledge Graph Integrity: The knowledge graph has been applied to many tasks, and there are also a large number of influential knowledge graph resources, including ConceptNet, WordNet, NELL, YAGO, Google Knowledge Map, Sogou Zhibo, Baidu Zhixin et al. [45]. However, there are more and more multi-language and multi-modal open domain knowledge graphs. Most of the knowledge graphs are far from being completed. Incomplete knowledge graphs will also directly affect the results of knowledge embedding. However, data is growing in a rapid rate, and the integrity and real-time nature of the knowledge graph will must be the problem to be faced.

At present, there are few related works about construct a knowledge graph based on knowledge base embedded. There is lots of research space about how to further optimize the neural network model and build a complete knowledge graph.

6 Conclusion

In this paper, the method of knowledge graph construction based on knowledge embedding is summarized. It mainly introduces the current mainstream methods and characteristics in terms of target text representation learning, knowledge embedding model and knowledge graph construction based on knowledge embedding. At the same time, this paper discusses the difficulties and hot issues in the construction of knowledge map. Knowledge graph construction is a very challenging problem. Now in depth study of research boom, how to use the knowledge graph based on knowledge embedded technology to better build domain-oriented knowledge graph is an at the early stages of study and challenging problems.

Acknowledgement. This article was supported jointly by the National Natural Science Foundation of China (No. F020807), Found Program of Ministry of Education of China for "Integration of Cloud Computing and Big Data, Innovation of Science and Education" (No. 2017B00030), Basic Scientific Research Operating Expenses of Central Universities (No. ZDYF2017006), Science and Technology Department Collaborative Innovation Program of Shaanxi Provincial (No. 2015XT-21) and Shaanxi Soft Science Key Program (No. 2013KRZ10). We would like to thank them for providing support.

References

1. Amit, S.: Introducing the knowledge graph. Official Blog of Google, America (2012)
2. Lenat, D.B.: CYC: a large-scale investment in knowledge infrastructure. Commun. ACM **38**(11), 33–38 (1995). https://doi.org/10.1145/219717.219745
3. Bizer, C., Lehmann, J., Kobilarov, G., et al.: DBpedia - a crystallization point for the web of data. Web Semant. Sci. Serv. Agents World Wide Web **7**(3), 154–165 (2009)
4. Suchanek, F.M., Kasneci, G., Weikum, G.: YAGO: a large ontology from wikipedia and WordNet. Web Semant. Sci. Serv. Agents World Wide Web **6**(3), 203–217 (2008)
5. Campbell, C.: Wikipedia: the free encyclopedia. Ref. Rev. **26**(16), 5 (2002)
6. Wang, Z., Li, J., Wang, Z., et al.: Xlore: A large-scale English-Chinese bilingual knowledge graph. In: Proceedings of the 2013 International Conference on Posters & Demonstrations Track-Volume 1035, pp. 121–124 (2013)
7. Carlson, A., Betteridge, J., Kisiel, B., et al.: Toward an architecture for never-ending language learning. In: Proceedings of the Twenty-Fourth AAAI Conference on Artificial Intelligence, vol. 5, pp. 1306–1313 (2010)
8. Bordes, A., Usunier, N., Garcia-Duran, A., et al.: Translating embeddings for modeling multi-relational data. In: Proceedings of NIPS, pp. 2787–2795. MIT Press, Cambridge, MA (2013)
9. Wang, Z., Zhang, J., Feng, J., et al.: Knowledge graph embedding by translating on hyperplanes. In: Proceedings of AAAI, pp. 1112–1119, Menlo Park, CA (2014)
10. Lin, Y., Liu, Z., Zhu, X., et al.: Learning entity and relation embeddings for knowledge graph completion. In: Twenty-Ninth AAAI Conference on Artificial Intelligence, pp. 2181–2187. AAAI Press (2015)
11. Ji, G., He, S., Xu, L., et al.: Knowledge graph embedding via dynamic mapping matrix. In: Proceedings of ACL, pp. 687–696. ACL, Stroudsburg, PA (2015)
12. Ji, G., Liu, K., He, S., et al.: Knowledge graph completion with adaptive sparse transfer matrix. In: Thirtieth AAAI Conference on Artificial Intelligence, pp. 985–991. AAAI Press (2016)
13. Xie, R., Liu, Z., Jia, J., et al.: Representation learning of knowledge graphs with entity descriptions. In: Proceedings of AAAI. AAAI, Mcnlo Park, CA (2016)
14. Xiao, H., Huang, M., Hao, Y., et al.: TransG: a generative mixture model for knowledge graph embedding. ArXiv Preprint arXiv:1509.05488, vol. 1509, p. 05488 (2015)
15. He, S., Liu, K., Ji, J., et al.: Learning to represent knowledge graphs with Gaussian embedding. In: Proceedings of CIKM, pp. 623–632. ACM, New York (2015)
16. Nickel, M., Tresp, V., Kriegel, H.P.: A three-way model for collective learning on multi-relational data. In: International Conference on Machine Learning, pp. 809–816. Omnipress (2011)
17. Yang, B., Yih, W.T., He, X., et al.: Embedding entities and relations for learning and inference in knowledge bases. Eprint Arxiv arXiv:1412.6575 (2014)

18. Trouillon, T., Welbl, J., Riedel, S., et al.: Complex embeddings for simple link prediction, pp. 2071–2080 (2016)
19. Nickel, M., Rosasco, L., Poggio, T.: Holographic embeddings of knowledge graphs. In: Thirtieth AAAI Conference on Artificial Intelligence, pp. 1955–1961. AAAI Press (2016)
20. Dettmers, T., Minervini, P., Stenetorp, P., et al.: Convolutional 2D knowledge graph embeddings (2017)
21. Zeng, D., Liu, K., Chen, Y., et al.: Distant supervision for relation extraction via piecewise convolutional neural networks. In: Conference on Empirical Methods in Natural Language Processing, pp. 1753–1762 (2015)
22. Lin, Y., Shen, S., Liu, Z., et al.: Neural relation extraction with selective attention over instances. In: Meeting of the Association for Computational Linguistics, pp. 2124–2133 (2016)
23. Wu, Y., Bamman, D., Russell, S.: Adversarial training for relation extraction. In: Conference on Empirical Methods in Natural Language Processing, pp. 1778–1783 (2017)
24. Frege, G.: Jber sinn und bedeutung. Wittgenstein Studien **100**, 25–50 (1892)
25. Hermann, K.M.: Distributed Representations for Compositional Semantics. Ph.D. Dissertation, University of Oxford (2014)
26. Socher, R., Karpathy, A., Le, Q.V., Manning, C.D., Ng, A.Y.: Grounded compositional semantics for finding and describing images with sentences. Trans. Assoc. Comput. Linguist. **2**, 207–218 (2014)
27. Socher, R., et al.: Recursive deep models for semantic compositionality over a sentiment treebank. In: Proceedings of the 2013 Conference on Empirical Methods in Natural Language Processing. Association for Computational Linguistics, Seattle, USA, pp. 1631–1642 (2013)
28. Socher, R., Huang, E.H., Pennin, J., Manning, C.D., Ng, A.Y.: Dynamic pooling and unfolding recursive autoencoders for paraphrase detection. In: Proceedings of Advances in Neural Information Processing Systems, vol. 24, Granada, Spain, pp. 801–809 (2011)
29. Socher, R., Pennington, J., Huang, E.H., Ng, A.Y., Manning, C.D.: Semi-supervised recursive autoencoders for predicting sentiment distributions. In: Proceedings of the 2011 Conference on Empirical Methods in Natural Language Processing, pp. 151–161. Association for Computational Linguistics, Scotland, UK (2011)
30. Mitchell, J., Lapata, M.: Composition in distributional models of semantics. Cognit. Sci. **34**(8), 1388–1429 (2010)
31. Elman, J.L.: Finding structure in time. Cognit. Sci. **14**(2), 179–211 (1990)
32. Hochreiter, S., Schmidhuber, J.: Long short-term memory. Neural Comput. **9**(8), 1735–1780 (1997)
33. Yao, Y., Huang, Z.: Bi-directional LSTM recurrent neural network for Chinese word segmentation. In: Hirose, A., Ozawa, S., Doya, K., Ikeda, K., Lee, M., Liu, D. (eds.) ICONIP 2016, Part IV. LNCS, vol. 9950, pp. 345–353. Springer, Cham (2016). https://doi.org/10.1007/978-3-319-46681-1_42
34. Chiu, J.P.C., Nichols, E.: Named entity recognition with bidirectional LSTM-CNNs. Comput. Sci. (2015)
35. Fukushima, K.: Neocognitron: a self-organizing neural network model for a mechanism of pattern recognition unaffected by shift in position. Biol. Cybern. **36**(4), 193–202 (1980)
36. Krizhevsky, A., Sutskever, I., Hinton, G.E.: Imagenet classification with deep convolutional neural networks. In: Advances in Neural Information Processing Systems, pp. 1097–1105 (2012)
37. Conneau, A., Schwenk, H., Barrault, L., Lecun, Y.: Very deep convolutional networks for natural language processing. arXiv preprint arXiv:1606.01781 (2016)

38. Kalchbrenner, N., Grefenstette, E., Blunsom, P.: A convolutional neural network for modelling sentences. arXiv preprint arXiv:1404.2188 (2014)
39. Kim, Y.: Convolutional neural networks for sentence classifcation. In: EMNLP (2014)
40. Zhou, G., et al.: Deep Interest Network for Click-Through Rate Prediction. arXiv preprint arXiv:1706.06978 (2017)
41. Friedman, L., Markovitch, S.: Recursive Feature Generation for Knowledge-based Learning (2018)
42. Annervaz, K.M., Chowdhury, S.B.R., Dukkipati, A.: Learning beyond datasets: knowledge graph augmented neural networks for natural language processing (2018)
43. Liu, Z.-y, Sun, M.-s, Lin, Y.-k, et al.: Knowledge representation learning: a review. J. Comput. Res. Develop. **53**(2), 1–16 (2016)
44. Kemp, C., Tenenbaum, J.B.: Structured statistical models of inductive reasoning. Psychol. Rev. **116**(1), 20–58 (2009)
45. Yan, J., Wang, C., Cheng, W., et al.: A retrospective of knowledge graphs. Front. Comput. Sci., 1–20 (2016)

Dynamic Detection Method of Micro-blog Topic Based on Time Series

Deyang Zhang[✉], Yiliang Han, and Xiaolong Li

Department of Electronic Technology,
Engineering University of the PAP, Xian 710086, China
zhangdy0126@126.com

Abstract. In order to monitor and detect hot topics in real time, this paper proposes a dynamic detection method for microblog topics based on time series. Firstly, according to the law of event development, a three-level topic evolution model based on time series is proposed, and the microblog text is divided into time slices. Then, a time decay function is introduced, regards it as time feature, use it and content feature together to calculate the text similarity. Finally, the traditional Single-pass algorithm has been improved to optimize its clustering center update strategy. Experiments show that the proposed method has certain improvements in both missed detection rate and false detection rate compared with the Single-Pass algorithm and the IEED algorithm, and can detect the microblog hotspot events in real time dynamically.

Keywords: Time series · Topic detection · Dynamic · Micro-blog topic
Single-pass

1 Preface

More and more people began to express their subjective ideas through the Internet. Internet users started from simply surfing the Internet, browsing information, and now everyone involved. With the emergence of some emerging social software such as WeChat and Weibo, the speed of the spread of hot topics has become even more unprecedented. With the advancement of the technology era, the world has fully entered the web3.0 era, and some hot issues have spread faster in such social media than in real life.

As an important part of online public opinion, hot topics are almost always associated with Weibo. The comments on different events, commodities, and public figures on Weibo include the expression of individual subjective emotions. At the same time, the emotional expressions of these Internet users have more or less influence on the emotions of other users; in addition, the government's decision-making behavior is often influenced by the public's opinions and behaviors. Therefore, how to obtain the current hot topics from Weibo It has become crucial.

© Springer Nature Singapore Pte Ltd. 2018
Q. Zhou et al. (Eds.): ICPCSEE 2018, CCIS 902, pp. 192–200, 2018.
https://doi.org/10.1007/978-981-13-2206-8_17

2   Related Work

The methods used in traditional topic detection algorithms are to vectorize texts and use different clustering methods to detect hot topics. Most of the current researches are directed at web page news on the Internet. There is less research on Weibo topics. However, with the rapid growth of Weibo users, the research on Weibo topics has become a hot topic, and has been achieved a lot of results at home and abroad. He Min et al. [1] proposed a hot topic clustering method based on meaningful string clustering. The method is relatively simple, but the complexity is high. Sayyadi et al. [2] proposed an algorithm for constructing keyword maps to detect events in blogs. The detection effect of the algorithm is significant, but the number of events detected by the algorithm depends on the setting of the threshold to achieve effective detection. Ye Shiren [3] proposed a method of topic detection combined with isolated point and Single-pass algorithm. Combined with the evolution of topics, hot spots have a certain degree of timeliness, but the existing results are less analyzed in conjunction with time factors.

With the delay of time, under the same event, new subtopics [4] maybe emerged, and existing results do not consider the time factor to study topic development. This paper introduces time series to divide the topic evolution process into multiple time slices, which constitutes a three-layer model of microblog-topic-event dynamic evolution. For the sensitivity problem of the cluster center in the Single-pass algorithm, the Single-pass is improved, the similarity of the Weibo text is defined, and the calculation method is given. Compared with the traditional Single-Pass algorithm and IEED algorithm, this method considers more relevant factors such as time features and semantic features. Through experiments, this method is more accurate in detecting hot topics.

3   Topic Evolution Model

The formation of lyrical topics on Weibo can be divided into three levels: the event level, the topic level, and the Weibo level, these three levels are all based on time series. Topics and hot topics dynamically change with time. An event is composed of microblogs with specific topics and specifies that a microblog has only one specific topic. Then the evolution process of the topic can be abstracted as a model of microblog topic evolution as shown in Fig. 1.

Define the events discussed by Weibo for a period of time as event sets are consists of different events $E = \{E_i\}$. Each of these events E_i consists of related topics T_i, $E_i \leftarrow \{T_i\}$, similarly, the topic T_i is composed of related microblogs m_i, $T_i \leftarrow \{m_i\}$. This formed a three-tier topic model. Since this article focuses on the formation of hot topics, our main research content is from the first layer of microblogging layer to the second layer of topic detection.

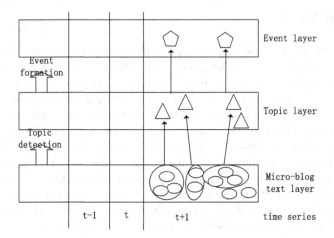

Fig. 1. Three-level topic evolution model

4 Micro-blog Text Processing

4.1 Text Preprocessing

In order to improve the operating speed and accuracy of the algorithm, we need to preprocess the microblog text, and represent a text with a length of less than 140 words as a space vector. The main steps are to remove noise, word segmentation, feature extraction and vector representation.

Denoising is the primary filter, such as some of the theme of the text contributes little punctuation ("[" and "]" and "#", "@", ":"), as well as some similar URL links. Word segmentation adopts the Chinese Academy of Sciences segmentation system NLPIR, which mainly includes Chinese word segmentation, named entity recognition, part of speech tagging, and new word recognition and user dictionaries. The specific usage is referred to its API. At present, the latest version of NLPIR has a precision of 98.45%, which is an excellent Chinese classifier. Remove stop words, and stop words are a kind of words that are not practical and cannot be used in the text. The method is to use an extended discontinuation word library. The preprocessed micro-blog text is mapped into a set of words $m_i \rightarrow \{t_1, t_2, t_3, t_4 \ldots \ldots t_j\}$, t_j is a preprocessed feature word.

For micro-blog, each word has different importance due to its different parts of speech. In order to highlight its importance, we introduce the concept of feature weight, and give each word a weight according to specific rules. So, the micro-blog text can also be expressed as $m_i = \{t_1, w_{i1}; t_2, w_{i2}; t_3, w_{i3}; \ldots \ldots t_j, w_{ij}\}$, the weight of the corresponding word of the characteristic word t_1 is w_{i1}.

4.2 Feature Selection

Currently there are three main feature selection methods: (1) based on the document frequency [5] (2) based on mutual information [6] (3) based on information gain [7]. This paper adopts the method of document frequency to extract the characteristics of microblog. This method can not only filter the noise information, but also reduce the data dimension, prevent the data from disasters, and help improve the accuracy and rate of the algorithm.

Under normal circumstances, nouns, verbs, adjectives, time, and number words are better able to express topics than prepositions and prepositions. Therefore, it is possible to increase the weight of these words and ignore other parts of speech. At the same time, it is necessary to set the threshold range of the text frequency to filter the feature. If the frequency is too high, the feature is not distinguishable. If the frequency is too low, the information carried by itself is not enough, so the feature word with the length less than 2 is excluded.

4.3   Calculate Feature Weights

The size of the weight has a great influence on the clustering result. Usually TF-IDF is used to calculate the feature weight, but there will be a case where the denominator is 0 after taking the logarithm. It is necessary to improve it by adding the smoothing to the logarithm portion with a factor of 0.01, the improved TF-IDF calculation is shown in Eq. (1):

$$w_{ij} = \frac{tf_{ij} \times \log\left(\frac{N}{n_{ij}} + 0.01\right)}{\sqrt{\sum_{j=1}^{M}\left(tf_{ij} \times \log\left(\frac{N}{n_{ij}} + 0.01\right)\right)}} \tag{1}$$

Among them, w_{ij} represents the weight of the jth word in the ith text. tf_{ij} represents the frequency of the jth word appearing in the ith text. n_{ij} represents the total number of the jth words, and N represents the total number of texts.

4.4   Similarity Calculation

Considering the strong timeliness of micro-blog, when calculating the similarity of micro-blog texts, we should consider not only the similarity between text contents, but also the time factors. Generally speaking, the public opinion about a particular hot topic will last for a period of time. The farther away the topic, the weaker the relevance of the topic. Therefore, in order to improve the precision of clustering, the time decay function [8] T is introduced in this paper.

$$T(m_i, m_j) = \begin{cases} 1 - t/ac, & if(t < ac) \\ 0, & else \end{cases} \tag{2}$$

$t = |t(m_i) - t(m_j)|$, $t(m_i)$ is the timestamp of the micro-blog m_i, represents the time when Weibom generated the topic, $t(m_j)$ means the time of the topic center. ac is a time decay factor, which represents the longest time to allow the interval. For a better topic clustering, the value of the ac is set to one month.

In this way, the micro-blog similarity between the text feature and the time feature is considered as follows:

$$sim(m_i, m_j) = d(m_i, m_j) + T(m_i, m_j) \tag{3}$$

$d(m_i, m_j)$ is the content similarity of micro-blog, expressed by the cosine formula, the calculation is as follows:

$$d(m_i, m_j) = \cos(m_i, m_j) = \frac{\sum\limits_{k=1}^{n} w_{ik} \times w_{jk}}{\sqrt{\sum\limits_{k=1}^{n} w_{ik}^2} \times \sqrt{\sum\limits_{k=1}^{n} w_{jk}^2}} \tag{4}$$

5 Single-Pass Clustering Algorithm Based on Time Series

How to extract "hidden" topics from microblog texts is a long-standing research issue. From the current research results, we mainly use clustering algorithms to discover hot topics of microblogs. Currently existing clustering algorithms include four, hierarchical-based clustering algorithms (AGENS [9] etc.), partition-based clustering algorithms (K-means [10] etc.), density-based clustering algorithms (DBSCAN [11] etc.), and incremental clustering algorithms. Combined with the actual situation of this article, since it is the detection of hot topic of microblog, the topic detection process is actually an incremental process. Therefore, the Single-Pass clustering algorithm [12] is adopted.

Traditional Single-Pass clustering algorithm can quickly and automatically obtain topics. It is an unsupervised clustering process. However, there is still a similarity calculation to the text content and ignores the time factor. Therefore, this paper is based on the original Single-Pass, the concept of time series is added, both the text feature and the time feature are considered in the text similarity calculation. There are several differences between the improved algorithm and the traditional algorithm: (1) Divided by time slice (2) In the text similarity judgment, introduce the time feature (3) Change strategy of cluster center update. The specific change of the update strategy is as follows: When the new text is added, the similarity between the original cluster center and each text in the new text and the cluster is calculated, and the average value is calculated, if the average value between new text and text in the cluster is larger, us it as a new clustering center, otherwise, remains unchanged.

The concrete steps of the algorithm are as follows:

(1) Arrange the collected microblog texts in time series;
(2) Initialize the model. In the interval of time series t1, consider the first microblog m_1 as a seed topic and set a threshold for similarity θ;
(3) Compare the new added text with the existing topic clustering center and calculate the similarity. If it is greater than the threshold, then join in the topic; otherwise, use the text as a new topic clustering center;

(4) Update the clustering center and follow the cluster center update strategy.

(5) Execute steps (3, 4, 5) for new text;

(6) Finally get the topic set $\{T1, T2, T3, \ldots\ldots\}$ in time series t1.

The improved algorithm is more in line with the development process of hot topics over time. It can autonomously perform incremental clustering and eventually achieve the purpose of topic clustering. Algorithm flow chart shown in Fig. 2.

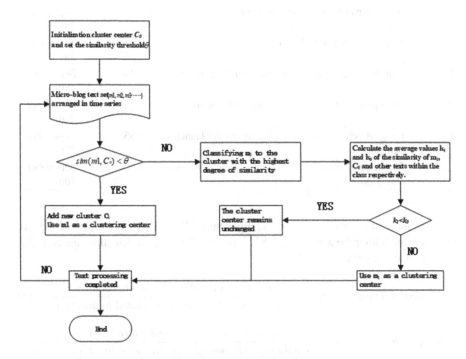

Fig. 2. The improved algorithm flow chart

6 Single-Pass Clustering Algorithm Based on Time Series

This paper uses the Sina Weibo API to randomly obtain the Weibo text within the month from September 2017 to October 2017. Since the Weibo topic has only a few days from the beginning to its demise, some topics will be quickly submerged with new topics. Taking into account the influence of time series interval on topic relevance, this paper sets the time interval Δt to 3 days. It is generally believed that in a topic clustering, the maximum number of microblogs is considered to be a hot topic. By extracting the microblog keywords, the hot events in each time interval can be finally obtained. The top 2 topics in the first 4 time intervals are shown in Table 1.

The comparison experiment between the algorithm in this paper and the IEED algorithm and the traditional Single-Pass algorithm in reference [9] is carried out. In order to better reflect the improvement of the algorithm in this paper, under the premise of detecting the above eight topics, uses of missed detection rate and false detection

Table 1. The top 2 topics in the first 4 time intervals

Time interval	Topic title	Number of reports	Start time
t1	A 6.3-magnitude earthquake in North Korea considers the nuclear bomb experiment	189	September 3rd
	Shaanxi girl name "king glory"	126	September 5th
t2	Xue Zhiqian Gao Leixin Compound	85	September 7th
	Earthquake 8.4 near Mexico	113	September 8th
t3	Remembrance Day of "September 18" Incident	148	September 18th
	Disapproval of plane release of mobile phone use	226	September 20th
t4	Long March 2 C rocket fired a successful launch of Samsung	155	September 27th
	The original secretary of Chongqing Sun Zhengcai was deposed	120	September 30th

rate to evaluate the performance of the algorithms. The smaller the value of the two indicators, the higher the accuracy of the algorithm. The missed detection rate and false detection rate are defined as follows:

$$missed\ detection\ rate = \frac{undetected\ topic - related\ microblogs}{total\ number\ of\ microblogs\ related\ to\ the\ topic} \quad (5)$$

$$false\ detection\ rate = \frac{detected\ topic - related\ microblogs}{total\ number\ of\ microblogs\ that\ are\ not\ related\ to\ the\ topic} \quad (6)$$

Figures 3 and 4 show the missed rate and false detection rate of different algorithms.

From Figs. 3 and 4, we can see that compared with other algorithms, the missed detection rate and false detection rate are the lowest, and the average missed detection rate is 14.5% and 12.8% lower than that of Single-Pass algorithm and IEED algorithm respectively. The average false detection rate is 5.4% and 2.4% lower than that of the Single-Pass algorithm and the IEED algorithm respectively. Compared with the two comparison algorithms, the algorithm has a certain improvement in the efficiency and accuracy of topic discovery. The reason is that the algorithm of this paper considers the similarity of text content, adds a time decay function, and considers the time factor. At the same time, it also improves the update strategy of the cluster center for the Single-Pass algorithm. However, due to the addition of a new time factor in the comparison of similarities, the efficiency of the algorithm is reduced.

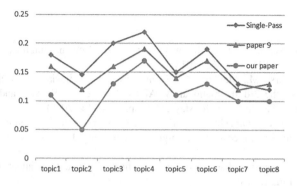

Fig. 3. Missed detection rate comparison of different algorithms

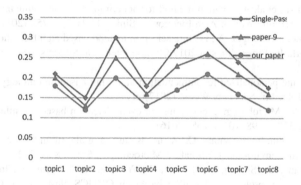

Fig. 4. False detection rate comparison of different algorithms

7 Conclusion

In the era of information explosion, a large amount of information are generated every day. How to obtain useful information from these information is an important research topic. Based on the existing results of topic detection, this paper proposes a method of dynamic detection of microblog topics based on time series, and designs a microblog-topic-event three-layer topic evolution model. In terms of clustering algorithms, The Single-Pass algorithm is improved and a new cluster center update strategy is proposed.

After the experimental analysis, this method has a certain improvement in the missed detection rate and the false detection rate compared to the Single-Pass algorithm and the IEED algorithm, and can better detect the microblog hotspot events in real-time and dynamically, but as compared to traditional algorithms increase the calculation of time-based similarity and the dynamic update of clustering centers. The running speed of the algorithm is reduced. With the development of cloud computing and other technologies, this paper will improve the algorithm's operating speed.

References

1. He, M., et al.: Weibo hot topic discovery method based on meaningful cluster clustering. Corr. Technol. **34**(S1), 256–262 (2013)
2. Sayyadi, H., Hurst, M., Maykov, A.: Event detection and tracking in social streams. In: Proceedings of the 3rd International AAAI Conference on Weblogs and Social Media (ICWSM 2009), San Jose, 17–20 May 2009, pp. 311–314 (2009)
3. Shiren, Y., Ying, Y., Changchun, Y., Mingfeng, Z.: Weibo topic detection method combined with isolated point preprocessing and single-pass clustering. J. Comput. Appl. **33**(08), 2294–2297 (2016)
4. Jiang, Z., et al.: Supervised search result diversification via subtopic attention. IEEE Trans. Knowl. Data Eng. (2018)
5. Ning, B., et al.: Low-cost message-driven frequency-hopping scheme based on slow frequency-hopping. IET Commun. **12**(4), 485–490 (2018)
6. Run-Nian, M.A., et al.: Evaluation method for node importance in communication network based on mutual information. Acta Electronica Sinica **45**(3), 747–752 (2017)
7. Wang, X., Zuo, M., Song, L.A.: Feature selection method based on information gain and BP neural network. In: Proceedings of 2017 Chinese Intelligent Systems Conference, pp. 23–30 (2018)
8. Li, H.: Dynamic personalized recommendation algorithm based on ontology similarity and time decay. Libr. Inf. Serv. pp. 95–98 (2017)(s1)
9. Li, H., Zhu, X.: Microblogging emerging hot event detection based on influence. Comput. Appl. Softw. **33**(05), 98–101 + 165 (2016)
10. Xuejuan, L., Jiabin, Y., Fengping, C.: K-means clustering algorithm for data distribution in cloud computing environment. Miniature Microcomput. Syst. **38**(4), 712–715 (2017)
11. Hong, W., Lina, G., Suqing, W., Liying, W., Yipeng, Z., Yucheng, L.: Improvement of differential privacy protection algorithm based on OPTICS clustering. J. Comput. Appl. **38**(01), 73–78 (2018)
12. Huang, J., Peng, M., Wang, H., Cao, J., Gao, W., Zhang, X.: A probabilistic method for emerging topic tracking in microblog stream. World Wide Web **20**(2), 325–350 (2017)

An Evaluation Algorithm for Importance of Dynamic Nodes in Social Networks Based on Three-Dimensional Grey Relational Degree

Xiaolong Li[✉], Yiliang Han, Deyang Zhang, and Xuguang Wu

Department of Electronic Technology, Armed Police Engineering University,
Xian 710000, China
463387834@qq.com

Abstract. The importance assessment of dynamic nodes in social networks has been a very challenging problem in the field of social network research. This paper creatively adopts the three-dimensional grey relational degree algorithm, introduces the dimension of time based on the attributes of nodes and nodes, and realizes the screening of important nodes in the dynamic social network. Finally, the algorithm is verified by Facebook data set for three consecutive months, and compared with the result of TOPSIS algorithm, which shows that this algorithm is more practical and accurate.

Keywords: Social networks · Dynamic nodes
Three-dimensional grey relational degree

1 Introduction

As an extension of the relationship between human beings in the real world, the social network is an important channel and carrier for sustaining human social relations and information communication in human society [1]. The important nodes in the social network are the basic indicators to characterize the structure and dynamics of the network. Evaluating the dissemination ability of important nodes is one of the fundamental issues in controlling the dissemination capabilities of social networks [2]. The importance evaluation algorithm of nodes in social networks has important applications in the field of fixed-point delivery of advertisements, defense of network viruses, and intervention of Internet public opinion. For the first time, the concept of node centrality was proposed by Shimbel [3], and he thought that the centrality of the node should be the number of the shortest path through the node. Freeman [4] proposed the Degree centrality indicator in 1978. This indicator has a simple and efficient calculation process, but it neglects the relevance of the nodes in the indicators caused by the global of the network. Closeness centrality [5] and Betweenness Centrality [4, 6, 7], proposed by Brandes have lower computational complexity, but the two indicators do not apply to

This work is supported by National Nature Science Foundation of China (No. 61572521), Research Project of Military Science (No. 16QJ003-097).

© Springer Nature Singapore Pte Ltd. 2018
Q. Zhou et al. (Eds.): ICPCSEE 2018, CCIS 902, pp. 201–212, 2018.
https://doi.org/10.1007/978-981-13-2206-8_18

large-scale networks. Harmonic Centrality [8] is also a centering algorithm based on the shortest path. Yannick [9] compares Harmonic's centrality with near-centrality, and points out that Harmonic's centrality can be used as an alternative to near-centrality and extends it to undirected graphs. Since then, many improved evaluation indicators have still failed to provide a comprehensive and objective evaluation of large-scale social networks. Yuxian Du [10] et al. used the technique of Similarity Solution method (TOPSIS) to implement multi-attribute evaluation of nodes in social networks, and the evaluation results were closer to the actual situation.

As is known to all, social networks [11] are a dynamic network that is constantly changing. For example, with the change of preferences, there will be mutual concern or attention cancellation on micro-blog. With the change of interpersonal relationship, there will be a discussion group joining and withdrawing from WeChat. Therefore, the nodes' continuous change will cause great changes in their importance degree in the network. so it is of great significance to evaluate the importance of nodes in dynamic social networks. This paper applies the three-dimensional grey relational degree algorithm [12] to the importance evaluation algorithm for dynamic nodes in social network for the first time, and realizes the screening of important nodes in dynamic nodes. This paper adopts Facebook's three-month data set to verify the algorithm. This data set, as a real social network relational data, has the characteristics of large amount of data and authenticity, and can effectively reflect the effectiveness of the algorithm in real social networks. Compared with the TOPSIS program's ranking results, The experimental results show that the algorithm is feasible, more in line with the actual situation, and has a high accuracy.

2   Theoretical Basis

2.1   Eigenvector Centrality

Social network can be represented by $G(V, E)$, where $V(v_1, v_2, \cdots, v_n)$ represents a set of all nodes in network G and $E(e_1, e_2, \cdots, e_m) \subseteq V$ represents a set of all edges in network G. $|V| = n$ Represents the number of all nodes in Network G.

2.2   Three-Dimensional Grey Relational Degree Algorithm

Gray relational analysis method uses gray relational space to realize quantitative analysis of system development trend. The basic idea is to measure the similarities between curves. The degree of correlation between factors is judged based on the changing proximity and similarity between the various curves. In this paper, nodes, node's attributes and time are used as three factors to measure the network changes. And The nodes in the network are analyzed by the three-dimensional grey correlation analysis method.

2.3   Attribute System

There are many attributes to evaluate a node. This paper selects four attributes as the attribute evaluation system of this algorithm.

2.3.1   Harmonic Centrality

$$c_H(x_i) = \frac{1}{n-1} \sum_{j \neq i} \frac{1}{dist(x_i, x_j)} \tag{1}$$

Harmonic is central, $dist(x_i, x_j)$ represents the distance between the two nodes i and j in the network. When two nodes are not connected, $dist(x_i, x_j)$ take infinite. $\sum_{j \neq i} \frac{1}{dist(x_i, x_j)}$ is defined as the sum of the inverted distances.

2.3.2   Closeness Centrality

$$CC_i = n / \sum_{j=1}^{n} d_{ij} \tag{2}$$

Where, d_{ij} indicates the number of edges on the shortest path formed by i and j as endpoints. It can be seen from the formula that the closeness centrality of node i is the summation of the distance between node i and other nodes except node i in the network. The value of CC_i indicates the distance between the location of the node and the center of the network. The larger the value of CC_i, the greater importance of node i is in the network.

2.3.3   Betweenness Centrality

$$BC_i = \sum_{j \neq i \neq k \in V} \frac{g_{jk}(i)}{g_{jk}} \tag{3}$$

Where $g_{jk}(i)$ represents the number of all shortest paths between node j and destination node k which passing through node i. g_{jk} represents the number of all the shortest paths formed between node j and node k. Betweenness centrality means that node i plays an important role in this network if it acts as a necessary path for any two other nodes in the network. Betweenness centrality is positively related to the node importance.

2.3.4   Degree

$$D_i = \sum_{j=1}^{n} e_{ij}(i \neq j)$$

$$e_{ij} = \begin{cases} 1 & v_i \text{ is } connected \text{ to } v_j \\ 0 & v_i \text{ is not } connected \text{ to } v_j \end{cases}$$

(4)

D_i represents the degree of node i, e_{ij} denotes the edge between i and j nodes. $e_{ij}=1$ indicates that there is a connection between the two nodes, $e_{ij}=0$ indicates that there is no connection between the two nodes.

3   Dynamic Node Importance Evaluation Algorithm

This article notes each time snapshot as $T_i(i = 1, 2, \cdots, m)$, whose corresponding weight is $e_i(0 < e_i < 1, \sum_{i=1}^{m} e_i = 1)$. Mark each attribute of the node as $p_j(j=1, 2, \cdots, n)$, whose corresponding weight is $w_{ij}(0 < w_{ij} < 1, \sum_{j=1}^{n} w_{ij} = 1)$. This article assumes that there are q nodes in the network in total. Consider each node as a scheme and record it as $S_k(k = 1, 2, \cdots, q)$. In summary, we will get a three-dimensional space. The three-dimensional space has q nodes, each scheme S_k has n attribute indicators p_j under m time snapshots T_i. This article notes each point in the three-dimensional space as a_{kij}, a_{kij} is a three-dimensional attribute value, indicating the attribute value of scheme S_k when it corresponds to T_i indexes at p_j moments.

(1) The decision matrix A_k of the kth scheme S_k:

$$A_k = \begin{array}{c} \\ T_1 \\ T_2 \\ \vdots \\ T_m \end{array} \begin{array}{c} p_1 \quad p_2 \quad \cdots \quad p_n \\ \begin{bmatrix} a_{k11} & a_{k12} & \cdots & a_{k1n} \\ a_{k21} & a_{k22} & \cdots & a_{k2n} \\ \vdots & \vdots & \cdots & \vdots \\ a_{km1} & a_{km2} & \cdots & a_{kmn} \end{bmatrix} \end{array} = (a_{kij})_{m \times n}$$

(1)

In the decision matrix for the scheme k, the jth column shows a time series of the index 1, and the i th row shows that the scheme k corresponds to the value of each index at time 2.

(2) The index can be divided into positive correlation index (The larger the node indicator value, the more important the node) negative correlation index (The larger the node indicator value, the less important the node), in order to facilitate the analysis, We want to apply the same trending to each indicator in A_k, the different indicators in the same dimension.

Positive correlation index:

$$b_{kij} = \frac{a_{kij} - \min_k \min_i a_{kij}}{\max_k \max_i a_{kij} - \min_i \min_i a_{kij}} \tag{2}$$

Negative correlation index:

$$b_{kij} = \frac{\max_k \max_i a_{kij} - a_{kij}}{\max_k \max_i a_{kij} - \min_i \min_i a_{kij}} \tag{3}$$

Formed normalization matrix B_k:

$$B_k = \begin{array}{c} \\ T_1 \\ T_2 \\ \vdots \\ T_m \end{array} \begin{array}{c} p_1 \quad p_2 \quad \cdots \quad p_n \\ \left[\begin{array}{cccc} b_{kij} & b_{kij} & \cdots & b_{kij} \\ b_{kij} & b_{kij} & \cdots & b_{kij} \\ \vdots & \vdots & \cdots & \vdots \\ b_{kij} & b_{kij} & \cdots & b_{kij} \end{array} \right] \end{array} = (b_{kij})_{m \times n} \tag{4}$$

(3) Determine the optimal scheme and the worst scheme according to the normalized decision matrix B_k:
Optimal scheme:

$$B^+ = (b_{ij}^+)_{m \times n} \tag{5}$$

The worst scheme:

$$B^- = (b_{ij}^-)_{m \times n} \tag{6}$$

Among them, the elements in matrix B^+, B^- are:

$$\begin{aligned} b_{ij}^+ &= \max_k (b_{kij}) \\ b_{ij}^- &= \min_k (b_{kij}) \end{aligned} \tag{7}$$

Among them, $i = 1, 2, \cdots, m, j = 1, 2, \cdots, n$.
(4) According to the normalized original data, the correlation coefficient between the Pending scheme and the decision matrix B_k can be obtained from the (8) formula:

$$\begin{aligned} r_{kij}^+ &= \frac{\min_k \min_i \min_j \left| b_{ij}^+ - b_{kij} \right| + \rho \max_k \max_i \max_j \left| b_{ij}^+ - b_{kij} \right|}{\left| b_{ij}^+ - b_{kij} \right| + \rho \max_k \max_i \max_j \left| b_{ij}^+ - b_{kij} \right|} \\ r_{kij}^- &= \frac{\min_k \min_i \min_j \left| b_{ij}^- - b_{kij} \right| + \rho \max_k \max_i \max_j \left| b_{ij}^- - b_{kij} \right|}{\left| b_{ij}^- - b_{kij} \right| + \rho \max_k \max_i \max_j \left| b_{ij}^- - b_{kij} \right|} \end{aligned} \tag{8}$$

Where ρ $(0<\rho<i)$ is the resolution, usually $\rho=0.5$. In the formula, $k=1,2,\cdots,q$, $i=1,2,\cdots,m$, $j=1,2,\cdots,$ n.

(5) Due to the difference in importance between time periods and indicators, the difference in the degree of correlation must take into account the corresponding weights. In the process of community development, the weight of each time period is equally important. However, the weight of each indicator is not the same. Therefore, when the weight of indicator w_{ij} is determined, each option S_k corresponds to the optimal solution B_k^+ and The relevance of the worst-case scenario B_k^- can be expressed as:

$$\xi_k^+ = \sum_{i=1}^m \sum_{j=1}^n w_{ij}r_{kij}^+$$

$$\xi_k^- = \sum_{i=1}^m \sum_{j=1}^n w_{ij}r_{kij}^-$$

(9)

(6) From the geometric meaning, we can see that the optimal scheme B_k^+ and the worst scheme B_k^- are equivalent to the optimal surface and the worst surface composed of the best and the worst discrete points in every direction, respectively. The correlation degree ξ_k^+ and ξ_k^- means the distance between the surface and the surface, which is equivalent to the degree of proximity between the surface of scheme S_k and the optimal surface and the worst surface. Similar to the two-dimensional grey relational theory, If ξ_k^+ is larger, the degree of association between scheme S_k and optimal scheme B_k^+ is greater, and if ξ_k^- is larger, the degree of association between scheme S_k and worst scheme B_k^- is greater.

(7) Calculate the relative closeness of each scheme to the optimal scheme B_k^+ and the worst scheme B_k^-.

$$d_k = \frac{\xi_k^+}{\xi_k^+ + \xi_k^-}$$

(10)

4   Experiment

Social networks are the extension of the human real world in the virtual world of the Internet. We regard people in the real world as one node in social networks. The edges in the social network represent the information interaction between people. Social networks Internetize the human relationships in real society. The experiment of the three-dimensional grey relational degree algorithm is mainly to verify the effect of this algorithm in specific data. The experiment of the three-dimensional grey correlation algorithm is mainly to verify the effectiveness of the algorithm in processing specific data. Therefore, we used Facebook's three consecutive months of data in the New Orleans area as a data set for verification. The first month contains 52,299 nodes, The second month contains 53,469 nodes, and the third month contains 56,621 nodes. The data set can reflect the status of interpersonal interactions in real society, which is of great practical significance. At the same time, this paper uses the Topsis algorithm based on the technique by Similarity Solution method and the algorithm of this paper for comparison experiments. The Topsis algorithm proposed by Yuxian Du [10] et al. is one of the main algorithms in the node importance evaluation algorithm of the static network at present, which is more suitable to the actual situation.

According to the data set, the indicators we selected are Harmonic Centrality, Closeness centrality, Betweenness centrality, Degree, and four positive correlation indicators. The indicator weights are the same, that is, $w_{ij} = 1/4\ (i = 1, 2, 3; j = 1, 2, 3)$. The time snapshot length is set to one month.

we use this algorithm to rank all the nodes that appear in three month data, and select the top 20 nodes with relatively strong importance to analyze.

After normalization of the three month data, we can get the optimal scheme B^+ and the worst scheme B^-.

$$B^+ = \begin{pmatrix} 0.93279022 & 0.98460221 & 0.98134834 & 1 \\ 0.93279022 & 0.99030428 & 0.98754257 & 0.98860872 \\ 1 & 1 & 1 & 0.98446663 \end{pmatrix}$$

$$B^- = \begin{pmatrix} 0 & 0 & 0 & 0 \\ 0 & 0.06004806 & 0.05435222 & 0 \\ 0 & 0.06078286 & 0.05505835 & 0 \end{pmatrix}$$

According to the optimal scheme and the worst scheme, the positive and negative correlation coefficient (ξ_k^+, ξ_k^-) can be obtained. Table 1 lists the top 20 data for all nodes with positive correlation coefficient ξ_k^+, and Table 2 shows the top 20 data for all nodes with negative correlation coefficient ξ_k^-.

The relative closeness of the positive and negative correlation coefficients can be calculated to obtain the importance of all nodes in the three months. Table 3 shows the top 20 data for relative closeness, that is, the top 20 nodes for importance of all nodes in three months.

Table 1. Positive correlation coefficient ξ_k^+

ID	Lable	Value	ID	Lable	Value
1	121	8.5082466	11	125	7.174769285
2	21	8.329090929	12	181	7.173830702
3	119	7.85669167	13	123	7.007364113
4	127	7.675562533	14	951	7.000579412
5	110	7.442575587	15	168	6.978164123
6	800	7.393631346	16	200	6.929343595
7	712	7.381283659	17	8	6.870547622
8	138	7.366097706	18	788	6.788186528
9	743	7.317150141	19	3223	6.756364019
10	136	7.308450645	20	114	6.725031927

Table 2. Negative correlation coefficient ξ_k^-

ID	Lable	Value	ID	Lable	Value
1	51686	10.56866649	11	51225	10.33046488
2	51145	10.51825472	12	51504	10.32965598
3	51399	10.51491144	13	49126	10.31059713
4	52191	10.50354694	14	52152	10.30270133
5	51685	10.4684224	15	49859	10.29509124
6	51402	10.43878879	16	51947	10.27666329
7	51144	10.37504107	17	52019	10.25961991
8	52182	10.3611426	18	51398	10.25501677
9	49860	10.35909102	19	51497	10.25497905
10	52153	10.33862687	20	52101	10.24869831

Table 3. Relative closeness

ID	Lable	Value	ID	Lable	Value
1	21	0.607103742	11	800	0.508237007
2	121	0.587296788	12	138	0.50688597
3	119	0.559388372	13	123	0.503515939
4	110	0.537567671	14	168	0.503038397
5	127	0.522089771	15	951	0.499222848
6	712	0.51670918	16	200	0.495242785
7	181	0.516412053	17	8	0.49129191
8	136	0.515963796	18	875	0.486825606
9	743	0.509404592	19	788	0.485346934
10	125	0.508516712	20	374	0.482998646

Using Topsis to rank the importance of nodes one month at a time, Table 4 shows the top 20 results for the importance of nodes in the first month. Table 5 shows the top 20 results for the importance of the node in the second month. Table 6 shows the top 20 results for node importance in the third month.

We know that nodes can be ranked high in the importance ranking of nodes for three months, and should be relatively high in each month's rankings or active in a month. For example, in the real-world network public opinion, the Internet giant will advocate an event in the beginning, and after a certain impact or under the supervision of the functional department, it will automatically disappear for a period of time, and then come out again to guide the direction of public opinion. For the above experimental data analysis, this article counts the rankings of the top 20 nodes in Table 3 in the Topsis algorithm for a single month. Since the nodes from the 10th to the 20th are not in the top 20 in the single month rankings. Therefore, Table 7 only lists the top 20 of the top 10 nodes in a single month's ranking. Those who are not in the top 20 are all represented by 0.

Table 4. The first month Topsis algorithm calculation results (Top 20)

ID	Lable	Value	ID	Lable	Value
1	121	0.999581	11	817	0.227734
2	119	0.604669	12	748	0.221957
3	706	0.584562	13	827	0.218931
4	110	0.427872	14	106	0.209009
5	762	0.374631	15	164	0.202712
6	181	0.338572	16	50	0.196152
7	21	0.302045	17	3635	0.177488
8	3014	0.272569	18	5203	0.170633
9	136	0.265761	19	1478	0.167942
10	115	0.230501	20	166	0.164965

Table 5. The second month Topsis algorithm calculation results (Top 20)

ID	Lable	Value	ID	Lable	Value
1	121	0.999537	11	823	0.224934
2	712	0.606649	12	833	0.217845
3	119	0.600863	13	755	0.216892
4	110	0.424421	14	106	0.205927
5	769	0.370813	15	166	0.19724
6	118	0.365611	16	51	0.193063
7	21	0.296337	17	3757	0.172717
8	3075	0.26613	18	5416	0.168251
9	817	0.262354	19	1488	0.165484
10	115	0.226825	20	168	0.163081

Table 6. The third month Topsis algorithm calculation results (Top 20)

ID	Lable	Value	ID	Lable	Value
1	127	0.99714	11	871	0.221052
2	743	0.634239	12	882	0.215252
3	125	0.595208	13	788	0.209392
4	114	0.420051	14	110	0.203077
5	124	0.386555	15	179	0.188315
6	807	0.361711	16	57	0.18638
7	21	0.325842	17	199	0.175844
8	865	0.264681	18	3955	0.167086
9	3223	0.25351	19	5749	0.164988
10	121	0.223107	20	181	0.15898

Table 7. Comparison of node importance rankings(Top 10)

Lable	Our algorithm (Ranking)	First month (Ranking)	Second months (Ranking)	Third months (Ranking)	In the top 20 (frequency)
21	1	7	7	7	3
121	2	1	1	10	3
119	3	2	3	0	2
110	4	4	4	14	3
127	5	0	0	1	1
712	6	0	2	0	1
181	7	6	0	0	1
136	8	9	0	0	1
743	9	0	0	2	1
125	10	0	0	3	1

From Table 7, we can see that the ranking of the top 10 nodes by our algorithm in the single month's ranking, the ranking is also relatively high. At least one month's ranking is in the top 20. After the top 10 nodes, the rankings in the single month are generally lower. This fully shows that the importance evaluation algorithm of nodes based on three-dimensional grey relational degree has rationality and objectivity.

Secondly, this algorithm does not simply superimpose the node's importance ranking in a single month. It is not possible to calculate the importance of a single monthly node by Topsis algorithm to calculate the node's ranking of node importance in the year. Table 8 is a discounted contrast diagram for node importance ranking, which is a clearer display of node ranking.

Table 8. Node importance ranking ranking comparison

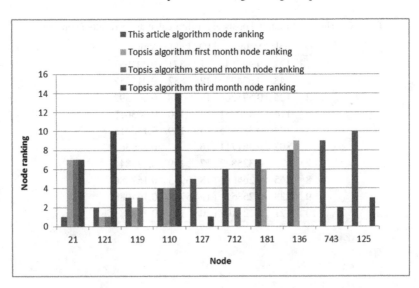

From Table 8, we can see that node 21 ranks first in the algorithm of this paper and ranks seventh in each month. Node 121 is ranked second in the algorithm of this paper, ranking 1, 1, 10 in each month respectively. Node 110 ranks fourth in the algorithm of this paper, ranking 4, 4 and 14 respectively in each month. If the ranking of the nodes from each month is considered, the importance of node 21 is weaker than that of node 121, which is similar to the degree of importance of node 110. If you look at the ranking of nodes from month to month, Node 21 is less important than node 121 and is similar to node 110 in importance. However, from the perspective of the overall three-month period, node 21 is more important. The importance evaluation of nodes based on three-dimensional grey relational degree is illustrated. The importance of nodes under each time segment can be well planned and the degree of importance of dynamic nodes in all current nodes can be analyzed accurately.

5 Conclusion

This paper creatively applies the three-dimensional grey relational degree algorithm to dynamic social networks and achieves the importance evaluation of dynamic nodes for the first time. The node's importance evaluation algorithm is extended from static multi-attribute evaluation to dynamic multi-attribute decision making. In this paper, we use four attribute indicators, Harmonic Centrality, Degree, Closeness centrality and Betweenness centrality as the four attribute indicators to verify Facebook's data of three consecutive months of in New Orleans and compare it with the Topsis algorithm in node importance evaluation under static conditions. Our algorithm of this paper is more reasonable and accurate in the evaluation of the importance of dynamic nodes, which has a strong practical significance and has obtained good effect.

References

1. Wang, L., Cheng, X.Q.: Dynamic community in online social networks. Chin. J. Comput. **38**(2), 219–237 (2015)
2. Wang, Z., Du, C., Fan, J., et al.: Ranking influential nodes in social networks based on node position and neighborhood. Neurocomputing **260**, 466–2477 (2017)
3. Shimbel, A.: Structural parameters of communication networks. Bull. Math. Biophys. **15**(4), 501–507 (1953)
4. Freeman, L.C.: Centrality in social networks conceptual clarification. Soc. Netw. **1**(3), 215–239 (1978)
5. Sabidussi, G.: The centrality index of a graph. Psychometrika **31**(4), 581–603 (1966)
6. Brandes, U.: A faster algorithm for betweenness centrality. J. Math. Sociol. **25**(2), 163–177 (2001). Social networks betweenness centrality algorithms
7. Newman, M.E.J.: A measure of betweenness centrality based on random walks. Soc. Netw. **27**(1), 39–54 (2005)
8. Bian, T., Hu, J., Deng, Y.: Identifying influential nodes in complex networks based on AHP. Physica A **479**(4), 422–436 (2017)
9. Yannick, R.: Closeness centrality extended to unconnected graphs: the harmonic centrality index ASNA, pp. 1–14 (2009)

10. Du, Y., Gao, C., Hu, Y., et al.: A new method of identifying influential nodes in complex networks based on TOPSIS[J]. Physica A **399**(4), 57–69 (2014)
11. Song, G., Li, Y., Chen, X., et al.: Influential node tracking on dynamic social network: an interchange greedy approach. IEEE Trans. Knowl. Data Eng. **29**(2), 359–372 (2017)
12. Wang, Z.X., Dang, Y.G., Shen, C.G.: Three-dimensional grey relational model and its application. Statist. Decis. **15**, 174–176 (2011)

Travel Attractions Recommendation with Travel Spatial-Temporal Knowledge Graphs

Weitao Zhang, Tianlong Gu, Wenping Sun, Yochum Phatpicha,
Liang Chang, and Chenzhong Bin[✉]

Guangxi Key Laboratory of Trusted Software,
Guilin University of Electronic Technology, Guilin 541004, China
binchenzhong@guet.edu.cn

Abstract. Selecting relevant travel attractions for a given user is a real and important problem from both a travellers's and a travel supplier's perspectives. Knowledge graphs have been used to conduct recommendations of music artists, movie and books. In this paper, we identify how knowledge graphs might be efficiently leveraged to recommend travel attractions. We improve two main drawbacks in existing systems where semantic information is exploited: fusion of multisource heterogeneous data and lack of spatial-temporal. Accordingly, we constructed a rich travel spatial-temporal knowledge graph from Baidupedia, Interactive Encyclopedia and Wikipedia. We proposed a methodnamed TSTKG4Rec to model the knowledge graph, it included two steps, attraction2Vec to model the feature attributes of attractions in the knowledge graph, and track2Vec to model the spatial-temporal semantics of the knowledge graph. Then, we obtained attraction vectors and user vectors fused with feature attributes of attractions and spatial-temporal semantics. At last we calculate the correlation between tourists and attractions with cosine similarity to give a list of recommendations. Our evaluation on real travel spatial-temporal knowledge graph showed that our approach improvement in terms of recall and MRR compared with the state-of-the-art approach.

Keywords: Spatial-temporal knowledge graph · Recommendation system
Network representation learning

1 Introduction

The rapid development of mobile Internet, artificial intelligence and other technologies have brought a lot of conveniences to people's life, but also brought the information overload problem. Search engine and recommendation system are representative technologies for solving problem of information overload. Traditional search engine can help people to find the useful information effectively and this approach satisfies the needs of most people but does not provide the personalized services. Compared to the traditional search engine, recommendation system can take individual requirements into account when solving the information overload problem. The purpose of recommendation system is to predict the user's preference for the new product based on

Q. Zhou et al. (Eds.): ICPCSEE 2018, CCIS 902, pp. 213–226, 2018.
https://doi.org/10.1007/978-981-13-2206-8_19

user's history preferences, personalized needs, and the characteristics of the product, so it can recommend the most suitable product for user, improve the user satisfaction, and also help the user to make a decision more quickly [1]. The value of the recommendation system is that it can provides the most suitable choice without asking user to explicitly provide the need for what they want. With the arrival of the era of big data, the performance of the traditional recommendation system is limited in data mining problems [2]. The emergence of the knowledge graph provides an effective way to design recommended system under the big data environment.

The core technology of using knowledge graph to solve recommendation problems is how to effectively establish the correlation between users and items from rich heterogeneous data. Therefore, how to accurately model the complex relationships between the nodes in network graph is a difficult technical problem in recommendation with knowledge graph area. In recent years, "network representation learning" has been paid more attention by academics and industry, and it's an effective way to represent the network. Network representation is a bridge to connect network data and the network application tasks. The network representation learning algorithm is responsible for learning the vector representation of each node in the network. And these node representations then can be used as the characteristics of the node for network application tasks, such as node classification, link prediction, and visualization.

This paper learns from the network representation learning methods Node2Vec [3] and Doc2Vec [4] to characterize the travel spatial-temporal knowledge graph. The spatial-temporal semantics in the knowledge graph can be learned with track2Vec and other attribute features can be learned based on the attraction2Vec. The feature vector of the tourists and attractions learned by the two methods are respectively summed and averaged to obtain the vector representation of the attractions and visitors which are fused with spatial-temporal semantics as well as the feature attributes. Finally, the relevance between the tourists and attractions is calculated by the method of cosine similarity, and then we can obtain the recommendation list according to the sorted relevance scores. The main contribution is as below:

(1) Integrate multi-source heterogeneous data to build a travel spatial-temporal knowledge graph.
(2) Propose a method attraction2Vec to model the feature attributes of attractions in the knowledge graph learned from the network representation learning method Node2Vec.
(3) Propose a method track2Vec to model the spatial-temporal semantics of the knowledge graph learned from the distributed representation method Doc2Vec.
(4) Combining the result of steps (2) and (3) to obtain the vectors of tourists and attractions which are fused with spatial-temporal semantics as well as the feature attributes.
(5) Propose a method to calculate the correlation between tourists and attractions with cosine similarity to give a list of recommendations.

2 Related Work

2.1 Spatial-Temporal Knowledge Graph

The concept of knowledge graph was proposed by Google in 2012. Knowledge graph are intended to describe the various entities or concepts that exist in the real world, and also describe the relationships between them. In the knowledge graph, each entity or concept is identified by a globally unique ID, each attribute-value pair describes the intrinsic attributes of the entity, and the relationship is used to connect two entities and to describe the relationship between them [5]. Spatial-temporal knowledge graph refers to a knowledge graph containing both time and space dimensions. In our work, knowledge graph is represented by a knowledge base formed by triplets and the interlinked between the triplets. Take the tourism spatial-temporal knowledge graph as an example. Nodes contain information of the attractions, tourists and the attribute value. Edges indicate attributes or relationships. These characteristics can describe the characteristics of entities and the spatial-temporal semantics in the knowledge graph.

2.2 Knowledge Graph Based Recommendation Algorithm

After the knowledge graph was proposed, relevant scholars applied the knowledge graph to recommendation systems in various fields and then achieved many great results. Oramas [6] applied the knowledge graph to recommendation for sound and music. Lu [7] used the data in DBpedia, Geonames, and Wikidata to build a world tourist attractions knowledge graph, to achieve the attractions recommendation. Noia et al. [8] completed the books recommendation with knowledge graph.

2.3 Network Representation Learning

The core of using knowledge graph to solve the recommendation problem is how to accurately extract the features of knowledge graph. Recently, network representation learning [9] has gradually become a popular research direction in machine learning. Network representation learning attempts to learn a low-dimensional representation vector for each node in the network while maintaining its original structural information. This method is proved to be an effective way to learning the graph feature. The word vector learning method represented by word2vec [10] is the most widely used. Doc2Vec adds segment vectors based on word2Vec, which can be used to extract the position features between words. Node2Vec joins random walk on the basis of word2Vec, it is more suitable for learning of network diagram. Based on Node2Vec, Entity2Vec [11] extracts attribute graph from the knowledge graph, so that the edge information in the knowledge graph can be extracted.

3 Construction of Travel Spatial-Temporal Knowledge Graph

3.1 Data Collection

This study takes Guilin as an example. The information for attractions, comments and tourist history trajectories are crawled from the Ctrip.com and the Baike websites(e.g. Baidupedia, Interactive Encyclopedia Wikipedia, etc.). A total of 395 attractions, 12,398 tourists, 28,477 attraction scores, 23,000 history trajectories were collected. Besides, some detail information of the attractions were also extracted from the Baike websites. Take the Lijiang River as an example, the detail information includes the attraction address, attraction level, average rating, attraction type, suitable play time, ticket price, tourist ID, and the time point for tourists visited.

3.2 Travel Spatial-Temporal Knowledge Graph Construction

Pre-processing of the collected information: delete the duplicate tourists' rating record; set the recently rating of the tourist as the tourist's rating to the attraction; select the attraction level, geographical location, average rating, attraction type, and the suitable playing time as the characteristics of the attraction; the relationship between the tourist and the attraction where the rating score is greater than 3 points is regarded as a favorite relationship; visitors' check-in of the attractions in a specific time is regarded as a time attribute; use the database tool to do the entity alignment(e.g., Guilin City and Guilin represent the same entity).

Construct the schema layer of the tourism spatial-temporal knowledge graph through the pre-processed information, including the attributes of attractions, the tourists entity and the attractions entity, as well as the relationship between tourists and attractions, which is shown in Fig. 1.

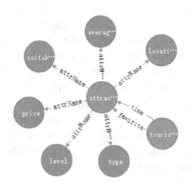

Fig. 1. Example of the pattern diagram.

Using the pre-processed Data and the designed schema layer, the tourism spatial-temporal knowledge graph is constructed. The travel spatial-temporal knowledge graph contains 383 attractions, 3,940 tourists, 19,724 rating records, 19,720 tourists history trajectories, and six attributes of various attractions, the total number of triple in the knowledge graph is 42,714. Figure 2 shows an example of a network structure stored in Neo4j which consists of some attractions, tourists and their attributes.

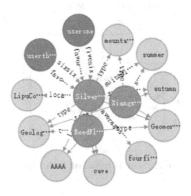

Fig. 2. Travel knowledge graph example.

4   Framework of System

The basic idea of the attraction recommendation based on travel spatial-temporal knowledge graph(TSTKG4Rec):

Use the method attraction2Vec to model the tourist attractions and the user characteristics in the travel spatial-temporal knowledge graph; then use the track2Vec to model the spatial-temporal semantics in the travel spatial-temporal knowledge graph, as a result, obtain the feature vectors of a attraction or tourists node that integrate both the semantics of attribute and the semantics of spatial-temporal; then we use the cosine similarity to calculate the correlation scores between the tourists and the attractions. At last, we normalize the correlation scores to generate a recommendations list. Figure 3 shows the recommend flow chart.

4.1   Fusion Attribute Semantic Feature Vectors

The constructed travel spatial-temporal knowledge graph is divided and stored as eight independent sub-graphs(location, price, type, suitable playing time, time-stamps, rate, favorite, level) through the SPARQL query and then use the attraction2Vec to model seven of them (non-time-stamps).

Take an attribute sub-graph as an example, for each node in the graph $u \in V$, define $N_s(u) \subset V$ as a set of network neighborhoods of nodes which are generated by random walks. Extend the skip-gram architecture to the network to optimize the following objective function, which maximizes the log-probability of the network neighbor $N_s(u)$. For a node u conditioned on its feature representation, given by f:

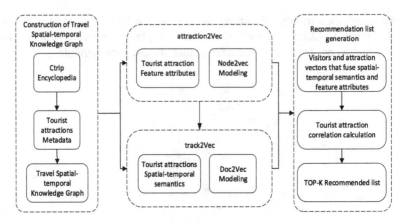

Fig. 3. Recommend flowchart of attractions based on travel spatial-temporal knowledge graph.

$$\max_f \sum_{u \in V} \log \Pr(N_s(u)|f(u)) \tag{1}$$

To make the optimization process easier, make two assumptions:

Conditional independence: We factorize the likelihood by assuming that likelihood of observing a neighborhood node is independent of observing any other neighborhood node given the feature representation of the source.

$$\Pr(N_s(u)|f(u)) = \prod_{n_i \in N_s(u)} \Pr(n_i|f(u)) \tag{2}$$

Symmetry of feature space: A source node and neighborhood node have a symmetric effect over each other in feature space. Accordingly, we model the conditional likelihood of every source-neighborhood node pair as a softmax unit parametrized by a dot product of their feature.

$$\Pr(n_i|f(u)) = \frac{exp(f(n_i).f(u))}{\sum_{v \in V} exp(f(v).f(u))} \tag{3}$$

With the above assumptions, the overall objective function can be summarized as:

$$\max_f \sum_{u \in V} [-log Z_u + \sum_{n_i \in N_s(u)} f(n_i).f(u)] \tag{4}$$

The per-node partition function, $Z_u = \sum_{v \in V} exp(f(u).f(v))$, is expensive to compute for large networks and we approximate it with negative sampling.

4.2 Fusion of Spatial-Temporal Semantic Feature Vectors

The method track2Vec, which extracts the spatial-temporal semantics features in travel spatial-temporal knowledge graph, not only considers the context of the node, but also takes the time sequence information between nodes into account. Each user's historical trajectory in the knowledge graph is sorted according to the time axis, then we can obtain a user-identified attraction trajectory sequence, and attractions in the attraction trajectory are represent as a vector by attraction2Vec. Use the track2Vec model as below in Fig. 4 to extract the features of obtained trajectory sequence to get the user's long-term preference.

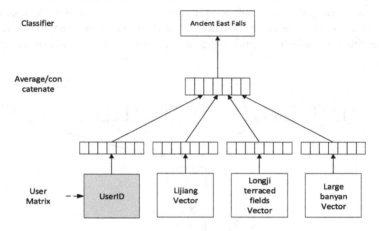

Fig. 4. A framework for track2Vec. In this model, the concatenation or average of this vector with a context of three words is used to predict the fourth word. The user vector represents the missing information from the current context and can act as a memory of the topic of the user.

Model the Contexts

In the travel spatial-temporal knowledge graph, the history of the visitor for a given visitor can be expressed as $T = <(u, t_1, v_1), (u, t_2, v_2), \ldots\ldots, (u, t_n, v_n)>$. Where u represents the tourist, t_n represents the time-stamps, and v_n represents the attraction. To express a tourist's historical preference information, based on the conditional probability formula can be drawn as the following formula:

$$P\{(u, t_1, v_1), (u, t_2, v_2), \ldots\ldots, (u, t_n, v_n)\} = P\{(u, t_1, v_1)\} \times P\{(u, t_2, v_2)|(u, t_1, v_1)\} \times \ldots$$
$$\times P\{(u, t_n, v_n)|(u, t_1, v_1), (u, t_2, v_2), \ldots\ldots, (u, t_{n-1}, v_{n-1})\}$$

$$(5)$$

By instantiating the semantic feature vector, the objective function of the final model is as follows:

$$\sum_{u \in E} \sum_{t \in T^u} \sum_{i=1}^{N_t} \log P(v_i | x^{u_i}) \tag{6}$$

Where E represents the collection of tourist nodes in the knowledge graph, T^u representing the history of the visitor, N_t indicating the length of the trajectory.

Parameter learning

Use the objective function to learn and optimize the parameters. In the sequence model, the parameters are trained in order to better use the context feature vector to learn the target vector. The parameters are generally learned by maximizing the logarithmic probability in Eq. (6). We use hierarchical softmax instead of softmax to optimize the objective function, which simplifies the problem of large computation complexity (Fig. 5).

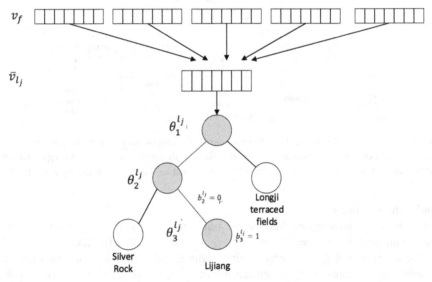

Fig. 5. An illustrative example for the binary tree in the hierarchical softmax ((l_j = "Lijiang" and L(l_j) = 3)). Every node is represented in a circle associated with an embedding vector θ. v_f denotes the embedding vector of the f-th context feature for location l_j and these vectors are averaged to v_{l_j}. The nodes on the path from the root to l_j are marked in orange and linked with red lines.

The sequence of historical trajectories for a visitor is:

$$T = \langle (u, t_1, v_1), (u, t_2, v_2), \ldots\ldots, (u, t_n, v_n) \rangle$$

The objective function is to maximize the probability of each attraction location based on given contextual information:

$$\sum_{j=1}^{N_t} \log P\left(l_j | x^{l_j}\right) \tag{7}$$

N_t shows the length of the historical trajectory, l_j represents the location of the target site, x^{l_i} represent the feature vector containing all contextual information.

Thus we can use the hierarchical softmax to defined the objective function as below:

$$\Pr\left(l_j | x^{l_j}\right) = \prod_{n=2}^{L(l_j)} \left(\left[\sigma\left(\bar{v}_{l_j}^T \theta_{n-1}^l\right)\right]^{1-b_n^{l_j}} * \left[1 - \sigma\left(\bar{v}_{l_j}^T \theta_{n-1}^l\right)\right]^{b_n^{l_j}} \right) \tag{8}$$

where

$$\sigma(z) = \frac{1}{1 + \exp(-z)} \tag{9}$$

$\theta_n^{l_j}$ represent the vector representation of the n-th non-leaf node in the path Ll_j. All parameters are trained using the gradient method. During the training process, all sequences are traversed. In this process, the target node l_j and its context node feature vectors are updated accordingly. After the calculation of hierarchical softmax, the gradient of errors can be obtained by back propagation. We use the gradients to update the parameters in the model and facilitate the update the θ in each step. We use the formula 12 to calculate the $\theta_{n-1}^{l_j}$ gradient

$$\frac{\partial_{k(l_j, n)}}{\partial \theta_{n-1}^{l_j}} = \left[1 - b_j^{l_j} - \sigma\left(\bar{v}_{l_j}^T \theta_{n-1}^{l_j}\right)\right] \bar{v}_{l_j} \tag{10}$$

Therefore, $\theta_{n-1}^{l_j}$ can be updated by the following formula:

$$\theta_{n-1}^{l_j} \leftarrow \theta_{n-1}^{l_j} + \mu \left[1 - b_j^{v_j} - \sigma\left(\bar{v}_{l_j}^T \theta_{n-1}^{v_j}\right)\right] \bar{v}_{l_j} \tag{11}$$

Where μ represents the learning rate. In order to facilitate the calculation of the feature vector to context nodes, the gradient of $\bar{v}_{l_j}$ is expressed as follows:

$$\frac{\partial_{k(l,n)}}{\partial \bar{v}_{l_j}} = \left[1 - b_j^{l_j} - \sigma\left(\bar{v}_{l_j}^T \theta_{n-1}^{l_j}\right)\right] \theta_{n-1}^{l_j} \tag{12}$$

Through the above calculations, the feature vector of context nodes the node l_j can be updated by the following formula:

$$v_f \leftarrow v_f + \mu \sum_{n=2}^{L(l_j)} \frac{\partial_k(l_j, n)}{\partial \bar{v}_{l_j}} \tag{13}$$

Recommended generation

In Sects. 4.2 and 4.1, the travel spatial-temporal knowledge graph of tourists and attractions have been learned into a same vector space based on the attributes of features and spatial-temporal semantics. We define v(attract) denoting the feature vector of the attraction and v(user) denoting the feature vector of the user. After we get the feature vector of the attractions and visitors in the same vector space, then can define the correlation score between the visitor and the attraction as below:

$$Rel(attract, user) = sim(v(attract), v(user)) \tag{14}$$

Where sim is the cosine similarity.

5   Experimental

5.1   Experimental DataSet

Experimental environment: Operating system Ubuntu 16.04, 64-bit, processor Intel Core i7-6700, memory size 8G, programming platform Pycharm, Python 2.7 version. Compared with the classic GRU4Rec [12] recommendation method on Recall@K and MRR@K. The data used in the comparison experiments and the real spatial-temporal knowledge graph for 42,714 triples.

5.2   Evaluation Criterion

Take 20% of the data in the database as a test set, and select the top-K attractions with high relevance scores as the recommendation list. Definition hit_r@k for a single test user, the recall rate for this user:

$$hit\_r@k = \frac{n}{u_{N_t}} \tag{15}$$

Where the n is the number of attractions in the recommended top-K that the tourist favorite, and u_{N_t} is the number of tourist favorite attractions.

The recall for all test visitors Recall@k indicates the probability of success for all test sets:

$$Recall@k = \frac{\sum hit\_@k}{\#all\_test\_user} \tag{16}$$

Randomly select 20% of the user's trajectory data as a test set. Defining hit_a@k for a single test visitor, if the user actually visited the sights included in the top-K list, set hit_a@k = 1 otherwise hit_a@k = 0. The MRR is:

$$MRR@k = \frac{\sum r(hit\_a@k)^{-1}}{\#all\_test\_user} \tag{17}$$

Where r() indicates the location index of the tourist's visited attractions in the top-K list.

5.3 Performance Comparison

The data used by the traditional recommendation algorithm is the user's history rating or the characteristics of the project, and there is no way to efficiently handle multi-source heterogeneous data. Travel spatial-temporal knowledge graph can be well integrated with multi-source heterogeneous data on the internet and also maintain spatial-temporal semantic information.

Comparing this model with the experiment of GRU4Rec, for each tourist u, based on the tourist's rating information and tourist history trajectories information in the spatial-temporal knowledge graph, vectors representation of the tourist is mined, and vectors representation of attractions can be mined according to the characteristics of the attractions. For any one visitor, a top-k recommendation list is generated by calculating the correlation between the visitor and the attraction vector. Figure 6 shows the model's recall rate and MRR.

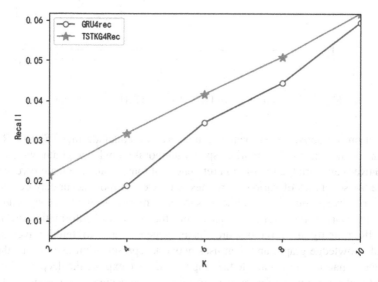

Fig. 6. Comparison of TSTKG4Rec and GRU4rec recall rates.

Figure 6 shows the recall results of two models when k falls in the range [2, 10], and Fig. 7 shows the results of MRR of two models when k falls in the range [2, 10]. By observing the two models in the recall rate and the MRR graph, we can conclude that with the increases of the length to the recommended list, both models show an increasing trend in terms of recall and MRR. Because the increase of recommended list means that the list of models recommended to tourists will increase, and the probability that the recommended attraction is the favorite attraction of tourists will increase at the same time, so the recall rate and MRR show an increasing trend. In the graphs of recall and MRR, although the curves of both models show an increasing trend, but our model TSTKG4Rec is more higher than GRU4rec, indicating that our model recommendation is better, which fully demonstrates that our model TSTKG4Rec can fuse more semantics and are more accurate to extract the preferences for tourists.

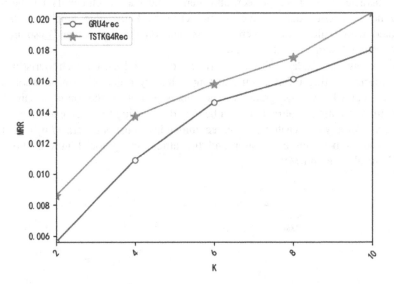

Fig. 7. Comparison of TSTKG4Rec and GRU4rec MRR rates.

Based on the above experimental results, we can conclude that TSTKG4Rec has fully learned the semantics of tourism spatial-temporal knowledge graph, which means TSTKG4Rec can learn the long-term preferences of tourists and the feature vectors that integrate the semantics of various attributes, then we can more accurately describe the correlation between tourists and attractions to make more reasonable recommendations. In real life, tourism information is sparse and the sources of information are messy, TSTKG4Rec firstly uses multi-source heterogeneous data to build travel spatial-temporal knowledge graph, and then uses network representation learning methods to extract the semantics in the knowledge graph, At last we explore the deeper preferences of tourist and the features of attractions to improve the accuracy of recommendations.

6   Conclusion and Future Work

Based on the travel spatial-temporal knowledge graph attractions recommendation, use the network representation learning methods to efficiently mine knowledge graph features information, so as to model the correlation between tourists and attractions more fully. We firstly collect the information on tourist attractions and tourist ratings from Guilin and Wikipedia websites to build a spatial-temporal knowledge graph of tourism, then use the network representation method to carry out feature mining on the spatial-temporal knowledge graph of tourism. At last, we use the cosine similarity to calculate the relevance of tourists and attractions to generate recommendation list. We design a comparative experiment program and use the recommendation system evaluation method to verify the effectiveness on the true travel spatial-temporal knowledge graph. The experimental results show that compared to the GRU4rec, TSTKG4Rec is more comprehensive and accurate for the feature modeling of tourists and attractions in the travel spatial-temporal knowledge graph, so as to make more accurate recommendations. In the future, we will consider to use more tourist information (such as photos and commentary information) to explore more fine-grained tourist preferences, and take more accurate model of attractions to further improve the accuracy of the recommendation.

Acknowledgment. This work was partially supported by the National Natural Science Foundation of China under Grant U1501252 and Grant 61572146, and partially supported by the Natural Science Foundation of Guangxi Province under Grant 2016GXNSFDA380006 and Guangxi Innovation Driven Development Project (Science and Technology Major Project) AA17202024.

References

1. Liang, C., Yuting, C., Wenping, S., et al.: Review of tourism recommendation system. Comput. Sci. **44**(10), 1–6 (2017)
2. Xiangwu, M., Weiyu, J., Yujie, Z.: Recommendation system in big data environment. J. Beijing Univ. Posts Telecommun. **38**(2), 1–15 (2015)
3. Grover, A., Leskovec, J.: node2vec: Scalable feature learning for networks. **2016**, 855–864 (2016)
4. Zhou, N., Zhao, W.X., Zhang, X., et al.: A general multi-context embedding model for mining human trajectory data. IEEE Trans. Knowl. Data Eng. **28**(8), 1945–1958 (2016)
5. Liu, Q., Li, Y., Duan, H., et al.: Knowledge graph construction techniques. J. Comput. Res. Dev. **53**(3), 582–600 (2016)
6. Oramas, S., Ostuni, V.C., Noia, T.D., et al.: Sound and music recommendation with knowledge graphs. ACM Trans. Intell. Syst. Technol. **8**(2), 21 (2016)
7. Lu, C., Laublet, P., Stankovic, M.: Travel attractions recommendation with knowledge graphs. In: Blomqvist, E., Ciancarini, P., Poggi, F., Vitali, F. (eds.) EKAW 2016. LNCS (LNAI), vol. 10024, pp. 416–431. Springer, Cham (2016). https://doi.org/10.1007/978-3-319-49004-5_27

8. Noia, T.D., Mirizzi, R., Ostuni. V.C., et al.: Linked open data to support content-based recommender systems. In: International Conference on Semantic Systems, pp. 1–8. ACM (2012)

9. Tu, C., Cheng, Y., Liu, Z., et al.: An overview of network presentation learning. Chin. Sci. Inf. Sci. (8) (2017)

10. Mikolov, T., Chen, K., Corrado, G., et al.: Efficient estimation of word representations in vector space. Comput. Sci. (2013)

11. Palumbo, E., Rizzo, G., Troncy, R.: entity2rec: Learning user-item relatedness from knowledge graphs for top-N item recommendation. In: Eleventh ACM Conference on Recommender Systems, pp. 32–36. ACM (2017)

12. Hidasi, B., Karatzoglou, A., Baltrunas. L., et al.: Session-based recommendations with recurrent neural networks. Comput. Sci. (2015)

Hierarchical RNN for Few-Shot Information Extraction Learning

Shengpeng Liu[(✉)], Ying Li, and Binbin Fan

Zhejiang University, Hangzhou 310027, China
lspvic@qq.com

Abstract. Web information extraction (IE) is the process of retrieving exact text fragments of record attributes from HTML web pages. Most of the existing approaches need to do a large amount of work on feature engineering, selecting or computing the underlying content, layout and contextual features from web pages. Another disadvantage is that a great number of human's labor on annotating training example is required. Methods via solving wrapper adaption drastically reduce the annotating work but still need to label many pages on the seed website. In this work, we present a hierarchical attention recurrent neural network, which is an end-to-end model and do not require traditional, domain-specific feature engineering. The network can be also trained with only a few pages in a site, i.e. Few-Shot learning. As the model automatically and deeply learns the semantics of text fragments in pages, we adapt the network to extract records from the previously unseen websites. Experiments on a publicly available dataset demonstrate that our networks for both wrapper induction and adaption showed competitive results compared against state-of-the-art approaches.

Keywords: Information extraction · Attention RNN
Few-Shot learning

1 Introduction

The World Wide Web has been growing dramatically and the web pages have reached several billion scales over the internet. The semi-structured HTML document is the most common format among web pages. The HTML pages displayed in the browser for humans contain explosive web information but provide no structured data for machines to store, index and read. Information Extraction (IE) is the process of retrieving exact text fragments of entity attributes from the semi-structured web data. This is crucial for handling continuously growing data published on the Web, especially in the era of big data. The IE tasks detect and retrieve structured records (or entity) that contain a collection of attributes from web pages, e.g., retrieving a *book*'s {*title, author, isbn_13, publisher* and *publish_date*} as shown in Fig. 1.

As many websites use dynamic web page techniques like JSP, PHP or ASP, the pages generated usually from a layout template and dynamic values from

© Springer Nature Singapore Pte Ltd. 2018
Q. Zhou et al. (Eds.): ICPCSEE 2018, CCIS 902, pp. 227–239, 2018.
https://doi.org/10.1007/978-981-13-2206-8_20

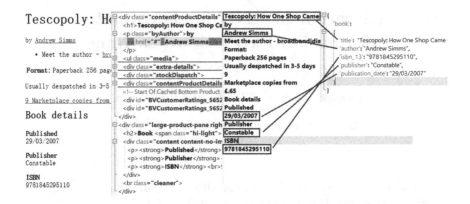

Fig. 1. A sample *book* record extraction. The images from left to right are: visual display, HTML source, text fragments sequence and extracted record. In our approach, we extract the record directly from text fragments sequence.

databases. Early IE methods are finding extraction rules and templates from manual labeled pages, which are referred as *Wrapper Induction* [14,20,25]. These methods are not general enough that one wrapper is only applied to one website and when the site updates its template, the wrapper will break. Great human efforts must be made for labeling new sites and maintaining wrappers. To reduce the labor of manually labeling, [12,22] both provided methods that learned wrappers can be applied to *previously unseen* Web sites, known as the problem of *Wrapper Adaption*. In that case, they only require labeled pages from one site for training to automatically deal with other sites in the same domain. Another category of approaches are totally completely unsupervised methods [21,23], they need no labeled pages in training phrase but require to clean and extract results manually. More extensive survey of IE approaches can be found in Sect. 2.

Although a great deal of research has been done in the area of IE, these methods still have room for improvement. On the one hand, manual work in IE process is still too heavy. Consider a web crawler, one of the most common tasks to extract information from a web site. In reality, a programmer is likely to write some regular expressions or XPath selectors with only analyzing one or a few pages to build it. Yet the state-of-the-art IE methods need a large amount of labeled example. [12,22] can avoid the labeling labor of previously unseen site, however, labeling many pages in the one source site is still required.

On the other hand, feature engineering is still a challenging task. Existing approaches highly rely on feature selection but the selected features are either domain specific or costly to compute. Roughly, these features can be divided into three categories:

– *Content Features* are the text values in page fragment, they represent the semantics of fragment in a specific domain and they are the most useful characters in wrapper adaption. However, text values in Web pages are not fully utilized when selecting content features – only some statistics on text e.g. word count, character count and predefined high-frequency token, are selected as content features because the word counts in text fragment are not fixed and traditional one-hot representation is too expensive to compute.

- *Layout Features* describe the visual properties of text fragment showed in a web browser. These features include text fragment position, size, font size, font color and so on. A disadvantage of the layout features is time-consuming and resource-consuming since web page and all its auxiliary resources (e.g., style files, script files, and image files) must be fully loaded and rendered.
- *Context Features* use surrounding text fragments as features, e.g. a indicating "*Title:*" fragment is always lying before the *title* value of a "*book*" record; and the *title* and *author* values of *book* are close in a web page. These content features need additional work to select and vary in different domains and sites.

Recently, deep learning approaches have obtained very high performance across many different Natural Language Processing (NLP) tasks (e.g. sentiment analysis, natural language translation and dialog systems). These models can often be trained with a single end-to-end model and do not require traditional, task-specific feature engineering. Recurrent Neural Network (RNN) is the most important component for these NLP tasks. It can handle the orders and dependencies in a sequence such as a sentence of words sequence.

In this work, we proposed hierarchical Attention RNNs for extracting text fragments from web pages. The model can automatically learn complicated underlying features including content features and context features. Word embedding layer transforms words to vectors. Words RNN with encoder layer encodes words sequence in text fragment to vector representation, i.e., *Text Fragment Representation*. The text fragment representations have deep semantic information and thus can be used to *Wrapper Adaption* problem. A second RNN layer to the fragment representation sequence in a web page is added to collect the contextual information. The two RNN layers are also extended with attention mechanism, which considers weights for each word in a fragment or each segment in a page, and then the words and fragments which are noisy or irrelevant have lower importance. Finally, a softmax layer classifies these contextual fragments with their attribute labels. A detailed description of each layer is presented in Sect. 3.

In training process, our network can be trained with only a few labeled pages, i.e., similar to the *One-Shot Learning* [8]. As the same with the process of constructing a web page crawler, an experienced programmer can easily extract rules from only a page. Few-Shot learning can dramatically reduce the annotation labor in automatically wrapper induction. There are two reasons why "Few-Shot Learning" is possible in our network: one is the high similarity in record data and DOM structure of pages generated from one template; the other is our network deeply exploits the semantics of text fragments and contextual dependencies. Experiments showed that even with the minimal labeling labor, our model would still get a high performance on different kinds of sites and domains.

Although the few-shot learning hierarchical RNN only works with web pages within one site, it is easy to adapt the model to previously unseen sites in the wrapper adaption problem. It is believed that the result of our model is credible in a site, and we use the whole predicted results as training data in wrapper adaption. In wrapper adaption process, we freeze the layers of trained text fragment representation layers and drop the fragments sequence RNN layers with contextual layer and train the rest layers to predict new site. The wrapper adaption process is presented in Sect. 3.3. The experiments showed appreciable performances.

In brief, our hierarchical attention RNN model have three characteristics for information extraction:

- a deep network eliminating the hard feature engineering
- an few-shot learning reducing the human's annotation
- easily adapting to previously unseen sites

The rest of this paper is organized as follows. We first refer to related work in Sect. 2. The proposed model details are introduced in Sect. 3. In Sect. 3.3, we adapt the model to new site. Experimental results are reported in Sect. 4. Section 5 concludes the paper.

2   Related Work

2.1   Information Extraction

Web information extraction has been studied extensively. Early research learnt a template that contains XPath rules to extract fragments in HTML tags, such as [14,20,25]. These approaches can't be used in large-scale extractions and long-term environment, because they're error-prone from noisy contents (e.g. advertise block), site template updates and various websites in the same domain. Supervised methods were proposed to learn similar structures in one site pages, but required manual annotation on quite a lot of example pages. [12,22] tried to learn the semantics of record attribute and then adapting the learned rules to previously unseen sites of the same domain. The wrapper adaption methods in these work drastically reduced the annotation labor to label example pages from one seed site. Our approach goes further to reduce the annotation labor to a few pages, i.e., the "few-shot learning".

There are also many works on completely unsupervised methods. These approaches usually cluster web pages to similar structures and then cluster text fragments to get record attributes. In training process, they do not require any labeled training data, however, an attributing cleaning and selection processes are left to the user. Another category methods exploit existing knowledge base to automatically generating training data and then adopt a supervised approach to learn wrappers. Recent works on this category are in [4,10].

We also noticed that there is a very recent work [11] which introduces convolutional neural networks (CNN) over web page visual image to extracting records and can be trained with only one labeled page. In their work, the authors send combined visual and textual data from web rendering engine to their deep CNN. By contrast, we only used the textual and their ordering information to train a deep RNN. The performance of our light-weighted approach is competitive with that reported in their work.

2.2 Related NLP Works

Natural Language Processing (NLP) has made breakthrough progress due to the deep learning. Many approaches in NLP can be used in web data extraction. Words embedding trains words vector representations with the neural network. [2] uses the concatenation of several previous word vectors to predict the next word. After the model is trained, the word vectors are mapped into a vector space that represents word semantically. [16] proposed two novel model architectures for computing continuous vector representations of words from very large data sets, i.e., the CBOW and Skip-Gram models.

Recurrent neural network [9] has been successfully applied to process sequence of words and sentences. RNN is able to encode an arbitrary sequence to a distributed representation. [15] proposed *Paragraph Vector*, an unsupervised algorithm that learns fixed-length feature representation from variable-length pieces of texts, such as sentences, paragraphs, and document. [24] presented sentence and document representations from hierarchical RNN, in addition to attention mechanism. Inspired by that, we learned a *Text Fragment Representation* for the variable-length of fragments in web pages.

Recently, Attention RNN has been the hot point of RNNs. Attentions are paid to the RNN process that each input in the past or future has different weights on constructing the whole sequence process. [24] adds two levels of attention mechanisms to words and sentence-level RNNs to classify documents. [7] integrate attention mechanism to build query-specific representations of tokens in the document for accurate answer selection. [17] pointed out that attention RNN can recognize and remove the noisy or irrelevant parts in a sequence.

3 Hierarchical Attention RNN for Few-Shot Learning

The overall architecture of the hierarchical attention RNN is shown in Fig. 2. It consists of several parts: a word embedding, a word sequence RNN encoder, a text fragment RNN layer and a softmax classifier layer, Both RNN layers are equipped with attention mechanism. In this section, we first prepare knowledge on Recurrent Neural Network (RNN), its variants we used in our model and the attention mechanism. And then we describe the details of different components in the following sections.

3.1 Recurrent Neural Network

Recurrent Neural Network (RNN) has been successfully applied to various tasks including natural language processing. The biggest advantage of the recurrent

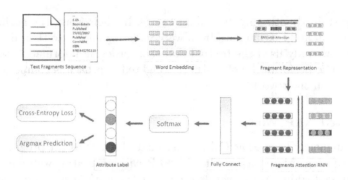

Fig. 2. Hierarchical attention networks.

neural networks lies in the capability of preserving information over time by the recurrent mechanism. Recurrent Neural Networks learn a representation for each time step by taking into account both the observation at current time step and the representation in the previous one. Commonly, the RNN transition function is as follows:

$$h_t = tanh(W x_t + U h_{t-1} + b)$$

$$y_t = tanh(W' h_t + b')$$

h_t and h_{t-1} are hidden states, y_t is the output. The hidden state of time t contains the information of current input x_t and last hidden state h_{t-1}, as h_{t-1} contains information of h_{t-2}, thus a recurrent transition is performed. Noticing that the output representation contains not only the input information but also the information from previous inputs with their order. e.g., y_4 contains the information from $\{x_4, x_3, x_2, x_1\}$, a arbitrary length of sequence can be represented as the final output, e.g. the sequence $\{x_1, x_2, x_3, x_4\}$ can be represented by y_4 or the mean of the output $(y_1 + y_2 + y_3 + y_4)/4$.

Long Short Term Memory (LSTM). Theoretically, RNN is able to process arbitrary length of a sequence. However, when dealing with long sequence, the plain-RNN faces the gradients vanishing and explosion problem. Long Short-Term Memory (LSTM) [5,13] and Gated Recurrent Unit (GRU) [6] are two of the most applied RNN variants to address the gradient vanishing problem. They both embed with gates to balance the information flow from the previous time step and current time step dynamically.

Bidirectional RNN. Basic RNNs only consider the past information while Bidirectional RNN (BRNN) [3,19] consider both the future and the past information to determine the output at any point in the sequence.

y_2 contains both the information from $\{x2, x1\}$ and $\{x2, x3, x4\}$.

Attention Mechanism. RNN incorporates the information from the contextual inputs. But not all input items contribute equally to the representation of the sequence. Attention mechanism has been studied to extract the important items to the semantic of the sequence and then aggregate those informative items to form a sequence vector. The mechanism we adopted is as in [24]. A softmax function uniforms the weights and a weighted sum of hidden representations s_i is gotten.

3.2   Hierarchical Model

Input Layer. We directly use the sequence of text fragments in a page as the input. A preprocessing program eliminates all the DOM-tree structure and HTML tags but keeps the order of the fragments in web pages. For a website in a domain (e.g. book), the web document collections are donated as d, page donated as p is preprocessed to a sequence of text fragments f_i, and each text fragment f is a sequence of word items (including punctuation marks and symbols) w_t split by spaces. A notation of the relation between the document collection, text fragments and words are as below, the capital D, P, T donate corresponding counts of the document collection, fragments in pages and words in fragments:

$$\{p_j \in d \, j \in [0, D]\}$$

$$\{f_i^{(j)} \in p^{(j)}, i \in [0, P]\}$$

$$\{w_i t \in f_i, t \in [0, T]\}$$

For each training or prediction process, we input a sequence of fragments in a page (Note that text fragment counts in a page P and the words counts in a fragment T is not fixed.):

$$[[w_0 0, w_0 1, ... w_0 T_0],$$

$$[w_1 0, w_1 1, ... w_1 T_1],$$

$$...,$$

$$[w_P 0, w_P 1, ... w_{PT_P}]]$$

Word Embedding Layer. We first create a vocabulary for all words in text fragments of all pages. Every symbol and number with the same number of digits are also considered as a word in the vocabulary, for some attributes have obvious symbol feature in the text. For every word in the vocabulary, a length n vector representation is assigned with random numbers initially. Thus for a size m of vocabulary, we get an embedding matrix W_E with shape $m \times n$. Then the input text fragment list can be translated to a list of sequences of vectors. For the i-th fragment with word sequence $w_{it}, t \in T_i$, where T_i is the numbers of words of the fragment, the translated vector sequence is:

$$x_{it} = W_{E w_{it}}$$

Text Fragment RNN Representation. When translated with embedding matrix, a text fragment becomes $x_{it}, t \in [0, T_i]$, where i is the index of fragment in fragment list, note that the T_i is not fixed for different fragment. We use a bidirectional LSTM to get annotations of words.

$$\overrightarrow{h}_{it} = \overrightarrow{LSTM}(x_{it})$$

$$\overleftarrow{h}_{it} = \overleftarrow{LSTM}(x_{it})$$

Every hidden state $[\overrightarrow{h}_{it}, \overleftarrow{h}_{it}]$ of the word at position t contains the context information. We put the attention to the hidden states of all words by

$$f_i = [Attention(\overrightarrow{h}_{it}), Attention(\overleftarrow{h}_{it})]$$

getting a text fragment vector representation f_i.

RNN for Text Fragment Sequence. Similarly, for the fragments f_i in a page, we conduct the same bidirectional LSTM:

$$\overrightarrow{u}_i = \overrightarrow{LSTM}(f_i)$$

$$\overleftarrow{u}_i = \overleftarrow{LSTM}(f_i)$$

to distinct with the hidden state notation h_{it} in last layer, we use u_i to notate the hidden state:

$$u_i = [Attention(\overrightarrow{u}_i), Attention(\overleftarrow{u}_i)]$$

Here, attentions are also introduced to address the fragments that contribute most for classify specified fragment to an attribute label.

Text Fragment Classification. The hidden states of text fragment sequence RNN can be used to representing the fragment with their context semantics. They're sent to the softmax function:

$$y_i = softmax(W_c u_i + b_c) \tag{1}$$

We use the cross-entropy loss function as training loss:

$$L = \frac{1}{TN} \sum_{i=1}^{T} \sum_{n=1}^{N} (y_{in} log\hat{y}_{in} + (1 - y_{in})log(1 - \hat{y}_{in})) \tag{2}$$

The predicted labels are:

$$label_i = argmax(y_i) \tag{3}$$

3.3   Adapting to Unseen Sites

It is easy to adapt the networks to previously unseen sites because we have learned the deep semantical information of the text fragments. We freeze the layers before we get the fragment representation vector, which means that all the trained parameters are used in the new sites. It is believed that same record attribute text fragments have similar representations. As the second RNN layer collect the contextual information to ensure the high precision in the same site, we replaced the second RNN with an MLP layer. In the training process, we use the whole seed site page as the training data to train the layers after the text fragments. The high precision in one site network showed in experiments make us believed that the predicted data in wrapper induction can be used as training data. The whole architecture of the adapted network can be seen in Fig. 3.

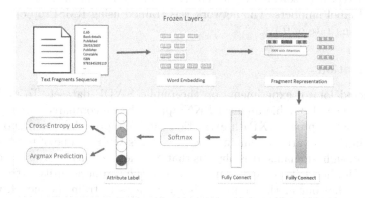

Fig. 3. Adapting to previously unseen site: the dark background means the fixed parameters in the model and they are not trainable.

4   Experimental Results

4.1   Experiment Setup

DataSet. We use the publicly available information extraction dataset presented in [12][1], which covers 8 domains, including Autos, Books, Cameras, Jobs, Movies, NBA Players, Restaurants, and Universities, with 10 websites per concept and around 2000 pages per website. The dataset is accompanied by ground truth values for 3 to 5 common attributes per concept.

Performance Metrics. Precision, recall and F-score values are evaluated as the performance. At the text fragment level, each fragment will be assigned

---

[1] The dataset with ground-truth information is publicly available at http://swde.codeplex.com/.

an attribute label after prediction. Precision is the number of fragments whose ground-truth attribute values are correctly labeled, called fragment hits, divided by the number of fragments that have extracted with a method. A recall value is the page hits divided by the number of fragments containing ground-truth attribute values. F1-Score is the harmonic mean of precision and recall.

Training Setup. We implement the hierarchical attention RNN with Tensorflow [1]. The hyperparameters are set as follows: the word vector in embedding layer has a length of 4, and word vocabulary size is dependent to the pages in a single site; for site-level transfer training, we use a shared of vocabulary with all the word items in the whole dataset. The LSTM cell unit size in text fragment RNN representation is 5, thus the fragment representation vector length is 10. In the fragment sequence RNN process, the LSTM cell unit size is 8, and send a 16-length vector to softmax layer. The softmax layer transforms that linearly to a size of label numbers. The network was trained using RMSProp optimizer with a learning rate of 0.1.

4.2   Evaluation

We conducted lots of experiments on the public SWDE dataset. To evaluate the performance of our hierarchical RNN approach, we compared our method with rule-based approach (XPath extraction) and Wrapper Adaption approach presented in [12] in various domains. Figure 4 shows the F1 value of total entity extraction in eight domains. It is obvious that our method outperforms the other two approach almost in every domain, except in the Carmera entity extraction. It is notable that our method only needs a few pages to train the model, while the XPath rules and Wrapper Adaption both use 80% pages of the dataset.

Unlike other networks that have a large proportion (usually 80%) of total dataset, we use the few-shot learning train process, a few pages (e.g. 10 pages) as the training data, and the other pages in the same site as test data. Figure 5 illustrate that only a few pages as training data can achieve excellent performance.

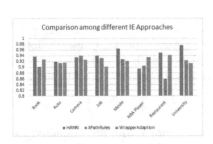

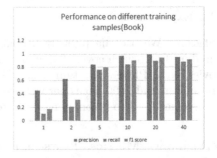

Fig. 4. Comparison among Hierarchical RNN, rule based and Wrapper Adaption

Fig. 5. Few-shot training: different page sizes of training dataset

With 5 pages training data, it can reach an 80% accuracy. 10 training pages can get about 90% accuracy and have no differences with more training pages.

For every experiment, we tried several times to compute the average of the precision, recall and F1 value. Figure 6 shows the result on the "waterstones" website of the *book* entity, which has the "title", "author", "isbn_13", "publisher" and "publish_date" attributes. There are 2000 pages on the site, we randomly select 10 pages as the training and predict the other 1999 pages. Figure 7 shows the results on all the domains presented in the dataset.

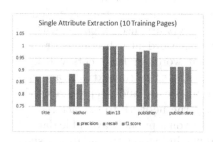

Fig. 6. Entity attribute extraction accuracy

Fig. 7. Domain accuracy comparisons

4.3 Analysis

We discard the DOM-tree structure and their tags which contain a lot of information. Most methods [18, 22, 23] utilize the DOM-tree layout features of web pages, these features play a roughly important role in their classification. However, we discover that with the aid of RNN, we could get a fairly good result from the plain text fragments. We discard some labor on feature selection and simplify the process.

Text Semantics are much considered in this research. Traditional methods usually use text statistics [12] as content features to train classifier. Some features content are domain-intensive, for instance, the symbol proportion in a text is a strong feature for a car price extraction, as the price text fragment has some symbols like '$', ',' and '.'. With the word embedding and word RNN for fragment vector representation, the word and fragment semantics are fully considered with no information loss.

Context features are naturally incorporated with RNN. A category of features is the relationship among attributes are adopted in some previous methods. These features are mainly used to select an optimized extract result; Since a page may contain multiple text similar fragments and these methods will get several results of an attribute extraction that only one is the exact result. The RNN layer for fragments sequence not only solves the above problem but also handles the case where a page has multiple extracting objects.

5   Conclusions

How to reduce the human's labor in web information extracting is a challenging problem. Most of the existing approaches need to do a large amount of work on feature engineering and training examples labeling. Inspired by the great success of deep learning on natural language processing, we present a hierarchical attention recurrent neural network, which is an end-to-end model and do not require traditional, domain-specific feature engineering. The network can be also trained with only one page in a site, i.e. Few-Shot learning. In the model, word embedding layer transforms words to vectors. Words RNN with encoder layer encodes words sequence in text fragment to vector representation. A second RNN layer to the fragment representation sequence in a web page is added to collect contextual information. The two RNN layers are also extended with an attention layer, which considers weights for each word in a fragment or each segment in a page, and then noisy or irrelevant words or fragments will have lower importance. Finally, a softmax layer classifies these contextual fragments with their attribute labels. As the model automatically and deeply learns the semantics of text fragments in pages, we adapt the network to extract records from the previously unseen websites. Experiments on a publicly available dataset demonstrate that our networks for both wrapper induction and adaption show competitive results compared against state-of-the-art approaches.

Acknowledgments. This work was supported by the national key research and development program of China under grant No. 2017YFC1001703 and the key research and development program of Zhejiang Province under grant No. 2017C01013.

References

1. Abadi, M., et al.: Tensorflow: large-scale machine learning on heterogeneous distributed systems (2016). arXiv preprint: arXiv:1603.04467
2. Bengio, Y., Ducharme, R., Vincent, P., Jauvin, C.: A neural probabilistic language model. J. Mach. Learn. Res. **3**, 1137–1155 (2003)
3. Berglund, M., Raiko, T., Honkala, M., Kärkkäinen, L., Vetek, A., Karhunen, J.: Bidirectional recurrent neural networks as generative models - reconstructing gaps in time series, April 2015. arXiv:1504.01575 [cs]
4. Bing, L., Lam, W., Wong, T.L.: Wikipedia entity expansion and attribute extraction from the web using semi-supervised learning. In: Proceedings of the Sixth ACM International Conference on Web Search and Data Mining, pp. 567–576. ACM (2013)
5. Cheng, J., Dong, L., Lapata, M.: Long short-term memory-networks for machine reading (2016). arXiv preprint: arXiv:1601.06733
6. Chung, J., Gulcehre, C., Cho, K., Bengio, Y.: Empirical evaluation of gated recurrent neural networks on sequence modeling (2014). arXiv preprint: arXiv:1412.3555
7. Dhingra, B., Liu, H., Cohen, W.W., Salakhutdinov, R.: Gated-attention readers for text comprehension (2016). arXiv preprint: arXiv:1606.01549
8. Fei-Fei, L., Fergus, R., Perona, P.: One-shot learning of object categories. IEEE Trans. Pattern Anal. Mach. Intell. **28**(4), 594–611 (2006)

9. Funahashi, K.I., Nakamura, Y.: Approximation of dynamical systems by continuous time recurrent neural networks. Neural Netw. **6**(6), 801–806 (1993)

10. Gentile, A.L., Zhang, Z., Ciravegna, F.: Self training wrapper induction with linked data. In: Sojka, P., Horák, A., Kopeček, I., Pala, K. (eds.) TSD 2014. LNCS (LNAI), vol. 8655, pp. 285–292. Springer, Cham (2014). https://doi.org/10.1007/978-3-319-10816-2_35

11. Gogar, T., Hubacek, O., Sedivy, J.: Deep neural networks for web page information extraction. In: Iliadis, L., Maglogiannis, I. (eds.) AIAI 2016. IFIP AICT, vol. 475, pp. 154–163. Springer, Cham (2016). https://doi.org/10.1007/978-3-319-44944-9_14

12. Hao, Q., Cai, R., Pang, Y., Zhang, L.: From one tree to a forest: a unified solution for structured web data extraction. In: Proceedings of the 34th International ACM SIGIR Conference on Research and Development in Information Retrieval, pp. 775–784. ACM (2011)

13. Hochreiter, S., Schmidhuber, J.: LSTM can solve hard long time lag problems. In: Advances in Neural Information Processing Systems, pp. 473–479 (1997)

14. Kushmerick, N.: Wrapper induction for information extraction. Ph.D. thesis, University of Washington (1997)

15. Le, Q.V., Mikolov, T.: Distributed representations of sentences and documents. In: ICML, vol. 14, pp. 1188–1196 (2014)

16. Mikolov, T., Chen, K., Corrado, G., Dean, J.: Efficient estimation of word representations in vector space (2013). arXiv preprint: arXiv:1301.3781

17. Pei, W., Baltrušaitis, T., Tax, D.M., Morency, L.P.: Temporal attention-gated model for robust sequence classification (2016). arXiv preprint: arXiv:1612.00385

18. Qiu, D., Barbosa, L., Dong, X.L., Shen, Y., Srivastava, D.: Dexter: large-scale discovery and extraction of product specifications on the web. Proc. VLDB Endow. **8**(13), 2194–2205 (2015)

19. Schuster, M., Paliwal, K.K.: Bidirectional recurrent neural networks. IEEE Trans. Signal Process. **45**(11), 2673–2681 (1997)

20. Soderland, S.: Learning information extraction rules for semi-structured and free text. Mach. Learn. **34**(1–3), 233–272 (1999)

21. Wong, T.L., Lam, W.: An unsupervised method for joint information extraction and feature mining across different web sites. Data Knowl. Eng. **68**(1), 107–125 (2009). https://doi.org/10.1016/j.datak.2008.08.009

22. Wong, T.L., Lam, W.: Learning to adapt web information extraction knowledge and discovering new attributes via a Bayesian approach. IEEE Trans. Knowl. Data Eng. **22**(4), 523–536 (2010)

23. Wong, T.L., Lam, W., Wong, T.S.: An unsupervised framework for extracting and normalizing product attributes from multiple web sites. In: Proceedings of the 31st Annual International ACM SIGIR Conference on Research and Development in Information Retrieval, pp. 35–42. ACM (2008)

24. Yang, Z., Yang, D., Dyer, C., He, X., Smola, A., Hovy, E.: Hierarchical attention networks for document classification. In: Proceedings of the 2016 Conference of the North American Chapter of the Association for Computational Linguistics: Human Language Technologies (2016)

25. Zheng, S., Song, R., Wen, J.R., Wu, D.: Joint optimization of wrapper generation and template detection. In: Proceedings of the 13th ACM SIGKDD International Conference on Knowledge Discovery and Data Mining, pp. 894–902. ACM (2007)

An Improved Collaborative Filtering Algorithm and Application in Scenic Spot Recommendation

Wanhong Bian[1] , Jintao Zhang[1] , Jialin Li[1] ,
and Lan Huang[2]([⊠])

[1] Software College, Jilin University, Qianjin Street 2699, Changchun, China
[2] College of Computer Science and Technology, Jilin University,
Qianjin Street 2699, Changchun, China
Huanglan@jlu.edu.cn

Abstract. To overcome the shortcoming of collaborative filtering algorithm based on item lacking of considering the attributes of items, an improved collaborative filtering is proposed. This paper defines a profile system composed of several tag-weight key-value pairs and implements it by means of text similarity analysis. It calculates the similarity of different items' profile to filter unrelated items in the recommendation process. Compared with other collaborative filtering algorithms based on item content to recommend, the proposed algorithm relies less on subjective experience from the design standpoint and can be applied at the industrial field. The experimental results on the real data set of scenic spots' ratings show that the proposed algorithm improves the performance of collaborative filtering algorithm. The problem of collaborative filtering algorithm based on item ignoring the attributes of items is effectively alleviated to some extent.

Keywords: Collaborative filtering algorithm · User profile · Similarity analysis
Travel recommendation

1 Introduction

As one of the leaders of tertiary industry, tourism plays an important role in the national economy. However, consumers' traveling satisfaction has not been significantly improved under the flourishment of the industry. Facing the massive scenic spot data, the public cannot make reasonable travel plans. With the news that two famous scenic spots of "Orange Island" and "Dragon Gorge" were delisted, the significance of traditional evaluation standard is gradually declining for the public, while the new scenic spot evaluation model together with the scenic spot recommendation algorithm will be imperative.

In the field of recommendation, collaborative filtering recommendation based on items [1] is a basic and classic algorithm. Its main idea is to use the interesting degree on the other scenic spots and the similarity between the target scenic spot and others to predict the user's rating on the target scenic spot. Aiming at the sparse user rating data,

Q. Zhou et al. (Eds.): ICPCSEE 2018, CCIS 902, pp. 240–249, 2018.
https://doi.org/10.1007/978-981-13-2206-8_21

domestic and foreign scholars have made many improvements on collaborative filtering algorithm, which were summarized as filling user rating data, improving similarity, fusing content to recommend and so on. The proposed algorithm belongs to the direction of "Applying item content to recommend". Numerous scholars have proposed good algorithms in this direction, such as "Collaborative filtering recommendation based on tags of scenic spots" [2], "Optimized Collaborative Filtering Recommendation Algorithm Integrating User Dynamic Tags" [3], etc. However, most of them are based on expert opinions and combine attributes and recommendations with a high degree of noise and rely too much on subjective experience to make it impossible to apply large-scale applications at the industrial level.

Nowadays, there are various Internet products in the tourism field. These travel websites produce huge amounts of user data. Providing these data could be used to accurately evaluate the scenic spot, the difficulties of selecting scenic spot set would be solved.

This paper introduces the profile technology which has matured in the field of social networking and e-commerce, and combines with the traditional collaborative filtering algorithm, proposes a more accurate recommendation algorithm.

2 Traditional Item-Based Collaborative Filtering Algorithm

The core thought of Item-based Collaborative Filtering Algorithm is to recommend issues similar to those they had affection for in the past time. In the view of rating evaluation framework in this passage, the algorithm predicts rating of target scenic spot according to rating of other scenic spots that the traveler has graded.

2.1 User-Rating Matrix Constructed

The first step of using Item-based Collaborative Filtering Algorithm is constructing user-rating matrix. The matrix provides data support for further calculation:

$$A_{p,q} = \begin{bmatrix} r_{11} & \cdots & r_{1q} \\ \vdots & \ddots & \vdots \\ r_{p1} & \cdots & r_{pq} \end{bmatrix} \tag{1}$$

In the user-rating matrix, p represents the quantity of users while q represents the quantity of scenery spots. r_{ij} represents the rating of scenery spot j given by user i. The rating would be 0 if the user did not grade the scenery spot.

2.2 Scenic Spots Similarity Calculated

We use Correlation-based Similarity method to calculate the similarity between two different scenic spots.

$$sim(s,t) = \frac{\sum_{u \in U}(r_{u,s}-\bar{r}_s)(r_{u,t}-\bar{r}_t)}{\sqrt{\sum_{u \in U}(r_{u,s}-\bar{r}_s)^2}\sqrt{\sum_{u \in U}(r_{u,t}-\bar{r}_t)^2}} \tag{2}$$

In the formula, I represents the scenic spots set of the certain user while s represents the target scenic spot whose rating needs to be predicted. And t represents one scenic spot rather than s in list I. U represents the set of users who have graded both s and t.$r_{u,s}$ represents the rating of scenery spot s given by user u. $\bar{r}_s$ represents the average rating of s given by all users in set U.

After the calculation, we obtain the similar scenic spots set I'. Each scenic spot in I' has positive similarity to target spot s.

2.3 Rating Predicted

Providing set I' existed, we use formula (3) to calculate the predicting rating of scenery spot s given by user u.

$$pred(u,s) = \frac{\sum_{i' \in I'} sim(s,i') * r_{u,i}}{\sum_{i' \in I'} sim(s,i')} \tag{3}$$

In the formula, i' represents certain spot belong to set I'. If set I' did not exist, the rating could not be predicted.

We discover that the traditional Item-based Collaborative Filtering Algorithm calculates the similarity between target spot and other spots in spots set according to users' grading performance rather than spots' intrinsic characteristics. In the other word, the crucial thought of the algorithm is that if item A's rating was similar to item B's, item A would be similar with item B.

However, spots that have little similarity with target spot might contribute unnecessary values to the predicting rating. The spots set including unrelated spots have negative effects on the predicting consequence. Thus, it is necessary to set several standards to eliminate unrelated spots in order to increase the precision of prediction.

3 The Proposed Algorithm

This paper defines a profile system composed of several tag-weight key-value pairs and implements it by means of text similarity analysis. It calculates the similarity of different items' profile, applying the result to recommendation.

3.1 Profile Technology

This paper applies the user profile [4, 5] technology, using the official introduction text data of the scenic spot on the mainstream tourism websites to describe the tourist resources. As the characteristic identifier of tourist resources, the profile is composed of several tag-weight key-value pairs, which is shown in Fig. 1. The tag represents the characteristics of the content and displays the personalized information of the target. The weight represents the characteristic degree of the target under this tag. The establishment of profile is divided into two stages: tag selection and weight determination.

Fig. 1. The profile of Nanjing Presidential Palace

3.2 Tag Selected

This paper quotes "Classification, investigation and evaluation of tourism resources" [6] and "Study on the Tourist Resources Classification System and Types Evaluation in China" [7], divides the tourist resource into six types: "Geographical type", "Waters type", "Climatic and biological type", "Historical type", "Modern human type", "Abstract modern type". Each type is divided into several patterns according to its own characteristics. The tag represents the characteristics of the content, considering from the aspect, it is appropriate to set "type" as the first level tag, set "pattern" as the second level tag. The tag system is shown in Table 1.

Table 1. Tag system

Type	Pattern
Geographical	(1) Mountains and valleys (2) Islands and beaches (3) Deserts and the wilds (4) Hiking grounds
Waters	(1) Lakes and rivers (2) Waterfalls
Climatic and biological	(1) Habitats of animals and plant (2) Steppes
Historical	(1) Military systems (2) Historic celebrities' mausoleums (3) Palace and government office (4) Classical gardens (5) Historical buildings (6) Religious sites (7) Famous bridges (10) Grottoes and statues (11) Historic water conservancy projects and transport projects (12) Historical blocks (13) Revolutionary commemoration sites
Modern human	(1) Modern large-scale bridge (2) Modern cities (3) Modern city squares (4) Modern city parks (5) Zoos and arboretums (5) Marine parks (6) Theme parks (7) Shopping malls and plazas (8) Sanatoriums (9) Museums and exhibition halls (10) Modern statues
Abstract modern	(1) Famous type spots (2) Folk villages and towns

3.3 Weight Calculated

For each pattern, several scenic spots are selected as the standard reference set. We name the standard reference set as Q. The result of similarity analysis between target scenic spots and spots in Q will be set as the corresponding weight value of the tag. The similarity analysis is based on the official introduction text data of scenic spots, specific steps are as follows:

(1) Use the jieba segmentation to process the official introduction texts of the target scenic spot and each scenic spot in the set Q especially remove the stop words in the Chinese word usage habits, such as "zhi", "yue" and so on. All the efforts are conducted to establish an initial corpus.

(2) Express text data in vector form: m scenic spots introduction document can be converted to m row vectors, and n words that appear in the corpus can be converted into n column vectors, establish a document matrix.

$$X = \begin{bmatrix} a_{11} & \cdots & a_{1n} \\ \vdots & \ddots & \vdots \\ a_{m1} & \cdots & a_{mn} \end{bmatrix} = \begin{bmatrix} a_1 \\ a_2 \\ a_3 \\ \vdots \\ a_m \end{bmatrix} = (a_1 \cdots a_n) \tag{4}$$

In the formula, $a_{i,j}$ represents the number of occurrences of the word j in the number i text.

(3) Apply TF-IDF [8] to weight words. The following formulas defines that the importance of a word increases proportionally with the number of times it appears in the file and decreases inversely with the frequency of its appearance in the corpus. TF, IDF, TF-IDF formulas are as follows:

$$TF_{i,j} = \frac{n_{i,j}}{\sum_k n_{k,j}} \tag{5}$$

$$IDF_i = \log \frac{|D|}{|\{j : t_i \in d_j\}|} \tag{6}$$

$$TF - IDF_{i,j} = TF_{i,j} * IDF_i \tag{7}$$

In the formula, D represents the total number of files in the corpus, $n_{i,j}$ represents the number of words to be tested, $\sum_k n_{k,j}$ represents the total number of words in file j, $|\{j : t_i \in d_j\}|$ represents the number of files containing this word.

TF-IDF excludes the interference of the common words to the semantic space and makes the importance of special word even more prominent.

(4) Establish the LSI (Latent Semantic Index) model [9]. LSI is an algorithm in order to solve word clustering and dimensionality reduction relying on singular value decomposition:

$$A_{t \times d} = T_{t \times n} S_{n \times n} (D_{d \times n})^T \tag{8}$$

In the formula, t represents original dimension, d is the number of documents, $n = min(t, d)$.

(5) Take only the first k columns of the matrix T, S, D to obtain the matrix $T_{t \times k}$, $S_{k \times k}$, $(D_{d \times k})^T$. The matrix after dimensionality reduction can be obtained:

$$B = T_{t \times k} S_{k \times k} (D_{d \times k})^T \tag{9}$$

The feature space is reduced from N dimension to K dimension. K represents the subject of the subjective division of the corpus. This paper sets the K value to 34 according to the number of scenic spots types.

(6) The cosine distance between vectors is used to calculate the degree of similarity between the target scenic spot and each scenic spot in the set Q. The result is returned as the weight value of the corresponding tag.

Through these five steps, the profile of scenic spot is established.

3.4   Similar Scenic Spots Set Selected

We make a perfection on the classic collaborative filtering algorithm based on item. Spots unrelated with the target spots are deleted in the recommendation calculation process. It is a fact that those unrelated spots contribute some ratings to the final result. Actually, those constitution of ratings are avoidable. Thus, we propose the following method to select valid similar scenic spots set to provide precondition for recommendation. Steps of determination of similar scenic spots set are defined as the following:

(1) Set Q as a set of scenic spots, set T as the target scenic spot.
(2) Establish the profile of T and each scenic spot in the set Q.
(3) For each profile, sort the tags in descending order of weight value, leaving only the top ten tags to form the unordered tag set H.
(4) For each scenic spot in the set Q, compare the degree of coincidence with T and rank them in decreasing order of the comparison result.
(5) Select the top N scenic spots of Q to establish the similar scenic spots set I.

Through these five steps, the determination of similar scenic spots set is accomplished. We could discover that the spots unrelated with the target spots are deleted in the recommendation calculation process. It is a fact that those unrelated spots contribute some ratings to the final result. Actually, those constitution of ratings are avoidable. Thus, we propose the following method to select valid similar scenic spots set to provide precondition for recommendation.

4   Application and Analysis

4.1   Data Acquisition

This paper gets 1030 active user rating data from the travel website. After prepro-cessing, 656 pieces of effective information are obtained, covering 42 distinctive scenic spots with domestic characteristics. The rating information is based on the five-star system and the corresponding meanings are as described in Table 2.

Table 2. Five-star rating system

Star	Meaning
★☆☆☆☆	Very disgusted
★★☆☆☆	Disgusted
★★★☆☆	Indifferent
★★★★☆	Satisfied
★★★★★	Very satisfied

4.2   Evaluation Indicators

(1) Root-Mean-Square Error, RMSE [10]: calculate the error between the predicted rating and the true rating:

$$RMSE = \sqrt{\frac{\sum_{(u_i, s_j) \in E} \left(r_{ij} - \bar{r}_{ij}\right)^2}{|E|}} \tag{10}$$

In the formula, E represents the sample set, $|E|$ represents the size of sample set, (u_i, s_j) represents a pair of user and scenic spot, $\bar{r}_{ij}$ represents the target user's prediction rating for the target scenic spot.

(2) Precision, this paper cites the RMSE mapping method proposed by the existing scholars [11] to define the precision:

$$Precision = 1 - \left(\frac{RMSE}{4}\right) \tag{11}$$

(3) Coverage [12], the proportion of the number of samples that can be predicted to the total number of samples:

$$Coverage = \frac{k}{M} \tag{12}$$

In the formula, k represents the number of samples that can be predicted, M represents the total number of samples.

(4) F-Measure, a comprehensive indicator that combines *Precision* and *Coverage*:

$$FMeasure = \frac{2 \times precision \times coverage}{precision + coverage} \tag{13}$$

In order to make the experiment more persuasive, this paper adopts the leave-one-out cross validation method, treating each piece of data separately as a test set, while treating other data as a training set, and summing up the results of each cycle test to calculate the evaluation indicators.

4.3 Analysis

Based on the same data sample, this paper set different radios (TOP-N values) for comparison experiments and observed changes in four evaluation indicators. The experimental results are shown in Table 3. We finally determine the best ratio of related scenic spots set by analyzing the variation of four parameters.

Table 3. Experimental result

Radio	RMSE	Precision	Coverage	F-Measure
15%	0.68708	0.82823	0.80018	0.82239
20%	0.68230	0.82942	0.81003	0.81961
25%	0.66027	0.83493	0.83772	0.83632
30%	0.67801	0.83049	0.88042	0.85473
35%	0.68100	0.82975	0.90176	0.86425
40%	0.69383	0.82654	0.91628	0.86910
45%	0.70662	0.82334	0.92197	0.86987
50%	0.74796	0.81301	0.92894	0.86711
55%	0.77573	0.80606	0.93087	0.86398
60%	0.80873	0.79781	0.93780	0.86216
65%	0.82187	0.79453	0.94892	0.86489
70%	0.82190	0.79452	0.94993	0.86530
75%	0.82589	0.79352	0.95479	0.86672
80%	0.88001	0.77999	0.96384	0.86222
85%	0.89345	0.77663	0.97488	0.86453
100%	0.89470	0.77632	0.98250	0.86732

In the sample data obtained in this experiment, the rating data given by the user are concentrated in the interval [2, 4], and the data is preprocessed, so the values of coverage and precision in the experimental results are always kept at a high level. The following figure show the change of each indicator with the radio.

Analyzing Figs. 2 and 3, we can see that the precision of the prediction shows a trend of increasing first and then decreasing with the increase of ratio and reaches a peak at 25%. When the preservation ratio reaches 100%, the algorithm degenerates into the original item-based collaborative filtering algorithm. The experimental results show that the proposed algorithm improves the precision by about 5.8% compared with the original algorithm in the optimal situation.

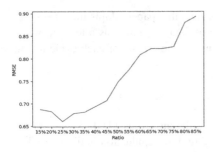

Fig. 2. The effect of ratio on RMSE

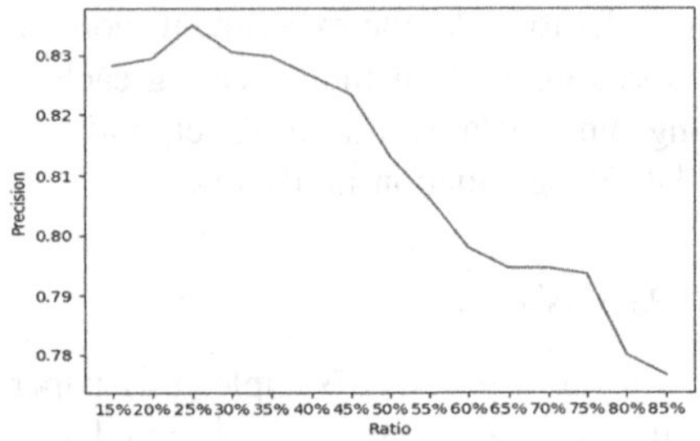

Fig. 3. The effect of ratio on Precision

Analyzing Fig. 4, we can see that the coverage shows a trend of increasing with the increase of ratio. Figure 5 shows that with the increasement in the ratio, the F-Measure gradually becomes stable at 40%. F-Measure is used as a comprehensive indicator to solve the problem of conflicting accuracy and coverage.

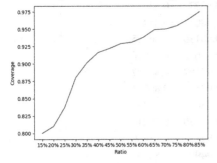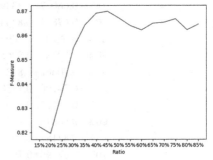

Fig. 4. The effect of ratio on Coverage

Fig. 5. The effect of ratio on F-Measure

After overall analysis, under the premise of ensuring that the F-Measure is at a reasonable level, it is more appropriate to set the ratio at 40%, which makes a good improvement compared to the traditional collaborative filtering algorithm.

The data required by the proposed algorithm comes from the official introduction text of the scenic spot's home page written by the experts, which has two characteristics of authority and accessibility. What's more, the establishment of the scenic spot's profile is based on text similarity analysis, compared to the subjective depiction, this kind of objective evaluation model is more convincing.

5   Conclusion

This paper introduces profile technology in the field of social networking and e-commerce into the tourism field and combines it with traditional collaborative filtering algorithms to propose a new type of recommendation algorithm. This algorithm combines the two-dimensional data of the user behavior record and the scenic spot's attributes to recommendation. The experimental results show that this algorithm effectively improves the performance of the original item-based collaborative filtering algorithm.

Due to the peculiarity of the tourist data, the scale of the appropriate data tested in this paper is not sufficiently convincing, while the type of data is limited to textual ones. The focus of future work will be on two aspects. First, we need to increase the number of data types that can be analyzed, such as pictures, travel notes, page visits, to enrich the source of profile data; Second, we need to expand the data size, test the performance of the algorithm under massive data which performed.

References

1. Deng, A., Zhu, Y., Shi, B.: A collaborative filtering recommendation algorithm based on item rating prediction. J. Softw. (9), 1621–1628 (2003)
2. Shi, Y., Wen, Y., Cai, G., Miao, Y.: Collaborative filtering recommendation based on tags of scenic spots. J. Comput. Appl. **34**(10), 2854–2858 (2014)
3. Jin, Z., Hu, G.: Optimized collaborative filtering recommendation algorithm integrating user dynamic tags. Mach. Des. Manuf. (2), 116–119 (2018)
4. Du, Q.: Personalized Service-Oriented User Profiling and Applications. South China University of Technology (2014)
5. Li, J.: A network datagram and big data based research on method of user profile. In: Proceedings of 2nd International Conference on Computer Engineering, Information Science and Application Technology (ICCIA 2017), p. 7. Wuhan Zhicheng Times Cultural Development Co., Ltd. (2017)
6. GB/T 18972-2003. Classification, investigation and evaluation of tourism resources
7. Guo, L., Wu, B., Liu, F., Fan, Y.: Study on the tourist resources classification system and types evaluation in China. Acta Geogr. Sin. (3), 294–301 (2000)
8. Tu, S., Huang, M.: Mining microblog user interests based on TextRank with TF-IDF factor. J. China Univ. Posts Telecommun. **23**(5), 40–46 (2016)
9. Shi, Q., Cong, S., Tang, X.: LSI_LDA: mixture method for feature dimensionality reduction. Appl. Res. Comput. **34**(8), 2269–2273 (2017)
10. Tang, J., Hu, X., Liu, H.: Social recommendation: a review. Soc. Netw. Anal. Min. **3**(4), 1113–1133 (2013)
11. Jamali, M., Ester, M.: Trustwalker: a random walk model forcombining trust-based and item-based recommendation. In: Proceedings of 15th ACM SIGKDD International Conference on Knowledge Discovery and Data Mining, pp. 397–406. ACM Press, New York (2009)
12. Victor, P., De Cock, M., Cornelis, C.: Trust and recommendations. In: Ricci, F., Rokach, L., Shapira, B., Kantor, P.B. (eds.) Recommender Systems Handbook, pp. 645–675. Springer, Boston (2011). https://doi.org/10.1007/978-0-387-85820-3_20

The Algorithms of Weightening Based on DNA Sticker Model

Chunyan Zhang$^{(\boxtimes)}$, Weijun Zhu, and Qinglei Zhou

Zhengzhou University, Zhengzhou, Henan Province, China
iecyzhang@163.com

Abstract. The algorithm of weightening serves as building blocks for the construction of more complex sticker algorithms. However, due to the previous weightening algorithms with single function, so the scope of solving problems is small. To this end, we propose three multifunctional weightening algorithms which running on the sticker machines. First, the basic operators of the algorithms consist of the pre-defined operations of the sticker model. Second, one can obtain the new algorithms by organizing these basic operations in a certain logical way. At last, running these new algorithms, we can obtain the corresponding functional results by reading the biochemical reaction products.

Keywords: Sticker model · DNA computing · Sticker machines
Weightening

1 Introduction

The concept of DNA Computing was not adopted until 1994, when Dr. Adleman, the Turing prize Laureate, published a paper in Science that solved a small-scale Hamilton path problem by standard biotechnology, which confirmed the feasibility of DNA molecular computing [1]. As an important model in DNA computing, the sticker model is mainly used to solve the problem of graph theory [2]. In recent years, many complex NP-complete problems have been solved under the circumstance of DNA computing, i.e., Satisfiability [3–5], vertex cover [6–8], clique [2, 9, 10], independent set [2, 11, 12], matching [13, 14], N Queen [10] and TSP (Traveling Salesman Problem) [15–17] and so on.

To solve some problems, the useful procedures are needed, which serve as building blocks (subalgorithms) for the construction of more complex sticker algorithms. In [18], some sticker algorithms take weightening as a procedure, such as set cover, vertex cover, clique, independent set and matching. It will be seen from these that the weightening plays an important role in many sticker algorithms. However, from the existing literature, there are two defects in the weightening algorithm: 1, the memory complexes in input test tube must be equal length; 2, we can extract the memory complexes in which exactly k are turned on. Obviously, these defects limit the problems of graph theory to be solved. To this end, we propose three new weightening algorithms based on literature [18], the new algorithms not only retain the original advantages, but also add two new functions, which can search for memory complexes

Q. Zhou et al. (Eds.): ICPCSEE 2018, CCIS 902, pp. 250–262, 2018.
https://doi.org/10.1007/978-981-13-2206-8_22

in which exactly k are turned off and do not require all the memory complexes of the initial test tube with equal length.

This paper is organized as follows. In Sect. 2, we introduce the related knowledge of weightening problem and DNA sticker machines. Section 3 describes the algorithms of weightening and the complexity analysis about new algorithms. In Sect. 4, the simulation experiments illustrate the feasibility of new algorithms. Section 5 highlights the advantages of new algorithms by comparing with the relevant work. Section 6, we give the conclusion of this paper.

2   The Sticker Model

As a non-autonomous DNA model, the sticker model provides 5 basic operations (merge, separate, set, clear, discard). Then, the researchers can call these basic operations and organize them in a certain logical way, to obtain the DNA encoding for solving a specific DNA calculation problem. In other words, the sticker model provides a common platform to isolate the underlying DNA molecules based biological implementation from high-level computational problems and solutions, making it unnecessary for DNA algorithm researchers to consider the underlying hardware (biology) implementation. The following are the structure introduction of the sticker model, the basic biochemical operations, and the library.

(1) The structure of sticker model

A random access memory of the sticker model consists of some so-called memory complexes which are partially double-stranded DNA. Each memory complex is formed by two basic types of single-stranded DNA molecules, referred to as memory strands and sticker strands, the detailed structure of sticker model can refer to [19].

A memory strand: one single-stranded DNA molecule which contains n non-overlapping substrands, and each substrand is m in length, so the whole memory strand is mn in length.

A sticker strand: one single-stranded DNA molecule which is m in length and required to be complementary to exactly one of the n substrands in a memory strand.

A memory complex: A DNA strand that is partially double-stranded and can be viewed as an encoding of a n bit number.

DNA encoding: Let each substrand of a memory strand is 1 bit, if a sticker strand is annealed to its matched substrand on a memory strand, the particular substrand is on (represented by "1"); otherwise, it is off (represented by "0"). In this way, we can skillfully make the memory complexes represent binary numbers.

Here's an example to understand the above concepts.

Example 1. A DNA molecule s is given

$$5'-TTTCCC \mid AAAAAA \mid GGGCCC-3'$$

$$3'-AAAGGG-5' \quad 3'-CCCGGG-5'$$

In s, $5'-TTTCCC \mid AAAAAA \mid GGGCCC-3'$ represents a memory strand, which contains 3 non-overlapping substrands, and each substrand is 6 in length. $3'-AAAG$ $GG-5'$ and $3'-CCCGGG-5'$ are both represent sticker strand, which are 6 in length. It will be seen that s is a memory complex which contains 3 substrands, so, the encoded 3-bit string is 101.

(2) Five basic biochemical operations

merge: Taking the union of the given test tubes and pouring the final product into the same test tube. In other words, the operation $merge(T_1, \cdots, T_n)$ produces a new tube T_n, which contains all the memory complexes from tube T_1 to T_n.

separate: the operation $separate(T, T^+, T^-, i)$ divides the test tube T into two new test tubes, T^+ and T^-, where T^+ consists of all memory complexes in T in which the ith substrand is "1", while T^- is comprised of all memory complexes in T, in which the ith substrand is "0", $1 \le i \le n$.

set: the $set(T, i)$ operation generates a test tube in which the ith substrand of each memory complex in T is "1", $1 \le i \le n$.

clear: the $clear(T, i)$ operation generates a test tube in which the ith substrand of each memory complex in T is "0", $1 \le i \le n$. Generally, this operation uses the method of heating to change a particular bit from "1" to "0".

discard: Empty all the contents in test tube T.

(3) Library

For a DNA sticker model, the input is an initial test tube (i.e. library) containing candidate solutions which are many DNA strands. While, the output are biochemical reaction products, and the encoding information contained in the molecule's products is the required result. The definition of library can see as follow:

Definition 1. [20] Let $n \ge k$ and $m \ge 0$. A $[m+n, \binom{n}{k}]$ library is a test tube given by a multiset of DNA molecules encoding as $m+n$ bit numbers where the first n characters are all k-combinations of n so that the last m characters are 0.

For instance, a $\left[7, \binom{4}{3}\right]$ library is given by a multiset of strings 1110|000, 1101| 000, 1011|000, and 0111|000, where vertical bars indicate the substrings.

3   The Improved Weightening Algorithms

3.1   The Weightening Problem

The algorithm weightening extracts from an input test tube those corresponding memory complexes based on the requirement. Usually, the weightening algorithm severs as a subalgorithm for the construction of more complex sticker algorithms, in order to solve NP problems. The definitions of improved weightening algorithms are as follows.

Definition 2. The first definition of weightening algorithm is that, we can extract all the memory complexes, in which exactly m are turned off (single-stranded DNA), from the initial test tube. Record as: WeighteningZ.

Definition 3. The second definition of weightening algorithm is that, we can extract all the memory complexes, in which exactly n are turned on (double-stranded DNA), from the initial test tube. Record as: WeighteningF.

According to the above two weightening algorithms, we can implement three new functions in the initial tube which consists of unequal length memory complexes:

If calling the WeighteningZ algorithm, we can get all the memory complexes in which exactly m are turned off.

If calling the WeighteningF algorithm, we can get all the memory complexes in which exactly n are turned on.

If calling both the WeighteningZ and WeighteningF algorithm, we can get all the memory complexes in which exactly there are the number of m are turned off and n are turned on.

To implement the above functional of the weightening problem, we have given the following three DNA sticker algorithms.

3.2   The Algorithm of WeighteningZ

According to Definition 2, we have given the algorithm of WeighteningZ (Table 1).

Table 1. The algorithm of WeighteningZ

Input: The initial test tube T , positive integer m
/* T contains all the memory complexes which are y in length, $y \geq m$ */

Output: The text tube T_m

/* T_m contains all the memory complexes which stored in T and exactly m are turned off */
Begin
1.for i:=0 to (y-1) do
2. for j:=min{i,m} down to 0 do

3.   separate $(T_j, T^+, T^-, i + 1)$

4.   merge (T^-, T_{j+1})

5.   merge (T^+, T_j)

6. end for
7.end for

8.return T_m
End

Let the input of the WeighteningZ algorithm is some memory complexes with a length of y. Due to y is an integer variable, it enlarges the range of solving the problem of weightening. It should be noted that in the following algorithms and complexity analysis, y represents the bit of the longest memory complexes.

3.3   The Algorithm of WeighteningF

Similarly, according to Definition 3, let the input of the WeighteningF algorithm is some memory complexes with a length of x. Due to x is an integer variable, it enlarges the range of solving the problem of weightening. It should be noted that in the following algorithms and complexity analysis, x represents the bit of the longest memory complexes (Table 2).

Table 2. The algorithm of WeighteningF

Input: The initial test tube T, positive integer n
/* T contains all the memory complexes which are x in length, $x \geq n$ */
Output: The test tube T_n
/* T_n contains all the memory complexes which stored in T and exactly n are turned on */
Begin
1. for i:=0 to (x-1) do
2. for j:=min{i,n} down to 0 do
3.   separate $(T_j, T^+, T^-, i + 1)$
4.   merge (T^+, T_{j+1})
5.   merge (T^-, T_j)
6. end for
7. end for
8. return T_n
End

3.4   Weightening Algorithm for Fixed Single and Double Strands

One can obtain a new algorithm, record as WeighteningZF, by organizing basic operations and subalgorithms (WeighteningZ and WeighteningF) in a certain logical way (Table 3).

Table 3. The algorithm of WeighteningZF

Input: The initial test tube T, positive integer n and m

/* T contains all the memory complexes which are y in length, $y \geq m$, $y \geq n$*/

Output: All the memory complexes which exactly n are turned on and are m turn off

Begin

1. WeighteningZ(T, m)

/* Take the test tube T as the input of the algorithm WeighteningZ and return tube T_m */

2. WeighteningF(T_m, n)

/* Take the test tube T_m as the input of the algorithm WeighteningF and return tube T_n */

3. if $\neg empty(T_n)$ then

4. report "The results are stored in tube T_n。"

5. else

6. report "No answer。"

End

3.5   The Workflow of the Algorithms

We explain the workflow of the algorithms and the running results through an example.

Example 2. An input test tube T_0 providing encoded DNA of memory complexes 0111, 10100, 01110, 01100 and 100011. Solving the following questions separately:

1. All the memory complexes which exactly 3 are turned off;
2. All the memory complexes which exactly 2 are turned on;
3. All the memory complexes which exactly 3 are turned off and 2 are turned on.

Seen from Example 2, the longest memory complex is 6 bits in T_0 which contains unequal length memory complexes. In addition, i represents the number of outer loops, while j means the number of inner loops, and the value of i and j begin from 0. The solutions to the above problems are as follows:

The first question can be solved by WeighteningZ.

From the above, the y is 6, so the outer loop is carried out 6 times. Take the first two loops as examples to describe the workflow of WeighteningZ.

The encoded DNA of initial test tube T_0 as shown the first line in Table 4, and the first outer loop is as follows:

When $i = 0$, then $j = 0$, that is once inner loop. The separate statement divides the test tube T_0 into T^+ and T^- according to the value of 1th bit of the memory complexes. When the value of 1th bit is "1", just as 10100 and 100011, put in T^+; When the value of 1th bit is "0", just as 0111, 01110 and 01100, put in T^-. The two merge statements separately implement the union of T^+ and T_0, T^- and T_1. That is to say, the memory complexes in test tube T^+ (10100 and 100011) are poured into T_0, while, the memory complexes in test tube T^- (0111, 01110 and 01100) are poured into T_1. See the second line in Table 4.

Table 4. Computation of WeighteningZ

	T_0	T_1	T_2	T_3	T_4	T_5
Initial	0111					
	10100					
	01110					
	01100					
	100011					
$i = 0$	10100	0111				
sep. on 1	100011	01110				
		01100				
$i = 1$		0111				
sep. on 2		01110				
		01100				
		10100				
		100011				
$i = 2$		0111	100011			
sep. on 3		01110				
		01100				
		10100				
$i = 3$		0111	01100	100011		
sep. on 4		01110	10100			
$i = 4$		0111	01110	100011		
sep. on 5				01100		
				10100		
$i = 5$		0111	01110	100011		
sep. on 6				01100		
				10100		

After the first outer loop, the encoded DNA of memory complexes, 10100 and 100011, are stored in test tube T_0. Meanwhile, 0111, 01110 and 01100 are stored in test tube T_1. Then, execute the second outer loop.

When $i = 1$, there are twice inner loops, $j = 1$ and $j = 0$.

When $j = 1$, the separate statement divides the test tube T_1 into T^+ and T^- according to the value of 2th bit of the memory complexes. When the value of 2th bit is "1", just as 0111, 01110 and 01100, put in T^+; When the value of 2th bit is "0", put in T^-. Now, T^- is empty, because there is no one memory complex which the value of the 2th bit is "0" in T_1. The two merge statements separately implements the union of T^+ and T_1, T^- and T_2. That is to say, the memory complexes in test tube T^+ (0111, 01110 and 01100) are poured into T_1, while, noting is poured into T_2, for T^- is empty.

When $j = 0$, the separate statement divides the test tube T_0 into T^+ and T^- according to the value of 2th bit of the memory complexes. When the value of 2th bit is "1", put in T^+. Now, T^+ is empty, because there is no one memory complex which the value of the 2th bit is "1" in T_0. When the value of 2th bit is "0", just as 10100 and 100011, put in T^-. The two merge statements separately implements the union of T^+ and T_0, T^- and T_1. That is to say, the memory complexes in test tube T^- (10100 and 100011) are poured into T_1, while, noting is poured into T_0, for T^+ is empty.

The second outer loop is over, at this time, T_1 contains all memory complexes, and other tubes are empty. See the third line in Table 4.

Similarly, when $i = 2$, there are three times inner loops, and the value of j are 2, 1 and 0 respectively....until $i = 5$, when the last bit of the longest memory complex is processed, the algorithm ends. Then, we can get all the memory complexes which exactly 3 are turned off from the initial tube, they are the products in T_3, 100011, 01100 and 10100, as shown in Table 4.

The second question can be solved by WeighteningF.

From the above, the x is 6, so the outer loop is carried out 6 times. Take the first two loops as examples to describe the workflow of WeighteningF.

The encoded DNA of initial test tube T_0 as shown the first line in Table 5, and the first outer loop is as follows:

Table 5. Computation of WeighteningF

	T_0	T_1	T_2	T_3	T_4	T_5
Initial	0111 10100 01110 01100 100011					
$i = 0$ sep. on 1	0111 01110 01100	10100 100011				
$i = 1$ sep. on 2		10100 100011 0111 01110 01100				
$i = 2$ sep. on 3		100011	10100 01100 0111 01110			
$i = 3$ sep. on 4		100011	10100 01100	0111 01110		
$i = 4$ sep. on 5			10100 01100 100011	0111 01110		
$i = 5$ sep. on 6			10100 01100	0111 01110 100011		

When $i = 0$, then $j = 0$, that is once inner loop. The separate statement divides the test tube T_0 into T^+ and T^- according to the value of 1th bit of the memory complexes. When the value is "1", just as 10100 and 100011, put in T^+; When the value of 1th bit is "0", just as 0111,01110 and 01100, put in T^-. The two merge statements separately implements the union of T^+ and T_1, T^- and T_0. That is to say, the memory complexes in test tube T^+ (10100 and 100011) are poured into T_1, while, the memory complexes in test tube T^- (0111, 01110 and 01100) are poured into T_0. See the second line in Table 5.

After the first outer loop, the encoded DNA of memory complexes, 0111, 01110 and 01100, are stored in test tube T_0. Meanwhile, 10100 and 100011 are stored in test tube T_1. Then, execute the second outer loop.

When $i = 1$, there are twice inner loops, $j = 1$ and $j = 0$.

When $j = 1$, the separate statement divides the test tube T_1 into T^+ and T^- according to the value of 2th bit of the memory complexes. When the value of 2th bit is "1", put in T^+, now, T^+ is empty, because there is no one memory complex which the value of the 2th bit is "1" in T_1; When the value of 2th bit is "0", just as 10100 and 100011, put in T^-. The two merge statements separately implements the union of T^+ and T_2, T^- and T_1. That is to say, the memory complexes in test tube T^- (10100 and 100011) are poured into T_1, while, noting is poured into T_2, for T^+ is empty.

When $j = 0$, the separate statement divides the test tube T_0 into T^+ and T^- according to the value of 2th bit of the memory complexes. When the value of 2th bit is "1", just as 0111, 01110 and 01100, put in T^+; When the value of 2th bit is "0", put in T^-. Now, T^- is empty, because there is no one memory complex which the value of the 2th bit is "0" in T_0. The two merge statements separately implements the union of T^+ and T_1, T^- and T_0. That is to say, the memory complexes in test tube T^+ (0111, 01110 and 01100) are poured into T_1, while, noting is poured into T_0, for T^- is empty.

The second outer loop is over, at this time, T_1 contains all memory complexes, and other tubes are empty. See the third line in Table 5.

Similarly, when $i = 2$, there are three times inner loops, and the value of j are 2, 1 and 0 respectively....until $i = 5$, when the last bit of the longest memory complex is processed, the algorithm ends. Then, we can get all the memory complexes which exactly 2 are turned on from the initial tube, they are the products in T_2, 01100 and 10100, as shown in Table 5.

The third question can be solved by WeighteningZF.

First, the algorithm calls the WeighteningZ, with the initial test tube T_0 and the integer 3 (the value of m) as an input, to get the needed test tube T_3, and the reaction process is shown in Table 4. Then, calling the subalgorithm WeighteningF which takes T_3 (see Table 6) and integer 2 (the value of m) as an input, the reaction process is shown in Table 7. At last, according to the conditions, the corresponding results are

Table 6. The input test tube of WeighteningF

	T_0	T_1	T_2	T_3	T_4	T_5
Initial	100011					
	01100					
	10100					

Table 7. Computation of the call subalgorithm WeighteningF

	T_0	T_1	T_2	T_3	T_4	T_5
Initial	100011 01100 10100					
$i = 0$ sep. on 1	01100	100011 10100				
$i = 1$ sep. on 2		100011 10100 01100				
$i = 2$ sep. on 3		100011	10100 01100			
$i = 3$ sep. on 4		100011	10100 01100			
$i = 4$ sep. on 5			10100 01100 100011			
$i = 5$ sep. on 6			10100 01100	100011		

given by algorithm WeighteningZF, they are the products in T_2. In other words, all the memory complexes which exactly 3 are turned off and 2 are turned on, they are 01100 and 10100.

The weightening algorithms's workflow and results are shown as above.

4 Complexity Analysis and Algorithm Purpose

4.1 The Complexity Analysis

We analyze the complexity of the new algorithms, in which the performance time of all biochemical and electronic operations are uniformly defined as standard CPU time, i.e. $O(1)$.

Proposition 1. The worst time complexity of the WeighteningZ algorithm is $O(3y(m+1) - m^2 - m)$.

Proof, There are 3 biochemical operations in each inner loop of the algorithm. When $i \leq m$, in each outer loop, the number of inner loop presents arithmetic progression with the tolerance of 1, i.e., $(1 + 2 + \cdots + m)$ times inner loop; When $i > m$, there remains $(y - m)$ times outer loop, and each outer loop with $(m + 1)$ times inner loop. So the time complexity of the WeighteningZ algorithm is $O(3[1 + 2 + \cdots + m + (y - m)(m + 1)]) = O(3y(m+1) - m^2 - m)$.

Proposition 2. The worst time complexity of the WeighteningF algorithm is $O(3x(n+1) - n^2 - n)$.

Proof, Empathy with Proposition 1, the time complexity of the WeighteningF algorithm is $O(3y(m+1) - m^2 - m + 3x(n+1) - n^2 - n + 2)$.

Proposition 3. The worst time complexity of the WeighteningZF algorithm is $O(3y(m+1) - m^2 - m + 3x(n+1) - n^2 - n + 2)$.

Proof, This algorithm calls the WeighteningZ and WeighteningF subalgorithms, and uses a conditional statement, so the time complexity of the WeighteningZF is

$$O(3y(m+1) - m^2 - m) + O(3x(n+1) - n^2 - n) + O(2)$$
$$= O(3y(m+1) - m^2 - m + 3x(n+1) - n^2 - n + 2)$$

4.2   The Algorithm Purpose

In view of the application of weightening to many sticker algorithms, the existing weightening algorithm requires the memory complexes equal length and can only extract k "1" memory complexes from the initial test tube. These defects limit the problems of graph theory to be solved, to this end, we propose three new weightening algorithms, which doesn't require the equal length of the DNA strands in the initial test tube and increase some new functions. For detailed advantages, see Sect. 5. We use the following platform to implement the simulation experiment: Intel(R) Core(TM) i7, Memory 8g, Windows 7.

5   The Comparison of Related Works

In [18, 21], the weightening algorithm as a useful subalgorithm is given, which serve as building blocks for the construction of more complex sticker algorithms. However, the weightening algorithm proposed in the two literatures can only deal with memory complexes with equal length. In contrast, the new algorithms do not require the memory complexes to be processed equal length. So, they enlarge the premise range of solving weightening problems.

In addition, the function of the weightening algorithm given in [18, 21] is to select the memory complexes which exactly n are turn on in the initial test tube. Compared with them, the new algorithms can not only select all the memory complexes which exactly m are turned off, but also all the memory complexes which exactly n are turned on, even can select all the memory complexes which exactly m are turned off and n are turned on.

It can be seen from the above that the new algorithms expand the scope of the premise to be solved, and adds some new functions. So, the range of graph theory problems that to be solved is naturally larger than that in [18, 21]. The specific comparison can be seen in Table 8.

Table 8. The comparison among the related DNA algorithms

	The weightening algorithm in [18]	The weightening algorithm in [21]	The three new algorithms
The memory complexes to be processed must be isometric?	Yes	Yes	No
Can obtain the memory complexes which exactly n are turn on?	Yes	Yes	Yes
Can obtain the memory complexes which exactly m are turn off?	No	No	Yes
Can obtain the memory complexes which exactly m are turn off and n are turn on?	No	No	Yes
The range of solving graph theory problems?	Small	Small	Large

6 Conclusion

This work presents weightening algorithms which running on the sticker machines. To the best of our knowledge, there is no DNA algorithm that implements the above functions, what is more, the new algorithms expand the range of graph theory problems that to be solved. Of course, When a graph is large, the algorithms need to deal with more data. Compared with the classical algorithm, the algorithms based on DNA can better deal with the large data in the graph theory because of its large-scale parallelism. This is the contribution of our work.

Acknowledgement. This work has been supported by NSFC under grant No. U1204608 and No. 61572444.

References

1. Adleman, L.M.: Molecular computation of solutions to combinatorial problems. Nature **369**, 40 (1994)
2. Fan, Y.-K., Qiang, X.-L., Xu, J.: Sticker model for maximum clique problem and maximum independent set. Chin. J. Comput. **33**(2), 305–310 (2010)
3. Lipton, R.J.: DNA solution of hard computational problems. Science **268**(5210), 542–545 (1995)
4. Braich, R.S., Johnson, C., Rothemund, P.W.K., Hwang, D., Chelyapov, N., Adleman, L.M.: Solution of a satisfiability problem on a gel-based DNA computer. In: Condon, A., Rozenberg, G. (eds.) DNA 2000. LNCS, vol. 2054, pp. 27–42. Springer, Heidelberg (2001). https://doi.org/10.1007/3-540-44992-2_3
5. Song, B.-S., Yin, Z.-X., Zhen, C., et al.: DNA self-assembly model for general satisfiability problem. J. Chin. Comput. Syst. **32**(9), 1872–1875 (2011)

6. Pan, L., Jin, X., Liu, Y.: A surface-based DNA algorithm for the minimal vertex cover problem. Prog. Nat. Sci. Mater. Int. **13**(1), 78–80 (2003)
7. Dond, Y.-F., Zhang, J.-X., Yin, Z.-X., et al.: An improved sticker model of the minimal covering problem. J. Electron. Inf. Technol. (4), 556–560 (2005)
8. Wu, F., Li, K., Sallam, A., et al.: A molecular solution for minimum vertex cover problem in tile assembly model. J. Supercomput. **66**(1), 148–169 (2013)
9. Al Junid, S.A.M., Tahir, N.M., Majid, Z.A., et al.: Potential of graph theory algorithm approach for DNA sequence alignment and comparison. In: 2012 Third International Conference on Intelligent Systems, Modelling and Simulation (ISMS), pp. 187–190. IEEE (2012)
10. Wu, F., Li, K.-L.: An algorithm in tile assembly model for N queen problem. Acta Electron. Sin. **41**(11), 2174–2180 (2013)
11. Xixu, W., Jing, L., Zhichao, S., et al.: A molecular computing model for maximum independent set based on origami and greedy algorithm. J. Comput. Theoret. Nanosci. **11**(8), 1773–1778 (2014)
12. Zhou, K., Duan, Y., Dong, W., et al.: A matrix algorithm for maximum independent set problem based on sticker model. J. Comput. Theoret. Nanosci. **13**(6), 3734–3743 (2016)
13. Zhou, X., Li, K.-L., Yue, G.-X., et al.: A volume molecular solution for the maximum matching problem on DNA-based computing. J. Comput. Res. Dev. **48**(11), 2147–2154 (2011)
14. Wu, X., Song, C.-Y., Zhang, N., et al.: DNA algorithm for maximum matching problem based on sticker computation model. Comput. Sci. **40**(12), 127–132 (2013)
15. Lee, J.Y., Shin, S.Y., Park, T.H., et al.: Solving traveling salesman problems with DNA molecules encoding numerical values. BioSystems **78**(1–3), 39–47 (2004)
16. Dong, Y.F., Tan, G.J., Zhang, S.M.: Algorithm of TSP based on sticker systems of DNA computing. Acta Simulata Syst. Sin. **17**(6), 1299–1387 (2005)
17. Kang, Z., Liu, W.B., Jin, X.U.: Algorithm of DNA computing of TSP. Syst. Eng. Electron. **29**(2), 316–319 (2007)
18. Martínez-Pérez, I.M., Zimmermann, K.H.: Parallel bioinspired algorithms for NP complete graph problems. J. Parallel Distrib. Comput. **69**(3), 221–229 (2009)
19. Roweis, S., Winfree, E., Burgoyne, R., et al.: A sticker-based model for DNA computation. J. Comput. Biol. **5**(4), 615–629 (1998)
20. Ignatova, Z., Martínez-Pérez, I., Zimmermann, K.H.: DNA computing models. Springer, Boston (2008)
21. Hasudungan, R., Rohani, A.B., Rozlina, M.: DNA computing and its application on NP completeness problem. Asian Pac. J. Trop. Dis. (2013)

Method and Evaluation Method of Ultra-Short-Load Forecasting in Power System

Jiaxiang Ou[1], Songling Li[2(✉)], Junwei Zhang[1], and Chao Ding[1]

[1] Electric Power Research Institute, Guizhou Power Grid Co., Ltd., Guiyang, China
[2] Shanghai University of Electric Power, Shanghai Yangpu District
Longchang Road No. 371, Shanghai 200090, China
18717831373@163.com

Abstract. This article describes the basic method of ultra-short-term load forecasting include the Linear Extrapolation, Kalman Filter Method, Time Series Method, Artificial Neural Networks and Support Vector Machine Algorithm. Then, it summarizes the commonly used methods to improve the accuracy of prediction from the comprehensive forecasting model and data mining technology. Finally, we divide the accuracy of the short-term load forecast into 5 min, 10 min, 30 min and 60 min and innovatively presented the concept of the accuracy in daily average load forecasting from different periods by giving the evaluation formula.

Keywords: Electric load forecasting · Ultra-short-term
Linear extrapolation · Kalman filter method
Time series method · Artificial neural networks
Support vector machine algorithm

1 Introduction

Power system load forecasting is based on historical data such as electricity load, weather, economy, society, time analyzing the influence of power load history data on future load to achieve the scientific forecast of future load. Ultra-short-term load forecasting often forecast the next 1 min to 4 h of the power load. Its significance lies in the realization of real-time power grid operation unit optimization control [1]. After the power into the marketed operation, its importance in the scheduling, marketing, market transactions and other departments will be more prominent [2].

Ultra-short-term load forecast has the features of high computational workload, high real-time and high accuracy requirements. The most important feature of the ultra-short-term load forecast is cyclical, that is to say, the changes of the corresponding time between different days are similar, the changes of different weeks but the same week type of days are similar, the changes in the respective

Q. Zhou et al. (Eds.): ICPCSEE 2018, CCIS 902, pp. 263–272, 2018.
https://doi.org/10.1007/978-981-13-2206-8_23

loads on the workday or weekend are similar, the changes in the load of different annual holidays (such as Spring Festival) are similar [3]. For the large city of grid on Monday morning and Friday afternoon, load changes are different from other days of the load features. Whats more, we can further separate historical data, which generally use the algorithm linear extrapolation method, Kalman filter method, time series method and modern intelligent prediction algorithm [4].

2   Basic Methods and Features of Ultra-Short-Term Load Forecasting

2.1   Basic Method Based on the Same Type of Daily Load Forecasting

When at a time to predict the future load, we need to determine the base date. Generally, taking the last similar type day as the base day and a few days before the base date for historical samples. If meets special holidays (Spring Festival, Golden Week, etc.), taking the recent years of the same day for the historical samples. We use the general working days as an example to illustrate as follow [5]. Starting from the base date, with a week for a cycle, separate the history sample as the first week, the second week, the third week etc. [6]. There must have a day in a single week is the same type day as the forecast day. Setting the load P_{et} on the day of e in time t, $e = 1 \cdots n$, $t = 1 \cdots T$ (T is the number of sampling points per day), $P_t = [P_{1t}, P_{2t} \cdots P_{ft}]$, P_{ft} is the closest history data in t time, then it is clear that we can calculate the data as follows.

- Historical maximum load data: $P_{t,max} = max(P_{1t}, P_{2t} \cdots P_{ft})$
- Historical minimum load data: $P_{t,min} = max(P_{1t}, P_{2t} \cdots P_{ft})$
- Historical average load data: $P_{t,ave} = average(P_{1t}, P_{2t} \cdots P_{ft})$

We can respectively use $P_{t,max}$, $P_{t,min}$, $P_{t,ave}$ as the base value processing the daily load curve to provide data for various prediction algorithms [7].

2.2   Linear Extrapolation

Linear extrapolation is the process of estimating, beyond the original observation range, the value of a variable on the basis of its relationship with another variable. It bases on the fitted load curve to calculate load curve in a specified time [8].

The prediction process is as follows:

Determine the predicted future day type: According to the default date type, we judge it as the general working day, weekends or holidays [9]. For special dates (such as Friday afternoon or after Spring Festival), we do the special handling.

Making the same type of date in history in history in normalization:

$$L_n(k, i) = \frac{L(k, i) - L_{min}(k)}{L_{max}(k) - L_{min}(k)}. \tag{1}$$

Where $L_n(k,i)$ is the normalized value of load data in the day of k and hour of i. $L(k,i)$ is the actual value of the load data in the day of k and hour of i. $L_{max}(k)$ is the maximum in the day of k. $L_{min}(k)$ is the minimum in the day of k.

According to the formula (1), we can take the average of the normalization coefficient in n days load data to calculate the forecast day with the load change coefficient $L_n(i)$.

Then, Import local weather forecast data [10].

Through the logical relationship between load and temperature, the maximum and minimum loads of the forecast day can be calculated using the least squares method.

When the maximum load of the day and the minimum load are calculated later, we can calculate the specify time load value on that day by formula (1). This method is generally suitable for regional power grids or larger provincial power grids, while for the small grid load or the grid for the traction stations or steel and other impact load accounted for a larger proportion, the effect of prediction often poor [11].

2.3   Kalman Filter Method

Kalman filtering, also known as linear quadratic estimation (LQE), is an algorithm that uses a series of measurements observed over time, containing statistical noise and other inaccuracies, and produces estimates of unknown variables that tend to be more accurate than those based on a single measurement alone, by using Bayesian inference and estimating a joint probability distribution over the variables for each timeframe [12].

State estimation is an important part of Kalman filter. It is important for mastering and analyzing a system. The most commonly used method is the statistic method including least squares estimation, the minimum variance estimate, the linear minimum variance estimate, the recursive least squares estimation, etc. No matter what kind of system model, for the different setting of the initial value, filter several times before the effect will be affected. But after several iterations it can be a good prediction of tracking the target [13]. This method does not require storage of historical observations and has more relaxed limit to the initial values [14]. When new input and output data are presented, using the latest time of the data and the previous moment of the estimated value and recursive formula, we can calculate a new estimate [15].

The biggest advantage of this method is that the computational complexity and the memory of the storage are reduced, which can be easily process online in real-time. But the Kalman filter method also has its shortcomings. It is difficult to measure the noise and system noise variance in using when require the satisfy of the Gaussian linear case [16].

2.4 Time Series Method

A time series is a series of data points indexed (or listed or graphed) in time order. Most commonly, a time series is a sequence taken at successive equally spaced points in time. Thus, it is a sequence of discrete-time data.

For a smooth time series, the core is to determine which parts of the change in data are made, which generally can be divided into three parts.

$$Y(t) = f(t) + p(t) + x(t). \tag{2}$$

Where, $f(t)$ is the trend item, $f(t)$, $p(t)$ are non-random items, $p(t)$ is the period item, which reflects $f(t)$ cyclical changes. $x(t)$ is the random item. It reflects the impact for $Y(t)$ with random factors such as noise.

The basic steps of time series analysis is according to the historical data of the load, obtaining the time series data of the observed system by use the means of observation, statistics, investigation and sampling [17]. Then, establishing a mathematical model and use it on the one hand to describe the historical power load changes in the statistical law. On the other hand, on the basis of the historical system model, determine the future load forecasting model to predict future power loads [18].

The advantage of the time series method is that the calculation is relatively simple, and the required power load history data is not too much. The drawback of this approach is that it does not take the relationship between power load and other factors into account [19]. Whats more, it also ignores account the causal relationship of the change in its numerical size. So, it is generally suitable for occasions where the power load changes smoothly.

2.5 Modern Intelligent Prediction Algorithm

Modern intelligent prediction algorithms include Gray system theory, Artificial neural networks (ANN) Algorithm, Support vector machine (SVM) algorithm etc. These algorithms involve biological evolution, artificial intelligence, mathematics and physical science, statistical mechanics and other science, which are based on a certain intuitive basis and structure [20]. Modern intelligent prediction algorithm starts since the early eighties of the last century, rapidly integrating with computer and operations research and other scientific, especially suit for solving the problem of model uncertainty, nonlinearity and so on.

Gray system theory describes the known degree of model information in color. The white model indicates that the information is all known, and the gray model is unknown. Load forecasting is to distinguish the multi-factors of the system by comparing the geometric relations of the statistical sequences [21]. The more similar the shape of the sequence geometry, the greater the degree of correlation between them. The advantage of gray system modeling is that the original quantity requirements are relatively small and with high precision. But there are some shortcomings: the greater the degree of data discretization will lead to the worse prediction accuracy. Its model generally need to through the correlation test, residual test and post-test differential test and other forms of precision. For the situation, the original data does not have good smooth performance, so the accuracy results are poor [22].

Artificial neural networks refers to the structure and function of the simulated human nervous system, using a lot of processing parts to build a network system manually which often with biological brain similar characteristics. ANN has achieved great success in applications such as pattern recognition, robot control, and network management. The back propagation (BP) network is the most commonly used algorithm for ultra-short-load forecasting neural networks. Its a multi-layer sensor feedforward network. The main disadvantage of the BP algorithm is that it uses the gradient method to train weights, and pays more attention on overcoming learning errors so that the generalization performance is not strong [23]. What's more, the number of hidden layers is difficult to determine, making the network training efficiency affected by the initial value.

Support vector machine algorithm can solve the problem of pattern classification and non-linear mapping as well as BP algorithm, which the main practice is to establish an optimal decision hyperplane, so that make the distance between the two sides of the plane is the largest to provide a good generalization ability for the classification problem [24]. In the case of a small amount of historical data, it can also get a good learning effect which can effectively avoid the other learning methods in the dimension of disaster, local minimal and so on.

This algorithm also has some problems in the application. Especially on how to set some key parameters, such as weight parameters, insensitive loss parameters, and shape parameters in nuclear parameters [25]. The selection of these parameters will directly affect the performance of the algorithm and the effect of prediction.

3   Methods to Improve the Precision of Ultra-Short-Term Load Forecasting

The error of load forecasting is mainly caused by two aspects: The first is due to the uncontrollable load, which means the randomness of load changes, especially the load mutation caused by weather or other causes seriously reduces the accuracy of load forecasting, and this part of the error is unavoidable [26]. In other words, there is an upper limit of accuracy. The second is due to the error caused by the prediction method itself. That is to say there is no algorithm can reach the upper limit of the prediction but just close to it. With the development of computer technology and modern mathematical methods, predictors continue to improve the prediction algorithm is to reduce the error caused by the method itself to maximize the approximation of load changes [27].

3.1   Comprehensive Forecasting Method

In the power load forecast, due to the different modeling mechanisms and objectives, the same problem can often have a variety of forecasting methods.

The selection of the method directly affects the accuracy of the prediction results. So, the more scientific approach is to combine several different principles of forecasting methods. This makes the combined prediction results less sensitive to a single poor prediction method. Meanwhile, it will not lose the key information, which can generally improve the accuracy and reliability of the forecast. In recent years, a number of related research using comprehensive forecasting models have been adopted. In this paper, a comprehensive forecasting model with time varying weights is taken as an example. The idea of the model is to distinguish the different periods of the day to establish a comprehensive model.

If the number of time periods for the forecast is T, we need to establish a total of T comprehensive forecasting models. Setting W_k^t as the weight of the kth method in the period t in the comprehensive prediction model. In this mode, the results of the load forecast for the final forecast date are:

$$\widehat{P_{0t}} = \sum_{k=1}^{q} (W_k^{(t)} \widehat{P_{0t}^{(k)}}). \tag{3}$$

The optimal weighting model can be obtained by predicting the residual sum of the residuals as the target optimization function.

3.2   Data Mining Technology

In the actual algorithm of ultra-short-load forecasting, we generally dont consider meteorological factors. Because the possibility of the weather mutation is small, while the actual load on the day of the situation has clearly reflected the impact of weather factors. But when something like thunderstorms happens, considering real-time meteorological factors will clearly make the forecast more accurate. At present, the power sector has achieved real-time access to weather forecast information.

Another approach is to separate the load day by season and week type. By comparing the historical load data, we can use the actual daily load variation rule from the similar day in the different seasons and different holidays.

In addition, for the summer climate anomalies, air conditioning load soared and other factors, there are predictive methods combined with weather forecasting, marketing and other data analyzing the cause and effect of abnormal data to improve the accuracy of prediction.

4   Evaluation Method for Accuracy of Ultra-Short-Term Load Forecasting

Taking a city power grid day power generation planning system as an example. Based on the ultra-short-term load forecasting generator set, the planned rolling update function starts at a frequency of 30 min to 60 min. But the ultra-short-term load forecast is constantly rolling to predict the future load of the grid, so the results of the prediction of a certain moment as the basis for the evaluation cant be sufficient to objectively evaluate the accuracy of its prediction. We present an evaluation method of short-term-load to forecast the accuracy based on forecasting time and applies to reality, which achieves a good result.

4.1 Ultra-Short-Term Load Forecasting Accuracy in Different Periods

Evaluation point load forecasting deviation rate:

$$E_i = \frac{L_{i,f} - L_i}{L_i} \times 100\% \tag{4}$$

Where E_i is load forecasting deviation rate in time i, $L_{i,f}$ is the value of load forecast in time i, L_i is the actual load value in time i.

Then the ultra-short-term load forecasting accuracy in different periods is:

$$E_{i,5} = \frac{L_{i,5} - L_i}{L_i} \times 100\% \tag{5}$$

$$E_{i,10} = \frac{L_{i,10} - L_i}{L_i} \times 100\% \tag{6}$$

$$E_{i,30} = \frac{L_{i,30} - L_i}{L_i} \times 100\% \tag{7}$$

$$E_{i,60} = \frac{L_{i,60} - L_i}{L_i} \times 100\% \tag{8}$$

Where $E_{i,5}$ is the ultra-short-term load forecasting accuracy in 5 min. $E_{i,10}$ is the ultra-short-term load forecasting accuracy in 10 min. $E_{i,30}$ is the ultra-short-term load forecasting accuracy in 30 min. $E_{i,60}$ is the ultra-short-term load forecasting accuracy in 60 min. The curve of the ultra-short-term load forecasting accuracy is shown in Fig. 1.

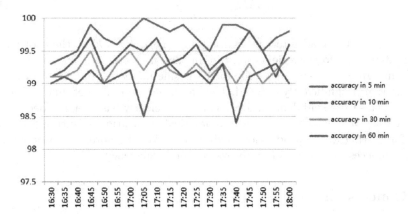

Fig. 1. Accuracy of the ultra-short-term load forecasting in different periods.

4.2   Accuracy of Daily Average Load Forecasting in Different Periods

The accuracy in daily average load forecasting from different periods mainly statistics the load forecast accuracy in 5 min, 10 min, 30 min and 60 min within a day. The histogram is shown in Fig. 2.

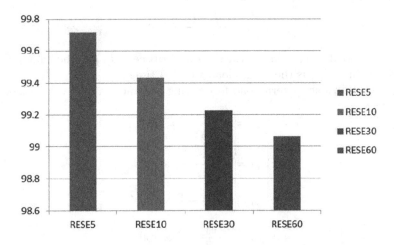

Fig. 2. Accuracy of the ultra-short-term load forecasting in different periods.

$$A_{d,i} = (1 - RMSE_i) \times 100\% \tag{9}$$

$$RSME_i = \sqrt{\frac{1}{n} \sum_{t=1}^{n} E_{t,i}^2} \tag{10}$$

Where $A_{d,i}$ is the load forecast accuracy in time i, $RSME_i$ is the root-mean-square daily load forecast deviation rate in time i.

From figures we can see, as the forecast time increases, the accuracy of the forecast is also significantly reduced. Therefore, the innovation of this assessment method is that the confidence interval can be set clearly based on the decrease of the accuracy rate, which can be used in practical applications of ultra-short-term load forecasting in the production environment. Whats more, it can be used to reduce unnecessary losses and improve economic efficiency.

5   Conclusion

Although there are already a variety of forecasting methods, scholars are still constantly exploring new predictive methods. Some theories have been applied to reality and have achieved remarkable results. Due to the random load of too many factors and with very strong nonlinearity, it is difficult to find a fast-online computing speed with high accuracy prediction method. In practical applications, it is necessary to take full account of the impact of abnormal data on the forecast. Subject to the real-time computing speed requirements, it has not been able to establish a set of different models of parameter estimation method analysis and comparison strategy. The ability of putting the system into operation of self-adaptation and self-learning needs to be further improved.

References

1. Toyoda, J., Chen, M.S., Inoue, Y.: An application of state estimation to short-term load forecasting, Part I: forecasting modeling. IEEE Trans. Power Appar. Syst. **89**(7), 1678–1682 (1970)
2. Vemuri, S., Huang, W.L., Nelson, D.J.: Online algorithms for forecasting hourly loads of an electric utility. IEEE Trans. Power Appar. Syst. **100**(8), 3775–3784 (1981)
3. Ziegel, E.R.: Time series analysis, forecasting, and control. In: Oakland, California, Holden-Day, 1976, vol. 37(2), pp. 238–242 (1976)
4. Vemuri, S., Hoveida, B., Mohebbi, S.: Short term load forecasting based on weather load models. IFAC Proc. Vol. **20**(6), 315–320 (1987)
5. Bengio, Y.: Deep learning of representations: looking forward. In: Dediu, A.-H., Martín-Vide, C., Mitkov, R., Truthe, B. (eds.) SLSP 2013. LNCS (LNAI), vol. 7978, pp. 1–37. Springer, Heidelberg (2013). https://doi.org/10.1007/978-3-642-39593-2_1
6. Christiaanse, W.R.: Short-term load forecasting using general exponential smoothing. IEEE Trans. Power Appar. Syst. PAS **90**(2), 900–911 (1971)
7. Sage, A.P., Husa, G.W.: Algorithms for sequential adaptive estimation of prior statistics. In: Adaptive Processes, p. 61 (1969)
8. Mehra, R.K.: On the identification of variances and adaptive Kalman filtering. IEEE Trans. Autom. Control **15**(2), 175–184 (1969)
9. Khalyasmaa, A.I., Dmitriev, S.A., Kokin, S.E.: Energy information model for power systems monitoring. Adv. Mater. Res. **732–733**, 841–847 (2013)
10. Coelho, I.M., Coelho, V.N., Luz, E.J.D.S., Ochi, L.S., Guimaraes, F.G., Rios, E.: A GPU deep learning metaheuristic based model for time series forecasting. Appl. Energy **201**, 412–418 (2017)
11. Dedinec, A., Filiposka, S., Dedinec, A., Kocarev, L.: Deep belief network based electricity load forecasting: an analysis of macedonian case. Energy **115**, 1688–1700 (2016)
12. Kazakov, E.G., Kirillov, A.V., Kutsin, V.V., Yasenev, N.D.: Educational process at electric drive and automation of industrial installations department of ustu-upi and its methodical and laboratory support. Russ. Electr. Eng. **80**(9), 473–476 (2009)
13. Fan, C., Xiao, F., Zhao, Y.: A short-term building cooling load prediction method using deep learning algorithms. Appl. Energy **195**, 222–233 (2017)
14. Feng, C., Cui, M., Hodge, B.M., Zhang, J.: A data-driven multi-model methodology with deep feature selection for short-term wind forecasting. Appl. Energy **190**, 1245–1257 (2017)
15. Li, L., Ota, K., Dong, M.: When weather matters: Iot-based electrical load forecasting for smart grid. IEEE Commun. Mag. **55**(10), 46–51 (2017)

16. Qiu, X., Ren, Y., Suganthan, P.N., Amaratunga, G.A.J.: Empirical mode decomposition based ensemble deep learning for load demand time series forecasting. Appl. Soft Comput. **54**(2), 246–255 (2017)
17. Abusaimeh, H., Yang, S.H.: Balancing the power consumption speed in flat and hierarchical WSN. Int. J. Autom. Comput. **5**(4), 366–375 (2008)
18. Ryu, S., Noh, J., Kim, H.: Deep neural network based demand side short term load forecasting. In: IEEE International Conference on Smart Grid Communications, p. 3 (2016)
19. Wang, L., Zhang, Z., Chen, J.: Short term electricity price forecasting with stacked denoising autoencoders. IEEE Trans. Power Syst. **1**(99), 1 (2017)
20. Chan, E.H.P.: Application of neural network computing in intelligent alarm processing (power systems). In: Power Industry Computer Application Conference, 1989, Conference Papers, pp. 246–251 (2002)
21. Wu, X., Shen, Z., Song, Y.: A novel approach for short term electric load forecasting. In: World Congress on Intelligent Control and Automation, pp. 1999–2002 (2016)
22. Moura, E.P.D., Souto, C.R., Silva, A.A., Irmão, M.A.S.: Evaluation of principal component analysis and neural network performance for bearing fault diagnosis from vibration signal processed by RS and DF analyses. Mech. Syst. Signal Process. **25**(5), 1765–1772 (2011)
23. Vaitheeswaran, N., Balasubramanian, R.: Stochastic model for optimal declaration of day ahead station availability in power pools in India. In: Power India Conference, p. 5 (2006)
24. Jain, A., Balasubramanian, R., Tripathy, S.C., Singh, B.N.: Power system topological observability analysis using artificial neural networks. In: Power Engineering Society General Meeting, pp. 497–502 (2005)
25. Nie, H., Liu, G., Liu, X., Wang, Y.: Hybrid of arima and SVMs for short term load forecasting. Energy Procedia **16**(5), 1455–1460 (2012)
26. Huang, G.B., Zhou, H., Ding, X., Zhang, R.: Extreme learning machine for regression and multiclass classification. IEEE Trans. Syst. Man Cybern. B Cybern. **42**(2), 513–529 (2012)
27. Shrivastava, N.A., Panigrahi, B.K., Lim, M.H.: Electricity price classification using extreme learning machines. Neural Comput. Appl. **27**(1), 9–18 (2016)

Ensemble of Deep Autoencoder Classifiers for Activity Recognition Based on Sensor Modalities in Smart Homes

Serge Thomas, Mickala Bourobou, and Jie Li[✉]

School of Computer Science and Technology,
Harbin Institute of Technology, 92, West Dazhi Street, Nangang District,
Harbin 150001, People's Republic of China
thomaserge@yahoo.fr, jieli@hit.edu.cn

Abstract. Over the past few years, a particular interest has been focused toward activity recognition domain. Indeed, human activity recognition pays more attention on the extraction of relevant and discriminative features whose the implementation facilitates the seamless monitoring of functional inhabitant abilities with the involvement of sensing technology in the smart home environments. However, despite the exponential efforts made by individual standard machine learning techniques, and recently by the remarkable breakthrough of deep learning methods, designing robust activity recognition architecture remains a major challenge in term of performance, due to a high degree of uncertainty and complexity caused by inherent behavior of human actions. For the former case, the drawbacks are essentially composed of heuristic and hand-crafted methods for features extraction, shallow features learning, and learning of low amount of well-labeled data. While the latter suffers from imbalanced datasets and problematic data quality in real-life datasets. In addition, the choice of suitable sensor types is also critical for successful human activity recognition. This paper proposes an ensemble of deep classifier techniques based on hybrid sensor types composed of wearable and environment interactive sensors to improve the prediction and recognition performance of activities of daily living in smart home environments. Indeed, this ensemble is designed by combining the both automatically learned features and hand-crafted features from Denoising Stacked Autoencoders (DSAE) and Random Forest (RF) algorithm respectively. Specifically, the combination involves both the features and outputs of the two techniques using stacking learning. The use of two public benchmark datasets has enabled to evaluate our approach. Furthermore, the experimental results show the accuracy improvement of the ensembles of deep autoencoders classifiers compared to denoising stacked autoencoder networks and random forest algorithm performed individually. Hence, our approach adaptability to ubiquitous environments and its effectiveness in the recognition of human activity applications.

Keywords: Activity recognition · Deep learning · Ensemble classifier
Sensor modalities · Shallow features · Smart homes · Stacked autoencoder

© Springer Nature Singapore Pte Ltd. 2018
Q. Zhou et al. (Eds.): ICPCSEE 2018, CCIS 902, pp. 273–295, 2018.
https://doi.org/10.1007/978-981-13-2206-8_24

1   Introduction

The convergence of ubiquitous computing and machine learning technologies has developed and fostered the emergence of sophisticated research fields in smart environment systems. Activity recognition constitutes an important consequence of that convergence over these past years. Indeed, it emphasizes the accurate detection of human activities based on a predefined activity model by exploring various kinds of sensing technologies to enhance the recognition rate while adapting to the generalized application techniques. In order to improve human activity recognition accuracy, we used a hybrid sensing technique that combines wearable and environment interactive sensors, as shown in Table 1. This hybrid sensor model takes advantages from strengths of each sensor type to construct a model that is able to accurately recognize complex and/or context-aware activities of multiple inhabitants. Given the effectiveness of each sensor type to recognize different activities, the wearable sensor model has the ability to capture the interaction between an inhabitant and objects through monitoring system in order to provide a solution to activity recognition in a smart home. It usually includes accelerometer, gyroscope, GPS, and RFID- readers/tags technology. On the other hand, the environment interactive sensor model has the ability to infer the activity of daily living through the interactions capture between an individual and a specific object. And it is essentially composed of contact sensors, temperature and pressure sensors, and motion detectors. Additionally, with the exponential evolution of information technology, the both types of sensor models have been widely used successfully for human activity recognition due to their easy deployment, low cost, less intrusiveness, and flexibility. Furthermore, that combination can allow a better capture of relevant information of human events. Therefore, elaborating information generated by hybrid sensors is critical for fine-grained activities recognition. The Fig. 1 introduces a human activity recognition system that combines shallow and deep features. It consists of five main steps as follow:

Raw activity signal: also called raw signal inputs, it generally includes data collection and data preparation and is acquired using various kinds of sensors. In our hybrid sensor model, the variation of data acquisition depends on the plurality of these sensor models. Therefore, data collection in human activity recognition systems requires the

Table 1. Hybrid sensor modalities for activity recognition

Sensor modalities	Wearable sensors	Environ. inter. sensors	Hybrid sensor
Description	Worn on different Parts of the body To recognize Activities	Applied in environment to reflect interaction user and Attached Objects	Cross sensor boundary
Illustration	Gyroscope, watch, Smartphone, wrist Glass band's accelerometer	Contact sensor, pressure, sound, temperature WiFi, RFID Bluetooth…	Combination of wearable and environment interactive types often deployed in smart environments

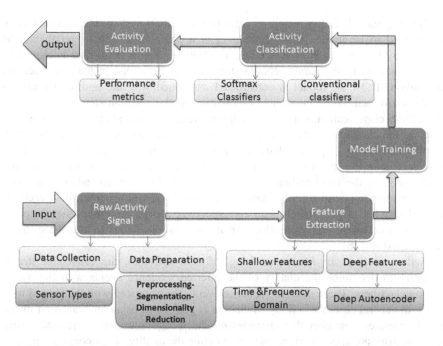

Fig. 1. Sensor based activity recognition system with shallow and deep features.

sensors [1]. On other hand, after collecting data from sensors and before making any computation, the preparation of acquired data is needed according to preprocessing, segmentation and dimensionality reduction steps for the features extraction. During the preprocessing step, the conversion of raw data to meaningful information meets the process composed of collection and conversion, mining, and clustering similar patterns. Subsequently, the collected data is processed in various ways. Almost always, high frequency noise from users in acceleration data is removed [2]. Then, the segmentation scheme is applied to it. The segmentation, also called spotting stage, ensures segments identification of preprocessed data which likely contains information about activities. Several segmentation methods are used to retrieve useful information from continuous streams of sensor data for continuous activity recognition [3]. After the segmentation, the dimensionality reduction is used to reduce the number of random variables considered by obtaining a set of principal variables. It aims to increase accuracy and reduce the computational efforts, since the high feature sets of dimensionality cause computation slowness and training difficulties [4].

Feature extraction: According to our model, two kinds of features including shallow and deep features are extracted from the raw data segmented. The former case is manually extracted from those readings based on human knowledge such as time and frequency domain in traditional machine learning architectures [5, 6], while the latter case is extracted automatically from the original low-level features through the network without any specific domain knowledge but with a general-purpose learning procedure.

Training model: Before using an activity recognition system, the classifier algorithms need to be trained. Indeed, training is a preparation step to get the model parameters for later use in classification. The training step is performed infrequently, and the resulting model parameters are stored for the future use in the actual activity recognition. Thus, the model is constructed and tuned with the optimal parameters usually performed offline in order to be used in the classification step.

Activity classification: After the training models, the selected attributes are used as input for the classification and recognition methods [7]. However, the activity classification includes two types of classifiers such as conventional classifiers and softmax classifiers referring to traditional machine learning and deep learning methods respectively. In the hand-crafted model, the trained classifiers are adopted to make activity inference in real activity recognition tasks. Moreover, each attribute vector is referred to a set of class labels with corresponding scores. And then, the computed scores are used by the end application to make a decision whether to trust the system's output transform [8].

Activity evaluation: After classifying different activities, the evaluation step is necessary to determine the quality of the inferred sequence events with the use of training classifiers. These inferred sequence events are compared with the ground truth labels. In general case, the activity evaluation is performed using an evaluation metric which generates a number that characterizes the recognition quality. Several performance metrics are applied. Those measures define the quality of a recognition method in order to accurately analyze their performance errors.

Therefore, this sensor-based activity recognition system showcases a brief idea of human activity recognition process for retrieving activity information from the sensors and combining the both traditional and deep learning classifiers. However, understanding the improvement of more accurate detection constitutes a challenge in activity recognition problems.

Instead of comparing traditional machine learning and deep learning techniques for sensor based activity recognition in smart environments, this work explores the way to combine the advantages of the both models considering their features and outputs using ensemble learning. Specifically, random forest (RF) algorithm is applied on the features learnt by the stacked denoising autoencoders. Moreover, a hybrid sensors model is used in order to accurately recognize various kinds of human activities of daily living. That hybrid sensors model consists of wearable and environment interactive sensors which can be able to capture very relevant information of human activities. The spectacular progress made on human activity recognition by applying traditional machine learning techniques such as Random Forest to sensor data further has favored the recognition of human activities. Those traditional techniques are more robust and better recognize simple activities which are represented with small number of examples. However, we note some drawbacks including the use of heuristic or hand-crafted approaches for feature extraction which is usually limited by human domain knowledge and result in information loss [9], shallow structures learning with difficulties in the discovery of latent non-linear relations inherent in features [10]. Indeed, feature extraction constitutes an essential step in constructing machine learning classification schemes, and it has great influence on the accuracy and the performance of those schemes.

Nevertheless, since this step is expensive, the features extracted manually can not guarantee a good generalization in different sensor data modalities.

Consequently, the performances concerning the classification accuracy and model generalization are restricted. On the other hand, deep architectures which are the state-of-the arts machine learning technology appear as a robust tool for learning complex and large scale problems in ubiquitous and sensor networks [11]. Those architectures enable automatic feature learning through the network from the original low-level features with a general purpose learning procedure. In other words, these models have the ability to include feature extraction, feature selection and classification steps into a single optimization scheme. Moreover, high level representation extraction in abstract layer makes the deep neural network more powerful and suitable for complex activity recognition tasks. Obviously, deep generative models are more effective in exploiting a large amount of unlabeled data for model training [12]. However, despite tremendous performances of deep learning methods, they are very computationally expensive compared to conventional machine classifiers, requiring huge training datasets for a good classification performance achievement. Therefore, the problematic data situation, and heavily imbalanced data distributions remain largely unaddressed for human activity recognition tasks.

In this paper, we introduce an ensemble of deep classifiers based on hybrid sensing data types for activity recognition in smart home environments. Unlike several previous works that compared traditional machine learning and deep learning techniques, our approach explores an alternative of merging the advantages of the both models. We combine the automatically learned features and hand-crafted features from Denoising Stacked Autoencoders (DSAE) and Random Forest (RF) algorithm respectively. More precisely, the combination involves both the features and outputs of the two techniques using stacking learning. However, two phases are needed to train the DSAE architecture, namely unsupervised pre-training and supervised phases. In the former phase, a greedy layer wise learning is performed to pre-train a deep network with one layer at time, and then each layer is trained as an autoencoder by minimizing reconstructing error of its inputs using unlabeled samples. While in the latter phase, fine-tuning is used to improve the results by tuning the parameters of all layers which are changed at the same time using the back-propagation. Additionally, this phase is called supervised since the labeled samples are needed to minimize the prediction error. Globally, the both phases unsupervised and supervised are helpful to better initialize parameters and minimize prediction error for classification respectively. In order to quantify how good a given scoring function is at correctly classifying data points in our datasets, the softmax layer is added on each hidden layer with fine-tuning. Finally, the accuracy and robustness of softmax are improved by considering the ensemble of classifiers and also the integration of the above classifier with different weight according to the corresponding accuracy rate [13].

The remainder of this work is organized as follows: In the next section, the background is introduced. In Sect. 3, a theoretical description of the proposed approach is presented. Section 4 provides experimental process description. In Sect. 5, experimental results are presented. Finally, the conclusion and future works are drawn in Sect. 6.

2   Background

Human activity recognition aims to understand human behaviors that enable the computing systems to proactively assist users based on their requirements [14]. In order to progressively develop and enhance the performance techniques for reliable human activity recognition and also to allow their applicability in the real world over the years, several remarkable works have been undertaken using pervasive and various sensory modalities [21]. It has shown that many works have combined different types of sensors for human activity recognition (HAR). In [22], combining body-worn, object and ambient sensors could foster a large number of find-grained and complex activities of multiple occupants to be recognized. Since HAR has become a leading broad and active research area of artificial intelligence, it is more than likely that the future innovations and developments will depend on robust and effective activity recognition from sensor data as well [23]. Thus, aiming for precision and conciseness, the synthesis of the global state-of-the-art in activity recognition are avoided from reiterating here. Therefore, nowadays HAR can be considered common knowledge referring to relevant works and tutorials such as [24–26]. More importantly, in the following we will highlight the specific algorithmic background for this paper, which consists of three main topic domains: (i) the stacked denoising autoencoders for HAR in pervasive and sensory modalities; (ii) random forest technique; and (iii) stacking ensemble classifiers.

2.1   Stacked Denoising Autoencoders

Architecturally, Autoencoder (AE) is a feed-forward neural network much similar to the multilayer perceptron (MLP). It consists of three layers: an input layer, one or more hidden layer(s) and an output layer. Moreover, autoencoder is considered as an unsupervised learning since it learns to reconstruct its input with a minimum of errors with the constraint that the target values of the output layer are equal or approximate to the inputs during training. AE is composed of two stages, namely encoder and decoder phases as depicted in the Fig. 2. The encoder stage is from input to hidden layer, while decoder stage is from hidden to output layer, and they can be formulated as in Eqs. (1) and (2) respectively. By assuming that the M and N represent the numbers of input neurons and hidden neurons respectively; an autoencoder transforms an M-dimensional vector $x = (x_1, x_2, \ldots, x_M)$ to a latent representation $h(x) = (h_1, h_2, \ldots, h_N) \in R^{N \times 1}$ according to Eq. (3) through a deterministic mapping also called encoding .... Equation 1 according to the following formulation:$h(x) = f(W^{(1)}x + b^{(1)})$, where $W^{(1)} \in R^{N \times M}$ is the weight matrix and b_1 defines the bias vector for encoding. On the other hand, $W^{(2)}$ and b_2 represent the matrix weight and bias respectively for the decoding, in Eq. 2. The $f(\cdot)$ (.) in (1) and (2) is the non-linear activation function of each neuron. In our case, we used sigmoid function

$$h = f(Wx + b_1) \tag{1}$$

$$\tilde{x} = f(W^T h + b_2) \tag{2}$$

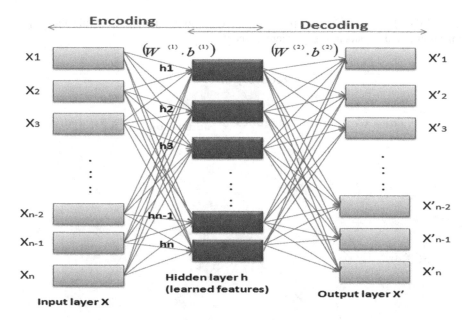

Fig. 2. Basic architecture of autoencoder.

$$h(x) = (h_1, h_2, \ldots, h_N) \in R^{N \times 1} \tag{3}$$

In order to motivate the hidden layer to learn good input representations, denoising autoencoder, which constitutes a variant of autoencoder has been proposed to achieve feature representations with better robustness and generalization performance. It tries to recover the original noise-free data from a corrupted input with a local denoising criterion. Indeed, that feature representations achievement is possible by adding random noises to the original inputs or randomly choosing a portion of them and setting them to be zero [15]. The process of denoising autoencoder can be described as follows: at first, the original input data X is corrupted into $\tilde{X}$ to create a partially destroyed version denoted $\tilde{X}$ using a stochastic mapping $\tilde{X} \sim qD(\tilde{X}|X)$, where $qD(\cdot)$ represents a modality of corrupted noise distribution and randomly sets a fraction of the input to be zero, while the other are remained unchanged. Thus, corrupted input $\tilde{X}$ is then mapped to a hidden representation network $y = f_\theta(\tilde{x}) = s(W\tilde{x} + b)$ from which $z = g_{\theta'}(y)$ should be recovered as showcased in the Fig. 3. Additionally, training of parameters θ and θ' leads to the minimization of the average recovery error over a training set in order to have a z as close as possible to the initial input data X. However, the essential difference is that z represents a deterministic function of $\tilde{x}$ rather than x. On the other hand, the decoder function is used for the translating hidden feature vector back to the original representation in a similar way as follows: $z = g_{\theta'}(h) = s(W'h + b')$. Finally, the cost function for minimizing the recovery error can be either the cross-entropy Loss $L_H(x, z) = H(B(x)||B(z))$, or the squared error

Loss $L_2(x, z) = \|x - z\|^2$. The parameters are randomly initialized while their optimization relied on stochastic gradient descent.

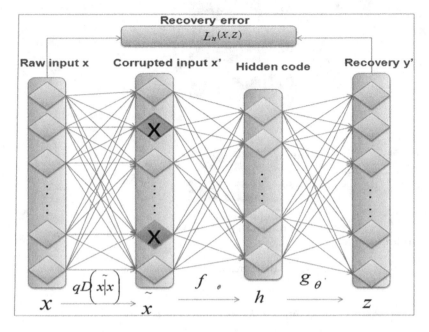

Fig. 3. Architecture of denoising autoencoder

On the other hand, stacked autoencoder that is a hierarchical architecture of deep learning, and in which an autoencoder is a building block, actively participates in the acquisition of more complex and non-linear relations underlying in data [16]. The Fig. 4 illustrates an activity recognition model based on stacked autoencoder and classifier layer. This architecture is formed by stacking the basic units (autoencoders) on the top of each other, and then by adding classifier layer on the top of the autoencoders. Generally, training stacked denoising autoencoders require two steps. The first step is called unsupervised pre-training where various autoencoder layers are stacked together. The latent representation obtained by an autoencoder is used as input to its successive autoencoder layer. During this process, the training is performed with one layer at a time in order to minimize reconstruction error of its input by using unlabeled samples. This reconstruction error can be computed using some models such as cross-entropy [17], while the error can be minimized by applying optimization methods such as a greedy layer-wise learning process that can considerably alleviate the issues related to gradient diffusion and poor local optimum on the network by improving the convergence rate [12]. The second step is called supervised fine-tuning. In this phase, the use of features learned with a set of labeled data in stacked autoencoder allows the building classifier layer. The classifier layer is stacked on top of the stacked autoencoder to classify an input and minimize the prediction error.

Furthermore, the feature vector encoded in the last hidden layer constitutes the input of a learning algorithm in the classifier layer. In a supervised manner, the activity recognition model is further optimized in order to improve its performance; then the whole network is fine-tuned. As we mentioned early in the literature, stacking denoising autoencoders is necessary for a deep network initialization and that process works in much the same way as basic autoencoders [27–29]. However, it is important to remind that input corruption is only used for the initial denoising-training of each individual layer with the ability to learn useful feature extractors. After learning the mapping f_θ, it will be automatically used on uncorrupted inputs; thus no corruption is applied to generate the representation that will be useful in cleaning input for training the next layer. Additionally, after constructing an encoders' stack, its highest level output representation is utilized as input to autonomous supervised learning algorithm such as Random Forest. Also, the parameters of all layers are then simultaneously fine-tuned using gradient-based optimization methods such as greedy layer-wise learning process. In addition to fine-tuning, the use of grid search strategy can enable to find the optimal learning parameters and the network layout, for example the number of hidden layers and neurons in each hidden layer, and also the choice of the best network structure as the final activity recognition approach via cross- validation.

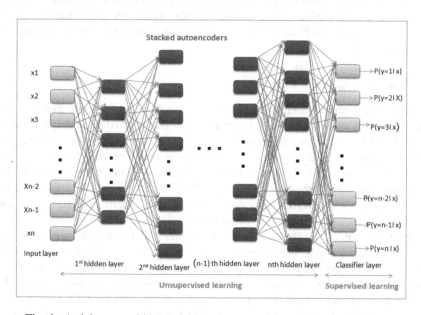

Fig. 4. Activity recognition model based on stacked autoencoders architecture

2.2 Random Forest Algorithm

Random Forest (RF) algorithm is one of the most popular traditional machine learning approaches that can be used for both classification and regression tasks. In this part, a particular accent has been put on the use of random forest for classification tasks.

Indeed, as its name suggests, this supervised algorithm creates the forest with a number of trees. Furthermore, the more trees in the forest, the more robust the forest looks like. On the other hand, in that classification technique, the higher the number of trees in the forest gives the high accuracy results. Compared with certain other classical classification algorithms, random forest meets the following advantages: (i) the random forest classifier can be used the both classification and regression tasks. (ii) The overfitting problem will never occur when the random forest algorithm is used in any classification problem. (iii) Random forest classifier has the ability to handle the missing values. (iv) The random forest algorithm can be used for feature engineering. In other words, identifying the most relevant features out of the available features from the training dataset. Like other conventional machine learning algorithms, a standard classification framework is used for random forest as follows: data segmentation, data filtering, feature extraction, feature selection, and building classification model. During the segmentation step, segments identification of preprocessed data is ensured using some techniques such as overlapping sliding window to retrieve useful information from continuous streams of sensor data for continuous activity recognition. After segmenting the sensor measurements, the preprocessing is made using filters. Afterwards, features are hand-crafted extracted from readings based on human knowledge. After features extraction, features vectors are formed and used by random forest algorithm to build an activity recognition model during training. Finally, during the classification step, the feature vectors extracted from test data are fed into the model in order to recognize the user activity.

2.3   Stacking Ensemble Classifiers

In artificial intelligence, ensembles learning are the meta-algorithms commonly used to combine various machine learning techniques into one predictive model. Indeed, the essential goal is to decrease variance, bias or to improve predictions by using bagging, boosting or stacking, respectively. Building robust and powerful ensemble models requires to each machine learning technique a significant contribution, and individual weaknesses and biases are offset by the strengths of other algorithms. In short, stacked ensemble models aim to combine various sets of individual models together via a meta-level classifier or meta-regressor in order to improvise on the stability and predictive power of the model. In stacking learning scheme, the predictions generated from various machine learning algorithms are used as inputs in a second-layer learning algorithm. This second-layer algorithm is trained to optimally combine the model predictions and to form a new set of predictions. In other words, the base level models are trained based on a complete training set, and then the meta-model is trained on the outputs of the base level model as features. That base level often consists of different learning algorithms and therefore stacking ensembles are often heterogeneous. The process of stacking learning can be summarized as follows: splitting the training set into two disjoint sets; training several base classification models on the first stage; testing the base classification models on the second stage as the inputs, and the correct responses as the outputs; finally, training a higher level classifier. In that combination process, the output of the base classification models (level0) is used as training data for meta-classifier (level1) to approximate the same target function. Therefore, the class predicted by each different model is given as input for a meta-level classifier whose output is the final class. In this work, the stacking learning has been applied using the multiple response model tree (M5') as meta- level classifier, as shown in the Fig. 5. Indeed, since these past years, stacking has been involved the use of top-level classification model which is able to learn from the prediction of based-level models in order to achieve greater classification power as underscored in [30–33]. Moreover, the ability to exploit the diversity in the predictions of based-level classifiers makes stacking more suitable than individual classifier based approaches or alternate fusion strategies that combine based-level classification by using a simple scoring or voting schemes as underlined in [34–38]. However, although the remarkable progress in the development of various data mining algorithms and also the testing of multiple meta-classifiers, overfitting remains an omnipresent issue in building predictive models because it is complex enough to perfectly fit the training data. In the case of stacking model, the overfitting partially caused by the colinearity between predictors, is a significant concern because of the combination of many predictors which predict the same target. Nevertheless, some efficient techniques such as cross validation and some regularization forms are used for training models [18]. Finally, stacked ensemble techniques such as regularized regression methods, gradient boosting, and hill climbing methods are combined with some meta-classifiers (e.g. logic regression, gradient boosted decision trees, factorization machine, forests...) to generate a various set of models. To the best of our knowledge, there is no general alternative solution in selecting models although tremendous efforts made in that field. However, the application of stacked models to real-world big data issues can produce greater prediction accuracy and robustness than doing individual models. Stacking approach is powerful and compels enough to alter initial data mining mindset from finding the single best model to finding a collection of really good complementary models. An ensemble learning meta-classifier for stacking is illustrated in the Fig. 5.

3 Datasets Description

Our experiments have been performed using two different smart home environment datasets. The first dataset was made by the Center for Advanced Studies in Adaptive Systems (CASAS), an integrated system of collecting sensor data and applied to machine learning in the smart home environments. The second dataset is the Human Activity Recognition using Smartphone dataset, one of the most commonly used benchmark datasets for activity recognition.

3.1 CASAS Smart Integrated System

CASAS Integrated System is a multidisciplinary research project at Washington State University (WSU) which focused on creating intelligent home environment. That smart home system is a three bedroom apartment located at university campus. It consists of three bedrooms, a living/dining room, a kitchen and a bathroom. In order to reflect the interaction between environment and objects, and also to enable the mobility of users, we used a hybrid sensor system composed of environment interactive and wearable

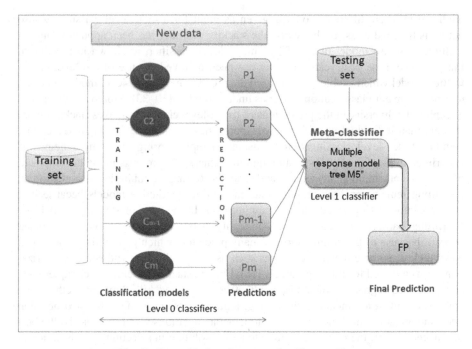

Fig. 5. Ensemble learning meta-classifier stacking

sensors such as motion detector sensors, contact switch sensors, temperature and pressure sensors, float sensors and mercury contact sensors disseminated in whole three bedroom apartment. Building sensor system in-house enables the capture of sensor data events that are stored in a database. These sensor data events are interpreted for activity of daily living that are used for training and testing the activity recognition algorithms. The collection of data in order to provide real training data for our algorithms has been done while two students in good health were living in the smart apartment. The both students shared the downstairs living areas but had a separate bedroom. All of our experimental data are produced by these students' day to day lives. Moreover, during several months of gathering training data, our dataset generated more than 100,000 sensor data events. Therefore, the analysis results are guaranteed to be real and useful. The sensor data collected by the dataset for our study is represented by the parameters shown in Table 2 as follow: Date, Time, Sensor ID, Message of sensors and activity Target Output, produced by the CASAS data collection system automatically. The sensor data events are interpreted with the corresponding activities which were been executed after gathering data from the smart three bedroom apartment while the sensor data events were produced. However, since learning algorithms performance strongly depends on the interpreted data quality, this data usage is very crucial for training machine learning algorithms. Furthermore, the seven activities performed by the both students in the smart three bedroom apartment and used for our experiment, are presented as follow: Sleeping, Watching TV, having Breakfast (Eating), Working at computer, Toileting, Cooking (Preparing meal), Grooming.

Table 2. Examples of raw data from sensor events.

Date	Time	Sensor ID	Message	Target Output
2009-2-6	17:17:36	M45	ON	Motion sensor start
2009-2-6	17:17:40	M45	OFF	Motion sensor end
2009-2-6	11:13:26	T004	21.5	Temperature reading
2009-2-5	11:18:37	P001	747 W	Current Power usage
2009-2-9	21:15:28	P001	1.929 KWh	Current Power usage

3.2   Human Activity Recognition Using Smartphone Dataset

The Human Activity Recognition using Smartphone dataset is a widely used public dataset in the UCI Machine Learning Repository and available following this link: https://www.elen.ucl.ac.be/Proceedings/esann/esannpdf/archive.ics.uci.edu/ml/datasets/Human+Activity+Recognition+Using+Smartphones. A series of experiments were conducted by a group of 30 volunteers aged between 19 and 49 years old. Each volunteer was wearing a Smartphone (Samsung Galaxy II) on the waist according to the activities protocol. In this activity recognition task, six classes are performed including lying, sitting, standing, walking, upstairs as well as downstairs. Thus, the volunteers conducted freely the activities sequence events for more naturalistic dataset in laboratory conditions. The produced dataset has been randomly divided in training and testing sets with 70% and 30% of selected data respectively. The training set contains 7352 activity instances while testing set contains 2947 activity instances. Therefore, all instances are labeled manually with each of the six activities. The dataset generates a total of 561 extracted features by making use of raw signals produced from sensors (accelerometer and gyroscope). The use of accelerometer and gyroscope allows the triaxial linear acceleration and angular velocity collection at a sampling rate of 50 Hz. In addition, median filter application to collected signals facilitates the noise removal. Then, the use of Butterworth low-pass filter permits the acceleration signals segregation into body gravity and acceleration. Furthermore, the time signals have been sampled in fixed-width sliding windows of 2.56 s and 50% overlap.

4   Proposed Ensemble Deep Autoencoder Learning Approach

4.1   Hybrid Sensor Model

As underlined earlier, some observations related to machine learning and deep learning techniques for sensor modalities based activity recognition in smart environment have predominately motivated our work. First, sensing modalities often cope with relevant challenges concerning both the reliability and amount of data. From this point of view, the individual sensor models can robustly infer only specific activities types. We can note for example wearable sensor models which use accelerometers, gyroscope, RFID-readers/tags, GPS technologies are embedded in Smartphone and other mobile devices to catch interaction between a subject and objects in order to recognize activity of daily living. While environment interactive sensor models which use sensors related to temperature, pressure, contact, motion, etc....have capacity to infer the activities of daily living performed by a human through the interactions capture between an individual and a specific object. Furthermore, this work combines the both sensor models types in a hybrid approach which has the ability to recognize more high-level activities of various inhabitants by providing more relevant and essential information for human activities. Thus, the choice of suitable sensor models is crucial for human activity recognition success. The output of an activity recognition system which constitutes a stream of sensor activations relies on a very powerful sensing model to handle the activity recognition as a time series analysis problem for a seamless quantity identification of sensor data stream attached to one of preselected activities. Thus, the hybrid sensor model used in this work can favor the deep high-level knowledge search about human activity with the combination of a large number of low-level sensor readings. Furthermore, supervised learning application extensively used to sensor based activity recognition with a specific training phase is composed of a sensor data stream segmentation, features extraction and raw signal data transformation into feature vectors followed by the classifier building, as well as the recognition step with the utilization of trained classifier in order to join a sensor data stream with a predefined activity, is then explored. However, obtaining activity recognition performance based on sensors strongly depends on some relevant parameters in ubiquitous computing and machine learning technologies such as relevant pattern representation of sensor data, adequate classifier choice and settings. Therefore, this work proposes to combine hand-crafted features with automatically learned features learning from random forest and denoising stacked autoencoder respectively in order to improve human activity recognition performance.

4.2 Feature Ensemble Learning

Instead of comparing traditional machine learning and deep learning techniques as commonly presented by many previous researches in activity recognition [39, 40], this work aims to apply a conventional classification algorithm on the features learned by a deep learning model. More precisely, we determine an ensemble of deep learning for activity recognition including trained stacked autoencoders by using different number of latent representation codes, and the random forest with inputs as the outputs of stacked autoencoders and the outputs as the final prediction. In other words, that combination involves essentially the features and outputs of the both models using stacking meta-learning. In order to maintain the substantial performance balance between accuracy and computational cost in activity recognition, in this work, the stacked autoencoder model is trained with three hidden layers, which lead to the combination of three layers into one feature set as shown in the Fig. 6.

Furthermore, the operation of our ensemble of deep stacked autoencoder algorithm is presented as follows: (i) stacked autoencoder training on the input data for highly compressed features learning; (ii) building and fine-tuning whole network of stacked autoencoders by using back-propagation algorithm for optimizing model parameters, and then to further improving the classification performance; (iii) stacking classifier layers on top of the stacked autoencoder for inputs classification with feature vectors in the last hidden layers as the inputs of a learning algorithm in the classifier layers;

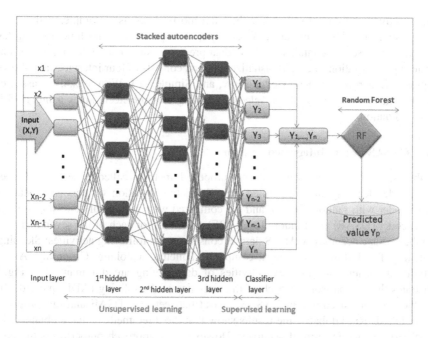

Fig. 6. Ensemble of proposed deep stacked autoencodes network

(iv) merging all the outputs of classifier layers into a matrix X*new* used for training a random forest with the expected prediction values Y. Therefore, in deep learning architectures and particularly in ensemble of deep stacked autoencoders, the model parameters optimization is crucial for high recognition accuracy generation. However, a standard optimization rule does not exist. In general case, some techniques such as choice of the best network architecture are largely applied. We can note for example the use of one hidden layer with several units (neurons), the use of many hidden layers (deep models) with a small number of units in each layer, as well as the use of several hidden layers with many units in each layer.

5 Experimental Process and Results

In order to evaluate performance of the proposed activity recognition model built on ensemble of deep denoising stacked autoencoder learning technique, we performed experiments on two different publicly available benchmark datasets for activity recognition equipped with multiple sensor types. Indeed, our experiments are based on a comparative analysis that aims at assessing activity recognition accuracy of the combination between a deep stacked autoencoder and a random forest algorithm in term of features and outputs. Furthermore, the accuracy of this ensemble of deep learning network is compared with the stacked autoencoder model and random forest algorithm respectively. Human Activity Recognition using Smartphone Dataset and CASAS Smart Integrated System represent our datasets. The former mostly consists of simple activities of daily living while the latter mostly consists of complex activities of daily living. To achieve this task, Weka, a popular open source machine learning test tool has been used. It contains tools for data pre-processing, classification, regression, clustering, association rules and visualization, and consists of four interfaces as follows: Explorer, Experimenter, Knowledge Flow, and Simple CLI. With Weka, we assessed the classification accuracy of the selected learning classifier techniques using 10-fold cross validation.

5.1   CASAS Smart Integrated System

In this sub-section, we evaluated our proposed method called ensemble of deep autoencoder learning network noted DSAE-RF (combination of denoising stacked autoencoder with random forest), and we compared with individual denoising stacked autoencoder (DSAE) and random forest (RF) algorithm for each of the seven activities and the average accuracy. CASAS dataset consists of the following activities: Sleeping, Watching TV, Eating, Working at computer, Toileting, Cooking, Grooming. As we introduced earlier, the complex activities of daily living are dominant. The Fig. 7 illustrates the comparison for each of the activities of daily living (ADL) individually and the averaged accuracy in CASAS dataset between RF, DSAE and our proposed model DSAE-RF. Globally, the results show that the three methods have obtained the good performance in term of accuracy. However, our approach outperforms the both other methods. Indeed, in carrying out certain activities, these both methods have substantially the same averaged accuracy. For the tasks of sleeping and working at computer, for example, we noticed the similar accuracy results. Thus, this similarity can be interpreted by the number of steps and the taken time to perform those tasks. On the other hand, for the activities having many steps for their execution, the averaged accuracy in DSAE method is higher than the averaged accuracy in RF model. In that case, it may be supposed that the longer the activity takes, the more the SDAE outperforms RF method. From these two cases, we can observe, on the one hand, the considerable efforts made by the traditional machine learning techniques in activity recognition for hand-crafted feature extraction, feature selection as well as the choice of the best algorithm. On the other hand, the robustness of deep autoencoder in automatically suitable features extraction and the accomplishment of better performance in term of accuracy. Overall, our method considers the strengths and weaknesses of the other both models by combining the features and outputs of the both techniques using stacking as meta-learning with multiple response model tree M5'' as meta-classifier. Furthermore, our approach yields the highest averaged accuracy for each of corresponding activities of daily living in CASAS dataset while RF algorithm yields the lowest.

Traditional machine learning method under RF algorithm, deep learning technique under DSAE and our approach DSAE-RF are compared in term of overall averaged accuracy. Thus, Table 3 shows that our approach outperforms the both denoising stacked autoencoder method and random forest algorithm by 3% and 7% respectively. Moreover, from Fig. 7, the similar accuracy results for some corresponding activities such as "sleeping" with 0.87 (87%) in random forest algorithm and denoising stacked autoencoders model and "working at computer" with 0.86 (86%) are improved in our

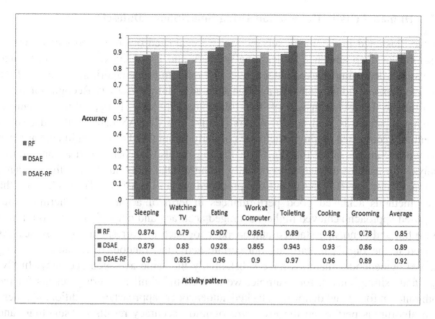

Fig. 7. Comparing RF, DSAE and DSAE-RF models for averaged accuracy with CASAS.

proposed approach by 0.03 (3%) and 0.04 (4%) respectively. In other hand, some other activities have very good averaged accuracy results in our proposed approach compared to random forest and denoising stacked autoencoders models for corresponding activities such as "eating", "toileting" and "cooking". For these above activities, our proposed method outperforms random forest algorithm and denoising stacked autoencoder by 0.06 (6%) and 0.04 (4%) for "eating" respectively, by 0.08 (8%) and 0.03 (3%) for "toileting" respectively, and by 0.14 (14%) and 0.03 (3%) for "cooking" respectively. We can notice a significant difference between random forest algorithm and our proposed method for the recognition of these corresponding activities, particularly "toileting" task by 14%. These activities tasks including "cooking", "toileting" and "eating" have many activity events and naturally take much time to be performed. Furthermore, it is difficult for traditional machine learning such as random forest which manually extracts features to achieve significant recognition accuracy results. Therefore, this analysis may be interpreted as follow: the longer the activity takes, better our proposed method performs and the more it outperforms the random forest algorithm and denoising stacked autoencoder methods.

Table 3. Comparing averaged accuracy of DSAE-RF with the both other algorithms in CASAS Dataset.

Approaches	Overall averaged accuracy %
Random Forests (RF)	85
Denoising Stacked Autoencoders (SDAE)	89
Our approach (DSAE-RF)	92

5.2   Human Activity Recognition Using Smartphone Dataset

As we illustrated in the above sub-section, we evaluated our proposed method called ensemble of deep autoencoder learning network and noted DSAE-RF and compared with single denoising stacked autoencoder and random forest algorithm for each of the six activities individually and the average accuracy. Human Activity Recognition using Smartphone Dataset consists of the following activities: lying, sitting, standing, walking, upstairs as well as downstairs. Unlike the CASAS dataset, this dataset is essentially composed of simple activities of daily living (ADL), as mentioned earlier in this work. Similar as for the Fig. 7, in the Fig. 8, also we performed a comparative analysis for each of the activities of daily living (ADL) individually and the averaged accuracy between RF, DSAE and our proposed model DSAE-RF. Thus, globally, the three methods achieved good performances in term of accuracy. Furthermore, our proposed approach achieved highest averaged accuracy and outperforms random forest algorithm and denoising stacked autoencoder for each of corresponding activities. As noticed in the Fig. 7, in carrying out certain activities, random forest and denoising stacked autoencoder methods have substantially the same averaged accuracy. In "lying" and "sitting" tasks, for example, we noticed the similar accuracy results for the both random forest and denoising stacked autoencoders approaches. Additionally, these both algorithms performed the lowest recognition accuracy results in "standing" and "sitting" tasks compared to other corresponding activities in the same dataset. Indeed, this low performance may be interpreted by a lower number of steps represented in those activities. However, the use of our method in activities with lower number of step has improved the recognition accuracy results and outperformed once again random forest and denoising stacked autoencoder models. However, the results show the best performances achieved in "downstairs" and "upstairs" tasks by all the three methods. Even though the recognition accuracy results in DSAE method is higher than accuracy results in RF algorithm, these performance results remain high. For these activities, it may be interpreted that there are many steps for their execution compared to other corresponding activities in the same dataset. Overall, our approach achieved the highest performance and outperformed the both other methods. From this dataset, we can observe the considerable efforts made by the traditional machine learning techniques in activity recognition for hand-crafted feature extraction, feature selection as well as the choice of the best algorithm. Thus, the random forest method performs good enough recognition accuracy results because the dataset is mostly composed of simple activities that are better for traditional machine learning techniques. However, the robustness of deep autoencoder to automatically extract suitable and relevant features enables to achieve good performance in term of accuracy in this dataset. Furthermore, our approach yields the highest averaged accuracy for each of corresponding activities of daily living while RF algorithm yields the lowest.

As mentioned from Table 3, traditional machine learning method under RF algorithm, deep learning technique under DSAE and our approach are compared in term of overall averaged accuracy. Thus, Table 4 shows that our approach outperforms DSAE and RF algorithm by 6.7% and 7.2% respectively. Moreover, as explained in 5.1, from Fig. 8, the similar accuracy results for some corresponding activities such as "lying" with 0.89 (89%) in RF algorithm and DSAE model and "sitting" with 0.782 (78.2%)

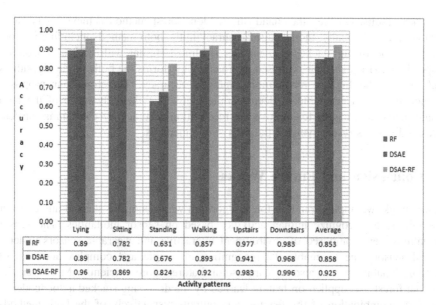

Fig. 8. Comparing RF, DSAE and DSAE-RF models for averaged accuracy with Human Activity Recognition using Smartphone Dataset.

are improved in our proposed approach by 0.07 (7%) and 0.087 (8.7%) respectively. Additionally, some activities have very good averaged accuracy results in our proposed approach compared to RF and DSAE for corresponding activities such as "lying", "sitting" and "standing". For these above activities, our proposed method outperforms random forest algorithm and denoising stacked autoencoder by 0.07 (7%) for "lying", by 0.087 (8.7%) for "sitting" and by 0.193 (19.3%) and 0.148 (14.8%) for "standing" respectively. We can notice a significant difference between random forest algorithm, denoising stacked autoencoder methods and our proposed approach for the recognition of these corresponding activities, particularly "standing" task by 19.3% and 14.8% respectively. It may be interpreted that these activities are represented with a few number of events which lead to a poor recognition accuracy.

Table 4. Comparing Averaged accuracy of DSAE-RF with the both other algorithms in Smartphone Dataset

Approaches	Overall averaged accuracy %
Random Forests (RF)	85.3
Denoising Stacked Autoencoders (SDAE)	85.8
Our approach (DSAE-RF)	92.5

Furthermore, it is difficult for deep learning techniques such as denoising stacked autoencoders model which needs more features to achieve significant recognition accuracy results. On the other hand, the random forest method achieved very satisfactory results. Globally, the averaged accuracy in the both methods including random forest and denoising stacked autoencoder is almost similar with 85.3% and 85.8% respectively. However, our proposed approach outperforms the both single algorithms and yielded the highest recognition averaged accuracy. Therefore, our proposed method that is an ensemble of deep learning network, takes advantage of the both single methods strengths to improve the recognition accuracy by combining the features and outputs respectively of the both single methods.

6   Conclusion and Future Works

In this work, we applied random forest algorithm with denoising stacked autoencoder pre-training to recognize human activities based on sensor modalities. The sensor modalities were composed of environment interactive and wearable sensors under a hybrid sensor type to enable the recognition of simple and complex activities of multiple inhabitants and also to learn useful features more efficiently. Moreover, the random forest was applied on feature learned by the denoising stacked autoencoder to allow the combination of the features and outputs respectively of the both methods under an ensemble of deep denoising stacked autoencoder approach using stacking learning with multiple response model tree M5″. Overall, on the both datasets, the experimental results revealed that our proposed approach achieved the best performances compared to other methods. On CASAS dataset which is composed in majority of complex activities, our approach yielded the highest averaged accuracy for each of corresponding activities despite the robustness of denoising stacked autoencoder model in automatically suitable features extraction and the accomplishment of better performance in term of accuracy in this dataset. On the other hand, on Human Activity Recognition using Smartphone Dataset which is composed in majority of simple activities of daily living, our approach also yielded the highest averaged accuracy for each of the corresponding activities despite the fact that random forest technique is more robust and better recognizes the basic activities which are represented with small number of examples. Therefore, by taking advantages of strengths of the both models, our method has considerably improved the recognition accuracy.

In the future work, we plan to consider the features of all layers, unlike the general case where only the more abstract features are represented, to further enhance the classification performance. Moreover, in order to improve the prediction of the activity recognition, the development of more advanced optimization algorithms should be taken into consideration. Therefore, the use of more ensemble of deep learning with traditional machine learning techniques should be more developed.

Acknowledgments. This work is partially supported by the National Natural Science Foundation of China (Grant No. 61471147), the Fundamental Research Funds for the Central Universities (Grant No. HIT.NSRIF.2017037). Natural Science Foundation of Heilongjiang Province (Grant No. F2016016), the National Key Research and Development Program of China (Grant No. 2016YFC0901905).

Authors Contributions. Serge Thomas Mickala Bourobou proposed the work and confirmed its efficiency through the experiments. Jie Li supervised the work and directed the implementation. All authors wrote the paper and discussed the revision together.

Conflicts of Interest. The authors declare no conflict of interest.

References

1. Smith, X.S., Tong, H., Ji, P.: Activity recognition with Smartphone sensors. Tsinghua Sci. Technol. **19**, 235–249 (2014)
2. Shoaib, M., Scholten, H., Havinga, P.J.: Towards physical activity recognition using Smartphone sensors. In: 2013 IEEE 10th International Conference on (UIC) Ubiquitous Intelligence and Computing, pp. 80–87. IEEE (2013)
3. Foster, N.C.K., Juillard, C., Colbry, D., Panchanathan, S.: Recognition of hand movement using wearable accelerometers. JAISE **1**(2), 143–155 (2009)
4. Yang, Y., Wang, J., Chen, Y.: Using acceleration measurements for activity recognition: an effective learning algorithm for constructing neural classifiers. Pattern Recogn. Lett. **29**(16), 2213–2220 (2008)
5. Bao, L., Intille, S.S.: Activity recognition from user-annotated acceleration data. In: Ferscha, A., Mattern, F. (eds.) Pervasive 2004. LNCS, vol. 3001, pp. 1–17. Springer, Heidelberg (2004). https://doi.org/10.1007/978-3-540-24646-6_1
6. Hu, L., Chen, Y., Wang, J., Shen, J., Jiang, X., Shen, Z.: Less annotation on personalized activity recognition using context data. In: UIC, pp. 327–332 (2016)
7. Avci, A., Bosch, S., Marin-Perianu, M., Marin-Perianu, R., Havinga, P.: Activity recognition using inertial sensing for healthcare, wellbeing and sports applications: a survey. In: Proceedings of the 23rd Architecture of Computing Systems Conference, Hannover, Germany, pp. 1–10 (2010)
8. Bulling, A., Blanke, U., Schiele, B.: Atutorial on human activity recognition using body-worn inertial sensors. ACM Comput. Surv. (CSUR) **46**(3), 1–33 (2014)
9. Bengio, Yoshua: Deep learning of representations: looking forward. In: Dediu, Adrian-Horia, Martín-Vide, Carlos, Mitkov, Ruslan, Truthe, Bianca (eds.) SLSP 2013. LNCS (LNAI), vol. 7978, pp. 1–37. Springer, Heidelberg (2013). https://doi.org/10.1007/978-3-642-39593-2_1
10. Bengio, Y.: Learning deep architectures for AI. Found. Trends Mach. Learn. **2**, 1–127 (2009)
11. Bengio, Y.: Learning deep architectures for AI. Found. Trends R Mach. Learn. **2**(1), 1–127 (2009)
12. Hinton, G.E., Osindero, S., The, Y.W.: A fast learning algorithm for deep belief nets. Neural Comput. **18**, 1527–1554 (2006)
13. Lu, Y., Zhang, L., Wang, B., Yang, J.: Feature ensemble learning based on sparse autoencoders for image classification. In: International Joint Conference on Neural Networks, pp. 1739–1745 (2014)
14. Bulling, A., Blanke, U., Schiele, B.: A tutorial on human activity recognition using body-worn inertial sensors. ACM Comput. Surv. (CSUR) **46**, 33 (2014)
15. Vincent, P., Larochelle, H., Bengio, Y., Manzagol, P.A.: Extracting and composing robust features with denoising autoencoders. In: Proceedings of the 25th International Conference on Machine Learning, pp. 1096–1103. ACM Press, New York (2008)

16. Hinton, G.E., Salakhutdinov, R.R.: Reducing the dimensionality of data with neural networks. Science **313**, 504–507 (2006)
17. Shore, J., Johnson, R.: Axiomatic Derivation of the Principle of Maximum Entropy and the Principle of Minimum Cross-Entropy IEEE
18. Gunes, F., Wolfinger, R., Tan, P.-Y.: Stacked ensemble models for improved prediction accuracy. Paper SAS-2017, SAS Institute INC
19. Stiefmeier, T., Roggen, D., Ogris, G., Lukowicz, P., Troster, G.: Wearable activity tracking in car manufacturing. IEEE Pervasive Comput. **7**(2), 42–50 (2008)
20. Polikar, R.: Ensemble based systems in decision making. IEEE Circuits Syst. Mag. **6**(3), 21–45 (2006)
21. Ordonez, F.J., de Toledo, P., Sanchis, A.: Sensors-based bayesian detection of anomaly living patterns in a home setting. Pers. Ubiquitous Comput. **19**, 259–270 (2014)
22. Vepakomma, P., De, D., Das, S.K., Bhansali, S.: A-wristocracy: deep learning on wrist-won sensing for recognition of user complex activities. In: 2015 IEEE 12th International Conference on Wearable and Implantable Body Sensor Networks (BSN), pp. 1–6. IEEE (2015)
23. Abowd, G.D.: Beyong Weiser-from ubiquitous to collective computing. IEEE Comput. **49** (1), 17–23 (2016)
24. Bulling, A., Blanke, U., Schiele, B.: A tutorial on human activity recognition using body-worn inertial sensors. Comput. Surv. **46**(3), 1–33 (2014)
25. Chen, L., Hoey, J., Nugent, C.D., Cook, D.J., Yu, Z.: Sensor-based activity recognition. IEEE Trans. Syst. Man Cybern. Part C: Appl. Rev. **42**(6), 790–808 (2012)
26. Avci, A., Bosch, S., Marin-Perianu, M., Marin-Perianu, R., Havinga, P.: Activity recognition using inertial sensing for healthcare, wellbeing and sports applications: a survey. In: Proceedings of the International Conference on Architecture of Computing Systems (2016a)
27. Bengio, Y., Lamblin, P., Popovici, D., Larochelle, H.: Greedy layer-wiser training of deep networks. In: Scholkopf, B., Platt, J., Hoffman, T. (eds.) Advances in Neural Information Processing Systems 19 (NIPS 2006), pp. 153–160. MIT Press (2007)
28. Ranzato, M., Poultney, C.S., Chopra, S., LeCun, Y.: Efficient learning of sparse representations with an energy-based model. In: Scholkopf, B., Platt, J., Hoffman, T. (eds.) Advances in Neural Information Processing Systems 19 (NIPS 2006), pp. 1137–1144. MIT Press (2007)
29. Larochelle, H., Bengio, Y., Louradour, J., Lamblin, P.: Exploring strategies for training deep neural networks. J. Mach. Learn. Res. **10**, 1–40 (2009)
30. Brazdil, P., Giraud-Carrier, C., Soares, C., Vilata, R.: Meta-learning: Applications to Data Mining. Springer, Berlin (2008)
31. Hansen, J.V., Nelson, R.D.: Data mining for time series using stacked generalizers. Neurocomputing **43**, 173–184 (2002)
32. Ting, K.M., Witten, I.H.: Stacked generalization: when does it work? In: Proceedings of the 15th Joint International Conference on Artificial Intelligence, pp. 866–871. Morgan Kaufmann, San Francisco (1997)
33. Wolpert, D.H.: Stacked generalization. Neural Networks **6**, 241–259 (1992)
34. Abbasi, A., Zhang, Z., Zimbra, D., Chen, H., Numamaker Jr., J.F.: Detecting fake websites: the contribution of statistical learning theory. MIS Q. **34**(3), 435–461 (2010)
35. Dzeroski, S., Zenko, B.: Is combining classifiers with stacking better than selecting the best one? Mach. Learn. **54**(3), 255–273 (2004)
36. Hu, M.Y., Tsoukalas, C.: Explaining consumer choice through neural networks: the stacked generalization approach. Eur. J. Oper. Res. **146**(3), 650–660 (2003)

37. Lynam, T.R., Cormack, G.V.: On-line spam filtering fusion. In: Proceedings of the 20th Annual International ACM SIGIR Conference on Research and Development in Information Retrieval, Seattle, WA, 6–11 August, pp. 123–130 (2006)
38. Sigletos, G., Paliouras, G., Spyropoulos, C.D., Hatzopoulos, M.: Combining information extraction systems using voting and stacked generalization. J. Mach. Learn. Res. **6**, 1751–1782 (2005)
39. Gjoreski, H., Bizjak, J., Gjoreski, M., Gams, M.: Comparing Deep and Classical Machine Learning Methods for Human Activity Recognition using Wrist Accelerometer

An Anomaly Detection Method Based on Learning of "Scores Sequence"

Dongsheng Li, Shengfei Shi[✉], Yan Zhang, Hongzhi Wang,
and Jizhou Luo

Harbin Institute of Technology, Xidazhi Street 92, Harbin 150001, China
shengfei@hit.edu.cn

Abstract. Anomaly detection is very important in the field of operation and maintenance (O&M). However, in O&M, we find that direct use of the existing anomaly detection algorithms often causes a large number of false positives, and the detection results are not stable. Nothing a data characteristics in O&M: Many anomalies are often anomalous time periods formed by continuous anomaly points, we propose a novel concept "Scores Sequence" and a method based on learning of Scores Sequence. Our method has less false positives, can detect anomaly timely, and the detection result of our method is very stable. Through comparative experiments with many algorithms and practical industrial application, it proves that our method has good performance and is very suitable for the anomaly detection in O&M.

Keywords: Scores Sequence · Operation and maintenance
Anomaly detection · False positives · Stability

1 Introduction

Operation and maintenance (O&M) are very important in industry, network security, finance and so on, it helps to provide efficient service, improve product quality and reduce losses.

Anomaly detection is widely used in O&M. We give an example, in cooperation with a railway equipment company, we have developed a system that detects anomaly of equipment set in different sections of the railway. A number of sensors are installed on the equipment and many metrics can be collected in real time, including equipment frequency, voltage, current and so on. The O&M personnel have collected a large amount of historical data, they mark a period of time before the equipment fails as a transition state, and the period from the occurrence of the anomaly to the restoration is regarded as an anomaly state. The anomaly detection system is expected to find anomalies as soon as possible to avoid railway accident. Our method is applied to this practical problem as one core algorithm and achieves good results.

There are two important requirements for anomaly detection system in O&M: (1) Anomaly detection system should minimize false positives and improve the accuracy of alarms. We find that in practical applications, the direct use or simple transformation of existing anomaly detection algorithms often causes a large number of false positives, which frequently disturbs O&M personnel, affects their work efficiency,

© Springer Nature Singapore Pte Ltd. 2018
Q. Zhou et al. (Eds.): ICPCSEE 2018, CCIS 902, pp. 296–311, 2018.
https://doi.org/10.1007/978-981-13-2206-8_25

and even allows them to abandon the anomaly detection system. (2) Emphasis on real-time nature. Once detecting the "beginning" of anomaly, anomaly detection system should alarm, rather than wait until the detection of a complete anomaly pattern.

We find that existing anomaly detection algorithms can't meet the O&M needs, they have many false positives, can't detect anomaly in time and can't find new anomaly in some cases. What's more, we find a big problem for these algorithms, that's the detection results are not stable at all in application of O&M. For example, in Fig. 1, the left chart shows this character of the existing algorithms, and even the very best-performing methods like ensemble learning algorithms have this problem too. We need an algorithm which behaves like the right chart in O&M.

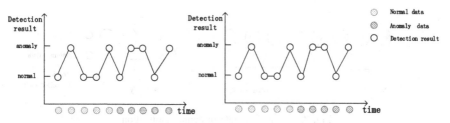

Fig. 1. The detection results of existing anomaly detection algorithms are not stable enough.

In practical problems, we first think of the transformation of the existing algorithm, and it is natural to introduce a sliding window and set the heuristic rule: only the number of anomaly points in window exceeds the threshold δ, an alarm is issued. But these heuristic rules do not work well because of the instability of detection results, we will elaborate on this issue in our experiments.

Motivated by these, we give a concept: Scores Sequence and propose our method.

Our method is based on two basic ideas: Each algorithm performs differently on different data, just like different human experts have their own areas of expertise. We use Scores Sequence to learn the judgments of different "experts".

We use Scores Sequence to focus on consecutive points instead of single point, which helps improve the stability of detection results and reduce false positives.

Our work has made the following contributions:

(1) We propose a novel concept: Scores Sequence and a method for anomaly detection in O&M: our method learns Scores Sequence instead of learning data directly.

(2) Our method has less false positives, can detect anomaly timely, and the detection result of our method is very stable with less fluctuation. In addition, our method is a flexible framework which can embed many algorithms. What's more, we also highlight the advantage to combine supervised algorithms with unsupervised algorithms.

(3) Compared with many other well-behaved algorithms including ensemble learning [14, 15] and distance-based algorithm [21] and so on, it proves that our method has a better performance and helps to realize intelligent O&M.

2 Problem Description

In this section, we describe the data characteristics in O&M and our target. The data type in O&M is data stream, it can describe as a possible infinite series of data points $\cdots, p_i, p_{i+1}, p_{i+2}, \cdots$, where data point p_i is received at time $p_i \cdot t$.

There is a data characteristics in the field of O&M: many anomalies are often anomalous time periods formed by consecutive anomalous points rather than many single anomalous points. For example, in [12, 13], the period that can't provide normal service is viewed as anomaly state. In actual application, there is a period of time between the normal state and the anomaly state, we call it transition state. We give an example in Fig. 2.

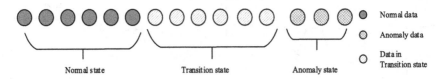

Fig. 2. Transition state in actual application

We use some variables to describe our target:

Alarm Point: the point which is regarded as anomaly by algorithms;

False Alarm: total number of false positives before anomaly occurs;

Delay Unit: the total number of points between the starting position of anomaly and the position of the first Alarm Point after anomaly occurs.

And we give a definition to quantify the fluctuations of detection results.

Definition 1 (Fluctuations Counter): A set of detection results $\mathbb{O} = \{o_1, o_2, \cdots, o_n\}$, for $\forall o_i \in \mathbb{O}, o_i = -1(anomaly) or \ 1(normal)$.

Fluctuations Counter $= \sum_{i=1}^{n-1} I(o^i \neq o^{i+1})$. $I()$ is indicator function. If x is true, then $I(x) = 1$, otherwise, $I(x) = 0$.

We give an example in Fig. 3, and we only do statistics in part of the data stream to illustrate these basic variables and definition.

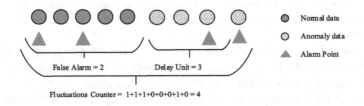

Fig. 3. An example in O&M data stream

We require False Alarm and Delay Unit as small as possible for algorithms in O&M. Besides, we expect Fluctuations Counter as small as possible.

3 Methodology

There are 2 important steps in our method: model training and online detecting, we first describe the workflow of online detecting to explain our core idea, then we will talk about how to realize model training.

3.1 Workflow of Online Detecting and Core Idea

In our method, we use a fixed-size window that contains some continues data, we use variable **Data Window** to represent it, and it is equivalent to sliding window that slides one cell at a time. We use variable **Winsize** to represent the size of Data Window.

We give a specific workflow of online detecting in Fig. 4 to show the main idea of our method. The models we use have two types: base models and top model. The base model set consists of many machine learning models, these models are built through the labeled training data and can output the anomaly scores for data. Top model is a higher level machine learning model, it process the detection results of base models, and the training method of top model is special, we will talk about it later. Base models is like many human experts who have their own area of expertise, and top model is like a leader, he can judge which expert's judgment to believe in different situations.

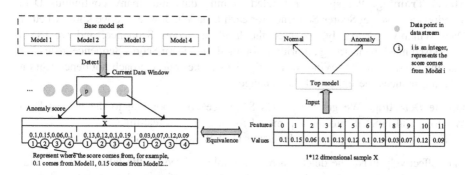

Fig. 4. A specific online detection workflow in O&M data stream

In Fig. 4, Winsize = 3, for each Data Window, the data points in Data window are scored by base models. There are 4 base models, so each data point has 4 anomaly scores corresponding to specific base models. For example, data point p has 4 regularized scores 0.1, 0.15, 0.06, 0.1, and 0.1 comes from Model 1, 0.15 comes from Model 2 and so on (the smaller the score, the more normal the data point is).

By splicing the scores of each point in Data Window, we get a 1 * 12 dimensional sample X that contains 4 * 3 = 12 scores. Each feature of X is a score corresponding to specific base model and specific data point in Data Window, for example, the seventh feature of X corresponds to Model 3 and the second point in Data Window. We input X to the top model to detect whether the current window is anomaly.

We use the above example to give the core definition: Scores Sequence.

Definition 2 (Scores Sequence): Let the number of base models = m, Winsize $\geq 1, m \geq 1$. The base models set $\mathbb{S} = \{Model_1, Model_2, \cdots, Model_m\}$, Data window $\mathbb{D} = \{p_i, p_{i+1}, \cdots, p_{i+\text{Winsize}-1}\}$. A Scores Sequence = $F(\mathbb{S} \times \mathbb{D})$, where $\mathbb{S} \times \mathbb{D}$ is Cartesian Product of $\mathbb{S}$ and $\mathbb{D}$, F is a function: F maps each pair $(Model_i, p_j)$ in $\mathbb{S} \times \mathbb{D}$ to the score that $Model_i$ gives to p_j.

For each Data Window, there are w data points, and there are m models to get anomaly scores for each data point in Data Window, we get a Scores Sequence for Data Window by splicing these scores in order, so each Scores Sequence contains m*w scores. For example, X is a Scores Sequence for current Data Window in Fig. 4.

We give two reasoning relations:

(1) Data Window is anomaly $\leftrightarrow$ Scores Sequence is anomaly. "$\leftrightarrow$" means "equivalence".
(2) Each data point in Data Window is anomaly $\rightarrow$ Data Window is anomaly. "$\rightarrow$" means "deduce".

The main idea of our method can summarize as follow:

Model Training: We split our labeled training data into many continuous Data Window, and we get Scores Sequence for each Data Window. Since we know whether Data Window is anomaly, we give the label of Scores Sequence, and we use a supervised machine learning algorithm (top model) to learn Scores Sequence instead of learning the data in Data Window directly. This makes our approach not dependent on data and reduces the impact of data changes.

Online Detecting: We get the Scores Sequence corresponding to the newest Data Window, and input it into our top model for detecting in real-time.

There are two main purposes for using Data Window and Scores Sequence:

(1) Effectively make the detection results stable and reduce the false positives. Traditional methods alarm as long as the detection of anomalous point, and these methods do not use the time continuity thus cause large number of false positives. With Scores Sequence, our method alarm only when there is enough confidence to judge that Scores Sequence is anomaly. Data Window is local data of data stream, this strategy based on local data can significantly reduce the fluctuation of detection result.

(2) There are some unusual anomalies that are not necessarily found using traditional methods (such as anomaly score is low), and our method learns this kind of special Scores Sequence. We give an example: in Fig. 5, current Data Window is anomaly, but by observing the Scores Sequence, we can find that Model 3 and Model 4 behave badly in this kind of data in Data Window, and Model 1 has a better performance. For example, for Model 3, the scores of 3 data points in Data Window is 0.42, 0.41, 0.42 respectively, and these scores are on the verge of normal and anomaly. In this condition, Model 1 plays a greater role and top model should more "believe" Model 1, since each point in Data Window is anomaly, we give Scores Sequence a label $Y = -1$, and get a sample (X, Y) to learn this special Scores Sequence.

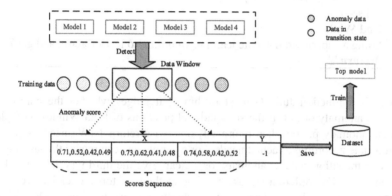

Fig. 5. Learn special scores sequence.

3.2 Model Training

The base models are trained in normal machine learning training method. But the training method of top model is special.

There are two requirements for the training data of top model:

(1) Training data should be time series data with label like Fig. 2, and training data can be divided into normal state, transition state and anomaly state.
(2) We require there are some new anomaly data that do not exist in the training data of base models. Since each base model does not see the new anomaly before, we encourage to use some unsupervised models in base model set, which have the ability to recognize new anomaly, and top model will learn the Scores Sequence when meeting new anomaly by this way. We give the algorithm 1 for building top model as follow.

Algorithm 1. Build Top Model

INPUT: (1) The training data of top model: $\{(p_1, y_1), (p_2, y_2), \cdots (p_m, y_m)\}$;
$\forall p_i, y_i = 1 \; if \; p_i \; is \; normal, otherwise \; y_i = -1$;

(2) Winsize;

(3) Base model set: $\mathbb{S}$;

RETURN: Top model: $\mathbb{T}$

1. Dataset = $\emptyset$; i = 1; Data Window $\mathbb{D}$ = $\emptyset$
2. **While** i+Winsize-1 $\leq$ m
3. $\mathbb{D} = \{(p_i, y_i), \cdots, (p_{i+Winsize-1}, y_{i+Winsize-1})\}$
4.   Get Scores Sequence x of $\mathbb{D}$
5.   y = **Judge**($\mathbb{D}$)
6.   Dataset = Dataset $\cup$ (x, y)
7.   i += 1
8. **End While**
9. Using a supervised machine learning algorithm to learn Dataset and get $\mathbb{T}$
10. **Return** $\mathbb{T}$

There is a function Judge() in algorithm 1, it judges whether the current Data Window is anomaly or not, if the start and end positions of Data Window completely fall in the anomaly period, then return -1, otherwise return 1 (We need to deal with transition state according to actual needs. For example, we can regard the last few Data Window in transition state as anomaly, or we can give the third type label for data in transition state. For different needs, training method of top model has very small difference in function Judge. There is a simple method: we just ignore Data Window which contains both normal data and anomaly data or data in transition state).

We can use the queue structure to realize algorithm 1 efficiently, we give an example in Fig. 6, there are 4 base models, and Winsize = 3. The current Data Window is drawn in solid line, and the previous Data Window is drawn in dotted line. In the previous Data Window, we get a 1 * 12 dimensional sample X. The previous Data Window is in normal state completely, so the label Y of X is +1, and we get a labeled sample (X, Y). Data Window slides, and we get another sample (X', Y').

As we can see, our method is a framework that can embed many existing anomaly detection algorithms flexibly.

4   Experimental Evaluation

We introduce the design of the Test Case Generator and test several representative anomaly detection algorithms, then transform these algorithms and compare them with our method. Finally, we will also discuss the features of our method.

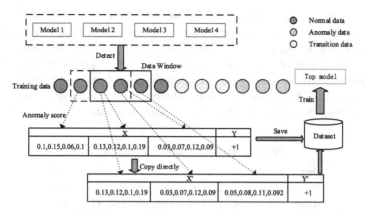

Fig. 6. An example for algorithm 1.

4.1 Test Case Generator

We use the UCI datasets often used in the anomaly detection including Forest cover, Shuttle [22] to generate test cases. We extract the categories of dataset firstly, and shuffle data in dataset so that data with the same category are grouped into the same group. By counting the number of data of each category, we consider the data of a larger category as normal and the data of a smaller category as anomalies. After getting different types of data, we splice the data of the normal type with the data of the anomaly type to generate many test cases. The whole process is shown in Fig. 7. Each test case consists of Size1 normal data and Size2 anomaly data.

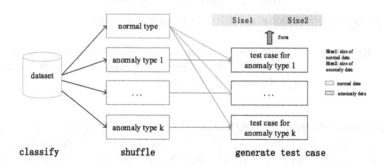

Fig. 7. Test case generator

In the next Sections, we take Forest cover as an example to give a detailed description of how to generate test cases and how to conduct comparative experiment.

4.2 Experiment Setting

To reflect the flexibility and usability of our method, we use a toy model: base models consist of 2 random forest (supervised) [20] and 2 isolation forest (unsupervised) [10], top model is a random forest model. We use isolated forest, random forest algorithm for contrast, in addition, we also compare with robust covariance estimation (statistical) [17], neural network algorithms, and well-established ensemble supervised learning algorithms [14, 15] and distance-based methods DODDS [21] proposed in recent years.

We generate test cases using data of categories 1, 4, 5 and 6 in Forest cover, treat data of categories 4, 5 and 6 as three different types of anomalies, and data of category 1 as normal data. Specific data information is shown in Table 1:

Table 1. Data information of forest cover

Dataset dimension	10
Number of data of Category 1 (normal data)	211940
Number of data of Category 5 (anomaly type 1)	9493
Number of data of Category 4 (anomaly type 2)	2747
Number of data of Category 6 (anomaly type 3)	17367
Number of data of category 1 used for training	10300
Number of data of category 5 used for training	300
Number of data of category 4 used for training	0
Number of data of category 6 used for training	100

As you can see, the number of samples we use for training is small and the proportion of anomalous data is small. Obviously the more data used for training model, the better results are. The information of training data of toy model is in Table 2:

Table 2. Training data of toy model

Training data of base models	9800 normal data, 200 anomaly type 1 data
Training data of top model	Two time series: one contain continuous 250 normal data and 100 anomaly type 1 data; another one contain continuous 250 normal data and 100 anomaly type 3 data

The data used to train other algorithms includes 10300 normal data, 300 anomaly type 1 data, and 100 anomaly type 3 data. For three types of anomalies, we generate several test cases respectively (determined by the number of samples). Set the parameters as: size1 = 100, size2 = 30.

Evaluation Metrics

Considering that algorithm can't detect anomalies in some cases, we define the metric *Miss Counter*. For each test case, *Miss Counter* is incremented if the algorithm can't detect any anomaly. In our experiment, we introduce some anomaly types that do not exist in the training data to test whether algorithm can find new anomaly. We use the following metrics for evaluation:

Total False Alarm (TFA): The sum of False Alarm in all test cases.

Total Delay (TD): The sum of Delay Unit in all test cases. In the experiment, in order to prevent that an algorithm can't find any anomaly, resulting in the delay is too large, the delay value does not count in the total delay in this case.

Miss Counter (MC): The total number of test cases where no anomaly is detected but anomaly exists.

Total Fluctuations Counter (TFC): The sum of Fluctuations Counter in all test cases.

We also use traditional evaluation metrics:

$$\text{Precision} = \frac{TP}{TP + FP}, \text{Recall} = \frac{TP}{TP + FN}.$$

4.3   Comparative Experiment

We conduct three experiments. Due to the randomness of the data shuffle operation and the randomness of the experiment, we conduct three experiments separately and average the final results. In experiment one, we use the existing anomaly detection algorithm directly. In experiment two, we set a heuristic rule for each anomaly detection algorithm to reduce false positives. In experiment three, we adjust the alarm threshold of sliding window.

Experiment One: Contrast Directly
Experiment one directly compares the various algorithms without introducing heuristic rules. Experiment results are shown in Tables 3, 4 and 5. It can be seen that our method behaves well using tradition evaluation metric including recall and precision, besides, our method has lower delay and lower false positives. We can also see that supervised algorithm can't detect anomaly in some cases and the unsupervised algorithm has the problem of too many false positives. This also highlights the advantage of combining the supervised algorithm with the unsupervised algorithm.

Table 3. Evaluation metrics for anomaly type 1

Model	TD	TFA	MC	Precision	Recall	TFC
Isolation forest	145	255	0	0.484	0.223	773
Robust covariance estimation	153	271	0	0.441	0.25	785
Ensemble supervised model	128	63	2	0.958	0.698	380
Random forest	133	65	0	0.9786	0.583	428
Neural network	196	107	0	0.9553	0.377	466
DODDS	328	320	0	0.323	0.61	337
Ours	**47**	**35**	**0**	**1**	**0.741**	**136**

By checking the results sequence for each algorithm, we highlight the issue of stability again. In Fig. 8, we enumerate the detection results of one test case, the

Table 4. Evaluation metrics for anomaly type 2

Model	TD	TFA	MC	Precision	Recall	TFC
Isolation forest	101	239	0	0.61	0.375	663
Robust covariance estimation	104	252	0	0.56	0.316	713
Ensemble supervised model	40	418	1	0.95	0.628	455
Random forest	120	85	0	0.98	0.732	490
Neural network	90	137	0	0.98	0.625	534
DODDS	229	278	0	0.25	0.521	276
Ours	**39**	**28**	**0**	**1**	**0.88**	**34**

Table 5. Evaluation metrics for anomaly type 3

Model	TD	TFA	MC	Precision	Recall	TFC
Isolation forest	121	250	0	0.63	0.423	609
Robust covariance estimation	114	254	0	0.55	0.312	611
Ensemble supervised model	78	312	3	0.95	0.678	433
Random forest	128	128	2	0.98	0.762	652
Neural network	221	321	3	0.98	0.625	928
DODDS	208	305	0	0.36	0.524	295
Ours	**64**	**39**	**0**	**1**	**0.87**	**40**

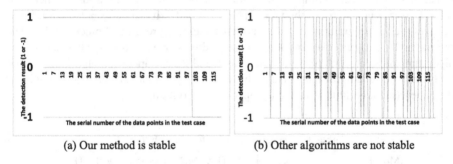

(a) Our method is stable   (b) Other algorithms are not stable

Fig. 8. The issue of stability of detection results

abscissa is the serial number of the data points in the test case, the ordinate is the detection result (1 or −1), and the points whose abscissa is after 100 are all anomalies.

As shown in Fig. 8(a), our method is very stable, with few false positives before the real anomaly occurs; whereas in Fig. 8(b), both before and after the real anomaly occurs there are frequent false positives and fluctuations, which seriously affects the detection results and leads to poor user experience.

Experiment Two: Introduce Heuristic Rule

In experiment two, we perform a heuristic rule on algorithms: only the number of anomaly points in Data Window exceeds threshold δ, an alarm is issued. Let

Table 6. Evaluation metrics (Winsize = 5, δ = 3)

Anomaly type	Anomaly type 1			Anomaly type 2			Anomaly type 3		
Model	TD	TFA	MC	TD	TFA	MC	TD	TFA	MC
Isolation forest	194	29	15	229	26	0	263	25	0
Robust covariance estimation	94	20	27	136	45	22	352	42	1
Ensemble supervised model	148	22	8	100	354	2	121	282	3
Random forest	131	22	3	120	31	1	118	20	1
Neural network	168	49	2	113	65	1	115	88	2
DODDS	211	43	0	181	27	0	128	24	0
Ours	**51**	**14**	**0**	**66**	**1**	**0**	**67**	**5**	**0**

Winsize = 5, δ = 3, the results are shown in Table 6 (due to space limitations, we only give total delay and total false alarm).

By comparing the results of experiment one and experiment two, we find that except for some algorithms (these algorithms are not effective enough because of the large fluctuation of detection results), it is in our expectation that total false alarm can be significantly reduced by introducing heuristic rules. At the same time, we must also emphasize the drawbacks arising from the introduction of heuristic rules: total delay is exacerbated. This is due to the fact that many algorithms do not continuously output alarms in the actual case of anomalies, and the results are similar to Fig. 8(b), which results in the introduced rules smoothing out actual true alarms, causing delay is longer.

We can see that simply using heuristic rule can't be competent for the actual O&M, we can also find our method is very stable, and even some of the false positives in experiment one are smoothed out and the result is better.

Table 7. Evaluation metrics (Winsize = 5, δ = 2)

Anomaly type	Anomaly type 1			Anomaly type 2			Anomaly type 3		
Model	TD	TFA	MC	TD	TFA	MC	TD	TFA	MC
Isolation forest	191	259	0	110	260	0	104	278	0
Robust covariance estimation	165	238	0	99	266	0	113	267	0
Ensemble supervised model	109	45	6	108	396	1	100	418	3
Random forest	126	85	2	116	156	1	126	90	2
Neural network	123	49	0	122	66	1	151	68	1
DODDS	206	291	0	235	276	1	235	219	0
Ours	**72**	**20**	**0**	**68**	**32**	**0**	**71**	**45**	**0**

Experiment Three: Adjust the Alarm Threshold

In experiment three, we set Winsize = 5 and δ = 2, that is, adjust the alarm threshold. The experimental results are shown in Table 7.

By comparing with experiment two, it can be found that the threshold has a great influence on the detection result, which reflects the importance of parameter adjustment in practical application, for example, the total delay decreases in anomaly type 3, but the number of false positives increases.

Based on the above three experiments, we find that our method has a good effect.

4.4   The Influence of Winsize

There is only one main parameter in our method: Winsize. We analyze the trade-offs of the Winsize setting. Consider the extreme case that Winsize = 1, our method can be considered as an ensemble method for detecting anomaly points like other algorithms.

In Fig. 9, we give the relationship between different Winsize and total delay with total false positives, the abscissa represents Winsize. Due to the limitations of the data and the experimental error, the law does not reflect sufficient, but we can still verify two laws: (1) in a certain range, the increase of Winsize will lead to increase of delay. (2) Appropriate increase of Winsize helps reduce the number of false positives.

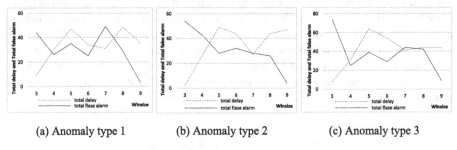

(a) Anomaly type 1   (b) Anomaly type 2   (c) Anomaly type 3

Fig. 9. Trade-off of winsize setting

When Winsize increases, the dimension of the training samples used to determine whether Data Window is anomalous increases, and the detection result should be more accurate, but at the same time the delay increases correspondingly. When Winsize increases to a certain extent, the risk of over-fitting increases. Therefore, the setting of Winsize is a trade-off.

4.5   Other Datasets and Experiments

In addition to the Forest cover dataset, we also use other datasets. We also use different types of data for experiment. For example, in Forest Cover, we use other types of data as normal data instead of just using data of category 1. In addition, we also carry out experiments with different Winsize and δ, and the test results all show the effectiveness of our method.

We also apply our method to the actual maintenance work of railway equipment. It proves that our method has little false positives, can detect multiple anomalies and has high detection accuracy.

Due to space limitations, we will not repeat them here.

5 Related Work

Due to the importance of anomaly detection, many methods of anomaly detection have been proposed in academia, including density-based approach [2], statistical-based approach [5, 17], distance-based approach [3, 4, 6], isolation-based approach [10], supervised learning approach [19], ensemble learning methods [14, 15] and so on. Research on anomaly detection of data stream is also very deep. The most widely used method is based on distance [7, 8], such methods need to introduce sliding windows and empirically set the threshold, the main drawback is the slightly higher computational complexity and over-reliance on experience for threshold setting. In addition there are statistical-based methods [9], window-based matching methods [18] and so on. However, these work attention to the detection of anomalous points, and do not consider the actual data situation in the field of O&M. More work about anomaly detection can refer to [16].

In practice, some anomaly detection methods are also proposed. For example, [12, 13] use an ensemble decision tree model to detect anomalous patterns in data streams in order to maintain the health of the machine cluster. [14] uses an ensemble learning model to maintain network security. [11] uses a statistical model to monitor the anomalous state of the hadoop cluster.

In addition, [1] gives a very effective method for human participation in the field of O&M, [1] emphasizes a very important issue in O&M: Many of the anomalies detected by unsupervised algorithms are not actual anomalies. The use of supervised algorithm enables the detection system to detect anomalies in accordance with the judgment of the O&M personnel, making the system more practical. However, this work only focuses on one-dimensional time series anomaly detection and can't find new anomaly. This inspires us to combine supervised algorithms with unsupervised algorithms.

6 Conclusions

In the actual O&M, we find that using the anomaly detection algorithm or simply transforming the existing algorithm (introducing heuristic rules) has many problems: there are large numbers of false positives, delays, and the detection results are not stable.

In this paper, in view of the data characteristic and requirements in O&M, we propose a novel method: we split the data stream into many continuous Data Window, we get Scores Sequence for each Data Window, and use supervised machine learning algorithm to learn Scores Sequence. We can detect anomaly using Scores Sequence.

Comparison experiments with many algorithms show the drawbacks of directly using the existing anomaly algorithms and highlight the practicality of our method. Our method can make anomaly detection algorithms even more Practical.

Acknowledgments. This work is supported by the National Key Research and Development Program (No. 2016YFB1000703).

References

1. Liu, D., Zhao, Y., Xu, H., et al.: Opprentice: towards practical and automatic anomaly detection through machine learning. In: Internet Measurement Conference, pp. 211–224. ACM (2015)
2. Breunig, M.M., Kriegel, H.-P., Ng, R.T., Sander, J.: LOF: identifying density-based local outliers. In: Proceedings of the 2000 ACM SIGMOD International Conference on Management of Data (SIGMOD), pp. 93–104 (2000)
3. Knorr, E., Ng, R.: Algorithms for mining distance-based outliers in large data sets. In: VLDB Conference (1998)
4. Knorr, E., Ng, R., Tucakov, V.: Distance-based outliers: algorithms and applications. VLDB J. **8**(3–4), 237–253 (2000)
5. Barnett, V., Lewis, T.: Outliers in Statistical Data. Wiley, Chichester (1994)
6. Tao, Y., Xiao, X., Zhou, S.: Mining distance-based outliers from large databases in any metric space. In: SIGKDD Conference, pp. 394–403 (2006)
7. Angiulli, F., Fassetti, F.: Detecting distance-based outliers in streams of data. In: CIKM Conference, pp. 811–820 (2007)
8. Kontaki, M., Gounaris, A., Papadopoulos, A.N., et al.: Continuous monitoring of distance-based outliers over data streams. In: IEEE International Conference on Data Engineering, pp. 135–146. IEEE (2011)
9. Zhu, Y., Shasha, D.: Statstream: statistical monitoring of thousands of data streams in real time. In: VLDB Conference, pp. 358–369 (2002)
10. Liu, F.T., Kai, M.T., Zhou, Z.H.: Isolation-based anomaly detection. ACM Trans. Knowl. Discov. Data **6**(1), 1–39 (2012)
11. Gupta, C., Bansal, M., Chuang, T.C., et al.: Astro: a predictive model for anomaly detection and feedback-based scheduling on Hadoop. In: IEEE International Conference on Big Data, pp. 854–862 (2014)
12. Gu, X., Wang, H.: Online anomaly prediction for robust cluster systems. In: IEEE International Conference on Data Engineering, pp. 1000–1011. IEEE Computer Society (2009)
13. Gu, X., Papadimitriou, S., Yu, P.S., et al.: Toward predictive failure management for distributed stream processing systems. In: The International Conference on Distributed Computing Systems, pp. 825–832. IEEE (2002)
14. Vanerio, J., Casas, P.: Ensemble-learning approaches for network security and anomaly detection. In: The Workshop, pp. 1–6 (2017)
15. Aggarwal, C.C.: Outlier ensembles: position paper. ACM SIGKDD Explor. Newslett. **14**(2), 49–58 (2013)
16. Gupta, M., Gao, J., Aggarwal, C., et al.: Outlier detection for temporal data: a survey. IEEE Trans. Knowl. Data Eng. **26**(9), 2250–2267 (2014)
17. Rousseeuw, P., Van Driessen, K.: A fast algorithm for the minimum covariance determinant estimator. Technometrics **41**(3), 212–223 (1999)

18. Yang, D., Rundensteiner, E.A., Ward, M.O.: Neighbor based pattern detection for windows over streaming data. In: Proceedings of the 12th International Conference on Extending Database Technology: Advances in Database Technology (EDBT), pp. 529–540 (2009)
19. Dasgupta, D., Nino, F.: A comparison of negative and positive selection algorithms in novel pattern detection. In: Proceedings of the 2000 IEEE International Conference on Systems, Man, and Cybernetics, vol. 1, pp. 117–125 (2000)
20. Liaw, A., Wiener, M.: Classification and regression with random forest. R News **23**(23), 18–22 (2002)
21. Luan, T., Fan, L., Shahabi, C.: Distance-based outlier detection in data streams. VLDB Endow. **9**, 1089–1100 (2016)
22. UCI datasets: http://archive.ics.uci.edu/ml

A Method of Improving the Tracking Method of CSI Personnel

Zhanjun Hao[1,2], Beibei Li[1(✉)], and Xiaochao Dang[1,2]

[1] School of Computer Science and Engineering, Northwest Normal University,
Lanzhou 730070, China
848275034@qq.com
[2] Gansu Internet of Things Engineering Research Center,
Lanzhou 730070, China

Abstract. In the trajectory tracking process of indoor environment, in order to reduce the communication overhead and reduce the complexity of the algorithm, a human trajectory tracking method for improving the CSI (Channel State Information) signal is proposed. Firstly, the AOA (Angle-of-Arrival) spectrum is extracted from the CSI to represent the probability of the target position (angle), and the Doppler shift obtained by the Music algorithm is combined with the AOA spectrum to determine the moving speed and position of the personnel. Finally, The neural network algorithm determines the position of personnel and simulates the movement trajectory of the personnel to achieve accurate tracking and positioning of indoor personnel. Compared with other algorithms and different people's movement speeds, simulation experiments show that the personnel tracking method proposed in this paper can greatly improve the accuracy and stability of positioning.

Keywords: CSI signal · Trajectory tracking · Doppler frequency shift
Neural network

1 Introduction

As an important part of pervasive computing and the Internet of Things, positioning technology has received more and more attention. The Global Navigation Satellite System (GNSS) plays a key role in outdoor precise positioning. However, due to the multipath effect, scattering and diffraction characteristics of GNSS in the indoor environment, GNSS It can no longer play its high-precision positioning advantages. At present, research on CSI-based positioning technology has achieved certain results, and it has even reached the positioning accuracy of meters or submeters.

In recent years, Massachusetts Institute of Technology, Stanford University, Washington University, Hong Kong University of Science and Technology, Tsinghua University and other universities have done a lot of research on CSI-aware applications. Document [1] uses the commercial WI-FI to realize indoor tracking without equipment, which can achieve high positioning accuracy even in the case of packet loss and delay, but does not consider multi-person tracking, real-time tracking and multipath effect Affect the problem. Document [2] uses software radio (USRP) to achieve gesture

© Springer Nature Singapore Pte Ltd. 2018
Q. Zhou et al. (Eds.): ICPCSEE 2018, CCIS 902, pp. 312–329, 2018.
https://doi.org/10.1007/978-981-13-2206-8_26

recognition across the entire experimental area and FFT to detect Doppler shifts. The literature [3] uses RFID to construct a virtual touch screen and uses AoA information for fine-grained tracking. Wi-Draw uses AoA information to provide a good idea of using phase information to determine location, but Wi-Fi transmitters are required to cover all directions in the environment [4]. In the actual environment by the multipath effect, there is a certain degree of difficulty. Wang et al. [5] used CSI to construct a low-cost and high-accuracy passive target localization method based on CSI model, LIFS, which effectively combines the features of CSI with the target localization without considering the relationship between the detection area and the detection rate. In [6], the phase information in CSI is applied to the movement of the oral cavity when speaking through the human body to achieve the effect of more fine-grained positioning, but the phase information is not easy to obtain and the special equipment USPR is used, which is stable and practical Sex is not high. The FIMD system in [7] exploits the stability of CSI to achieve finer granularity detection in static environments but does not achieve high detection rates and can impact overall performance due to changes in the experimental environment. The BFP system in [8] makes use of CSI to conduct behavior-independent motion detection, the overall performance is better, but the overall efficiency of the algorithm is not high.

Most existing trajectory tracking methods improve precision at the expense of higher communication overhead and higher algorithmic complexity. The proposed method based on CSI signals can well avoid the above problems. The first extracted CSI signal is processed with a smoothing algorithm to form an enhanced CSI algorithm. The enhanced CSI algorithm compares the value of the previous CSI with the current CSI value, reducing the fluctuation of the kinetic energy of the moving personnel. Finally, the improved three-sided positioning centroid algorithm is used to determine each activity point, and all the activity points are plotted as trajectory images to be visually displayed so as to achieve more accurate tracking and positioning for indoor personnel.

2 Related Theory

2.1 Channel Status Information

CSI is channel state information (Channel State Information) which is used to measure channel conditions and belongs to the PHY layer, and is derived from subcarriers decoded by the OFDM system [9]. CSI is fine-grained physical information and is more sensitive to the environment. It is used in motion recognition, gesture recognition, keystroke recognition, and tracking [10]. Firstly, it needs to know the received power of the measured node and the signal power of the transmitting node as a reference, and then uses the signal propagation attenuation model to convert the distance between the nodes. Finally, using the improved trilateration algorithm, the position of the unknown node can be simply determined. The relationship between the transmission power and the received power of a wireless signal can be expressed by formula (1):

$$P_R = P_T/d^n \tag{1}$$

In the formula, the received power of the wireless signal is P_R, the transmitting power of the wireless signal is P_T, the distance between the transceiver units is r, the propagation factor is n, and the value depends on the environment in which the wireless signal propagates.

Take the logarithm of the two sides of the formula (1) to get the formula (2),

$$10 * n \lg r = 10 \lg \frac{P_T}{P_R} \tag{2}$$

The transmit power of a node is known, and the transmit power is substituted into Eq. (3) in Eq. (2):

$$10 \lg P_R = A - 10 * n \lg r \tag{3}$$

The left half of Eq. (3) is $10 \lg P_R$ the expression of the received signal power converted into dBm a formula, which can be directly written as formula (4). In formula (4), A can be seen as the power of the received signal when the signal is transmitted for 1 m away.

$$P_R(dBm) = A - 10 * n \lg r \tag{4}$$

From (4) $P_R(dBm)$ is the CSI constant, A and n the value of the sum determines the relationship between the received signal strength and the signal transmission distance.

Because each reference point has a different degree of signal attenuation, only one value n is not generic and its accuracy is hard to tell. From (4) to obtain the formula (5), so as to get a set of n_i values,

$$n_i = -\left[\frac{CSI_i - A}{10 \lg r_i} \right] \tag{5}$$

The value of n_i a set $n_1, n_2, n_3 \cdots n_n$ obtained from Eq. (5), $\bar{n}$ is the average of the number of sets, $\bar{n}$ is the desired propagation factor.

2.2 CSI Signal Doppler Shift Extraction

In indoor environments with a pair of transmitters and receivers [11], multipath propagation involves the signal not only propagating along the direct path [12], but also being obstructed by other objects such as people and walls [13]. The signal received at the receiver is therefore a superposition of the signals from all the paths. When a person is moving in the environment, the length of the path is changed correspondingly by the reflection signal of the human body. In this case, the Doppler shift reflection signal of the carrier frequency is changed, and the carrier frequency is as shown in Eq. (6):

$$f_{Doppler} = f \frac{v_{path}}{c} \qquad (6)$$

f is the original carrier frequency of the signal, the speed at v_{path} the path length changes, c is the propagation speed of the Wi-Fi signal in the air. When a person is moving, the frequency conversion that incorporates a Doppler Wi-Fi signal is only a few tens of Hertz for a 5 GHz channel. Obviously, it is very difficult to detect such a small Doppler carrier frequency shift with fine granularity.

3   CSI - Based Trajectory Tracking Method

3.1   Doppler Estimation Based on MUSIC

When the status information is collected, packet loss/delay caused by actual environmental noise and interference easily causes the problem that the sample data packet can not be normally sent or received. In order to solve these problems, how to use Wi-Fi equipment to obtain accurate Doppler frequency shift estimation becomes the key. In this paper, a MUSIC-based algorithm is proposed to obtain accurate Doppler frequency shift estimation algorithm.

Let's collect the first CSI sample at t_0, with the sample interval for each sample for the first sample is $[0, \Delta t_2, \ldots, \Delta t_M]$, where $\Delta t_1 = 0$. In a short sample window, the speed of change of the path v_{path} is treated as a constant because different i CSI samples have different attenuation crossovers. The first sample of phase difference samples between the first CSI is $e^{-j2\pi f \frac{v \Delta t_i}{c}}$, where f is the original carrier frequency of the signal. Therefore, Mth the difference between the phase CSI sample and the first CSI sample can be expressed as follows:

$$\vec{a}(v) = [1, e^{-j2\pi f \frac{v\Delta t_2}{c}}, e^{-j2\pi f \frac{v\Delta t_3}{c}}, \cdots, e^{-j2\pi f \frac{v\Delta t_M}{c}}]^T \qquad (7)$$

The Doppler vector $\vec{a}(v)$ and M CSI sample matrix with one sample are expressed as:

$$X(f) = [x(f, t_0), x(f, t_0 + \Delta t_2), \ldots, x(f, t_0 + \Delta t_M)]$$

$$= [1, e^{-j2\pi f \frac{v\Delta t_2}{c}}, e^{-j2\pi f \frac{v\Delta t_3}{c}}, \cdots, e^{-j2\pi f \frac{v\Delta t_M}{c}}]^T x(f,t_0) + n(f) = \vec{a}(v)x(f,t_0) + n(f) \qquad (8)$$

$n(f)$ is the noise. When only one path signal exists, the Doppler shift can be easily calculated from the phase measurements of the CSI samples. In the real world with multipath, the L path signal will reach the receiver. From (7) and (8) CSI sample matrix can be expressed as:

$$X(f) = \sum_{i=1}^{L} \vec{a}(v)x(f,t_0) + n(f) = [\vec{a}(v_1), \vec{a}(v_2), \cdots, \vec{a}(v_L)][s_1(f, t_0), s_2(f, t_0), \cdots, s_L(f, t_0)]^T x(f,t_0) + N(f) = AS(f) + N(f)$$

$$(9)$$

v_i is the path change speed of the signal of the firs i path,

$A = [\vec{a}(v_1), \vec{a}(v_2), \cdots, \vec{a}(v_L)]$ is the total matrix $M * L$ Doppler shift The Doppler vector is the CSI of the first i path signal measured at the first sampling time t_0, the matrix $s(f) = [s_1(f, t_0), s_2(f, t_0), \cdots, s_L(f, t_0)]^T$ is the signal matrix and $N(f)$ is the noise matrix. In order to obtain the Doppler shift of each path signal. This article takes multiple snapshots in each CSI sample frequency domain, as shown in Fig. 1. The device provides CSI on multiple subcarriers. Let K be the number of CSI subcarriers and have CSI K snapshot CSI samples. For the first CSI sample, it can be expressed as:

$$\vec{x}(f, t_0 + \Delta t_i) = [x(f_1, t_0 + \Delta t_i), x(f_2, t_0 + \Delta t_i), \ldots, x(f_K, t_0 + \Delta t_i)] \qquad (10)$$

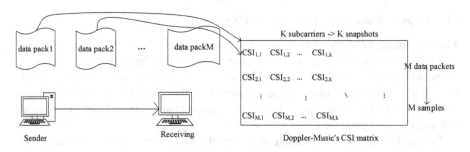

Fig. 1. Extracts K CSI snapshots of K subcarriers per sample from M groups of packets

f_k is the original carrier frequency of the k subcarrier. The signal of the first i path in the signal matrix $S(f)$ can be expressed as:

$$\vec{s}_i(f, t_0) = [s_i(f_1, t_0), s_i(f_2, t_0), \ldots, s_i(f_K, t_0)] \qquad (11)$$

$s_i(f_K, t_0)$ is the CSI of the first i snapshot of the first path signal at the first sampling time (t_0). In order to improve the accuracy of the Doppler shift estimation, CSI is selected from all the subcarriers so as to avoid selecting a specific subcarrier with a low signal to noise ratio. From (10) and (11), (9) can be expressed as:

$$X(f) = [\vec{x}(f, t_0), \vec{x}(f, t_0 + \Delta t_2), \ldots, \vec{x}(f, t_0 + \Delta t_M)]^T$$
$$= [\vec{a}(v_1), \vec{a}(v_2), \cdots, \vec{a}(v_L)] * [s_1(f, t_0), s_2(f, t_0), \cdots, s_L(f, t_0)]^T x(f,t_0) + N(f) = AS(f) + N(f) \qquad (12)$$

The MUSIC algorithm is used to estimate Doppler shift using CSI samples. The correlation matrix with M eigenvalues R_X can be expressed as:

$$R_X = E[XX^H] = AE[SS^H]A^H + E[NN^H] = AR_SA^H + \sigma^2I \qquad (13)$$

R_S is the correlation matrix of the signal matrix, I is the identity matrix, and δ^2 is the noise variance. The correlation matrix R_X has M eigenvalues. The smallest eigenvalue $M - L$ corresponds to noise while the other L eigenvalues correspond to L-path signals. The eigenvector constructs a noise subspace $E_N = [\vec{e_1}, \ldots, \vec{e_{M-L}}]$ corresponding to the smallest eigenvalue $M - L$. The signal and noise subspaces are orthogonal, so the Doppler velocity spectral function can be expressed as:

$$p(v)_{MUSIC} = \frac{1}{a^H(v) - E_N E_N^H a(v)} \tag{14}$$

Where the spike corresponds to the maximum of the spectral function. Since the original carrier frequency and propagation velocity of the signal are known, the Doppler velocity can be obtained.

3.2 Improved Three-Sided Positioning Centroid Algorithm

In complex indoor locations, the detected trajectories of people are momentarily changing [14], together with the influence of external uncertainties [15]. As a result, the value of the CSI signal fluctuates greatly, resulting in a large error on the positioning of the personnel, which can not achieve the expected effect [16]. In this paper, a smoothing algorithm is introduced to compare the adjacent CSI signal values to remove large fluctuations Value, to achieve precise positioning [17].

This paper proposes to improve the smoothing algorithm is divided into two phases: (1) distance estimation; (2) position estimation.

3.2.1 Estimated Distance

The CSI localization mechanism mentioned in this paper consists of two parts, the reference node and the unknown node. In the two-dimensional coordinate system, the position coordinates of the reference node in the fixed position and the CSI values received by the unknown node are calculated. Seek both the standardized Euclidean distance.

Euclidean distance calculation:

(1) Take out the coordinates of the position of the reference point as the horizontal axis x_i;
(2) CSI values collected for unknown nodes each time are recorded as y_i;

The normalized Euclidean distance between the two is as shown in formula (15):

$$d(x,y) = \sqrt{(x_1 - y_1)^2 + (x_2 - y_2)^2 + \cdots + (x_n - y_n)^2} = \sqrt{\sum_{i=1}^{n} (x_i - y_i)^2} \tag{15}$$

In order to avoid accidental errors, we obtain Eq. (16) from Eq. (15):

$$d(x,y) = \frac{1}{n}(d(x_1, y_1) + d(x_2, y_2) + \cdots + d(x_n, y_n)) \tag{16}$$

Data smoothing definition:

$R_{est(n)}$ is the range of the first n smoothed CSI;
$R_{pre(n)}$ is the range of the first n predicted CSI;
$R_{mea(n)}$ is the range of the first n measured CSI;
$R_{rat(n)}$ is the range ratio of the first n smooth estimate;
$R_{prerat(n)}$ is the first n prediction range ratio;
λ, ξ said the gain constant, T_n is a time of change.

Processing steps:

Step1: Calculate $R_{est(n)}$ and $R_{rat(n)}$, as shown in Eqs. (17), (18).

$$R_{est(n)} = R_{pre(n)} + \lambda(R_{mea(n)} - R_{pre(n)}) \tag{17}$$

$$R_{est(n)} = V_{pre(n)} + \frac{\xi}{T_s}(R_{mea(n)} - R_{pre(n)}) \tag{18}$$

Step2: By comparing the estimation of the range and ratio of the first n CSI value and the estimation of the third $n + 1$ CSI range and ratio, Eqs. (19), (20).

$$R_{pre(n+1)} = R_{est(n)} + V_{est(n)}T_s \tag{19}$$

$$V_{pre(n+1)} = V_{est(n)} \tag{20}$$

Step3: The value of the first $n + 1$ reference point is obtained from Eq. (20), and then the value of CSI is calculated by Eq. (5). Finally, the calculated CSI is processed and sent to the unknown node.

3.2.2   BP Neural Network Algorithm

Traditionally, the neural network algorithm uses an arbitrary set of free weights, and then establishes a set of linear equations based on the transfer function. The linear arithmetic algebra and mathematical thinking are used to obtain the right to be calculated [18]. The improved algorithm of the BP neural network proposed here is to calculate the neuron by replacing the traditional inverse algorithm using the transfer function by directly giving the given target output as the linear equation algebra and establishing the linear equation set. Of the net output, so that the solving steps to further optimize [19].

In principle, the principle of the error back propagation algorithm of the traditional neural network is complied with. Therefore, even the results of the simulation of the improved BP neural network algorithm are consistent with those obtained by the conventional neural network algorithm. The basic principle of the BP improved algorithm is as follows: First, establish a linear equation group by acting on the input and output patterns of the neural network; then use the Gaussian elimination method to solve the linear equations to obtain the position weights; finally, it is different from the traditional BP neural network. The nonlinear function error feedback optimization idea [20].

3.2.3 BP Implementation Steps to Improve the Algorithm

Solving process: Select a certain number of sample patterns. The fixed weight between the output layer and the hidden layer is determined by a set of free weights given randomly. The actual output of the hidden layer is calculated by the transfer function, and then the output layer is The hidden layer weight as the amount of demand, the target output as an unknown amount, so as to establish equations to solve the demand.

The symbols appearing in the improved algorithm are defined as follows:

The number of input neurons i; the number of output neurons o; the number of hidden neurons h; the actual input vector of the input layer is $x(p)$; the output vector of the output layer is $y(p)$; the target output vector is $t(p)$; the output layer and the hidden layer Of the weight matrix V; hidden layer and the input layer between the weight matrix W.

The improved BP algorithm implementation steps are as follows:

Step1: Give the initial value of the neuron between the hidden layer and the input layer a random weight w_{ij}.

Step2: The actual output value $a_j(p)$ is calculated from the values entered for the known sample $x_i(p)$. In order to simplify the calculation, the threshold value in the network is written into the connection weight. In this case, let the hidden layer threshold θ_i be equal to w_{nj} the actual input amount $x(i) = -1$ of the i-th input layer.

$$a_j(p) = f(w_{ij}x_i(p))(j = 1, 2 \cdots m - 1) \tag{21}$$

Step3: Through the above data to calculate the output layer and the hidden layer between the weights v_{jr}. Taking the 0th neuron of the output layer as an object, a set of equations is established by using the given output target value $t_r(p)$ as a polynomial of the equation, which is represented by a linear equation group:

$$a_0(1)v_{1r} + a_1(1)v_{2r} + \cdots + a_m(1)v_{mr} = t_r(1)a_0(2)v_{1r} + a_1(2)v_{2r} + \cdots + a_m(2)v_{mr} = t_r(2) \tag{22}$$

$$a_0(p)v_{1r} + a_1(p)v_{2r} + \cdots + a_m(p)v_{mr}p = t_r(p) \tag{23}$$

Summarizing the law of equations is not difficult to find:

$$Av = T \tag{24}$$

In order to ensure that the system of equations has a unique solution (to prevent multiple root-caused objections), it is easy to derive from the relationship between the linear algebraic value and the rank only when the equation matrix A is a non-enterprise matrix and its rank is equal to the rank of its augmented matrix. The equations in this case have the only solution. That is $r(A) = r(A:B)$, when, and the number of equations is equal to the number of unknowns, when h = p, the only solution to the equation is:

$$V_r = [v_{0r}, v_{1r}, v_{2r} \cdots v_{mr}] \tag{25}$$

Step4: Repeated the third step can be obtained by the output layer h neuron's weight, by calculating the weight of the output layer of the weight matrix and random fixed layer hidden layer and output layer weight can be obtained BP neural network training The matrix. Figure 2 is a 3-layer BP neural network structure.

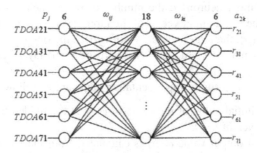

Fig. 2. Three-layer BP neural network structure

3.2.4 Modification of TDOA Based on BP Neural Network

Under the multi-path and non-line-of-sight environments, the BP neural network model with the revised TDOA measurement estimates of the more classic 7 base stations is often used. The BP neural network consists of the input layer, the hidden layer and the output layer. The input layer consists of six TDOA measurements provided by seven related base stations

So the input is:

$$INPUT = [INPUT21, INPUT31, INPUT41, INPUT51, INPUT61, INPUT71] \tag{26}$$

Hidden layer neurons array N_2 by the empirical formula, that is $N_2 > = lbT$ (T is the number of training samples). Although increasing the number of hidden neurons can improve the positioning accuracy, but the amount of computation will be greatly increased. However, as a system with fewer samples and only improved accuracy, the hidden layer function uses Sigmoid model functions, that is,

$$f_1(x) = 1/(1 + \exp(-x)) \tag{27}$$

Its input is any value and the output value is in the interval $[-1, +1]$.

The output layer is composed of 7 neurons and the linear transfer function Purelin is used in the output layer.

$$f_2(x) = kx \tag{28}$$

The output is the corrected TOA value. Output is:

$$O = [r_{21}, r_{31}, r_{41}, r_{51}, r_{61}, r_{71}] \tag{29}$$

3.2.5 TDOA Location Algorithm Based on BP Neural Network

It is known from past experience that the Chan algorithm has a small error in TDOA, especially under the ideal zero-mean Gaussian random variable, and the accuracy of the TDOA measurement has a direct impact on the performance of this algorithm [6]. Here, we use the characteristics of the nonlinear infinite curve approach of the BP neural network to correct the final measurement results of the TDOA so as to reduce the NLOS error in the TDOA measurement values. Finally, we use the Chan algorithm to further improve the positioning accuracy [7].

Positioning steps are as follows:

Step1: There are N group of TDOA values in NLOS environment, establishing a BP network for correcting NLOS errors and training. BP neural network is trained with TDOA of NLOS error of MS as the target sample vector;

Step2: In the experimental environment, the BP neural network that has already been trained is compared to the simulated TDOA measurement data;

Step3: The corrected TDOA values are further estimated by the Chan algorithm to improve the accuracy.

3.3   CSI - Based Personnel Tracing Method

Based on the analysis of the above theoretical model, this paper proposes a trajectory tracking method based on the above improved neural network algorithm. The specific implementation of the algorithm is shown in Fig. 3.

Step1: The CSI signal value is extracted through the deployed experimental environment.

Step2: First, the Doppler frequency shift (that is, the proposed Doppler-music algorithm) is estimated by using the Music algorithm, and then the AOA spectrum is estimated to determine whether the position of the personnel is within the preset range. If yes, go back to step one.

Step3: The smoothing algorithm is used to remove the redundant points from the active points obtained in the second step. Finally, an improved trilateration algorithm is used to determine the active point position to determine whether the result is within the reference range. If yes, step 4 is performed, otherwise, it is returned.

Step4: The obtained points within the range of motion are processed, enough points are collected to simulate the trajectory image, and the program ends.

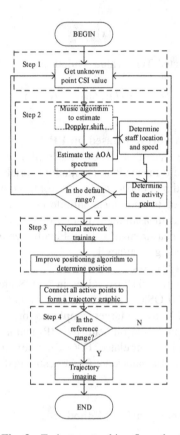

Fig. 3. Trajectory tracking flow chart

3.4 Communication Overhead and Algorithm Complexity Analysis

There are several parameters that need to be introduced in the calculation of communication overhead: N is the number of nodes in the network; A is the number of anchor nodes in the network; G is the average degree of connectivity of the network; C is the average number of neighbor nodes; and K is the number of anchor nodes participating in certain multilateral measurement and positioning.

When the unknown node obtains a distance from 3 or more anchor nodes, trilateration positioning is performed. Suppose the coordinates of unknown nodes are $A(x, y)$, the coordinates of anchor nodes are $L_1(x_1, y_1), \ldots, L_k(x_k, y_k)$ and the distances between unknown nodes and anchor nodes are $r_1, r_2, \ldots, r_k$ respectively, then the system of linear equations can be established and expressed as the form of Eq. (26):

$$Cx = b \tag{30}$$

among them:

$$C = -2 \times \begin{bmatrix} (x_1 - x_k) & (y_1 - y_k) \\ (x_2 - x_k) & (y_2 - y_k) \\ \vdots & \vdots \\ (x_{k-1} - x_k) & (y_{k-1} - y_k) \end{bmatrix}, x = \begin{bmatrix} x \\ y \end{bmatrix} \tag{31}$$

$$B = \begin{bmatrix} r_1^2 - r_k^2 - x_1^2 + x_k^2 - y_1^2 + y_k^2 \\ r_2^2 - r_k^2 - x_2^2 + x_k^2 - y_2^2 + y_k^2 \\ \vdots \\ r_{k-1}^2 - r_k^2 - x_{k-1}^2 + x_k^2 - y_{k-1}^2 + y_k^2 \end{bmatrix} \tag{32}$$

After the establishment of a linear system of equations, the location estimate of unknown nodes can be solved by least square method:

$$\hat{x} = (C^T C)^{-1} C^T b \tag{33}$$

Communication overhead: As a result of controlled flooding in the network to send messages, each anchor node sends broadcast packets, the intermediate node only sends unsolicited packets, due to go through two flood process, the average number of nodes in the network Send a $2A$ packet, the communication overhead algorithm is $2AN$.

Algorithm complexity: The number of anchor nodes and the number of anchor nodes participating in a certain multilateral measurement and positioning are required for matrix multiplication in the process of least squares processing. The time complexity is $T(AK)$, i.e., approximate $T(n^2)$, and the spatial complexity is $O(n)$. The time complexity of K-means algorithm is $O(n \log_2 n)$, the space complexity is $O(n^2)$.

4 Simulation Experiment and Analysis

4.1 Lab Environment

At present, the CSI signal can be obtained through the Intel 5300 and Atheros 9380 wireless network adapters. This article uses the Atheros 9380 network card solution. The positioning algorithm requires two desktop computers with an Atheros 9380 NIC and the CPU model is Intel Core i3-4150. The operating system is Ubuntu10.04LTS, and the kernel and wireless network card drivers are all customized. One of the desktop computers is used as a signal transmitter and the other is used as a receiver. The experimental site selects a 9 m × 6 m office area and deploys 25 In the square area, each square area is 0.8 m × 0.8 m, the distance between the receiver and the transmitter is 4.5 m, and the antenna height is 1.2 m. The layout and detailed area layout are shown in Fig. 4, and Fig. 5 shows the test site for testers.

The office environment floor plan as shown, the actual test chart shown in Fig. 6. The size of the tracking sensing area in this paper is 9 m by 6 m. Place some markers on the floor and use your camcorder to record when people have walked through the markers. In the experiment, five students were selected to walk in different lines,

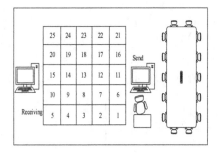

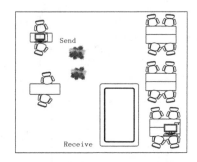

Fig. 4. Experimental environment plan

Fig. 5. Office environment plan

rectangles and circles within 2 weeks (the same time of the day). This article collected 100 tracks for each person and reported tracking errors to show tracking performance.

Figure 7 shows the tracking results achieved using the proposed algorithm in the real environment. Figure 7 shows the accuracy of the human body's velocity amplitude. Figure 8 shows the accuracy of the human body in velocity direction.

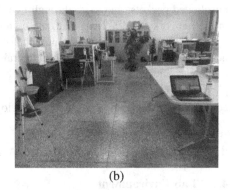

(a)   (b)

Fig. 6. (a) Real Environment Test Chart (laboratory environment), (b) Real Environment Test Chart (Office Environment)

From Fig. 7 it can be seen that when the measured person's trajectory is circular, the tracking effect is closest to the expected result, indicating that the tracking effect is best at this time. From Figs. 8, the amplitude error of the medium speed motion is as small as 14%, and the direction error is only 7°. In addition, the Droppler-Music method proposed in this paper is compared with the K-means algorithm based on the partition. Since the K-means algorithm estimates the Doppler speed based on the amplitude of the CSI, no direction information is provided. In the comparison process, only Use the K-means method to estimate the magnitude of the Doppler velocity and use the Droppler-Music method to provide direction information. As shown in Fig. 8, with the addition of direction information to improve performance, the medium-speed error of the K-means method is still 44% with a directional error of 17° which is much

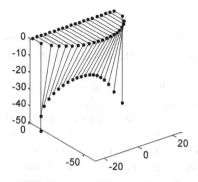

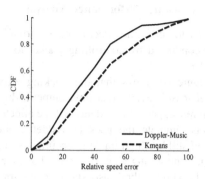

Fig. 7. Trajectory tracking results

Fig. 8. Speed accuracy of the direction

larger than the Droppler-Music method due to the Droppler The -Music method estimates the Doppler velocity's CSI phase, which is more stable than the CSI amplitude used in the K-means method. In addition, compared with other methods, this article's Droppler-Music method can simultaneously estimate the magnitude and direction of Doppler velocities.

The effects of antenna power (PA) and static components (RS) on the trajectory tracking accuracy are shown in Fig. 9:

As can be seen from Fig. 9, when the power of the antenna is not adjusted and the static component is not removed, the speed error is 37% and the direction error is 26°; only the power of the antenna is adjusted, the relative speed error is 19%, and the direction error is 17°; only the static component is removed, the relative speed error is 17% and the median direction error is 8°; the antenna power is adjusted to remove the static component, the relative speed error is 11% and the median direction error is 7°. Visible, remove the strong interference static component, the tracking performance improved significantly. At the same time, the power of the two antennas is adjusted, and the direction information of the Doppler speed obtained is more accurate.

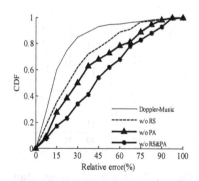

Fig. 9. Related error chart

4.2   Tracking Performance Analysis

In order to verify the effect of packet loss/delay on tracking performance, experiments were conducted by controlling packet loss and delay size. The results are shown in Fig. 10:

Figure 10 shows that the tracking error of this algorithm is about 8% without packet loss/delay. In real environment, due to the impact of environment noise and interference, packet loss/delay will be affected. Therefore, some random discarding CSI packets simulate the packet loss/delay situation. Even with 50% packet loss, the method still has high tracking accuracy similar to no packet loss. Figure 10 shows that tracking the wrong CDF in the lab has more multipath effects due to more shelters in the lab. This method also shows good tracking performance, demonstrating significantly higher tracking performance under static conditions In a dynamic environment.

To further investigate the impact of speed on tracking performance, the target was walking at three different speeds: slow (<1 m/s), medium (1–1.5 m/s), and rapid (1.5–3 m/s). The impact of speed on tracking performance is shown in Figs. 11 and 12.

Figure 11 illustrates the effect of the target velocity on the tracking performance of the method. The error is about 8% at the slowest and fastest speeds, indicating that the slower the speed, the better the tracking performance. Figure 12 illustrates the effect of the speed of the people moving at different speeds on the tracking performance. The impact of tracking performance indicates that the error in the speed direction increases with the increase of the moving speed.

This method tracks five targets. The error analysis is shown in Fig. 13:

It can be seen from Fig. 13 that the proposed method achieves consistent accuracy across different targets.

To sum up, the experimental results show that the proposed algorithm has great improvement in positioning with the loss rate, speed difference, direction difference and multi-person effect. Effectively alleviating the tracking accuracy caused by multipath weakness and non-line-of-sight environment. Especially in the direction of speed error, this algorithm proposes a good solution.

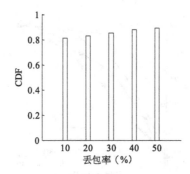

Fig. 10. The impact of loss rate chart

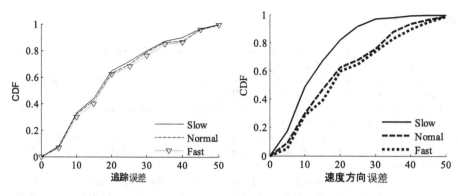

Fig. 11. Influence of speed difference   **Fig. 12.** Speed direction influence diagram

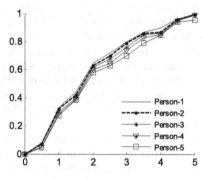

Fig. 13. Tracking error chart

5 Conclusion

In this paper, we aim to design a method that uses the improved neural network algorithm to track indoor personnel based on CSI signals. The experimental results show that the algorithm can effectively improve the accuracy of indoor personnel tracking and reduce the positioning error. To achieve a person's track accurate tracking. The problems that need to be solved in future work include multi-person tracking, real-time tracking, etc. The application of indoor personnel methods has very broad prospects. In addition an important difficulty is the current experimental environment is 2.4 GHZ, the next step is to transplant the development environment to 5 GHZ environment, in order to obtain more accurate positioning performance.

Acknowledgements. This work was supported by the National Natural Science Foundation of China under Grant No. 61363059, No. 61762079, and No. 61662070, Key Science and Technology Support Program of Gansu Province under Grant No. 1604FKCA097 and No. 17YF1GA015, Science and Technology Innovation Project of Gansu Province under Grant No. 17CX2JA037 and No. 17CX2JA039.

References

1. Xu, Y., Xu, X., Li, C., et al.: People's detection based on SVM classifier and HOG feature extraction. Comput. Eng. **42**(1), 56–60 (2016)
2. Xie, J., Wang, Y.: K-means algorithm for optimizing initial cluster centers with minimum variance. Comput. Eng. **40**(8), 205–211 (2014)
3. Li, C., Qin, P., Zhang, J.: Image denoising based on deep convolutional neural network. Comput. Eng. **43**(3), 253–260 (2017)
4. Jue, W., Deepak, V., Dina, K.: RF-IDraw: virtual touch screen in the air using RF signals. In: Proceedings of the 2014 ACM Conference on SIGCOMM (SIGCOMM 2014), pp. 235–246. ACM, New York (2014)
5. Li, S., Sen, S., Koutsonikolas, D., Kim, K.-H.: WiDraw: enabling hands-free drawing in the air on commodity Wi-Fi devices. In: Proceedings of the 21st Annual International Conference on Mobile Computing and Networking (MobiCom 2015), pp. 77–89. ACM, New York (2015)
6. Wang, W., Liu, A.X., Sun, K.: Device-free gesture tracking using acoustic signals. In: Proceedings of the 22nd Annual International Conference on Mobile Computing and Networking (MobiCom 2016), pp. 82–94. ACM, New York (2016)
7. Yun, S., Chen, Y.-C., Qiu, L.: Turning a mobile device into a mouse in the air. In: Proceedings of the 13th Annual International Conference on Mobile Systems, Applications, and Services (MobiSys 2015), pp. 15–29. ACM, New York (2015)
8. Sen, S., ChSoudhury, R.R., Radunovic, B., et al.: Precise indoo localization using PHY layer Sen information. In: International Conference on Mobile Systems, Applications, and Services, DBLP, pp. 1–6 (2011)
9. Qian, K., Wu, C., Yang, Z., et al.: Decimeter level passive tracking with Wifi. In: Proceedings of the 3rd Workshop on Hot Topics in Wireless. ACM, pp. 44–48 (2016)
10. Zheng, Y., Wu, C.S., Liu, Y.: Location Calculation: Wireless Network Positioning and Localization. China University Press, Beijing (2014)
11. Peng, Y., Yang, Y.: A Bayesian indoor location algorithm based on RSSI. Comput. Eng. **38**(10), 237–240 (2012)
12. Qi, W.: Indoor location technology based on RSSI ranging. Electron. Technol. **25**(6), 64–66 (2012)
13. Shi, X., Yin, A., Chen, X.: Multi-scale indoor positioning algorithm based on RSSI. J. Instrum. Chin. **35**(2), 261–268 (2014)
14. Yong, Z., Jie, H., Xu, K.: WLAN indoor positioning based on PCA-LSSVR algorithm. Chin. J. Sci. Instrum. **36**(2), 408–414 (2015)
15. Wang, X., Gao, L., Mao, S., et al.: CSI-based fingerprinting for indoor localization: a deep learning approach. IEEE Trans. Veh. Technol. **66**(1), 763–776 (2017)
16. Xiao, J., Wu, K., Yi, Y., et al.: FIFS: fine-grained indoor fingerprinting system. In: International Conference on Computer Communications and Networks, pp. 1–7. IEEE (2012)
17. Ma, C., Klukas, R., Lachapelle, G.: Time-of-arrival based localization under NLOS conditions. IEEE Trans. Veh. Technol. **55**(1), 19–23 (2006)
18. Li, J., Geng, L., Cao, M., et al.: Super-resolution delay estimation algorithm based on channel frequency domain model and OFDM technology. J. Sens. Technol. **19**(3), 733–736 (2006)

19. Voltz, P.J., Hernandez, D.: Maximum likelihood time of arrival estimation for real-time physical location tracking of 802.11a/g mobile stations in indoor environments. In: PLANS 2004: Proceedings of Position Location and Navigation Symposium, pp. 585–591. IEEE Press, Piscataway (2004)
20. Ni, H., Ren, G., Chang, Y.: A new TOA estimation algorithm for OFDM wireless networks. J. Xidian Univ. **36**(1), 17–21 (2009)

Design and Implementation of the Forearm Rehabilitation System Based on Gesture Recognition

Dexin Zhu[1（✉）], Zhiling Li[1], Kui Huang[1], and Sato Reika[2]

[1] College of Computer Science and Technology, Changchun University,
Changchun, China
38925023@qq.com
[2] Osaka Electro-Communication University, Shijonawate, Osaka, Japan

Abstract. With the development of human computer interaction technology, the computer virtual reality technology makes people's body language signals by simulating the human body language and other natural means of communication, so as to drive instruction of human-computer interaction. Therefore, demand for such a natural and harmonious interaction, the voice, face, gesture recognition is an important issue, identify the effect of experience is critical for interactive applications. The paper presents the forearm rehabilitation system based on vision-based gesture recognition, including pre-processing of images collected by the camera, gesture segmentation based on skin color information, gesture recognition based on hand shape feature. This system can effectively extract the gesture area, combined with the capture sphere of action to achieve rehabilitation exercise on the forearm.

Keywords: Gesture recognition · Forearm rehabilitation
Human computer interaction

1 Introduction

Rapid development of science and technology, human computer interaction has become increasingly close, from the 1980s and 1990s of the keyboard, mouse, until now the touch screen, improved human-computer interaction not only reduces the interactive difficulty, enhances the interactive speed and stability, but also the people's life has brought great convenience. At present, the gesture recognition in virtual reality is the study of the gesture as a human computer interaction interface, using intuitive gestures as interactive instruction, which is more in line with the interactive habits. In addition, gesture recognition can also be used in 3D animation, interactive visualization, virtual reality, medical treatment and other fields. However, due to the ambiguity of the gesture and the gesture is affected by the temporal and spatial variation, thereby gesture recognition research has a great challenge. Compared with traditional gesture recognition based on data glove, vision-based gesture recognition on the implementation has the advantage such as natural intuitive, so as to attract more scholars to conduct research and application.

© Springer Nature Singapore Pte Ltd. 2018
Q. Zhou et al. (Eds.): ICPCSEE 2018, CCIS 902, pp. 330–336, 2018.
https://doi.org/10.1007/978-981-13-2206-8_27

In recent years, some scholars have studied the vision-based gesture recognition. For dynamic gesture recognition, Yoon et al. developed a gesture recognition system based on gesture position, angle and speed [1], Liu et al. use different HMM model for identifying the 26 letters [2]. For the hand gesture recognition is used to describe different gestures, a descriptor based on geometry, SIFT descriptors based bag of words method [3], and the orientation histogram descriptor [4].

In the paper, image segmentation method in gesture segmentation stage uses the skin color model and the depth image, after skin segmentation, face skin color region are also included, then use the depth information of all the color points, according to the histogram model can effectively extract the gesture area, through the capture ball, realizing rehabilitation exercise on the forearm.

2 Skin Color Segmentation

The skin color is an important feature of the human body, it can be used as a gesture segmentation features from image sequence. After a lot of research and statistical analysis, the colour in the exclusion effect of light condition has good clustering. The color represents the use of color space, common color spaces are: RGB, YCbCr, HSV and so on. Many researchers, such as: Zarit [5], Jayaram [6], Terrilion [7] compared the effects of different color space applications in skin color modeling.

2.1 RGB

RGB color is the primary colors, R represents Red, G represents Green, B represents Blue. RGB color space can also be used to describe a three-dimensional cube. As shown in Fig. 1.

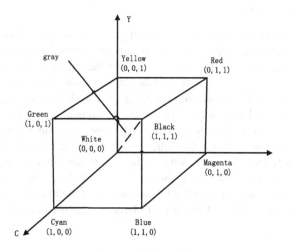

Fig. 1. RGB model

Red, green and blue in the RGB model closely related, it is the most common color space, which commonly used in image acquisition device and color display device. Because the components of RGB model red, green and blue do not remove the luminance information, the skin color segmentation is serious interfered by light skin.

2.2 YCbCr

The human eye is more sensitive to the luminance information matching hue, so many colors space separate the luminance component and the hue component, Y represents luminance component, Cb for blue color, Cr is red color, it is more in the digital TV, video codecs and other applications; RGB conversion equation it follows:

$$
\begin{bmatrix} Y \\ Cb \\ Cr \end{bmatrix} = \begin{bmatrix} 0.2990 & 0.5870 & 0.1440 \\ -0.1687 & -0.3313 & 0.5000 \\ 0.5000 & -0.4187 & -0.0813 \end{bmatrix} \begin{bmatrix} R \\ G \\ B \end{bmatrix} \tag{1}
$$

$$
\begin{bmatrix} R \\ G \\ B \end{bmatrix} = \begin{bmatrix} 1.0 & 0 & 1.371 \\ 1.0 & -0.3336 & -0.698 \\ 1.0 & 1.7320 & 0 \end{bmatrix} \begin{bmatrix} Y \\ Cb \\ Cr \end{bmatrix} + 128 \begin{bmatrix} -1.731 \\ 1.034 \\ -1.732 \end{bmatrix} \tag{2}
$$

Seen by the conversion formula to calculate the amount of the two is small.

In the paper, gesture recognition color separation uses the image segmentation channel cvSplit function of OPENCV, which displays the image in the Cb single channel of YCbCr color space.

3 Image Preprocessing

3.1 Image Binarization

Image binarization is to set the value of pixel grayscale image to 0 or 255, in digital image processing, image binarization occupies a very important position. First, image binarization is conducive to further processing of the image, it can make the image becomes simple, reduces the amount of data and can shows the outline of the interest target. Second, if the binary image processing and analysis, from gray-scale image thresholding, obtain a binary image. However, the amount of pixel gray value distribution calculated from the 0 images to 255 of is large, by a suitable general grayscale threshold binarization, that is, the image gray value within the threshold range belonging to the target, do not belong to the background or non target, reduce the amount of image data; In the paper, the cvThresholdBinarization function is used to process Image binarization, it uses the manual setting of the threshold, this method can greatly gray histogram of good two level difference. The pseudo code as follows:

```
void cvThresholdBidirection(IplImage* img)
{
    for (int h = 0; h<img->height; h++) {

      for (int w = 0; w<img->width; w++) {
          unsigned char* p = (unsigned char*)(img->imageData + h*img->widthStep
+ w) ;

          if (*p<=Threshold_Upper&&*p>=Threshold_Lower)
                 *p=255;
          else
                 *p=0;

      }
    }
}
```

3.2 Advanced Morphological Transformation

Morphological processing in OpenCV, it mainly removes isolated noise points in the image and connecting adjacent the image area, including dilation and erosion in two ways.

CvDilate expansion processing effect makes the two value images in the neighboring area has been expanded and connection.

CvErode corrosion processing the two value of the adjacent region in the image is isolated and narrow.

4 Experimental Results and Analysis of Gesture Recognition

The camera captures images, and after image preprocessing and skin color segmentation, image display is shown in Fig. 2. Among them, Ts_Lower and Ts_upper are used to manually adjust the threshold. Through multiple experiments, Ts_Lower = 0 and Ts_upper = 115. The image shows that the non-target region with skin color is the least.

In order to separate the gesture area from the background, we use convexityDefects in OpenCV to calculate the convex contour defect of the gesture contour, and the convex contour defect of the gesture contour is shown in Fig. 3. The calculation process is as follows:

(1) Find the mask of the hand
(2) Find the profile of the mask
(3) Find the multi-distortion curve of the contour

(4) find convex hull sets of polygonal fitting curves and find out convex points.
(5) find the concave set of the multi deformation fitting curve and find out the concave point

The image outside the gesture area is further processed, as shown in Fig. 4.

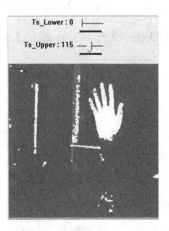

Fig. 2. Image preprocessing and image segmentation

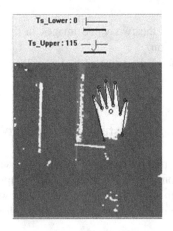
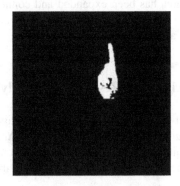

Fig. 3. Gesture area   **Fig. 4.** Gesture area

5 Forearm Rehabilitation System Based on Gesture Recognition

Flow chart of the system is shown in Fig. 5, the graph contains the whole process system. After the system is running, the interface is shown in Fig. 6. When the center of the gesture area and crosses the ball, the ball will disappear.

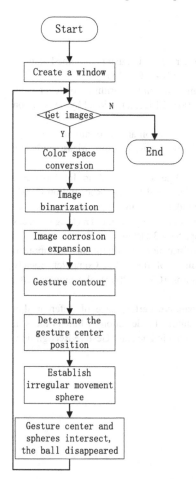

Fig. 5. System flow chart

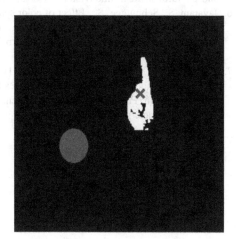

Fig. 6. System interface

6 Conclusion

The paper presents forearm rehabilitation system based on gesture recognition, first of all, the system implementation process uses the adaptive threshold value method and the morphological transformation to preprocess the image; Then in the process of skin color segmentation, the image is displayed in Cb on a single channel of the YCbCr color space; Finally, according to the gesture space position, use the breadth iteration method to part the gesture area.

Acknowledgment. This study was supported by the Science and Technology of Jilin province development plan projects with grants No. 20170204023GX, the Education Department of Jilin Province with grants No. 2016292, No. JJKH20170496KJ, the spring plan of Ministry of Education with grants Z2016013.

References

1. Yoon, H.S., Soh, J., Bae, Y.J., Yang, H.S.: Hand gesture recognition using combined features of location, angle and velocity. Pattern Recogn. **34**(7), 1491–1501 (2001)
2. Liu, N., Lovel, B., Kootsookos, P.: Evaluation of hmm training algorithms for letter hand gesture recognition. In: American: Proceedings of the 2003 IEEE International Symposium on Signal Processing and Information Technology (2003)
3. Dardas, N.H., Georganas, N.D.: Real-time hand gesture detection and recognition using bag-of-features and support vector machine techniques. IEEE Trans. Instrum. Measur. **60**(11), 3592–3607 (2011)
4. Freeman, W., Roth, M.:. Orientation histograms for hand gesture recognition. In: American: Proceedings of International Workshop on Automatic Face and Gesture Recognition (1994)
5. Zarit, B.D., Super, B.J.: Comparison of five color models in skin pixel classification. In: Proceedings of the International Workshop on Recognition, Analysis, and Tracking of Faces and Gestures in Real-time Systems, Corfu, Greece, pp. 58–63 (1999)
6. Jayaram, S., Schmugge, S.:. Effect of color space transformation, the illuminance component and color modeling on skin detection. In: Proceedings of the IEEE Computer Society Conference on Computer Vision and Pattern Recognition (CVPR 2004), Washington, DC, USA, pp. 813–818 (2004)
7. Tenilion, J.C., Shirazi, M.N., Fukamachi, H.: Comparative performance of different skin chrominance models and chrominance spaces for the automatic detection of human faces in color images. In: Proceedings of the Conference of Automatic Face and Gesture Recognition, Grenoble, France, pp. 54–61 (2003)

Research on Traffic Passenger Volume Prediction of Sanya City Based on ARIMA and Grey Markov Models

Xia Liu[1(✉)], Fang Wan[2(✉)], Lei Chen[1], Zhao Qiu[3],
and Ming-rui Chen[3]

[1] Sanya Aviation and Tourism College, Sanya 572000, Hainan, China
paolo_lx@qq.com
[2] College of Information Engineering, Hainan Institute of Science
and Technology, Haikou 570228, Hainan, China
36851357@qq.com
[3] Information and Technology School, Hainan University,
Haikou 570228, Hainan, China
5454734@qq.com, 1607885098@qq.com

Abstract. It is crucial to the economic and social development of the entire city to accurately predict its traffic passenger volume. For Sanya, as a coastal tourism city, it is rather vital to analyse the traffic volume to its tourism development. This paper, based on the tourism statistical data of monthly passenger volume from 2012 to 2017 in Sanya, applied ARIMA prediction model and grey Markov model to fit and predict the passenger volume. Upon empirical analysis, the result indicates that the mean absolute percentage error (MAPE) of such two models is 4.42% and 3.78% respectively with high prediction precision. Finally, the grey Markov model was utilized for trend extrapolation prediction and it has been found that the passenger volume in 2018 in Sanya would be expected to reach almost 36 million. Such prediction result would play an active role in policy formulation in the tourism and transportation industries, etc. in Sanya City.

Keywords: ARIMA model · Grey prediction · Grey Markov model
Passenger volume · Prediction

1 Introduction

It is vital to study the traffic passenger volume for the tourism development of a coastal city. In the course, several approaches can be adopted to predict the traffic passenger volume. In terms of time series, Reference 1 constructed GDP prediction model based on ARIMA Model. This GDP prediction model featured sound prediction precision but can be further improved. Reference 2 applied the grey system theory prediction model and the time series ARIMA prediction model to predict the traffic volume respectively. On this basis, this paper proposed the combined prediction model to predict the traffic volume. Upon case comparison and analysis, it has been found that the combined prediction model featured higher prediction precision than the individual grey prediction model or the time series analysis model. In this sense, it can be adopted as an

© Springer Nature Singapore Pte Ltd. 2018
Q. Zhou et al. (Eds.): ICPCSEE 2018, CCIS 902, pp. 337–349, 2018.
https://doi.org/10.1007/978-981-13-2206-8_28

effective measure to predict the short-term traffic volume. Reference 3 constructed ARIMA prediction model and analysed ARIMA model identification, model test and model prediction in a systematic manner; meanwhile, such model was applied to predict the monthly traffic volume of a highway. As verified by the result, ARIMA prediction model can be better adapted to predict the monthly traffic volume of highways. Reference 4 established the ARIMA Model and found upon verification that the prediction result of ARIMA Model highlighted higher degree of fitting with the short-term passenger volume in the future. Thus it can be adopted to predict the development trend of passenger volume and warn in advance the prevention and control of passenger aggregation to a certain extent. In Reference 5, based on the passenger volume data of Sanya Airport from 2008 to 2016, ARMA Model, Grey Prediction GM (1, 1) Model and ARMA-improved Regression Model were adopted for data fitting. Upon verification, the mean absolute percentage error of such three models was 4.19%, 4.20% and 1.97% respectively with high prediction precision. In terms of grey theory, Reference 6, upon verification, indicated that the grey prediction method featured higher prediction precision in respect of the airline passenger volume. It would be of direct guiding significance for the airline companies to estimate the passenger volume and formulate the sales policy. Considering GM (1, 1) Model would be easily affected by the stochastic disturbance of modeling data and the model had inferior stability, Reference 7 proposed the optimized GM (1, 1) Model based on Markov theory. Such optimized grey Markov prediction model was superior to the traditional GM (1, 1) prediction model and the optimized GM (1, 1) prediction model for its stronger applicability and better stability. Reference 8 firstly established the unbiased grey model to modify the inherent deviation of the traditional grey model, then utilized the particle swarm optimization to compute the optimal whiting coefficient value to form a new model with higher prediction precision—the particle swarm unbiased grey Markov chain model. Upon computation, this model could achieve higher prediction precision in comparison with the traditional grey Markov chain model. Considering the heavy computation of the traditional grey Markov chain model, Reference 9, in combination with the accumulation method and the grey Markov prediction model, proposed a grey Markov prediction model based on the accumulation method, which has conquered the drawbacks of the traditional prediction model. Meanwhile, this grey Markov Prediction model was applied to predicted to analyse and predict the tourism data of Guizhou Province in previous years. The result indicated that such model can reduce computation and increase the prediction level. Reference 10, considering the heavy computation of the grey Markov prediction model, combined the accumulation method and the grey Markov prediction model, and proposed a grey Markov prediction model based on the accumulation method, which has conquered the drawbacks of the traditional prediction model. Moreover, this model as proposed was also applied to analyse and predict the tourism data of Guizhou Province in previous years. The result indicated that such model can reduce the computation burden and increase the pre-diction level. Reference 11, based on the basic theories of grey model and Markov chain, constructed the grey Markov model for prediction. The result indicated that, upon two-order weakening, grey model construction, grey metabolism and grey Markov prediction, this grey Markov model can prominently increase the prediction precision. Reference 12, on the basis of grey prediction theory, carried out Markov prediction of residual sequence with great random fluctuation, thus has realized the advantage complementation of the classic grey theory and Markov chain with the deficiency of both conquered in the meantime. Therefore, this paper mainly adopted ARIMA model and grey Markov model to predict the traffic volume of Sanya from January 2012 to December 2017 respectively, and compared the prediction precision of such two models. The result showed that the prediction precision achieved by grey Markov prediction model was superior to that made by ARIMA model. Then the former was adopted to predict the passenger volume of Sanya of each month in 2018.

2 Data Sources and Method Description

2.1 Data Sources

This paper adopted the monthly tourism statistical data of Sanya City as publicized on Sanya Tourism Administration website from January 2012 to December 2017. Upon data purification, the total passenger volume of the airport and the train station was adopted as the research object.

2.2 Method Description
2.2.1 Description of ARIMA
ARIMA model, known as Autoregressive Integrated Moving Average Model in full, is a renowned time series prediction method proposed by Box and Jenkins in early 1970s [1], therefore, it is also known as Box-Jenkins model or Box-Jenkins method. In ARIMA (p, d, q), AR refers to auto-regression, p is the regression item; MA refers to the moving average, q refers to the moving average number and d refers to the difference frequency when the time series is stable. The stated ARIMA model refers to the model as established that transfers the non-stable time series to the stable time series, then regresses the dependent variable's lagged value, the current value and the lagged value of the random error. Based on the stability of the original sequence and the different parts included in the regression, ARIMA model includes the moving average (MA), the auto-regression (AR), the auto-regressive moving average (ARMA) and ARIMA process.

For the basic procedures of ARIMA model prediction:

- To identify the sequence's stability according to the scatter diagram, the autocorrelation function and the partial autocorrelation function diagram of the time series, check the variance, tendency and seasonal change rules with the roots of unity of ADF. Generally speaking, the time series of economic operation is unstable.
- To stabilize the unstable sequence. In case the data sequence is unstable and even indicates the increase or decline tendency to a certain extent, the data have to be subject to differential treatment. If the data have heteroscedasticity, the data have to be treated technically until the data's autocorrelation function value and partial correlation function value upon treatment are not substantially different from zero.
- To establish the corresponding model according to the identification rules of time series model. In case the stable sequence's partial correlation function indicates

truncation while the autocorrelation function indicates trailing, AR model shall be applicable; if the stable sequence's partial correlation function indicates trailing while the autocorrelation function indicates truncation, MA model shall be applicable; if both the stable sequence's partial correlation function and partial correlation function indicate trailing, ARMA model shall be applicable (the truncation refers to the nature that the time series' autocorrelation function (ACF) or partial autocorrelation function (PACF) is zero after a certain order (such as the PACF of AR); the trailing is the nature that neither ACF nor PACF is 0 after a certain order (such as AR's ACF)).

- To estimate the parameters and check whether such parameters have statistical significance;
- To perform hypothesis testing to diagnose whether the residual sequence is white noise;
- To utilize the models that have been tested for predictive analysis.

2.2.2 Description of Grey Markov

According to previous studies, the grey prediction features higher prediction precision for data sequences with certain laws and smaller rate of change, but it is difficult for grey prediction to limit the prediction precision within a relatively small range for the data sequences that change irregularly. Not only that, the grey prediction model also requires the exponentially increasing nature of the accumulative generated sequence. Only such a sequence is suitable for fitting with the differential equation, but the accumulative generated sequence from the non-negative and disordered time-series may have no exponential law. In this sense, it is very likely to generate great errors by using the exponential equation of the grey prediction model. At the same time, it is difficult to predict the errors due to great randomness. The Markov chain, with a dynamic system with random changes as its prediction object, predicts the future change tendency in the data according to the transition probability of the existing status, which reflects the influence of each influencing factor on the forecasting result from the side. Therefore, the Markov chain is applicable for predicting the system with fluctuating original data. The combination of the grey model with the Markov chain model can make up for the drawbacks of the grey prediction model and combine the advantages of both to improve the prediction precision. The grey Markov model is divided into grey transition probability Markov model and grey transition state Markov model. In this paper, the latter is applied based on the general steps shown as follows:

- To divide the prediction state. The residual errors are divided to the status of r categories according to the difference value between the predicted value and the actual value achieved by the grey model. There's no strict regulations on the number of state division, which is determined by comprehensive consideration of the number of samples and the error of fitting. It is generally classified into 3 to 5 categories.
- To establish the state transition probability matrix, shown in Formula (1) and (2).

$$P(\mathrm{k}) = \begin{bmatrix} P_{11}(k) & \cdots & P_{1n}(k) \\ \vdots & & \\ P_{n1}(k) & \cdots & P_{nn}(k) \end{bmatrix} \tag{1}$$

$$P_{ij} = \frac{M_{ij}}{M_i} \tag{2}$$

Where P_{ij} refers to the one-step transition probability, M_{ij} refers to the one-step transition quantity from State i to j, and M_i refers to the quantity of State i.

- To calculate the predicted value. Assuming that the state sequence is in State j at the moment of (k), according to the median of the residual error in State j, $[\varepsilon_{j^-}, \varepsilon_{j^+}]/2$, and the grey prediction value $\hat{x}^{(0)}(k)$, the predicted value achieved by the grey Markov model is:

$$\hat{y}_{(k)} = \frac{(e_{j^-} + e_{j^+})}{2} + \hat{x}^{(0)}(k) \tag{3}$$

3 Traffic Volume Fitting Based on ARIMA Model

Fit and predict the monthly passenger volume as mentioned above with the traditional time series ARIMA model. Work out the sequence chart of such airline before the ARIMA model is established. The result is shown in Fig. 1:

Figure 1 indicates prominent seasonal factor of the monthly passenger volume, which is characterized by less passengers in July, August and September and obvious more passengers in January, February and March. X-12 seasonal adjustments were then adopted to adjust the monthly passengers seasonally. The monthly seasonal index result is shown in Table 1.

Carry out unit root test of the monthly passenger volume upon adjustment. See Table 2 for the test result.

The monthly passenger volume upon adjustment is an integrated process according to Table 2; therefore, establish ARIMA (p, 1, q) model.

3.1 Model Recognition

Draw ACF and PACF diagrams of NUM to recognize p and q. See Fig. 2 for the result.

As seen from ACF and PACF diagrams in Fig. 2, the corresponding PACF diagram of D (passenger_num) indicates "truncation" of Phase 1 and ACF diagram shows "trailing". In combination of the principle of the minimum AIC value, initiate the ARIMA (1, 1, 0) model of traffic passenger volume finally. Upon repetitive debugging, see Table 3 for the result of ARIMA (1, 1, 0) model as finally established.

passenger_num series

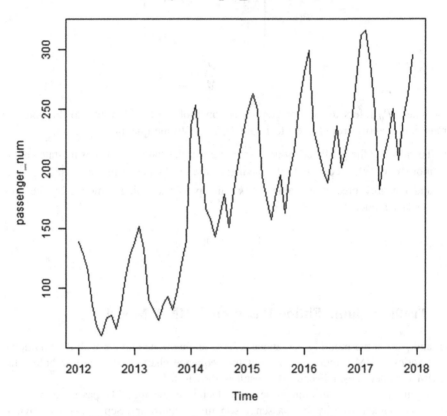

Fig. 1. Time series chart of monthly passenger volume

Table 1. Monthly seasonal index

Time	Seasonal index	Time	Seasonal index	Time	Seasonal index
January	1.3481	May	0.8237	September	0.7477
February	1.4072	June	0.7776	October	0.8661
March	1.2090	July	0.8390	November	1.0012
April	0.9493	August	0.8937	December	1.1373

Table 2. Unit root test result of passengers upon adjustment

Variable	P value	Test result	D(num)	P Value	Test result
Num	0.2399	Non-stationary	−3.4524	0.01	Stationary

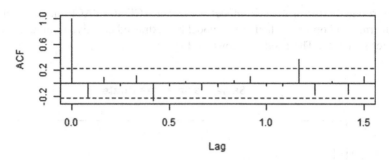

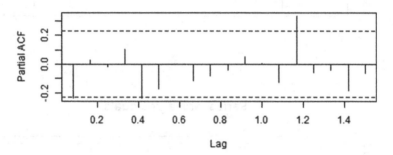

Fig. 2. ACF and PACF diagrams of D (passenger_num)

Table 3. ARIMA result

Variable	Coefficient	Std. error
C	189.3814	35.7818
AR(1)	0.9191	0.0451
Log likelihood	−608.7546	
AIC	692.6	

Under the significant level of 0.05 in Tables 2 and 3, the general likelihood-ratio test of the model is significant. All coefficients of ARIMA (1, 1, 0) are significant, as shown in the specific equation:

$$D(\text{passenger\_num})_t = 189.3814 - 0.9191D(\text{passenger\_num})_{t-1} \qquad (4)$$

3.2   Model Test

Test the residual error of the said model, work out ACF and PACF diagrams of such residual error, and observe whether the model has extracted effective information of the passenger sequence. The result is shown in Fig. 3:

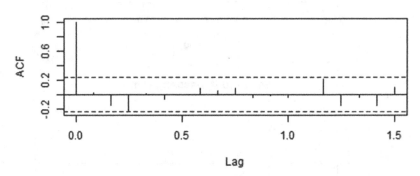

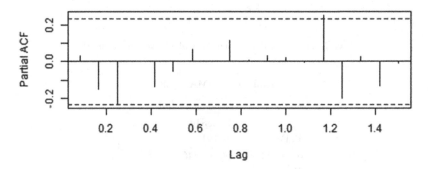

Fig. 3. ACF and PACF diagrams of residual error

Figure 3 indicates that ACF and PACF diagrams in respect of the residual error, upon two standard deviations, have explained the said model and effectively extracted all the sequence information. Then ARIMA (1, 1, 0) model as established was utilized to predict the monthly passenger volume.

3.3   Model Prediction

On the basis of a series of model establishment and test as mentioned above, ARIMA (1, 1, 0) model as established was utilized to predict the monthly passenger volume within the samples to achieve Sequence Y_1', which thus multiplied with the monthly seasonal index to acquire the final prediction sequence:

$$YF_1 = Y_1' * S_j \qquad (5)$$

In which, YF1 indicates the prediction sequence of monthly passenger volume and Sj indicates the monthly seasonal index. Then compute the mean absolute percentage (MAPE) of the predictions.

$$MAPE = \frac{1}{n} \sum_{i=1}^{n} \left| \frac{YF_{1,i} - Y_i}{Y_i} \times 100 \right| \qquad (6)$$

Based on the said ARIMA (1, 1, 0) Model as established above to predict the prediction precision of the monthly traffic passenger volume, MAPE = 4.42%.

4 Fitting Based on Grey Markov Model

Figure 1 shows prominent seasonal trend of traffic passenger volume from 2012 to 2017. Establish the grey Markov prediction model with the corresponding passenger volume of each month from 2012 to 2017 as the original sequence to carry out grouping prediction. This approach can effectively eliminate the influence of seasonal effect on prediction.

4.1 Model Establishment

Carry out grey prediction firstly and then establish the specific grey Markov model with the passenger volume from 2012 to January 2017 as the cases.

Firstly, order

$$X_1^{(0)} = \{139.1203, 138.4494, 235.6057, 247.1665, 280.7489, 311.25\}$$

Secondly, solve the differential equation $\frac{dX^{(1)}}{dt} + aX^{(1)} = \mu$, utilize the least square method to solve parameter vector $\hat{\alpha} = (B^T B)^{-1} B^T Y_n$. The prediction sequence from 2012 to 2017 by means of grey prediction is (Table 4):

Table 4. Grey prediction value

Year	True value	Predicted value	Predicted error	State
January 2012	139.1203	139.1203	0	E2
January 2013	138.4494	174.8933	−36.4439	E1
January 2014	235.6057	203.7775	31.8282	E3
January 2015	247.1665	237.4322	9.7343	E2
January 2016	280.7489	276.645	4.1039	E2
January 2017	311.25	322.3339	−11.0839	E2

Thirdly, divide the residual error state achieved from grey prediction and classify the predicted errors to three different state spaces: E1, E2 and E3 respectively shown as follows:

$$E1 = (36.44, -13.69), E2 = (-13.69, 9.07), E3 = (9.07, 31.82) \qquad (7)$$

According to the said division, one E1 state, three E2 states and one E3 state from 2012 to January 2017 were achieved.

Finally, compute the predicted value, obtain the class mid-value of residual error of the predicted value, then add the preceding grey predicted value to obtain the predicted value based on grey Markov prediction value featuring higher precision. Taking the predicted value of February 2012 as the case:

$$Y = 138.4484 + (36.44 - 13.69)/2 = 149.8283 \qquad (8)$$

4.2   Model Prediction

Similarly, obtain the predicted value of grey Markov prediction model from 2012 to January 2017:

Table 5 indicates the prediction precision of MAPE = 8.13% from January 2012 to January 2017 by grey Markov model and the prediction precision of MAPE = 3.70% by the grey prediction model. In this sense, the prediction precision of the former grey Markov model is higher than that of the later grey model; therefore, the grey prediction model was adopted to predict the traffic passenger volume from 2012 to February-December 2017. The prediction precision upon computation is MAPE = 3.78%.

Table 5. Grey Markov predicted value

Year	True value	Grey prediction	Markov prediction	Grey prediction error	Markov prediction error
January 2012	139.1203	139.1203	139.1203	0.00%	0.00%
January 2013	138.4494	174.8933	149.8283	26.32%	6.51%
January 2014	235.6057	203.7775	224.2275	13.51%	5.58%
January 2015	247.1665	237.4322	235.1222	3.94%	5.07%
January 2016	280.7489	276.645	274.335	1.46%	2.32%
January 2017	311.25	322.3339	320.0239	3.56%	2.72%

5 Comparison of Two Models and Trend Extrapolation Prediction

5.1 Model Comparison

Compare ARIMA model and grey Markov model to predict the traffic passenger volume from January 2012 to December 2017. The prediction sequence chart as obtained is shown as follows (Fig. 4):

The prediction precision of two models are shown as follows (Table 6):

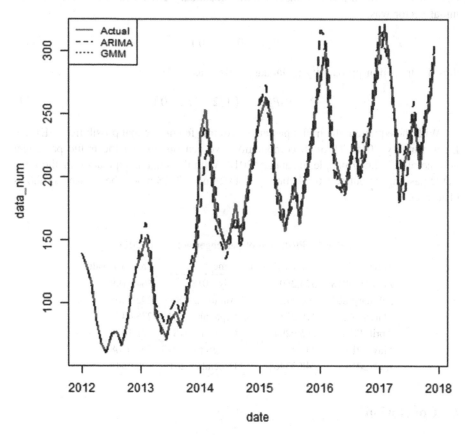

Fig. 4. Predicted value of two models

Table 6. Comparison of prediction precision of two models

Model	ARIMA Model	Grey Markov Model
Prediction precision	4.42	3.78

5.2   Trend Extrapolation Prediction

On the basis of comparison of the prediction precision of the said two models, the grey Markov model highlighting better prediction precision was adopted to predict the traffic passenger volume from January to December 2018.

Firstly, compute the state transition probability matrix from 2012 to January 2017, shown as:

$$P(1)=\begin{pmatrix} 0 & 0 & 1 \\ 1/2 & 1/2 & 0 \\ 0 & 1/2 & 1/2 \end{pmatrix} \qquad (9)$$

through which, the traffic passenger volume in January 2012 was in E2 state, thus the initial vector was:

$$P_0 = (0 \quad 1 \quad 0) \qquad (10)$$

thus the transition probability in January 2018 was:

$$P = P_0P(1) = (1/2 \quad 1/2 \quad 0) \qquad (11)$$

With the equation, the traffic passenger volume featured equal possibility of E1 and E2 in January 2018. Through computation, we can obtain that the traffic passenger volume of 3,732,607 people in January 2018. With the same computation method, the traffic passenger volume from February to December 2018 would be shown as follows (Table 7).

Table 7. Prediction of passenger volume in 2018

Time	Predicted value	Time	Predicted value
January 2018	373.2607	July 2018	264.0199
February 2018	401.3155	August 2018	292.7408
March 2018	310.258	September 2018	275.0423
April 2018	285.9294	October 2018	273.1624
May 2018	247.4187	November 2018	291.0208
June 2018	248.3346	December 2018	330.6543

6   Conclusion

Establish the model and make analysis based on different models. Upon comparison of the prediction precision of various models, the grey Markov prediction model based on sample prediction and highlighting higher precision stood out for its capability of predicting the traffic passenger volume in the future one year in a more precise manner. The prediction result indicated February would be the peak season of travel in Sanya and its passenger volume may break through 4 million, which would be of great significance for the local traffic and tourism departments to make preliminary work arrangement. Besides, it is simple and straightaway to master the methods and result of grey Markov model, therefore, it can be promoted to other applications in terms of passenger volume prediction.

Acknowledgements. This research was financially supported by Hainan Provincial Natural Science Foundation of China (618QN258). Thanks to Professor PhD Zhao Qiu and Professor Ming-rui Chen, as correspondents of this paper.

References

1. Sun, S., Li, S., Fan, C., Liu, H.: Construction and application of GDP prediction model based on ARIMA. J. Univ. Sci. Technol. Liaoning **37**(04), 337–342 + 349 (2014)
2. Song, Z.: Study on short-term traffic volume prediction model. Sci. Decis. **4**, 83–94 (2014)
3. Rui, S., Kuang, A.: ARIMA model of expressway traffic volume monthly forecastingj. J. Chang'an Univ. (Natural Science Edition) **30**(04), 82–85+91 (2010)
4. Zhang, J., Du, M.: Application of ARIMA model to risk analysis of urban passenger aggregation. Telecommun. Bull. **12**, 22–28 (2016)
5. Liu, X., Huang, X., Chen, L., Qiu, Z., Chen, M.: Prediction for passenger flow at the airport based on different models. In: Chen, G., Shen, H., Chen, M. (eds.) Parallel Architecture, Algorithm and Programming. PAAP 2017. CCIS, vol. 729. Springer, Singapore (2017)
6. Xia, L., Jie, Y., Lei, C., Ming-rui, C.: Prediction for air route passenger flow based on a grey prediction model. In: 2016 International Conference on Cyber-Enabled Distributed Computing and Knowledge Discovery (CyberC), Chengdu, pp. 185–190 (2016)
7. Li, K., Li, Z., Zhao, L.: An optimized grey model deformation forecasting modeling based on Markov theory. Sci. Surv. Mapp. **41**(08), 1–5 (2016)
8. Fan, D.: Research on railway passenger volume prediction methods based on optimized grey markov chain model. Chongqing Jiaotong University (2015)
9. Zhang, G.: Grey Markov prediction model and its application based on accumulation method. Stat. Decis. **8**,157–158 (2011)
10. Nian, Zhua, Haizhong, Gan, Weiwei, Feng, Lishan, Huang: The prediction of guangxi border trade based on grey Markov Chain forecasting model–the third paper in series article of grey system theory and application. J. Math. Practice Theory **47**(03), 270–277 (2017)
11. Liu, Z., Jia, Z., Li, X.: Traffic volume prediction based on grey Markov chain model. J. East China Jiaotong Univ. **29**(01), 30–34 (2012)
12. Liu, Y.: Application of citrix application virtualization to telecom IT supporting system. Guangdong Commun. Technol. **32**(3), 75–78 (2012)
13. Chen, X., Ding, B., He, G., et al.: Probe into construction of uniform access platform among mobile operators. Guangdong Sci. Technol. **8**, 89–90 (2009)

Predictive Simulation of Airline Passenger Volume Based on Three Models

Han-Tao Yang and Xia Liu$^{(\boxtimes)}$

Sanya Aviation and Tourism College, Sanya 572000, Hainan, China
371750371@qq.com, paolo_lx@qq.com

Abstract. It is of great significance to predict the airline passenger volume accurately no matter for the transport capacity arrangement, the airline adjustment or the planning and development. Considering so many uncertainties and insufficient data in terms of the passenger volume prediction of civil aviation, this paper, based on the daily passenger data of the airline from Beijing to Sanya for the period from 2010 to 2017, applied the random forest prediction model, the support vector regression model and the neural network model to fit the airline data. Upon verification, the mean absolute percentage error (MAPE) of the said three models was 4.18%, 6.87% and 12.38% respectively. In this sense, the random forest prediction model featured the highest prediction precision and the optimal simulation effect in passenger volume prediction.

Keywords: Random forest · Support vector regression · BP neural network
Prediction · Simulation

1 Introduction

As it is of crucial significance to predict the airline passenger volume accurately no matter for the transport capacity arrangement, the airline adjustment or the planning and development, many scholars have carried out fruitful attempts in terms of passenger volume prediction. In terms of the random forest prediction, Reference 1 introduced the application of random forest regression algorithm to the seismic reservoir prediction and achieved sound prediction effect. Reference 2 put forward the random forest regression annual wastewater discharge prediction model which based on random interpolated structure samples, which featured high prediction precision and strong generalization capability upon verification. Reference 3 utilized the random forest regression method to predict the military equipment demand and achieved sound prediction effect. In terms of the support vector regression prediction, Reference 4 would apply the support vector regression to fit the data of some airline in the latest six years and achieved high fitting precision. Reference 5 adopted the support vector regression algorithm. With the key factors of spares requirement as the input and the spares requirement as the output, a SVR prediction model of spare part was established. Upon simulation verification through matlab software, such model highlighted high prediction precision. Reference 6 put forward a regression tree prediction model based

© Springer Nature Singapore Pte Ltd. 2018
Q. Zhou et al. (Eds.): ICPCSEE 2018, CCIS 902, pp. 350–358, 2018.
https://doi.org/10.1007/978-981-13-2206-8_29

on support vector, which can also achieve accurate prediction; in addition, as its performance was superior to the standard analysis method, it could be widely applied to the engineering field. In terms of the neural network prediction, Reference 7 applied BP neural network algorithm to predict the actual passenger volume of Shanghai rail transportation and then provided the prediction result data and analysis. Reference 8 utilized BP and Elman neural networks respectively to establish a model for predicting the passenger volume of subway escalators. Upon analysis and comparison of the predicted data, the BP neural network was verified to be more suitable in this regard.

2 Data Sources and Descriptive Analysis

2.1 Data Selection

This paper selected the passenger volume of the airline from Beijing to Sanya for the period from January 2010 to December 2017 as the research object, and the data were mainly the actual daily data in respect of the airline from Sanya to Beijing provided by the airline company. The random forest regression, the support vector regression and the neural network were adopted respectively for prediction. And the prediction precision of such three models was finally compared.

2.2 Descriptive Analysis

Based on the sequence chart of passenger volume of the airline from Beijing to Sanya to observe the variation trend of passenger volume, the result as achieved is shown in Fig. 1:

In Fig. 1, the passenger volume showed a rising trend for the airline from Beijing to Sanya during the sample period and it was obviously affected by seasonal factors. The passenger volume reached the peak in 2014 and descended in 2017. Afterwards the random forest regression, the support vector regression and the neural network would be respectively utilized for prediction and final comparison.

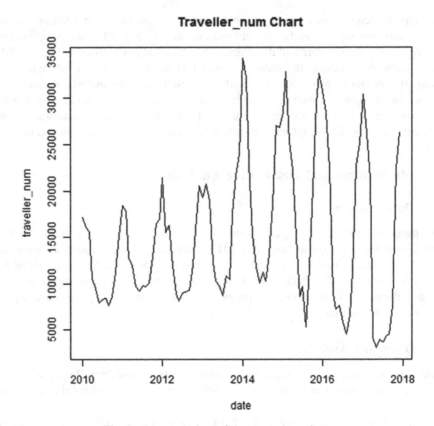

Fig. 1. Sequence chart of the passenger volume

3   Predictive Simulation Based on Random Forest Regression

As the integrated algorithm of the decision-making tree, the random forest includes several decision-making trees to reduce the risk of over-fitting. Similarly, the random forest also has the problems of understandability, processible characteristics of types, scalable to multi-classification and regression. As the random process is added to the algorithm, each decision-making tree has some minor distinctions. Therefore, the prediction result of each tree is merged to reduce the prediction variable and improve the performance of test set.

The random forest has to integrate the prediction result of each decision-making tree in prediction, but the way of integrating regression and classification problems differ slightly. For the classification problems, the voting system is adopted. In the course, each decision-making tree will vote to a classification and the classification with the majority votes will be the final result. For the regression problems, the prediction result of each tree is a real number, and the final prediction result will be the mean value of each decision-making tree's prediction result.

3.1   Model Parameter Optimizing

In the random forest model building, 500 decision-making trees shall be firstly selected with the variable number of 7 to establish the model initially, and then confirm the optimal parameter. First of all, observe the relationship between the quantity of decision-making trees and the model's prediction error. The result arising therefrom is shown in Fig. 2:

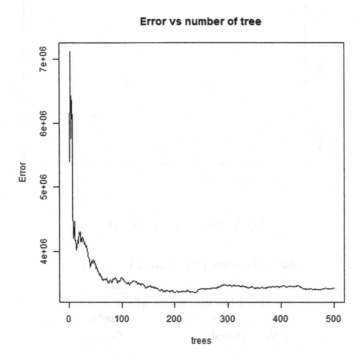

Fig. 2. Error vs number of trees

As indicated by Fig. 2, the model has the minimum prediction error in case of 230 decision-making trees as selected in the random forest model. Then explore the relationship between the variable number of each tree and the prediction error. The result arising therefrom is shown in Fig. 3:

As indicated by Fig. 3, the model has the minimum prediction error when the decision tree has 4 variables. Therefore, 230 trees were selected in this paper and each tree had 4 variables to establish the random forest model. The model parameters as obtained are shown in Table 1.

3.2   Model Building

The prediction result of the passenger volume from Beijing to Sanya within the samples by the said model as established is shown in Fig. 4.

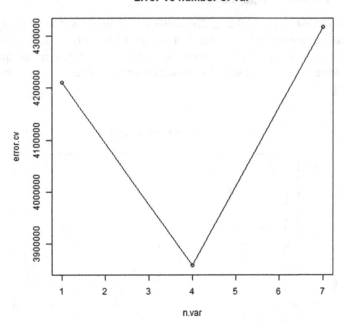

Fig. 3. Error vs number of VAR

Table 1. Estimated result of prediction parameters

Number of trees	230
No. of variables tried at each split	4
Mean of squared residuals	3489303
% Var explained	94.44

Figure 4 indicates the actual passenger volume from Beijing to Sanya Airport based on the random forest regression model almost overlaps with the predicted passenger volume. The computation formula of the mean absolute percentage error (MAPE) is:

$$MAPE = \frac{1}{n}\sum_{i=1}^{n}\left|\frac{\hat{y}_i - y_i}{y_i} \times 100\right| \tag{1}$$

Upon computation, MAPE = 4.18%. The correlation coefficient between the predicted passenger volume and the actual passenger volume is cor = 0.9951, indicating the random forest regression can well predict the passenger volume from Beijing to Sanya with ideal prediction effect.

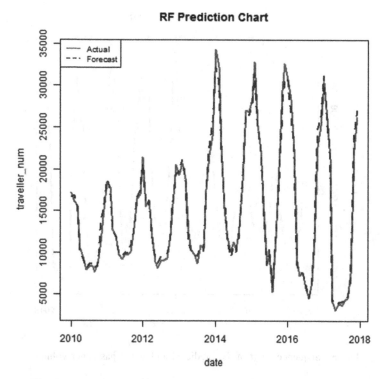

Fig. 4. Sequence chart of original sequence and prediction sequence

4 Predictive Simulation Based on Support Vector Regression (SVR)

This paper adopted the departure airline from Beijing to Sanya, the flight time, the passenger load factor, the total revenue per available seat kilometre (RASK), the average price and the average discount rate as the characteristic attributes of the passenger volume. Then SVR model was utilized to predict the airline passenger volume and the prediction result is shown in Table 2:

Table 2. The optimal parameters of SVR model

Parameter	SVM-Kernel	Cost	Gamma	Epsilon	Number of support vector	MAPE
Parameter value	Radial	1	0.1429	0.1	59	6.87%

In Table 2, SVR model adopted the radial basis function and achieved the related optimal parameters, showing the model's bonding violation cost $c = 1$, the gamma function factor in radial basis function $g = 0.1429$, the estimated error $\varepsilon = 0.1$ and 59 support vectors. The sequence chart of the predicted and actual passenger volume is shown in Fig. 5:

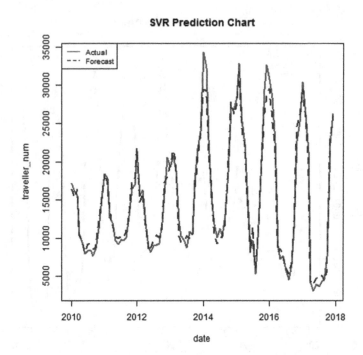

Fig. 5. Sequence chart of the predicted and actual passenger volume

Upon computation, MAPE = 6.87%, the correlation coefficient of the predicted passenger volume and the actual passenger volume is $cor = 0.9907$, indicating the support vector regression can well predict the passenger volume from Beijing to Sanya.

5 Predictive Simulation Based on Neural Network

The neural network is a kind of computing model composed by a great amount of variables and neurons, among which, each variable represents a specific output function, known as the activation function. The connection between each variable all represents a weighted value passing through the connecting signals, known as the weight which is equivalent to the memory of artificial neural network. The network

Table 3. Weight among various variables in the model

Output layer	b->h1	i1->h1	i2->h1	i3->h1	i4->h1	i5->h1	i6->h1	i7->h1
Weight	−1.21	0.91	1.01	0.2	0.46	0.29	−0.64	0.26
Input layer	b->h2	i1->h2	i2->h2	i3->h2	i4->h2	i5->h2	i6->h2	i7->h2
Weight	0.04	−0.2	−0.22	−0.05	−0.11	−0.09	0.09	−0.04
Input layer	b->h3	i1->h3	i2->h3	i3->h3	i4->h3	i5->h3	i6->h3	i7->h3
Weight	0.05	−0.23	−0.25	−0.05	−0.12	−0.1	0.11	−0.04
Output layer	b->o	h1->o	h2->o	h3->o				
Weight	−0.07	1.84	−0.38	−0.44				

BP Prediction Chart

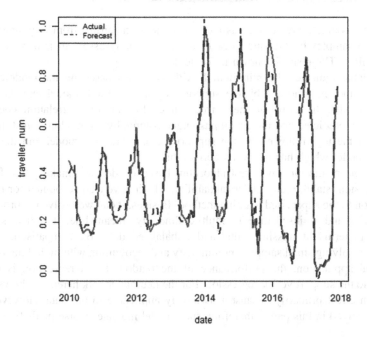

Fig. 6. Sequence chart of predicted passenger volume by BP network

output depends on the way of connecting and differs because of different weights and activation functions. Besides, the network itself is approaching some external algorithm or function. At present, the neural network is usually applied to solve the classification and regression problems.

This paper adopted the back propagation neural network (BP neural network) to predict the passenger volume from Beijing to Sanya and established the single-layer neural network including only one hidden layer. In the network, the input layer had 7 variables, the hidden layer had 3 variables and the output layer had 1 variable. The weight among various variables is shown as in Table 3:

The prediction result of the passenger volume from Beijing to Sanya within the samples by the neural network model as established is shown in Fig. 6.

Upon computation, MAPE = 12.38%, the correlation coefficient between the predicted passenger volume and the actual passenger volume is cor = 0.9836, indicating the neural network can well predict the passenger volume from Beijing to Sanya.

Table 4. Comparison of the prediction result

Prediction method	Random forest regression	Support vector regression	Neural network
MAPE	4.18%	6.87%	12.38%
COR	0.9951	0.9907	0.9836

6 Comprehensive Comparison and Conclusion

Build the model and predict the passenger volume of the airline from Beijing to Sanya within the samples by the said three models, and finally assess their respective prediction effect. The result is shown in Table 4.

The table indicates the minimum MAPE is 4.18% based on the random forest regression model, followed by the support vector regression model and the neural network model. In terms of the correlation coefficient, the correlation coefficient between the predicted and the actual passenger volume by the random forest model is ranked on the top, followed by the support rector regression model and the neural network model which has the worst correlation.

To sum up, the random forest algorithm based on decision-making tree features high precision, stability and understandability, which have also accounted for its wide application to the supervised learning method. Besides, it can well solve the problem of nonlinearity and is insensitive to multi-collinearity; meanwhile, it indicates sound stability in terms of missing value and unbalanced data. By comparison, SVR is superior in solving small samples, nonlinearity and recognition with high dimension. In the actual application, the performance of the random forest regression is usually superior to the support vector regression. For the neural network hereof, it shows worse prediction effect primarily because it is mostly applicable to big data. However, less data is involved in this paper; therefore, such model indicates worse predictive ability.

Acknowledgements. This research was financially supported by Hainan Provincial Natural Science Foundation of China (618QN258). Thanks to associate professor Xia Liu, correspondent of this paper.

References

1. Song, J., Gao, Q., Li, Z.: Application of random forest regression to seismic reservoir prediction. Oil Geophys. Prospect. **51**(06), 1202–1211 (2016)
2. Cui, D.: Application of random forest regression model to the wastewater discharge prediction. Water Technol. **8**(01), 31–36 (2014)
3. Gu, J., Lin, M., Ju, G.: Prediction of military equipment demand based on random forest regression. Autom. Appl. **9**, 24–25 (2017)
4. Liu, X., Huang, X., Chen, L., Qiu, Z., Chen, M.: Improving the forecasting accuracy of civil aviation passengers based on machine learning models. In: 2017 International Conference on Cyber-Enabled Distributed Computing and Knowledge Discovery (CyberC), Nanjing, pp. 298–304 (2017)
5. Liu, Y., Chen, D.: Research on spare part prediction of steel enterprises based on support vector regression algorithm. Guangxi J. Light Ind. **33**(06), 100–103 (2017)
6. Xianglan, W., Mingjun, F.: The trend prediction of engineer vibration signals based on SVM-RT. Ship Sci. Technol. **37**(05), 67–69 (2015)
7. Cheng, H., Xu, X.: Short-term prediction of rail passenger volume based on BP neural network. Electron. Technol. Softw. Eng. **22**, 15–16 (2016)
8. Mingxu, L., Yinzhong, Y., Xianghua, M.: Application of neural network in the subway passenger flow prediction. Mech. Res. Appl. **25**(03), 86–89 (2012)

Research on Monitoring Methods for Electricity Hall Staff Based on Autonomous Updating and Semi-supervising Model

Yao Tang[1]([✉]), Zhenjuan Qiao[2], Rui Zou[3], Xueming Qiao[1], Chenglin Liu[1], and Yiliang Wang[1]

[1] State Grid Weihai Power Supply Company, Weihai 264200, Shandong, China
tangyao201801@163.com
[2] School of Computer Science and Technology, Shandong University, Jinan 250100, Shandong, China
[3] Weihai No. 1 Middle School, Weihai 264200, Shandong, China

Abstract. With the continuous improvement of the concept of grid marketing and service awareness, Whether the electricity hall staff is on the job has become the basic guarantee for monitoring service levels. In this paper, a self-updating and semi-supervised model is used to optimize the existing monitoring system. First of all, the combination of significant facial features and classical key point positioning has been used to construct feature extraction methods. And this method is used to extract the features of the random monitoring acquisition image, so as to realize the high frequency detection of "human" in the monitoring area. Subsequently, the person's batch-acquisition image is used to identify his identity information within a fixed period. The feature extraction method is also used to extract features from standard marker images with the features of partial random monitoring acquisition image, which are used as input data for the semi-supervised k-NNM identification module. And the non-linear least-squares optimization algorithm was integrated to realize the weight distribution of the facial features. The above method builds a one to one recognition model. With the development of monitoring work, the correct identification sample data of the previous monitoring period is input into the model as a self-updated marking training sample to continuously improve the identification accuracy of the model. Under the condition of variable posture, the experiment shows that this method can meet the requirements of high timeliness for personnel on-the-job monitoring, and can also identify the appointed staff on the basis of facial features mining data in batches. This method can also update the sample to adapt to the change of electricity hall staff and provide non-staff warning.

Keywords: In-service time detection · Feature extraction · Match weight
Semi-supervised k-NNM identification module

1 Introduction

With the development of staff service monitoring work in the Power Supply Business Office, the monitoring of personnel at work has become a prerequisite for the development of other monitoring services. However, the current face recognition technology

© Springer Nature Singapore Pte Ltd. 2018
Q. Zhou et al. (Eds.): ICPCSEE 2018, CCIS 902, pp. 359–376, 2018.
https://doi.org/10.1007/978-981-13-2206-8_30

is mainly aimed at the frontal image recognition of the personnel, and if the single face recognition fails, the identified person will actively report, so as to adjust and update the person identification feature. The source of the images monitored by the staff at the post is random shooting. First of all, when the images are captured, the employees are unknowable, so that it is impossible to update the database storage information when the personnel image changes. Secondly, the change of the angle of the captured image leads to a decrease in the recognition rate.

In response to the above problems, this paper draws on the relatively mature image recognition technology [1], combined with the actual application requirements of the monitoring business of the staff for the power supply business office, that is, high-frequency acquisition and detection, and identification of fixed posts, to achieve effective processing of personnel on-site monitoring acquisition images. Especially, in recent years, for the image processing such as face recognition and gesture recognition become a hot research issue [2]. In this paper, the feature extraction of the monitoring image is realized by reference to the positioning of key points, and the method is improved on the basis of the existing research. In addition, through the measurement of many factors, a combination of machine learning algorithms is used to achieve the identification of on-site personnel monitoring images, thereby improving the management service level of the power supply business hall. In addition, through the measurement of multiple factors, the identification of monitoring images of on-the-job personnel is realized through a combination of machine learning algorithms, thereby improving the management service level of the power supply business hall.

2 Design and Analysis

In order to improve the service level and responsibilities of staff in the power supply business office, this paper conducts research on staff monitoring. On the one hand, monitoring staff attendance, on the other hand, as the basis for handling business, and pushing non-service personnel (including non-updated employee data). Through image acquisition, a personnel monitoring and identification model is built to monitor the situation of personnel in posts in the case of a fixed position for personnel, which satisfies the demand for smart grid construction with reform and innovation to promote service improvement. The method design is as follows.

The description of problems in the process of constructing on-site staff monitoring methods:

Question 1: In order to determine the on-duty time of personnel, it is necessary to collect high-frequency images of on-duty personnel based on existing equipment. Due to the large amount of images, the processing method must have high timeliness.

Question 2: With the reform of the power system, the mobility of employees has increased. Therefore, when the old and new employees alternate, the monitoring model must have the ability to self-renew.

Question 3: The application of machine learning methods is becoming more and more complicated, such as deep learning [3]. The model training of this method requires a large number of data samples, and the parameter adjustment process is complicated and difficult, so this required to deploy high-performance equipment to meet its computational overhead. For small dataset processing, deep learning may not be possible with the same equipment and time conditions. The application background of this article is that the position of the employee is fixed, so it is a one-to-one person identification with known information, thus broadening the range of machine learning methods.

Question 4: Due to the change of the angle of the human face, the accuracy of image region positioning and recognition is significantly reduced. For example, the key points in the identification of key points are 68, but the side images can only be positioned 30 to 40, and when the angle changes significantly, the degree of discrimination of personnel is insufficient.

In order to solve the above problems, On the one hand, this paper uses a simple method to achieve high-frequency detection of personnel. On the other hand, the way of coarse positioning to fine matching was used to achieve personnel identification. The improvement is described as follows:

Improvement 1: For the monitoring needs of personnel at work time, the overall model is divided into two parts: the time identification process of personnel at work and the identification process of personnel. This meets the timeliness and accuracy requirements of business applications.

Improvement 2: For the requirement of self-renewal of human characteristics, this paper uses the key points to achieve coarse localization of facial features, and then uses the way of local features to local features for fine matching facial features. The two methods constitute a facial feature extraction method. When person updated, new image of the human can be obtained. Perform self-extraction to assist in the development of sample update data. When the personnel are updated, self-extracting images of new people can also be performed to assist in making sample update data.

Improvement 3: In the algorithm selection process of the personnel identification module, the problem of sample updating with monitoring as the target is fully considered. In practice, a large number of acquired images are in an unlabeled state, so the semi-supervised learning algorithm is used to achieve the latest classification using unlabeled samples. Therefore, the identification module does not rely on external interactions and automatically uses unlabeled samples to improve model learning capabilities. Because the semi-supervised learning of complex algorithms has difficulties in balancing the computational complexity and the learning accuracy, a relatively simple and effective algorithm that a weighted k-Nearest Neighbor Means algorithm [6] is used in this paper, which can match weights to meet the recognition accuracy requirements.

Improvement 4: For the problem of changing angles, facial panorama images are used instead of traditional facial images to determine the location of the key points of the panoramic human face. This method can lock the effective facial features and extract high-availability facial features, thereby improving the accuracy of the model.

In addition, based on face recognition technology has achieved results, refer to CLM algorithm [4], this article uses the local feature concept to replace features of the facial features with part facial features. And PCA (Principal Component Analysis) [5] algorithm is used to reduce features, which reduces the complexity of the model. This also improves the timeliness and accuracy of the monitoring model.

3 Construction of Semi-supervised Model Based on Autonomous Update

In the model building process, first, the image is collected for feature data extraction. Then, by using the semi-supervised k-NNM (k-Nearest Neighbor Means) [7] combined with the nonlinear least squares optimization algorithm, this paper constructs a personnel identification module that can realize the identification of the same person in the phase period.

3.1 Feature Extraction Method

This paper uses feature extraction as the input data for the personnel identification module. To solve the problem of changing angles and extract features of regional features, significant feature extraction and classical key point location combined to build feature extraction method. The specific process is as follows:

(1) Existing equipment is used to capture and store images of on-duty personnel;
(2) Screening clear images from historical image databases and graying the images to extract features of facial features as significant facial features;
(3) Based on the standard Chinese facial proportion image, the key points of facial features were acquired by manual calibration. Then, the key points were globally expanded in combination with the facial panorama to obtain the classic key points;
(4) The combination of significant facial features and classical key point positioning constitutes an image feature extraction method.

The specific process of feature extraction of the facial features is described as follows:

3.1.1 Significant Feature Extraction

For the characteristics of the facial features, such as clear black and white eyebrows, eyebrows texture, outline of the nose, etc. These feature areas are the main basis for distinguishing different people. Therefore, this paper extracts information with obvious features.

The specific process is described as follows:

(1) The acquisition of images from the historical image database. The features to be included in the image are clear, such as facial features, clear facial features. Select multiple images that meet the above requirements, and both are frontal/90 ° side view;
(2) The above image is calibrated and divided into facial features;
(3) The HOG features were extracted from the image region. That is, the each regional $[hog(x, y)]$ are obtained. (x, y) represents a pixel in the image. If the detection target is a 64×64 image, the image is divided into cells of 3×3 pixels, and 2×2 cells constitute a block. When the HOG feature is extracted, there are 9 bins in the gradient direction in each cell, and the gradient amplitude $m(x, y)$ and the gradient direction $\theta(x, y)$ of each cell of the image are calculated as follows:

$$I_x(x,y) = I(x+1,y) - I(x-1,y) \tag{1}$$

$$I_y(x,y) = I(x,y+1) - I(x,y-1) \tag{2}$$

$$m(x,y) = \sqrt{I_x^2(x,y) + I_y^2(x,y)} \tag{3}$$

$$\theta(x,y) = \arctan\left[I_y(x,y)/I_x(x,y)\right] \tag{4}$$

The characteristics of all the cells in a single block are concatenated to obtain the HOG feature value of the block. The HOG features of all blocks are concatenated to obtain the HOG feature vector of the block. The characteristics of all the cells in a single block are concatenated to obtain the HOG feature value of the block. The HOG features of all blocks are concatenated to obtain the HOG feature vector for the area;

(4) For different regions, PCA algorithm is used to screen the high-differentiation feature matrix so as to obtain the characteristics of maximizing and distinguishing facial features. Taking the eye area image (including a part of peripheral skin) as an example, a total of h images of an employee's eye area are set, and the sample feature set extracted through the HOG feature is recorded as:

$$B = \{b_1, b_2, \cdots, b_i, \cdots b_h\}^T \tag{5}$$

The above b_i is the d-dimensional row vector, d is the HOG feature dimension of each image, and the covariance matrix is:

$$S = \frac{1}{h}\sum_{i=1}^{h}(b_i - \bar{b})(b_i - \bar{b})^T \tag{6}$$

In the formula, $\bar{b} = \frac{1}{h}\sum_{i=1}^{h}b_i$ represents the average eigenvalue.

Through the feature decomposition of the covariance matrix, the eigenvectors corresponding to the m larger eigenvalues are obtained which constructed the matrix.

$$W = [w_1, w_2, \cdots, w_m] \tag{7}$$

Eye characteristics with the greatest differentiation are as follows:

$$\left[hog_{eye}\right] = W^T(b - \bar{b}) \tag{8}$$

In the same way, high-degree HOG features of other features are extracted. Finally, for eyebrows, eyes, ears, nose and mouth, the eight types of facial features were acquired. The characteristics of each area are recorded as:

$$[hog(x,y)] = \begin{bmatrix} hog_{eb}^l & hog_{eb}^r & \cdots & hog_m \end{bmatrix}^T \tag{9}$$

3.1.2   Classical Key Point Location

Considering the change of face angle and the problem of staff updating, this paper uses standard Chinese facial proportion map combined with face panorama to construct the classical key point location (Fig. 1).

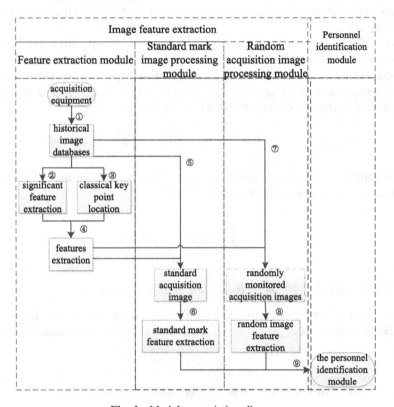

Fig. 1. Module association diagram

In this paper, the historical image database is used as the data source in the process of significant feature extraction, so as to ensure the consistency of the acquired image specifications in the training and testing stages. However, in the classical key point location extraction stage, the purpose is only to estimate the position of the five official positions through the distribution of key points. Based on the standard Chinese human face ratio image, manually calibrated 102 key points are used to locate facial region contours. Based on Face++'s key point detection algorithm in 2013 [8], 68 key points are used to achieve the initial positioning of the face contour. At the same time, in order to better extract facial image features, two key points are added on both sides of the nasal bone. To achieve multi-angle recognition, there are 32 key points in the contour of the ear. Specific calibration as shown below (Fig. 2):

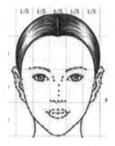

Fig. 2. Key point distribution based on standard facial proportions

Considering the change of face angles, combined with the face panorama, the position of the key points in this paper is displaced and scaled, so that it more meet the actual estimated position of the acquired image.

This article translates the scaled keypoints in Fig. 3 to face panoramas. After the eye position aligned, the other key points are panned and zoomed to achieve the classical key point position.

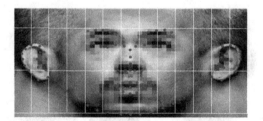

Fig. 3. Person's face panorama

Referring to the part ideas of the CLM algorithm, this paper uses the classical key points to achieve the initial calibration of the facial sensory regions. The difference is that the purpose of this section is to estimate the facial features position of the captured image. Therefore, the accuracy requirement is reduced, and at the same time, employees' changes are prevented from causing the update of positioning key points.

3.2 Feature Extraction of Standard Marker Images

In the monitoring identification module, in order to improve the recognition accuracy of the model, this paper uses a semi-supervised algorithm. In the training process, in order to update the sample as much as possible to include the latest image of the personnel, the model will automatically update the marked sample data. This may lead to the loss of the basic information of the person. To avoid this problem, features are extracted from the frontal facial features image as the standard invariant recognition feature of the model. The feature extraction process of the standard marker positive facial features is as follows:

(1) The person's face rotates horizontally +90 °, 0 ° and −90 °. Each of the on-the-job personnel collected three images of the above-mentioned angles. Subsequently, Viola-Jones face detection [9] is used to extract facial regions of the image;

(2) The two-dimensional Gabor filter is used to detect the position of the eyes and mouth. The position of the facial features of the standard marker sample is estimated by the classical key point position.

(3) The characteristics of facial features marked by officers were extracted by extracting significant features, thus obtaining the invariant features of the standard mark sample.

3.3   Feature Extraction of Randomized Monitoring Acquisition Image

For the randomly monitored images, Viola-Jones and two-dimensional Gabor filters were also used to obtain the position of the eyes and mouth of the image. The Platts analysis was performed on the positions of the above three detection points and the corresponding positions of the classical key points, and parameters such as rotation, scaling, and displacement of the collected facial images were obtained. According to the above parameters, the classical key points are positioned by rotating, scaling and displacing, thereby determining the estimated position of the facial region (Fig. 4).

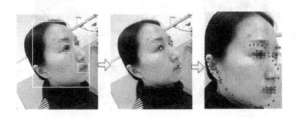

Fig. 4. Initial positioning of regional in captured images [10]

From Fig. 5, the accuracy of the estimated position is inaccurate. Therefore, in order to improve the robustness of the locked five regional features, this paper uses significant features to match and adjust. In this paper, the classical optimization algorithm in the ellipse fitting, that is, the nonlinear least squares optimization algorithm is used to achieve the maximum fitting for significant features and the features of the randomly acquired images. Accurately locate facial features and extract facial features. After precise positioning of five regional areas, features are extracted. Take the eye area as an example, the specific process is as follows:

(1) With the unified cell and block partition, the HOG features are extracted from the estimated sensory regions of randomized monitoring acquisition image. Subsequently, the PCA algorithm is used to obtain the down dimension feature matrix;

(2) In the same way as shown in Fig. 5, the estimated area of the sensory perception is obtained for the randomized monitoring acquisition image. The constrained fitting of features was performed in the area of its initial localization with significant eye region HOG features. The fitting algorithm is described as follows:

$$f_h(\hat{x}, \hat{y}) = \min\left\{\sum_{i=1}^{u} \|q_i - z[P(\hat{x}, \hat{y}) \times p_i + t]\|^2\right\} \tag{10}$$

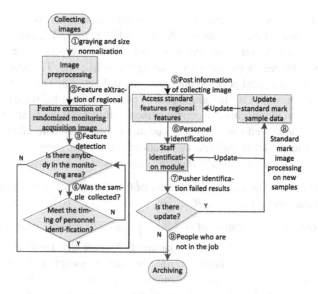

Fig. 5. Overall identification process

In the formula, $(\hat{x}, \hat{y})$ is the two-dimensional movement of the face. u is the number of HOG feature points in the eye region. q_i is the HOG feature point of the eye region of the random image. p_i is the HOG feature point of the eye region of significant feature. P is the displacement matrix. t is the plane offset vector. z is the scaling factor. In the stage of experimental evaluation, this paper uses Matlab's optimization function lsqnonlin to solve the optimal solution for the nonlinear least-squares function, so as to obtain the precise eye region of the acquired image.

In the same way, the above process accurately acquires the facial features of the randomly monitored images. This paper extracts precise sensory regions in the constrained prediction region, rather than fitting the entire facial region. At the same time, in order to avoid severe facial or facial area disturbances caused by facial angle changes, an optimal solution threshold α is set. When the fitted value exceeds α, the extraction area is discarded as shown the right eyebrow/eye in Fig. 5. Finally, there are

A extraction areas. Then, based on the optimal solution $f(\hat{x}, \hat{y})$, the regions are weighted by the normalization method. The weight value ω_a is calculated by stereo matching adaptive weight algorithm. That is, the Gaussian kernel is used to determine the weights of each regional features based on the optimal solution determined by the HOG feature comparison. Calculated as follows:

$$\omega_a = k \cdot \exp(-f_h(\hat{x}, \hat{y})/\delta_h) \tag{11}$$

In the formula, $\delta = \sum f(\hat{x}_a, \hat{y}_a)/A$. ω_a can well reflect the contribution of each region to face recognition. The ω_a value is proportional to the regional weight.

3.4   Personnel Identification Module Based on Semi-supervised k-NNM

The data input of the model mainly includes three parts: one part is the standard mark invariant feature data of fixed position staff obtained from Sect. 3.2, and this part data is used as the cluster's various classification points; The other part is the random monitoring images collected by the personnel during the audit cycle. The feature data of the unlabeled samples for fixed position are generated in Sect. 3.3. This part is the sample to be tested; The third part is the sample data of pre-review results, which is usually the characteristics from identifying the correct person within 5–10 verification cycles. This section serves as a labeled sample that is updated in real time. The initial phase is not available, and the model is updated automatically at a later stage. The specific classification process of the monitoring identification model is as follows:

(1) In the initialization phase, the sample set of five feature regions obtained from the random monitoring image is $B' = \{b'_1, b'_2, \cdots, b'_i, \cdots, b'_n\}$. The marked person identification feature sample set is $\{b'_j \cdots\} \in B$. The labeled person sample containing the identification feature is classified. Assume that there are k individuals and it is divided into 2 k clusters in total, so that the sample clusters are clustered into $C = \{c_1, c_2, \cdots, c_k, \cdots, c_{2k}\}$. Among them, from the k+1th class to the 2kth class, which is the class prepared for failing to recognize the person;

(2) For the fixed position regional characteristics data, the distance between each unmarked person sample b'_i and each labeled person sample b'_j is calculated. The calculation method also uses a nonlinear least-squares optimization algorithm:

$$l_{ij} = \left\| b'_i - b'_j \right\|_2$$
$$= \min \left[\sum_{a=1}^{A} \sum_{i=1}^{u} \sqrt{(hog\|q_i - p_i\|_2)/\omega_a} \right] / A \tag{12}$$
$$\text{precondition}: \text{only calculate non}-\text{zero feature areas}$$
$$\text{of unlabeled samples}$$

Sample b'_i which l_{ij} less than the set threshold L is assigned to the corresponding fixed-post personnel cluster r.

$$C_r = C_r \cup \{b'_i\}, r \in \{1, 2, \cdots, k\} \tag{13}$$

For samples that failed assigned to the corresponding fixed category, they are assigned to the k + r class. At this time, the number of the k + r class sample is smaller than the overall sample. Subsequently, the above-mentioned failure categorization samples are identified by mean clustering;

(3) After the first categorization of all unlabeled samples, the squared error is calculated as follows:

$$E = \sum_{r=1}^{k} \sum_{b' \in C_r} \|b' - u_r\|_2^2 \tag{14}$$

In the formula, $u_r = \frac{1}{|C_r|} \sum_{b' \in C_r} b'$ is the mean vector of the cluster C_r. The E more smaller, and the higher similarity within the cluster;

(4) For minimize the E, the semi-supervised k-NNM uses the greedy algorithm. Iterative optimization is used to achieve the approximate solution of the E, The iteration threshold is set to W and initializes $w = 1$. The distance $l'_{ir} = \|b'_i - u_r\|_2$ between each unmarked person sample b'_i and the cluster mean value vector u_r is calculated for the feature recognition of the five regions. Selecting the smallest value of l, which is $r = \arg\min_{j \in \{1,2,\cdots,k,\cdots,h'\cdot k\}} l_{ij}$. Then sample b'_i is placed in this cluster $C_r = C_r \cup \{b'_i\}, r \in \{1, 2, \cdots, k\}$;

(5) After this round of updates is completed, the mean vector u_r of each cluster is calculated. And the squared error E is calculated and updated. If u_r are not updated for all, the iterative process stops. At this time, the categorized of the sample is considered as the output result and goes to (7);

(6) If u_r is updated data, $w = w + 1$,

①If $w < W$, then return to (4),

②If $w = W$, for the first categorization and the iterative process, the minimum value of the squared error E is considered as the sample categorization result;

(7) According to the results of sample categorization, batch identification of unmarked personnel is completed.

The sample distance calculation of the semi-supervised k-NNM algorithm, first of all, K-nearest neighbor clustering achieves a one-to-one comparison of the samples. This achieves relative accuracy of the initial classification and reduces the number of later iterations. Subsequently, the K-means clustering, that mode is, single sample comparison with the mean, which improves the participation of the latest unlabeled sample data. When the appearance of a person changes greatly, this method avoids the problem that the feature deviates from the recognition failure. In this paper, iterative optimization is used to achieve clustering with the largest similarity between samples. At the same time, in order to avoid falling into an iterative loop, an iterative threshold is set so that the category division is made according to the squared error E. For example, only one unlabeled sample jumps back and forth between two clusters.

4 Overall Identification Process

Based on the method of coarse positioning to fine matching, this paper implements personnel monitoring at work and continuously updates historical samples with feedback samples. The specific process is shown in the figure:

(1) Image preprocessing is performed on randomly collected images;
(2) As the way of feature extraction of randomized monitoring acquisition image, the image region features are extracted and the weights are matched;
(3) It is judged whether the image is successfully extracted from the random monitoring acquisition feature. If the feature extraction fails, the result of the employee's absence is output directly to (9); if the extraction is successful, the process proceeds to the next step;
(4) Determine whether the captured image is stored for personal identification. According to the actual business requirements, for the time interval, the on-the-job time detection is less than the person identification in the monitoring area. Therefore, if the time interval requirement is not satisfied, return (2), and if it is satisfied, the process proceeds to the next step;
(5) Based on the fixed posts attached to the captured images, the standard marked features of the staff of the fixed posts are acquired;
(6) Set the person identification period, and the feature extraction of randomized monitoring acquisition image and the standard mark regional features as the data input for the identification module of the fixed staff, and output the identification failure data of the person identity;
(7) For personnel identification failure data, manually determine whether there is an employee update status, if there is, enter (8), if not, enter (9);
(8) Invoking the standard mark image processing module to realize the collection of personnel standard information and update the historical image database information, thereby updating the characteristics of the standard mark area of the staff;
(9) Output the non-employed person, which generate assessment records and stored.

The on-duty staff monitoring method is divided into two parts. The first is based on the feature detection of random monitoring acquisition images which can improve the timeliness of on-duty personnel monitoring. The second is based on the identification of fixed staff, which can identify fixed-site personnel, and identify the actual business process personnel of the position when the personnel substitutes. In addition, the acquisition information of the unidentifiable person (including collection time, job position, and image acquisition) is sent to the auditor as failure result of identification. The auditor extracts information that updates the feature of standard marker image, when there is an employee replacement or the appearance of the person changes greatly.

5   Experiments and Analysis

5.1   Experimental Environment and Image Preprocessing

This article takes a power supply business hall of the State Grid as the on-site experimental environment. The image acquisition device is a computer camera when the staff is working. Experimental equipment is intel i3 2.2 GHz CPU, with 2G memory and win10 operating system. Simulation software is MATLAB2012b. The image collected by the on-site laboratory equipment is shown in Fig. 6:

Fig. 6.   The image collected by the on-site laboratory equipment

The specific experimental process is:

(1) Based on laboratory equipment, random monitoring images and standard marker images are collected for on-the-job personnel. Within 12 weeks, 28 employees of the power supply hall were collected for on-the-job monitoring images. Shooting interval is 20 s. One shot captures 2 images. That is, the daily collection of images is 8640 (excluding low-quality retakes) for each person. The time interval for personnel identification is 20 min. That is, the 6 randomized monitoring acquisition image were collected at this point, what as the sample of the semi-supervised k-NNM personnel identification module. In the 11th week, two employees are alternated;

(2) The image is normalized and grayscale processed with 364 × 480 pixels. The pixels in each region of face area are as follows (Table 1):

The image is divided into cells with 16 × 16 pixel, and 2 × 2 cells form a block. Each cell has 9 bins in the gradient direction, and 8 pixels as one step size. Taking the eye area image as an example, there are 16 and 6 scanning windows in the horizontal and vertical directions, respectively. Therefore, the initial HOG feature of a single eye region image is 3456. After the PCA algorithm reduces the dimension, the HOG feature is reduced to 1944 [11].

Table 1. Pixels of each area

Face area	The average size of the picture
Eye	136 × 56
Eyebrow	144 × 40
Nose	112 × 214
Mouth	168 × 64
Ear	96 × 200

(3) Based on the above data, the experiment of personnel working time detection and fixed station personnel identification was implemented. The recognition rate and time consumption are calculated and analyzed.

5.2   Experimental Process and Results

The experimental phase is divided into two parts. The first stage is from 1 to 10 weeks. At this stage, the staff of the business hall is fixed and its appearance does not make specific requirements. In the 11th week, two people were replaced. The second phase is the 11th and 12th weeks.

For personnel on-the-job time detection, in the process of feature extraction of randomized monitoring acquisition image, if there are any facial features in the captured image, that is, $A \geq 1$, then it is determined that there are people in the region. During the experiment, different thresholds α were set for different regions. This article uses the nose extraction feature as an example. Based on the experimental results of the extracted features of the nose, definition that $0 \leqslant f_h(\hat{x}, \hat{y}) \geqslant 0.7$ indicates successful feature extraction and $f_h(\hat{x}, \hat{y}) > 0.7$ indicates unsuccessful. In the course of high-frequency on-duty testing, if there are $A = 0$ for three consecutive times, the person is determined not to be on duty. As shown in Fig. 6, in the image 2 and the image 7, it is determined that the person is not in the post (including the staff sabotage state). In the Fig. 6, even if there is partial occlusion of the face, it is still effective to identify the presence of a part of the facial features so as to determine that the person is on the job. The detection time for each image is approximately 0.12 ms, and the recognition accuracy is approximately 100%. This method satisfies the application requirements of on-duty personnel time detection.

In the identification process of fixed positions, the sample identification period is in units of days. In the identification process of the semi-supervised K-NNM algorithm, positive samples of the previous 10 cycles of the test time continuously update the marker samples. In the experimental process, we select facial images of three angles (that is, direct look, low head, and right side face) to compare the features. The specific examples are left eye and mouth (Figs. 7 and 8).

Fig. 7. The image from left to right is right side face, low head, direct look for left eye

Fig. 8. The image from left to right is right side face, low head, direct look for mouth

From the above image, when the face angle changes, the HOG characteristics of the same person change greatly, especially the front image and the side image. However, in contrast, the lateral image of the ear and nose changes less. Therefore, the acquisition of ear images has been added as part of facial feature extraction. The facial organ weights were determined based on the fitted similarity values. This article uses semi-supervised k-NNM personnel identification module to achieve personnel identification. In the staff update phase of the 11th week, taking the identification process of an updater as an example, the results of the first identification class division and the final iterative identification class division are shown in the following figure:

As shown in the figure, the standard sample is the standard mark image of the employee. The un-updated sample is the most recent positive sample of the previous 10 cycles before the test date, which enables continuous updating of labeled sample data. The updated sample is the acquisition sample of the new employee. From Fig. 9, we

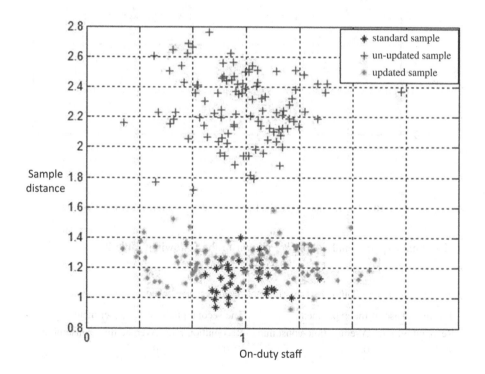

Fig. 9. The results of the first identification class division

can see that the samples collected by new employees do not meet the criteria for classification because they do not match the features of the original personnel. Therefore, the samples are classified into unrecognizable categories. As can be seen from Fig. 10, a new category is eventually formed due to the data distribution characteristics of the updated samples themselves. The personnel identification module pushes this category as part of the recognition failure, and in the case of the presence of personnel updates, the new marker personnel features are created to update the characteristics of the fixed position personnel.

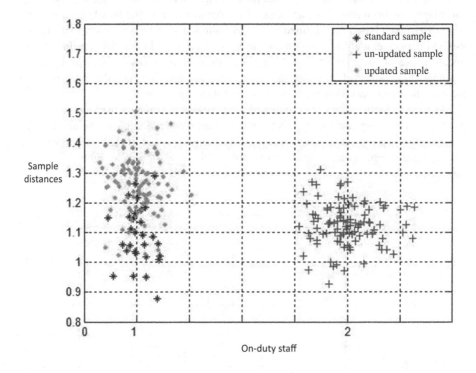

Fig. 10. The results of the final iterative identification class division

The final experimental results are shown in the following Table 2:

Table 2. Experimental results for five schemes

The first phase of the experiment		The second phase of the experiment	
Recognition rate	Average time-consuming	Recognition rate	Average time-consuming
99.14%	0.54 s	96.02%	0.61 s

5.3   Result Analysis

Based on the above experimental results, in terms of timeliness of detection and personnel identification rate, the methods described in this paper can meet the business needs of on-duty personnel monitoring for power office. In the first stage of the experiment, the semi-supervised K-NNM algorithm can improve the model recognition rate based on the data distribution information of unlabeled samples. In the second stage of the experiment, when there are people updating, feature extraction methods based on significant feature extraction and classical key point location can adaptively extract and update the region features and then update the fixed input data of the identified model sample. Only at the 11th week, the recognition rate is reduced. At the 12th week, the recognition rate of personnel increased, so the average recognition rate decreased less in this period. In summary, the method of monitoring on-site staff in this paper is effective. The recognition rate and timeliness of the monitoring service are satisfied, the model can be updated independently, thereby reducing manual maintenance costs.

6   Conclusion

Taking a power supply business hall of the State Grid as an experimental environment, this paper constructs a method for monitoring on-the-job staff based on autonomous updating and semi-supervising model. Monitoring is divided into two parts. On the one hand, real-time inspection personnel obtain working time records; on the other hand, periodical batch identification of fixed-post personnel. First, the collected images are processed in a unified manner to prepare input data for the model. It is divided into three stages. On the one hand, the positioning of facial features is achieved by combining significant feature extraction and classical key point location [12]. On the one hand, the feature extraction of the standard mark image of the on-the-job personnel is used as a basic mark feature of the input data, so as to ensure that the personnel sample does not change the on-the-job staff in the self-renewal phase; On the one hand, for the random monitoring images, the features of the five regional image are extracted, which is used as a test data input for working time detection and personnel identification module. There is no unlabeled sample in the initial stage. Then, the non-linear least squares optimization algorithm is used to match the weights for the local area of the face. The semi-supervised k-NNM algorithm is used to build the personnel identification module, which ultimately completed the monitoring and identification of personnel. In addition, with the development of on-the-job monitoring work, the sample data that was correctly identified in the previous monitoring cycle was input into the model as self-updated marker sample, thereby improving the recognition accuracy of the model. Experimental data show that this method is effective and feasible.

References

1. Zhao, S., Pan, H., Xie, X., Zhang, Z., Feng, X.: A range-threshold based medical image classification algorithm for crowdsourcing platform. In: Zou, B., Li, M., Wang, H., Song, X., Xie, W., Lu, Z. (eds.) ICPCSEE 2017. CCIS, vol. 727, pp. 433–446. Springer, Singapore (2017). https://doi.org/10.1007/978-981-10-6385-5_37
2. Abouyahya, A., El Fkihi, S., et al.: Features extraction for facial expressions recognition. In: 2016 5th International Conference on Multimedia Computing and Systems (ICMCS), pp. 46–49. IEEE (2016)
3. Zhang, K., Zhang, Z., Li, Z., et al.: Joint face detection and alignment using multitask cascaded convolutional networks. IEEE Signal Process. Lett. **23**(10), 1499–1503 (2016)
4. Wang, N., Gao, X., Tao, D., et al.: Facial feature point detection: a comprehensive survey. Neurocomputing **275**, 50–65 (2017)
5. Wu, Q., et al.: Android malware detection using local binary pattern and principal component analysis. In: Zou, B., et al. (eds.) ICPCSEE 2017. CCIS, vol. 727, pp. 262–275. Springer, Singapore (2017). https://doi.org/10.1007/978-981-10-6385-5_23
6. Bicego, M., Loog, M.: Weighted K-nearest neighbor revisited. In: 2016 23rd International Conference on Pattern Recognition (ICPR), pp. 1642–1647. IEEE (2016)
7. Guo, Y., Cao, H., Han, S., et al.: Spectral-spatial hyper spectral image classification with K-nearest neighbor and guided filter. IEEE Access **6**, 18582–18591 (2018)
8. Zhou, E., Fan, H., Cao, Z., et al.: Extensive facial landmark localization with coarse-to-fine convolutional network cascade. In: 2013 IEEE International Conference on Computer Vision Workshops (ICCVW), Sydney, Australia, pp. 386–391 (2014)
9. Vikram, K., Padmavathi, S.: Facial parts detection using Viola Jones algorithm. In: 2017 4th International Conference on Advanced Computing and Communication Systems (ICACCS), (08), pp. 1–4 (2017)
10. Kowalski, M., Naruniec, J., Trzcinski, T.: Deep alignment network: a convolutional neural network for robust face alignment. In: 2017 IEEE Conference on Computer Vision and Pattern Recognition Workshops (CVPRW), pp. 2034–2043 (2017)
11. Ahmadkhani, S., Adibi, P.: Face recognition using supervised probabilistic principal component analysis mixture model in dimensionality reduction without loss framework. IET Comput. Vis. **10**(03), 193–201 (2016)
12. Wu, Y., Tal, H.: Facial landmark detection with tweaked convolutional neural networks. IEEE Trans. Pattern Anal. Mach. Intell. (TPAMI) **PP**(99), 1 (2017). https://doi.org/10.1109/TPAMI.2017.2787130

Research on Electricity Personnel Apparel Monitoring Model Based on Auxiliary Categorical-Generative Adversarial Network

Xueming Qiao[1(✉)], Yiping Rong[2], Yanhong Liu[1], and Ting Jiang[1]

[1] State Grid Weihai Power Supply Company, Weihai 264200, Shandong, China
qxm_3351@163.com
[2] State Grid Shandong Electric Power Company,
Jinan 250000, Shandong, China

Abstract. This article aims at the problems that exist in the monitoring process of employees' standard clothing. Adopting the reverse way of thinking to carry out the normative identification work, that is, the classification ability depends on the feature fitting ability. First of all, building an employee work card template library, which is used to realize personnel authentication in combination with template matching method. The recognition result is given as a label to the real captured image. Then, the construction of the auxiliary categorical-generative adversarial network was done with a few real images. The training of generator G and discriminator D is completed through iterative interactive update, so as to realize the fitting of the real clothing image. The trained ACat-GAN is used as a standard clothing monitoring model. Respectively, based on the discriminant image and parameter adjustment data, discriminator D and generator G push out the list of well-dressed people and Image of non-standard. Based on a power supply business hall as the background platform, this article collected personnel clothing images for experiments. The experimental results show the feasibility and practicability of the method which described in this paper.

Keywords: Template matching method · Personnel authentication
Generative Adversarial Network · Clothing standardization recognition
Feedback iterative optimization

1 Introduction

With the development of a new round of power system reforms [1], and the power supply business hall has served as a direct service window for power companies and power users to contact, and its normative clothing intuitively demonstrates the cultural soft power of the entire power group. The traditional standard clothing monitoring work mainly depends on manual identification. However, due to internal and external factors, the manual identification of standard clothing is difficult to carry out, so it is necessary to rely on the processing and analysis of image data to realize the automatic identification of the person's clothing specifications [2].

© Springer Nature Singapore Pte Ltd. 2018
Q. Zhou et al. (Eds.): ICPCSEE 2018, CCIS 902, pp. 377–388, 2018.
https://doi.org/10.1007/978-981-13-2206-8_31

2 Design Principles and Ideas

In order to solve the problem of standard clothing recognition, this paper aims at the specific problems in the process of identification, determines the model construction ideas through the design principle requirements, constructs the clothing specification recognition model. The design principles and ideas as follows.

(1) Imbalance in sample categories. Traditional machine learning algorithms often rely on positive and negative samples to construct the standard clothing recognition model, even if the final obtaining a high accuracy rate is difficult to meet actual needs. For example, the 100 captured images in the actual test, only 2 are unqualified, and the test result is 100. The accuracy of the model is still 98% [3].

(2) Data differences between samples. Because each person's posture is different, even in the face of the camera, the personnel maintain the same state, their clothing images themselves are also very different [4].

(3) Data fitting. Mining data based on the algorithm, we must always focus on the fitting of the entire data, so the overall grasp of the higher requirements, this article needs to be the clothing business norms of the entire staff of the power supply sales office to conduct clothing testing, so the data fitting degree has higher requirements.

Based on the principle requirements of the above design, this paper adopts the Generative Adversarial Network (GAN) [5] to implement the construction of the clothing specification identification model. The model is constructed using positive sample data, through the generator G and discriminator D continuously iteratively update the positive sample data to globally fit the data distribution of the clothing specification sample [6]. In addition, this paper applies the license plate recognition technology that is relatively mature and applies it to the identification of employee badges. Based on this, it also conducts the clothing standard identification and builds the clothing specification recognition model of the auxiliary classification generation confrontation network.

3 Standard Clothing Recognition Model

The purpose of this method is to solve the problem of clothing specification recognition for personnel in the power supply business hall. Firstly, unified image preprocessing is performed on the collected training sample images to extract part of the input data into the training model. Then the template matching method is used to achieve personnel identity authentication. And based on this to build an auxiliary classification to generate the confrontation network, complete the construction of the clothing specification identification model.

3.1 Image Preprocessing and Feature Extraction

In order to achieve standardized testing of standard clothing by the staff of the power business hall. The specific ways are as follows:

(1) This paper uses skin color feature threshold to determine whether there is a person in the captured image. This article first converts RGB images to YC_bC_r images, and sets thresholds as follows:

$$s_{skin} = \begin{cases} 1, & (85 \le C_g \le 135) \&\& ((-C_g + 210) \le C_r \le (-C_g + 190)) \\ 0, & other \end{cases} \tag{1}$$

Among them, if s_{skin} is 1, there are people in the area, else no people.

(2) Perform image normalization processing on images with skin color features, including image graying, normalized size, removal of background noise, and image morphological operations to obtain a standard upper body clothing grayscale image, as shown in the following Fig. 1:

Fig. 1. Image normalization results

(3) Perform gradient calculation on the image and perform gradient calculation on the fixed point (x, y) of each pixel in the processed image. The formula is as follows:

$$G_x(x,y) = f(x+1,y) - f(x-1,y) \tag{2}$$

$$G_y(x,y) = f(x,y+1) - f(x,y-1) \tag{3}$$

$$m(x,y) = \sqrt{G_x^2(x,y) + G_y^2(x,y)} \tag{4}$$

$$\theta(x,y) = \arctan\left[G_y(x,y)/G_x(x,y)\right] \tag{5}$$

Statistical gradient direction projection, that is, the image is divided into uniform cells, there are 9 directions in each cell, multiple cells composed of block, scans the image based on the block, and normalizes the cells between the overlapping blocks, thereby realizing the normalization of the HOG feature, and the HOG feature vector of the image is obtained.

3.2 Personnel Authentication

The current license plate recognition template matching method is used to realize the identification of the working employee badge. The specific process is as follows (Figs. 2, 3 and 4):

Fig. 2. Unified extraction of employee badge information

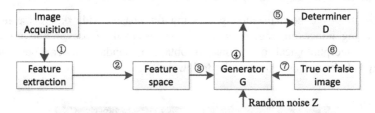

Fig. 3. ACat-GAN training process

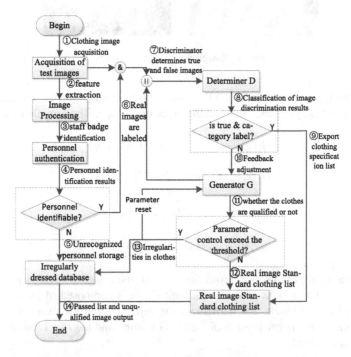

Fig. 4. General flow chart of standard clothing recognition

(1) First, badge template library is built. Through the unified collection of employee badges, normalization of the size, and locking of the number area, obtain the HOG feature vector ($f[hog]$) of the number area. As shown below:

(2) Then according to the brand's HOG feature vector, lock the employee badge area to be identified captures images. According to the method of constructing the employee badge template library, extract the HOG feature vector $g[hog]$.

(3) Match the employee badge area information with badge template library, select the best match result as the sample category attribution, and use the coincidence degree function to measure the match value. The formula is as follows:

$$C_{fg} = \frac{100}{100} \sum_{i=1}^{M} \sum_{j=1}^{N} \frac{|f_{ij} \cap g_{ij}|}{T_f + T_g} \tag{6}$$

Among them, f is template binary image, g is corresponding employee badge binary image, i and j represent the coordinates, the size of the two images are the same as M × N, T_f and T_g represent the number of pixels which value is 1.

3.3 Construction of ACat-GAN

This article is based on the comprehension of the generative anti-network GAN [7], constructing ACat-GAN. In this paper, the G and the D are constructed using a batch-normalized layer convolutional neural network [8]. The G tries to generate the image acquisition information by simulating the characteristics of the real image, so as to maximize the probability that the discriminator will make an erroneous judgment. The D constantly corrects the parameters of the generator by interpreting the real sample and the generated sample so as to maximally simulate the actual image information. Therefore, the final G captures the general distribution of the actual captured image data. At this point, the D is difficult to distinguish between the authenticity of the input image data, and when the input is a situation where the actual collection image is unsatisfactory, the D will launch a real image that does not conform to the general distribution of the features. The ACat-GAN training phase as follows:

(1) Based on the collected image of the standard clothing of the staff of the power supply business hall, image preprocessing is implemented according to Sect. 3.1, so as to realize feature extraction;

(2) Extract the feature data of all the images collected by all personnel to form a feature space as the input data of the generator G;

(3) The G has 4 convolution layers. The input data is feature-normalized, scaled to [0,1], which pushed to the G for transposition convolution operation of the upsampling layer with sliding s step. The picture size is N1 × N1, the convolution kernel is N2 × N2, and after once transposition convolution is (N1 + N2−1) × (N1 + N2−1), and it is filled with 'same'. After the convolution operations, the narrow and deep input feature vectors become wide and shallow;

(4) The G puts the generated false image and the real collected image into the D;

(5) The D also has 4 convolution layers. The obtained image (real/false) is subjected to conventional convolution operation in the upsampling layer, and the image size is [(N1–N2)/s + 1] × [(N1–N2)/s + 1] after once convolution. The last layer uses the sigmoid activation function, and sets the number of iterations K, and updates the D by increasing its stochastic gradient.

$$\nabla_{\theta_d} \frac{1}{m} \sum_{i=1}^{m} \left[\log D\left(x^{(i)}\right) + \log\left(1 - D\left(G\left(z^{(i)}\right)\right)\right) \right] \tag{7}$$

Among them, m is the number of false and real images, and x is real image, and z is false image;

(6) After the iteration is stopped, the discrimination result of the D is output, and the difference between the false image and the real image is transmitted to the G;
(7) The G receives the backward propagation gradient of the D, and adjusts its parameters, and updates the G by decreasing its random gradient.

$$\nabla_{\theta_d} \frac{1}{m} \sum_{i=1}^{m} \log\left(1 - D\left(G\left(z^{(i)}\right)\right)\right) \tag{8}$$

As the training progresses, the G starts outputting a false image closer to the real image, thereby extracting the data distribution of the real image.

The G and the D continue to confront each other and eventually reach the global optimum of the GAN. The objective function is as follows:

$$\min_{G} \max_{D} V(D, G) = E_x[\log D(x)] + E_z[\log(1 - D(G(z)))] \tag{9}$$

In this paper, the GAN model is constructed to achieve the extraction of the clothing standard specification rules for the images collected.

3.4 The Overall Process

This article through the twice standard screening which to achieve the identification of clothing norms. The specific process is as follows:

(1) Set up an image acquisition device in the workbench area of the employee to achieve the collection of clothing images for the employee to storage of captured images, and provide real image for subsequent the D;
(2) Through the image preprocessing process, the HOG features of the acquired image information are acquired;
(3) Based on the HOG feature, the badge template library is established. And the template matching method is used to realize the identification of the employee;
(4) According to the results of the identification of work cards, if the personnel identity is not recognized, enter (5), otherwise enter (6);

(5) Because the identity of the personnel is not recognizable, so it does not meet the requirements for badge wear, and the image is input into the non-standard storage of personnel clothing, which is used as a source of evidence for subsequent clothing and sample optimization;

(6) Assign the identifiable person label to the corresponding real image. The relationship between them is 'and'. Only when the collection of the real image is successful, the calculation can enter the next step;

(7) The G generates a virtual image, and the D selects the tagged real image or the virtual image for discrimination. Since the GAN after trained generates an approximate optimal state, When the G parameter is fixed, the discrimination result of the D is:

$$D_G^*(x) = \frac{p_r(x)}{p_r(x) + p_v(x)} \approx \frac{1}{2} \tag{10}$$

Among them, $p_r(x)$ is real image, and $p_v(x)$ is false image, and $D_G^*(x)$ is the rate that the D recognizes the real image;

(8) According to the discrimination result of the D, the secondary classification of the passing and failing of the standard clothing is performed. If the image is judged to be true and the person tag is attached, enter (9), otherwise enter (10);

(9) Because the image is tagged with personnel label, its essence is real image and meets the standard clothing feature, so it outputs a list of standard clothing personnel;

(10) The image information that is judged to be false by discriminator D is used as a feedback parameter to adjust the parameters of the generator G;

(11) The G is divided into two parts. One part records the parameter value before the unadjusted parameter, and the other part adjusts the parameter according to the feedback data provided by the D, and uses the parameter change as judgment basis. If the parameter change does not exceed the threshold, enter (12), otherwise enter (13);

(12) The part is judged as "false" image and meets the requirements of the standard clothing. Since the D is difficult to distinguish between true and false images at this time, so the "false" image contains the real image with some people tags, and the list of personnel is output as a list of standard clothing;

(13) Due to parameter adjustment exceeds the threshold, it indicates that the "false" image leads to a larger change in the characteristics of the standard clothing retained by the G. If the image contains a person label, it means that it is a real image and does not meet the requirements of the standard clothing, and the image is input into a non-standard storage repository, and a reset of the parameters is performed;

(14) The list of personnel standard clothing and the person's irregular clothing storage image are taken as the identification result of the standard clothing, which output the result to the staff.

The method uses a small number of positive samples to complete the training of the model, and which can simulate and retain the standard characteristics for the maximum extent, without considering the problem of negative sample shortage and class imbalance. In addition, the GAN model is continuously self-adaptively optimized so as to always maintain the best recognition of the standard clothing.

4   Experimental Process and Analysis

In order to verify the applicability and superiority of this method, this paper is based on a power supply business hall of the State Grid as the experimental implementation environment, In addition, because of the difference in clothing for women and men, this experiment only selected female staff.

4.1   ACat-GAN Training Process

In the training process of ACat-GAN, in order to reduce the training difficulty of the G, and the regional gray features of the standard clothing are preserved, the feature space is constructed with the standard clothing image. The standard clothing collection image size is 288 × 288. For the selection of the connected area of the HOG feature, a comparison is made with the cell as an object. As shown in the following Fig. 5, the cells are 4 × 4 pixels, 8 × 8 pixels and 16 × 16 pixels from left to right:

Fig. 5. Example of a gradient magnitude map

Since the input of the G is added to the image extraction feature, its purpose is to obtain a relatively stable boundary. In the experimental process, the extraction of the image features of the small connected regions leads to the problem that the feature range is solidified and the G is over-fitted, which leads to the collapse. So the image is divided into 16 × 16 pixel cells, and 2 × 2 cells constitute a block, and each cell has 9 gradient direction bin, with a 16 pixel step, a scanning window 17 of the horizontal and vertical direction, and the initial HOG characteristics of image is 10404. Data normalization is performed on all acquired image HOG features, and normalized HOG features together with noise data are taken as input data of the G. As shown in the Fig. 6 below, the data for the feature space is normalized.

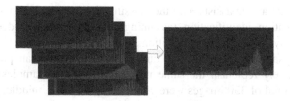

Fig. 6. Feature space data normalization based on standard samples

In the initial stage, the G and the D are alternately optimized. After iterating 200 times, according to the output result of the G, the respective optimization times are adjusted several times before and after, and finally the GAN that meets the standard clothing characteristics is obtained. As shown in the Fig. 7 below, the virtual image generated by the trained G is compared with the real image:

Fig. 7. Comparison between real and virtual images

As shown in the Fig. 7, the left image is a real image, and the right image is a virtual image generated by the G. It can be seen that although the virtual image has a pixel blurring condition, it retains the characteristics of the standard clothing area to the greatest extent and does not over-fit a person at the same time, resulting in the detection being too harsh.

4.2 Experimental Results and Analysis

This article collects images of the summer clothing of 22 women employees in the power supply business office. Due to the normal light conditions, the personnel authentication technology is relatively mature and stable, and the identification rate can reach over 97%. So this experiment focuses on the detection of the personnel's standard clothing as a goal, and adopts a comparative method to prove the superiority of the method described in this article. The specific method description is as follows:

Method 1: super-pixel merging wearer's clothing segmentation + HOG feature + SVM classification to achieve standard clothing recognition [9];

Method 2: Deep convolutional neural network recognized of standard clothing [10];

Method 3: The image feature extraction stage uses two-layer convolutional CNN. The GAN model implements the clothing specification identification (assuming that the person's identity has been identified and only its clothing specification is tested).

Method 4: This article assists with the classification of the GAN model to achieve clothing specification identification (assuming that the person's identity has been identified and only its clothing specification is tested).

The training samples were 22, 200, and 2 000, which were all positive samples. Method 1 separately replenish the same number of negative samples as part of the training data. A total of 330 images were collected at the early, middle, and late nights for the week (5 working days) under the same image acquisition conditions for the test sample, including 66 non-standard samples, and the following are the results of three kinds of model training and testing:

As shown in the above table, TP (True-Positive) indicates that the actual specification is detected as a specification, FP (False-Positive) indicates that the actual non-standard detection is a specification, TN (True-Negative) indicates that the actual specification is detected as non-standard, and FN (False- Negative) indicates that the actual non-standard detection is non-standard. According to the experimental results, the correct rate and recall rate of the three methods are calculated.

The result of the calculation is shown in the figure below:

As shown in Table 1 and Fig. 8, although the method described in this article takes more time in the training phase, once the training is completed, the model can quickly implement the inspection of the personnel's standard clothing. In the case of a small number of samples, this method does not need to rely on positive and negative samples (such as method 1) to achieve the refinement of the standard clothing, and it does not need to rely on a large number of sample data to achieve the extraction of sample features (such as method 2 and method 3), and it can rely on only a few positive sample data to realize the construction of the GAN. And in the follow-up work depends on the increase in the number of samples continuously optimize the model. As shown in the experimental results, method 3 uses CNN in the feature extraction stage. In the case of

Table 1. Comparison of three model training and test results

	Training time	Test time	TP	FP	TN	FN
Training samples	22 sheets					
Method 1	209 s	1062.6 s	166	25	98	41
Method 2	–	12.54 s	140	31	124	35
Method 3	–	11.75 s	232	11	34	53
Method 4	–	11.55 s	253	0	11	66
Training samples	200 sheets					
Method 1	1448 s	1108.8 s	197	18	67	48
Method 2	–	12.55 s	192	19	72	47
Method 3	–	11.84 s	243	4	23	60
Method 4	–	11.53 s	262	0	2	66
Training samples	2000 sheets					
Method 1	12052 s	1095.6 s	216	12	48	54
Method 2	–	12.55 s	227	9	37	57
Method 3	–	11.73 s	260	2	4	64
Method 4	–	11.54 s	263	0	1	66

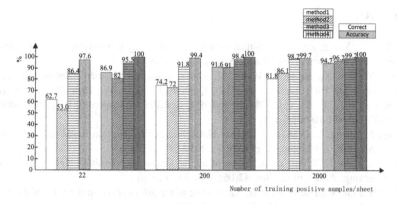

Fig. 8. Contrast diagram of experimental results

a small number of training samples, the feature extraction effect is not ideal and when the training sample is increased, there is no advantage compared to the HOG feature extraction method. In the case of a small sample of the training set, the HOG feature can maintain good invariance to the geometric and optical deformation of the image, and is more suitable for the requirement of describing the boundary of the image in this paper. The generator and discriminator self-extracted the data distribution character-istics of the clothing specification image. Therefore, when the clothing does not meet the requirements of the specification, it can be identified 100%, to meet the monitoring needs of the staff of the power supply business office on the dress.

5   Conclusion

This paper uses a power supply business hall of the State Grid as the background for data collection environment, and bases on the existing job area to achieve the collection of clothing images of on-the-job staff. On the one hand, the template matching method is used to realize the identity authentication of the personnel, and on the other hand, the GAN is constructed, and ACat-GAN is generated in combination with the identifiable person information to realize the discrimination of the dress code.

In addition, GAN relies entirely on convolutional neural network algorithms, so it takes more time in the training phase, and the virtual image boundaries generated by the self-extraction method of the clothing characteristics of generator G and discrim-inator D are still not clear enough. With the advancement of personnel standard clothing testing work and the continuous development of science and technology, in the later period, the method mainly increases the key semantic features of the different dressing areas of the GAN, and completes the adaptation from the virtual to the real domain, so as to continuously reduce the training difficulty of the model and improve the definition of the virtual image generated by the model.

References

1. Kaibao, F., Xiang, Z.: Current situation analysis and policy recommendations of electricity market reform. Macroecon. Manag. **33**(01), 49-54 (2018)
2. Chen, Y., Cai, X., Zeng, Y., Wang, M.: Crossing-scene pedestrian identification method based on twice FAS. In: Zou, B., Han, Q., Sun, G., Jing, W., Peng, X., Lu, Z. (eds.) ICPCSEE 2017. CCIS, vol. 728, pp. 483–491. Springer, Singapore (2017). https://doi.org/10.1007/978-981-10-6388-6_41
3. Dong, J., Zheng, B., Yang, Z.: License plate character recognition based on convolutional neural network. J. Comput. Appl. **37**(07), 2014–2018 (2017)
4. Zhu, Y., Zhao, J., Wang, Y.: A review of human behavior recognition algorithms based on deep learning. Acta Autom. Sin. **42**(06), 848–857 (2016)
5. Wang, K., Gu, C., Duan, Y., et al.: Generative adversarial networks: the state of the art and beyond. Acta Autom. Sin. **43**(03), 321–332 (2017)
6. Springenberg, J.T.: Unsupervised and Semi-supervised Learning with Categorical Generative Adversarial Networks. In: International Conference on Learning Representations (2016). arXiv:1511.06390v2
7. Goodfellow, I.J., Pouget-Abadie, J., Mirza, M., et al.: Generative adversarial nets. In: Proceedings of the 2014 Conference on Advances in Neural Information Processing Systems 27, Montreal, Canada, pp. 2672 − 2680. Curran Associates (2014)
8. Shi, L., Wang, Y., Cao, Y., Wei, L.: Vehicle type recognition based on deep convolution neural network. In: Zou, B., Han, Q., Sun, G., Jing, W., Peng, X., Lu, Z. (eds.) ICPCSEE 2017. CCIS, vol. 728, pp. 492–502. Springer, Singapore (2017). https://doi.org/10.1007/978-981-10-6388-6_42
9. Dong, Y., Chen, Z., Song, H.: Application of improved feature matching algorithm in bank card number recognition. J. Ji lin Univ. (Sci Edit.) **56**(01), 126–129 (2018)
10. Ma, Y.: Research on Clothing Identification Algorithm for Downstream People in Video Surveillance Scenes. Anhui University (2015)

Optimization Method of Suspected Electricity Theft Topic Model Based on Chi-square Test and Logistic Regression

Jian Dou[✉] and Ye Aliaosha

China Electric Power Research Institute, Beijing, China
krauser3a@163.com

Abstract. In recent years, the electricity theft has presented characteristics of high-tech and covert. Therefore, the factors that reflect the existence of stealing electricity become varied and complex. It brings the problems such as low efficiency and poor accuracy for power grid enterprises to identify the customers who had been stealing electricity. In this paper, Chi-square test and logistic regression are used to optimize the suspected electricity theft topic model. Chi-square test is used to determine the factors interrelated with the electricity theft firstly, and then the logistic regression algorithm is used to optimize the weights of the interrelated factors, and finally constructed a prediction function that can predict the customers who had been stealing electricity. Experiments show that the method proposed in this paper can help the power grid enterprises to identify the customers who had been stealing electricity, on account of having high accuracy rate, precision rate and recall rate.

Keywords: Anti-electricity theft · Chi-square test · Logistic regression
Power consumption inspection

1 Introduction

For a long time, power grid enterprises have been committed to investigating and punishing the behavior of electricity theft. With the continuous development of technology, the electricity theft has presented characteristics of high-tech and covert. It brings the problems of low efficiency and poor accuracy for power grid enterprises to identify the customers stealing electricity [1, 2]. The traditional methods to identify electricity theft users such as artificial analysis have been difficult to meet the requirements of current anti-electricity theft. Modern techniques need to be used to screen the factors interrelated with electricity theft, and help power grid enterprises to accurately identify the electricity theft [2, 3].

Due to the widely application of various modern techniques in electricity theft, the concealment of electricity theft is improving, and the factors that can reflect the behavior of electricity theft are becoming more and more complicated [4–7]. As a result, the requirement for the ability of the power consumption inspector has become stricter. Some stealers even use kinds of modern equipment such as interference unit, to

© Springer Nature Singapore Pte Ltd. 2018
Q. Zhou et al. (Eds.): ICPCSEE 2018, CCIS 902, pp. 389–400, 2018.
https://doi.org/10.1007/978-981-13-2206-8_32

mislead the power consumption inspector, which made more difficult for the inspector to identify the thieves from various factors [8].

In 2015, the State Grid proposed a set of models for the on-line monitoring and intelligent diagnosis of power metering. The electricity theft model contained in these models could help the power consumption inspector to identify the electricity theft users.

However, this model is based on rule matching, and the rule is dependent on the subjective experience heavily. With the increase of new artifice of electricity theft, it is difficult for this method to capture the characteristics of diverse behavior of electricity theft accurately. The existing identification methods of electricity stealing behavior are the same as this model, which can not meet the requirements of electricity theft users and behavior recognition in current stage.

Therefore, it is necessary to optimize the model based on data analysis, in order to find the characteristic factors of electricity theft quickly and effectively, and predict the probability of customers appear the acts to steal electric power accurately. The optimization model proposed in this paper can reduce the economic losses caused by electricity theft, improve the accuracy of power consumption inspection, and reduce the working pressure of inspector greatly.

2 Research on Chi-square Test and Logistic Regression

Chi-square test is a commonly used hypothesis testing method. The most common use of this algorithm is to investigate whether the distribution of disordered categorical variables is consistent between two or more groups. In addition, it can also be used to compare the relevance between two or more samples and the classified variables. Chi-square test is not restricted by the overall distribution, and has many advantages such as wide scope of application, easy to operate, and has much superiority in practical applications [9]. Based on the Chi-square test, this paper calculates the main factors that can determine whether the customers have been stealing electricity, thus building a topic model of the electricity theft.

The regression model is a mathematical model for the quantitative description of the statistical relationship. Regression analysis is the method of studying the specific dependence of independent variables on dependent variables. Based on a set of sample data, the regression analysis determines the mathematical relationship between variables, and then carries out reliability of the relationship by statistical test, and finds out the significant variables from the variables that affect a specific variable finally. With the aid of obtained relationship, the value of another specific variable would be predicted or controlled according to the value of one or several variables, and the accuracy of prediction or control is also given [10]. The logistic regression analysis method is often used for classifying variables [11]. The predictive result of logistic regression is a probability between 0 and 1, which is easy to use and explain. In practical applications, logistic regression plays an important role in the classification problem, such as predicting the probability of a disease, predicting the probability of commodity purchase, or judging the sex of a user [12–14]. However, this analysis method is rarely used in the field of anti-electricity stealing. This paper will screen out the factors that are significantly associated with electricity theft, and optimize the weight of factors based on logistic regression, in order to get a prediction model with high accuracy, accuracy and recall rate.

3 Topic Model of Electricity Theft Based on Chi-square Test

Chi-square test can be used to compare the association between two or more samples and the classified variables. The factors and results that affect the electricity theft can be considered as the classified variables. In this paper, the correlation of factors and results is obtained by Chi-square test, so as to eliminate irrelevant factors.

3.1 Screen Out Interrelated Factors of Suspected Electricity Theft

Suppose that the topic model is built on the basis of the electricity theft users sample set c and the normal users sample set m.

The data set of factors from the electricity theft users and the normal users is used as the original data set. Define the factors set as $q_i, i \in (1, 2, \ldots, n)$. The fourfold table of actual values for building the Chi-square test is shown in Table 1.

Table 1. Fourfold table of actual values

Group	Electricity theft users	Normal users	Total
q_i occurred	c_1	m_1	$c_1 + m_1$
q_i not occurred	c_2	m_2	$c_2 + m_2$
Total	c	m	$c + m$

c_1 and c_2 are the sample number of factors q_i that occurred and didn't occur in electricity theft users samples respectively. m_1 and m_2 are the sample number of factors q_i that occurred and didn't occur in normal users samples respectively. $c = c_1 + c_2$, $m = m_1 + m_2$

First, suppose that q_i occurs or not is independent of whether the user has been stealing electricity or not. Select a sample from user data randomly. The probability of this sample belongs to electricity theft users is $\mu = \frac{c}{c+m}$.

Then, according to the independence hypothesis, a new fourfold table of theoretical values is generated as shown in Table 2.

Table 2. Fourfold table of theoretical values

Group	Electricity theft users	Normal users	Total
q_i occurred	$(c_1 + m_1) \cdot \mu$	$(c_1 + m_1) \cdot (1 - \mu)$	$c_1 + m_1$
q_i not occurred	$(c_2 + m_2) \cdot \mu$	$(c_2 + m_2) \cdot (1 - \mu)$	$c_2 + m_2$

Obviously, if the two variables are linearly independent, the difference between the theoretical values and the actual values in the fourfold table is very small.

The formula of Chi-square is

$$\chi^2 = \sum \frac{(A-T)^2}{T} \tag{1}$$

A is the actual values, which is the data shown in Table 1. T is the theoretical value, which is the data shown in Table 2.

After calculating the value of χ^2, determine whether the independence hypothesis is reliable by querying the critical value table of the Chi-square distribution. The degree of freedom (DF) of the fourfold table is 1. At this time, the critical probability of the Chi-square distribution (part) is shown in Table 3.

Table 3. Chi-square distribution critical value table (part)

Group	P					
DF	0.975	0.2	0.1	0.05	0.025	0.02
L	$9.82*10^{-4}$	1.642	2.706	3.841	5.024	5.412

By querying the whole table, the probability of all factors interrelated with electricity theft can be obtained. Suppose that the threshold is ε, when the correlation probability $P > \varepsilon$, this factor is identified as the interrelated factor. At last, a set of interrelated factors $x_i, i \in (1, 2, \dots, n)$ could be screened out by this way.

3.2 Eliminate High Interrelated Factors

In order to ensure the independence between the interrelated factors, we need to calculate the correlation degree between the interrelated factors. If the correlation degree between the two factors ρ is greater than the set threshold φ, the factor which has the lower correlation degree will be eliminated.

Test the linear dependence between the consecutive and ordinal factors by the Pearson correlation coefficient. The calculation formula is as follows.

$$\rho_{x_i, x_j} = \frac{\sum (x_i - \overline{x_i})(x_j - \overline{x_j})}{\sqrt{\sum (x_i - \overline{x_i})^2 (x_j - \overline{x_j})^2}} \tag{2}$$

$$j \in (1, 2, \dots, n), j \neq i, x_i = \begin{cases} 1, x_i \text{ occurred} \\ 0, x_i \text{ not occurred} \end{cases} \tag{3}$$

if $\rho_{x_i, x_j} > \varphi$, the factors that are less associated with the results of the two factors are eliminated. Suppose the set of factors which have been eliminated high interrelated factors is $y_i, i \in (1, 2, \dots, k)$, then $y_i \subseteq x_i$.

Test the dependence between the disorderly scalar factors by the Apriori algorithm, which is the nonlinear dependence between any two factors in $y_i, i \in (1, 2, \ldots, k)$.

Calculate the support degree σ_{y_i} of each factor in $y_i, i \in (1, 2, \ldots, k)$ based on the electricity theft sample.

Define the threshold of support degree as η.

If $\sigma_{y_i} \leq \eta$, eliminate the i'th factor from $y_i, i \in (1, 2, \ldots, k)$, to generate a new set $z_i, i \in (1, 2, \ldots, l)$, and $z_i \subseteq y_i$.

Constitute the set $\{z_i, z_j\}$ by selecting two random factors from $z_i, i \in (1, 2, \ldots, l)$. And on this basis calculate the support degree $\sigma_{z_i z_j}$ of each set in $z_i, i \in (1, 2, \ldots, l)$.

If $\sigma_{z_i z_j} \geq \eta$, we could judge that there are correlation between factor z_i and factor z_j. Eliminate the factor which has higher entropy after discretization from y_i. At last, we generate a new set $u_i, i \in (1, 2, \ldots, r)$, and $u_i \subseteq z_i$.

3.3   Determine the Combination Correlation

Combined with the factors which were eliminated in Sect. 3.1 based on the maximum combined threshold δ. Verify the correlation degree of each pair of combinations and results by Chi-square test in the same way as described in Sect. 3.1, excepting the groups were change to the combination factors occur simultaneously or not.

If the association probability between the combination factors and the results $P > \varepsilon$, this combination factors will be regarded as interrelated factors.

Finally, we generate the set of interrelated factors as $x_i, i \in (1, 2, \ldots, h), h \leq n$.

4   Factor Weight Optimization by Logistic Regression

On the basis of all the interrelated factors, we construct the loss function by logistic regression algorithm. And then, calculate the optimal solution of each factor weight by gradient descent method. At last we get the final prediction function.

4.1   Construction of Logistic Regression Function

Logistic regression is mainly used in the binary classification, so we use Sigmoid function which form is as follows.

$$g(z) = \frac{1}{1 + e^{-z}} \tag{4}$$

The Sigmoid function plot is shown in Fig. 1.

It is seen that the Sigmoid function transforms the z value into a value close to 0 or 1, and its output value changes significantly in the vicinity of $z = 0$.

The structural prediction function is as follows.

$$h_\theta(x) = g(\theta^T x) = \frac{1}{1 + e^{-(\theta_0 + \theta_1 x_1 + \ldots + \theta_k x_k)}}, k \leq n \tag{5}$$

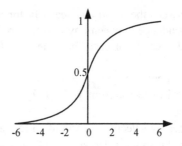

Fig. 1. Sigmoid function plot

Therein, $\theta^T = \sum \theta_i$, θ_i represents the weight of the interrelated factor $x_i, i\epsilon(0, 1, \ldots, k)$.

4.2 Optimize the Weights Based on Gradient Descent Algorithm

Prediction function $h_\theta(x)$ is an algorithm which inducts the sigmoid function to realize classification boundary fitting on the basis of linear regression. This function has the following properties.

$$P(y = 1 | x; \theta) = h_\theta(x) \tag{6}$$

$$P(y = 0 | x; \theta) = 1 - h_\theta(x) \tag{7}$$

When we calculate the optimal solution of corresponding parameters through optimizing function, a non-convex loss function would be got if we applied the loss function of linear regression directly. This loss function could not meet the application requirements of gradient descent algorithm. Thus, a monotone convex function need to be rebuilt as the loss function of logistic regression factor weight optimization.

Therefore, logistic regression loss function based on the log-likelihood loss function is constructed as follow.

$$\cos t(h_\theta(x), y) = \begin{cases} -\log(h_\theta(x)) & \text{if } y = 1 \\ -\log(1 - h_\theta(x)) & \text{if } y = 0 \end{cases} \tag{8}$$

Based on maximum likelihood estimation, the loss function can be deformed as

$$J(\theta) = \sum_{i=1}^{k} -y_i \log(h_\theta(x)) - (1 - y_i) \log(1 - h_\theta(x)) \tag{9}$$

$J(\theta)$ is convex function in this situation. The concrete steps to update the weight by the gradient descent method are as follows.

Initialize: $\theta_0, \theta_1, \ldots, \theta_k$, threshold η and learning rate α.

(1) Determine the gradient of the loss function of the current position. The gradient expression of θ_i is $\frac{\partial}{\partial \theta_i} J(\theta)$.

(2) Multiply the gradient of the loss function by the learning rate, to get the distance from the current position called step is $\alpha \frac{\partial}{\partial \theta_i} J(\theta)$.

(3) Confirm that the distance of the gradient descent of each θ value is less than η. If this is the case, algorithm will terminate. Otherwise, enter step (4).

(4) Update θ value according to the following formula:

$$\theta_i' = \theta_i - \alpha \frac{\partial}{\partial \theta_i} J(\theta) \tag{10}$$

Then turn back to the step (1).

By the iteration of all above steps, the optimal weight would be calculated, the predictive function would be obtained to predict the samples.

As described above, the topic model is built and optimized based on the electricity theft user sample set c and the normal user sample set m. The critical value of the prediction ξ is calculated according to the following formula.

$$\xi = \frac{m}{c + m} \tag{11}$$

If the predictive value belongs to $(\xi, 1]$, the user would be predicted as an electricity theft user. If the predicted value belongs to $(0, \xi]$, the user would be predicted as a normal user.

5 Experiment Design and Result Analysis

5.1 Experiment Environment

In this experiment, we chose 2000 electricity theft users with their related electricity utilization data during the period of stealing electricity, as well as 2000 normal users with their related electricity utilization during a period of time from a provincial power grid enterprise.

5.2 Experiment Procedures and Results

We selected the initial interrelated factors as shown in Table 4.

Using the method described in Sect. 2, we calculated the correlation degree between each of the suspected factors and the result that the electricity theft occurred or not by Chi-square test. Defined the threshold $\varepsilon = 0.8$. When the calculated correlation probability $P > \varepsilon$, the factor would be regarded as interrelated factors. Thus the conclusion was that the factors with serial number 3–5, 7–12, 15–16, 22, 26, 28, 30–35, 37–38 were interrelated factors, and others were irrelevant factor.

Then, we used the Pearson correlation coefficient to calculate the correlation degree between the interrelated factors, so as to ensure the independence between the factors. Defined the threshold $\varphi = 0.9$. If the correlation degree between the two factors was greater than 0.9, we removed the factor which P value is less.

Table 4. Initial relevance factor

Sn.	User behavior factors	Sn.	User behavior factors
1	Energy Meter Value Unbalanced	28	Current Return
2	Energy Meter Overspeed	29	Fee Control Order Failure
3	Energy Meter Reversed	30	Surplus Money Exception
4	Energy Meter Stopped	31	Single Phase Meter Shunted
5	Energy Meter Fee Rate Exception	32	Secondary Circuit Shorted (Shunted)
6	Voltage Phase Failure	33	Secondary Circuit Opened
7	Voltage Overlimit	34	Primary Circuit Shorted
8	Voltage Unbalanced	35	Energy Meter Shorted
9	Phase B Exception of High Voltage Supply & High Voltage Metering	36	Energy Meter Value Error
10	Current Lost	37	Circuit Series Semiconductor
11	Current Unbalanced	38	Magnetic Exception
12	Energy Meter Covering Opened	39	Electricity Address
13	Meter Terminal Covering Opened	40	Power Category
14	Measuring Door Opened	41	Contract Capacity
15	Magnetic Interference	42	Running Capacity
16	Energy Differential	43	User Status
17	Power Differential	44	Power Supply Organization
18	Power Failure Exception	45	Electricity Price Category
19	Load Overcapacity	46	Electricity Price & Industry Category
20	Requirement Overcapacity	47	Voltage Level
21	Current Overload	48	Measurement Method
22	Power Over Offline Continuously	49	Transformer Type
23	RTU Clock Exception	50	Energy Meter Comprehensive ratio
24	Power factor Exception	51	Last Year Energy Status
25	Energy Meter Clock Exception	52	Last Year Fee Status
26	Reversed Energy Exception	53	Last Year Arrears Times
27	Phase Sequence Exception	54	Last Year Arrears Amount

Afterwards, we calculated the correlation degree of combined factors and results from the irrelevant factors based on Chi-square test. $\delta = 2$. If the calculated combined correlation probability $P > \varepsilon$, the combined factor would be regarded as interrelated factors. Thus, loss of voltage phase and transformer type is a pair of interrelated combined factors, and other factors is eliminated.

And then, we substituted all interrelated factors and interrelated combined factors into the logistic regression model, as followed the steps in Sect. 4 for calculations. Initialized all θ values as 0.5, $\eta = 0.1$, and defined the learning rates were $\alpha = 0.1$ and $\alpha = 0.5$ respectively. When getting the updated θ value in iterations, the predictive function would be updated, into which the samples would be substituted to forecast.

Calculated the error rate of electricity theft users as the prediction result.

$$\sigma = \sqrt{(h_\theta(x) - 1)^2} \tag{12}$$

The error rate curve was shown in Fig. 2.

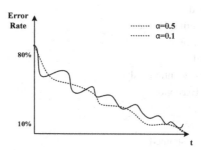

Fig. 2. Error rate curve

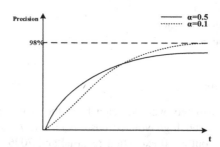

Fig. 3. Precision curve

Substituted the result of each iteration into the model to calculate the prediction precision of samples. The precision curve was shown in Fig. 3.

The result of the solution didn't change after 152 iterations. The results of the final model factors weights were shown in Table 5.

Finally, we substituted the samples into the prediction function. The average accuracy rate was 94.1%, the average precision rate was 96.6%, and the average recall rate was 95.8%.

Table 5. Model factors weight

Sn.	Factors	Weight
1	Energy Meter Reversed	0.1
2	Energy Meter Stopped	0.1
3	Energy Meter Fee Rate Exception	0.1
4	Voltage Phase Failure (Dedicated Transformer)	0.1
5	Voltage Phase Failure (Public Transformer)	0.7
6	Voltage Overlimit	0.2
7	Voltage Unbalanced	0.1
8	Phase B Exception of High Voltage Supply & High Voltage Metering	0.2
9	Current Lost	0.4
10	Current Unbalanced	0.1
11	Energy Meter Covering Opened	0.9
12	Magnetic Interference	0.5
13	Energy Differential	0.5
14	Power Over Offline Continuously	0.1
15	Reversed Energy Exception	0.5
16	Current Return	0.2
17	Surplus Money Exception	0.1
18	Single Phase Meter Shunted	0.5
19	Secondary Circuit Shorted (Shunted)	0.9
20	Secondary Circuit Opened	0.5
21	Primary Circuit Shorted	0.9
22	Energy Meter Shorted	0.9
23	Circuit Series Semiconductor	0.5
24	Magnetic Exception	0.9

5.3   Model Verification

100 thousand typical customers were selected from the provincial power grid enterprise, including 30 thousand high voltage customers and 70 thousand low voltage customers. We screened out 8110 exception records, and 7096 records of them were ascertained as electricity theft or measure abnormal. The precision rate was 87.5%.

The experimental results show that the prediction model constructed in this paper can achieve the prediction of the results accurately, which can effectively help the power grid enterprises identify the electricity theft users, reduce the economic losses of the power trade, and ensure the stable development of the power trade.

6   Conclusion

In view of the difficulties of power grid enterprises in inspecting electricity theft, this paper proposes an optimization method of suspected electricity theft topic model based on Chi-square test and logistic regression.

Based on the analysis of the factors that affect suspected electricity theft, Chi-square test is used to calculate the factors and combined factors that have high correlation degree with the results hat the electricity theft occurred or not. Screen out the interrelated factors and eliminate the irrelevant factors. The logistic regression algorithm is used to optimize the weights of the interrelated factors iteratively, to get the final prediction function. It can predict whether the users have been stealing electricity through the values of interrelated factors.

To verify the optimization model presented in this paper, we chose some electricity theft users and normal users with their related electricity utilization data from provincial power grid enterprise as the initial data samples. First, we screened out interrelated factors. Then, we constructed logistic regression function to optimize the weights of each factor. Finally got the predictive function. Substituted the experimental samples into the prediction function to get the prediction results which had a good performance in accuracy, precision, and recall rate.

Experimental results show that Chi-square test and logistic regression algorithms have a good applicability in selecting electricity theft interrelated factors as well as predicting whether the users have been stealing electricity. This method can inspire the power grid enterprises in anti-electricity theft, and improve the accuracy of power consumption inspection efficiently, and promote the stable development of power grid enterprises, and maintain well social power using order.

References

1. Wang, J., Meng, Y., Yin, S., Zhang, Y.: The present situation and development trend of anti electric stolen function of power demand information acquisition system. Power Syst. Technol. **12**(S2), 177–178 (2008)
2. Cheng, C., Zhang, H., Jing, Z., Chen, M., Jiao, L., Yang, L.: Study on the anti-electricity stealing based on outlier algorithm and the electricity information acquisition system. Power Syst. Prot. Control **43**(17), 69–74 (2015)
3. Hu, S., Guan, J., Yang, Z., Yu, H.: Research on electricity quantity metrology and acquisition system based on embedded system. Modern Electron. Tech. **39**(22), 163–166 +170 (2016). https://doi.org/10.16652/j.issn.1004-373x.2016.22.040
4. Wang, Q., Li, S.: Technology analysis and preventive measures of electric larceny prevention technology based on electric energy data acquisition system. Electr. Meas. Instrum. (2016)
5. Ren, S.: Strengthen the supervision and management of electric power to combat the theft of electricity. Global Mark. Inf. Guide **45**, 156 (2014)
6. Zhuang, C., Zhang, B., Hu, J., Li, Q., Zeng, R.: Anomaly detection for power consumption patterns based on unsupervised learning. Proc. CSEE **36**(2), 379–387 (2016). https://doi.org/10.13334/j.0258-8013.pcsee.2016.02.008
7. Zhao, L., Luan, W., Wang, Q.: Accurate line loss analysis of LV distribution network using AMI data. Power Syst. Technol. **39**(11), 78–83 (2015). https://doi.org/10.13335/j.1000-3673.pst.2015.11.026
8. Ma, S.: Supervision and management of electricity and measures for preventing electricity theft. Theor. Res. Urban Constr. **11**, 2440 (2016)

9. Xu, C., Lu, G., Ye, Y., Mi, Y.: Cooperative spectrum sensing using Chi-square test for multi-antenna cognitive radio. Chin. High Technol. Lett. **26**(7), 650–656 (2016). https://doi.org/10.3772/j.issn.1002-0470.2016.07.005

10. Wu, D.: Electricity theft identification method based on curve similarity. Electr. Power **50**(2), 181–184 (2017). https://doi.org/10.11930/j.issn.1004-9649.2017.02.181.04

11. Chen, A., Xia, F., Zhong, Y.: A new independence test of four grid table. Stat. Decis. **13**, 85–88 (2017). https://doi.org/10.13546/j.cnki.tjyjc.2017.13.020

12. Xu, J., Su, W., Wu, S., Wu, X.: Modeling user reliability based on logistic regression in micro-blog. Comput. Eng. Des. **3**, 772–777 (2015). https://doi.org/10.16208/j.issn1000-7024.2015.03.042

13. Guo, J., Sun, J., Liang, T., Tan, R.: Evaluation model of disruptive design scheme based on logistic regression. Comput. Integr. Manuf. Syst. **21**(6), 1405–1416 (2015). https://doi.org/10.13196/j.cims.2015.06.001

14. Wang, Z., Liu, K., Zheng, Z., Li, C.: Prediction retweeting of microblog based on logistic regression model. J. Chin. Comput. Syst. **37**(8), 1651–1655 (2016)

A Comparison Method of Massive Power Consumption Information Collection Test Data Based on Improved Merkle Tree

Enguo Zhu[1(✉)], Fangbin Ye[2], Jian Dou[1], and Chaoliang Wang[2]

[1] China Electric Power Research Institute, Beijing, China
krauser3c@163.com
[2] State Grid Zhejiang Electric Power Company Electric Power Research Institute, Hangzhou, China

Abstract. It is necessary to conduct comparison test of massive electricity data on the unified interface platform for ensuring that the power consumption information collection system can provide data for kinds of business systems stably and accurately. In this paper, a method of data comparison for mass data of power energy data acquire system based on improved Merkle Tree is proposed to solve the problem that the traditional one to one comparison method cannot adapt to massive data test. Merkle Tree is improved by forming all the subtrees into complete binary tree. Based on the improved algorithm mentioned above, the improved Merkle Tree is constructed for the source data from power consumption information collection system and the publish data from unified interface platform. Through the comparison of Merkle Root and its sub nodes, a fast consistency check is achieved, and the optimal data blocking strategy and multithreading optimization method are put forward. This method greatly improves the efficiency of the unified interface platform for the consistency of mass data.

Keywords: Power consumption information collection · Hash algorithm
Improved Merkle tree · Data consistency

1 Introduction

As an important data source of power marketing, power consumption information collection system provides business and data support for many application systems. As the range of data acquisition expands, the volume of data provided to the other systems also increases greatly [1–4]. The State Grid Corporation proposed to establish the unified interface platform for power consumption information collection system, so as to achieve the interface data transmission between the power consumption information collection system and other business systems. At this stage, the volume of data transferred is huge, and this data may be used in the important business areas such as electricity settlement. In order to avoid business anomalies caused by data variation, it is necessary to ensure data consistency between source data and interface data to pass data consistency test for unified interface platform [5, 6]. Because the data provided by the power consumption information collection system has the characteristics of large

© Springer Nature Singapore Pte Ltd. 2018
Q. Zhou et al. (Eds.): ICPCSEE 2018, CCIS 902, pp. 401–415, 2018.
https://doi.org/10.1007/978-981-13-2206-8_33

scale and wide varieties, the traditional verification mode that contrasts data one to one is difficult to meet the requirements of this test scenario. It is necessary to investigate an efficient data comparison method for massive data, which could support data integrity and consistency test of power consumption information collection system unified interface platform.

The consistent hash algorithm is usually used to verify the integrity and consistency of data. However, simple consistent hash computation can only be used to verify the consistency of data, but cannot meet the needs of precise locating of inconsistent data during testing [3–5]. In this paper, we build Merkle Tree for the data should be compared on the basis of data partitioning, and verify the data consistency by comparing the hash values of each node in Merkle Tree. By retrieving the path of the different hash value in the Merkle Tree, the data block with consistency differences would be located. Then, compare the whole data in block, so as to locate inconsistent data accurately.

In the first chapter of this paper, we introduce the Merkle Tree, analyze its technical characteristics, and put forward the advantages that applied Merkle Tree to comparison of massive power consumption information collection test data. The main content of the second chapter is the detailed procedure of the comparison method of massive power consumption information collection test data based on improved Merkle Tree. In this chapter, we select the appropriate Hash algorithm based on application scenarios firstly, then expound the data comparison process based on Merkle Tree as well as the data partitioning strategy, and then propose an improved Merkle Tree constructing method, and finally propose a multithreading optimization strategy based on this improved method. In the third chapter, correlated comparative experiments are carried out to verify the effectiveness of the comparison method proposed in this paper. We make contrast tests on the comparison method based on improved Merkle Tree, the original Merkle Tree, and the traditional one to one comparison method, and the multi-threaded and single-threaded mode. The results prove that the comparison method based on improved Merkle Tree can significantly improve data execution efficiency, and could be more suitable for using multi-threaded optimization compared with the other two algorithms.

2   Merkle Tree Based on Hash Algorithm

Hash algorithm maps the binary value of any length to a shorter fixed length binary value, which is called a hash value. The hash value is a unique and compact numerical representation of a piece of data [7]. Consistent hash algorithm was proposed by Massachusetts Institute of Technology in 1997. Its original design goal is to solve the problem of hot spot in the Internet. Later, this algorithm is widely used in the comparison of data consistency [8, 9].

Merkle Tree, also known as Hash Tree, is a tree that stores hash values. The leaves of the Merkle Tree are the hash values of the data blocks. A non-leaf node is the hash value of its subnode series string. First, the data is divided into small data blocks with a corresponding hash values. Then, merge the two adjacent hash values into a string. And then, calculate the hash value of merged string, so as to obtain a sub-hash. If the total number of hash value in this layer is singular, the far right-hand single leaf carries out the hash value of itself as its sub-hash. With the same way in the upper layer, fewer new hashes can be obtained. Finally, a binary tree is formed, and the root node is called Merkle Root [10–12]. At present, Merkle Tree is mostly used for comparison and verification in the field of computer. When used in the comparing or validating application scenarios, especially in distributed environment, Merkle Tree will reduce data transmission and computation complexity greatly [13–16].

In this paper, if the Merkle Root consistency contrast is passed, there is no need for a detailed comparison to prove the consistency of the contrast data. When Merkle Root is different, the data would be compared with the hash value of its subnodes. Finally, we can find out all the inconsistent data blocks by traversing through the tree, and then make a one-to-one comparison to locate the discrepancy. This method could reduce the cost on time and resources of data comparison substantially.

3   Data Comparison Based on Merkle Tree

3.1   Selection of Hash Algorithm

Hash algorithm can be divided into encrypted hash algorithm and unencrypted hash algorithm. The encrypted hash algorithms include the algorithms of MD5 and the SHA family such as SHA-1, SHA-224, SHA-256 and so on. Unencrypted algorithms include the algorithms such as CRC32, MurmurHash [17–20]. Because the unencrypted hash algorithm is simple and irreversibility, the efficiency of the unencrypted hash algorithm is higher than that of the encrypted hash algorithm. In this paper, the hash algorithm is used to data verification, so the privacy protection is not considered in this application scenario. Therefore, we select the more efficient unencrypted hash algorithm.

The CRC algorithm, or cyclic redundancy check algorithm, has the longest history in unencrypted hash algorithm. Now commonly used CRC algorithm is CRC32, which can hash a string into a value of 32 Bit. Under the same condition of 32 Bit, the performance of MurmurHash is higher than that of CRC32, but the collision probability of CRC32 is lower than that of MurmurHash.

Because MurmurHash supports 128 Bit, the collision probability is greatly reduced, and there are almost no collisions within the range of 0–10^8. According to the test data status of the power consumption information collection system, the number of data blocks for a single test will not exceed 10^8, so MurmurHash could completely meet the requirement of collision probability. And 128 Bit MurmurHash is better than CRC32 in terms of performance. Therefore, the MurmurHash is selected in this paper to be the core hash algorithm for the construction of Merkle Tree.

3.2 Process of Data Comparison and Discrepancy Analysis Based on Merkle Tree

Process of Data Comparison

The process of data comparison based on Merkle Tree is as follows.

(1) The measured data is partitioned into blocks according to data partitioning strategy. The data blocks are numbered and sequenced by its numbers.
(2) According to the data content of each block, generate the instruction set within data request from the unified interface platform and the mapping relation of instructions and data blocks.
(3) Hash the numbered data blocks in order to get a sequenced hash value list.
(4) Construct the Merkle Tree from the hash value list.
(5) According to the instruction set, request block data from the unified interface platform in multi-threaded.
(6) After a data block is downloaded, identify the block according to the mapping relation of instruction set and data block.
(7) Get the hash value of downloaded data block.
(8) After the adjacent two data blocks are downloaded, generate and combine the hash value of the data blocks to get the sub-hash of it. Repeat the steps above to build the Merkle Tree of test data.
(9) Compare the Merkle Root of source data and test data. If the two are equal, the test is passed, otherwise it is not passed.

When using this method for data comparison, if the depth of Merkle Tree is d, after the last block is downloaded, it is only had to hash d times to structure the whole Merkle Tree to get the Merkle Root for comparing. It is much more efficient than compare data one to one.

Data Discrepancy Retrieval Method

When there is discrepancy between the Merkle Root of source data and test data, the Merkle Trees will be retrieved and compared in order to find the discrepancy. The retrieval process is as follows.

(1) Compare the hash value of subnode owned by the Merkle Root of the source data and test data.
(2) If the hash values of the subnode of the source data and test data are same, there's no need for comparing the nodes in their subtree.
(3) If they are different, it is necessary to compare their subnode until all of the leaf node is compared.
(4) Locate all the leaf nodes which hash value are different, and elect the related data blocks from source data and test data.
(5) Compare data items in the blocks of source data and test data in order to find the discrepant data.

The node localization process is shown in the following diagram (Fig. 1).

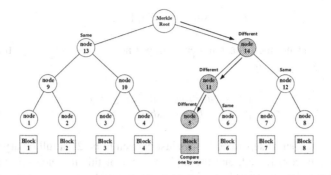

Fig. 1. Data discrepancy localization process based on Merkle Tree

3.3 Data Partitioning Strategy

The greater the depth of the Merkle Tree, the lower the efficiency of its retrieval location. In the condition of the same volume of data, the more data items in a single block, the fewer volume of block, the longer the matching time after block has been located.

For any constructed binary tree, when the tree is a full binary tree, the number of leaf nodes k and depth d have the following relation.

$$k = 2^d \tag{1}$$

At this point, the leaf node number depth ratio of the binary tree is the biggest.

Thus, in order to improve the retrieval location efficiency, the volume of data blocks should be nearly or slightly less than the number which is integer power of 2.

According to the calculation, the single data length of the unified interface platform of power consumption information collection system which is expressed in XML is between 64 Bytes and 128 Bytes. In this paper, we choose 128 Bytes as a single data estimated length.

For example, in case of the test data which contains 3,000,000 data items, the estimated volume of it is 384,000,000 Bytes $\approx$ 366.21 Mbytes. If the volume of expected data block is less than 10 M, the data block could be partitioned into $64(2^6)$ parts. At this point, the volume of data block should be between 5.81 Mbytes and 5.72 Mbytes for the optimal search performance.

3.4 Improved Merkle Tree Constructing Method

The unimproved Merkle Tree constructing method is as follows.

If the node number of lower layer is m, the nodes of the lower layer are $a_1, a_2, \ldots, a_m$.

If m is an even number, the number of upper nodes is $n = \frac{m}{2}$, the nodes of upper layer is $b_1, b_2, \ldots, b_n$.

$$b_i = hash(a_{2i-1}, a_{2i}), 1 \leq i \leq n \tag{2}$$

If m is an odd number, the number of upper nodes is $n = \frac{m-1}{2} + 1$, the nodes of upper layer is $b_1, b_2, \ldots, b_n$.

$$\begin{cases} b_i = hash(a_{2i-1}, a_{2i}), 1 \leq i \leq n \\ b_n = hash(a_m) \end{cases} \tag{3}$$

When the number of separated blocks just happen to be 2^n, a full binary tree which depth is $n + 1$ is constructed, and the node number of this tree needed to traverse is $2^{n+1}-1$. At this point, the Merkle Tree has an optimal performance on construction and retrieval.

When the number of separated blocks just happen to be $2^n + 1$, a binary tree which depth is $n + 2$ is constructed, and the node number of this tree needed to traverse is $2^{n+1}+n + 1$. The left subtree is a full binary tree whose depth is $n + 1$, and the right subtree is a right oblique tree. At this point, the Merkle Tree has a worst performance on construction and retrieval.

As it is seen, the cost of constructing and retrieving tree rises sharply when the numbers of separated data blocks are not matched 2^n.

In order to resolve the problems above, we improved the constructing method of Merkle Tree which is as follows:

Suppose the number of nodes in the lower layer is m, $m > 1$, the nodes are $a_1, a_2, \ldots, a_m$.

When $m = 2$, the number of upper nodes $n = 1$, the nodes are b_1.

$$b_1 = hash(a_1, a_2) \tag{4}$$

When $m \bmod 4 = 0$, the number of upper nodes $n = m/2$, the nodes are b_i.

$$b_i = hash(a_{2i-1}, a_{2i}), 1 \leq i \leq n \tag{5}$$

When $m \bmod 4 = 3$, the number of upper nodes $n = (m - 1)/2 + 1$, the nodes are $b_1, b_2, \ldots, b_n$.

$$\begin{cases} b_i = hash(a_{2i-1}, a_{2i}), 1 \leq i \leq n - 1 \\ b_n = a_m \end{cases} \tag{6}$$

When $m \bmod 4 = 2$, the number of upper nodes $n = (m - 2)/2 + 1$, the nodes are $b_1, b_2, \ldots, b_n$

$$\begin{cases} b_i = hash(a_{2i-1}, a_{2i}), 1 \leq i \leq n - 2 \\ b_{n-1} = a_{m-1} \\ b_n = a_m \end{cases} \tag{7}$$

When $m \bmod 4 = 1$, the number of upper nodes $n = (m - 3)/2 + 1$, the nodes are $b_1, b_2, \ldots, b_n$.

$$\begin{cases} b_i = hash(a_{2i-1}, a_{2i}), 1 \leq i \leq n-3 \\ b_{n-2} = a_{m-2} \\ b_{n-1} = a_{m-1} \\ b_n = a_m \end{cases} \tag{8}$$

When the number of leaf nodes is 2, 5, 6, 7, 8 respectively, the Merkle Tree improved construction status is shown in figures as follows (Fig. 2, 3, 4, 5 and 6).

Fig. 2. Construction process of the tree when the number of leaf nodes is 2

Comparing to the unimproved Merkle Tree constructing method, this method has the following advantages:

(1) The number of summary points that need to be traversed is reduced. The single value on the far right, needs to be used one or more hashes to maintain the balance of the tree, but it also increases the cost of retrieval. The improved algorithm no longer carries out its own hash, but regards the leaf nodes as the upper node directly. It reduces the total hash times.

(2) For the tree generated by the improved algorithm, all subtrees under the root node are complete binary trees which have a good performance on retrieval, and therefor this algorithm could improve the efficiency of discrepant data retrieval. Especially in multi-threaded mode, all of the subtrees searched by each thread are complete binary tree, so the search efficiency of each single-threaded is relatively close. It improves the overall search performance under the same resource conditions rather than the traditional methods.

3.5   Multi-threaded Optimization Based on Improved Merkle Tree

Because the left subtree of Merkle Tree is a complete binary tree, when the depth of the tree is d, the depth of left subtree should have to be $d_1 = d - 1$, and the depth of right subtree should be $1 \leq d_r \leq d - 1$. This tree is not a balanced tree. If we use single-threaded to retrieve, the retrieval efficiency would be affected by its non-equilibrium. In this paper, a multi-threaded method is proposed to compare the discrepancy nodes of tree. The process of it is as follows.

(1) Check whether Merkle Roots are consistent or not. If they are inconsistent, start two threads for retrieving left and right subtrees separately. The processes of retrieving those two subtrees are the same.

(2) Check whether the roots of subtrees are consistent. If they are consistent, return, otherwise, start two threads for its left and right subtrees retrieving.

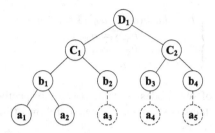

Fig. 3. Construction process of the tree when the number of leaf nodes is 5

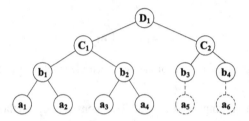

Fig. 4. Construction process of the tree when the number of leaf nodes is 6

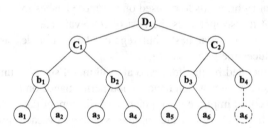

Fig. 5. Construction process of the tree when the number of leaf nodes is 7

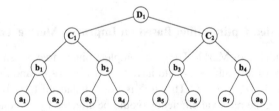

Fig. 6. Construction process of the tree when the number of leaf nodes is 8

(3) When the left and right subtrees are all consistent, return.

(4) When only one of the left and right subtrees is inconsistent, the root node of that subtree is continued to be retrieved as a new root node, then return to step 2.

(5) When the left and right subtrees are all inconsistent, start a new thread for retrieving left subtree whose root is regard as the new root node. The current thread

is used for retrieving right subtree whose root is regard as the root node. The retrieval method of the subtrees follows the step 2 and its following steps.

(6) Threads are managed in a thread pool, and the maximum number of active threads can be determined according to the machine resource situation used for testing. When the number of threads exceeds the maximum number of active threads, the extra thread goes into the waiting queue until one thread is completed, and the thread runs one to one according to the order of the queue.

The advantages of this method are: the left and right subtrees are complete binary trees, that is to say, any subtrees of non-root node is a leaf node or a complete binary tree. When retrieving discrepancies of the subtrees by multi-threaded mode, the retrieval efficiency of each thread is relatively optimal. By the way of thread independent retrieval, it effectively solves the problem of Merkle Tree imbalance. It transforms the retrieval problem of an imbalanced tree into multiple balanced tree retrieval problems, and effectively balanced the retrieval efficiency and resources by controlling the thread pool.

4 Experiment Design and Result Analysis

4.1 Experiment Environment

In order to verify the validity of the test data comparison method based on the improved Merkle Tree for power consumption information collection system, the test data of the power consumption information collection system unified interface platform are used to carry out the simulation experiment. The experimental data are generated by simulating the actual data. During the testing process, both the test data and the source data are stored in XML format, and the document format follows the unified interface platform interface protocol.

The test data set contains about 36.6 million data items, and the volume of data set is about 1.57 G. The status of data set is as follow table (Table 1).

Table 1. Test data quantity

Sub catalog	Test data scale
Daily freeze energy value	3,150,625
Daily freeze energy	3,150,625
Load curve	7,008,000
Voltage curve	2,628,000
Current curve	2,628,000

4.2 Experiment Procedures and Results

In order to verify the performance of the test data comparison method of the massive power consumption information collection system based on the improved Merkle Tree, the experiment is set up as follows.

Experiment 1: According to the volume of test data, the Merkle Tree with a depth of 9 is constructed. The size of each data block is 6.3 M, and there is 256 blocks totally. Test the time expended by the traditional one to one comparison method and the comparison method based on the Merkle Tree. The experiment is carried out 10 times. Before the test, 150 discrepant data are randomly added to the test data and distributed randomly in 32 data blocks. Finally, the average consumption time of 10 experiments is obtained. The results are in figures as follows (Fig. 7).

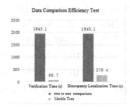

Fig. 7. Data comparison efficiency test result

According to the experiment, we can find that all data must be compared, so the verification time and discrepancy localization time are same.

The use of improved Merkle Tree only needs to compare the Merkle Root of the built tree, so it is far more efficient than one to one comparison. In the process of data validation, Merkle Tree is a full binary tree which has 511 nodes within 256 leaf nodes. According to the data retrieval method proposed in this paper, there is 32 discrepant data blocks are located from the 192 nodes retrieved, and the time spent on retrieving is 1/256 of one to one method. Thus it can be seen that the comparison method based on Merkle Tree can effectively improve the efficiency of data validation and discrepancy localization.

Experiment 2: According to the test data, the improved Merkle Tree with a depth of 9 is constructed. The size of each data block is 6.3 M, and there is 256 blocks totally. Test the time expended by the traditional one to one comparison method and the comparison method based on the Merkle Tree. Each experiment is carried out 10 times. Before once test, 10, 50, 100 and 150 discrepant data are randomly added to the test data and distributed randomly in the No. 1, 8, 16, 32 data blocks. Test the efficiency of discrepancy localization in multi-threaded and single-threaded under different distribution of discrepant data. Finally, the average consumption time of 10 experiments is obtained. The results are in figures as follows.

It is seen in Fig. 8, when the discrepant data is distributed to one data block, the retrieval efficiency of multi-threaded and single-threaded is almost the same. When the number of distributed blocks is more, and discrepant data is relatively equally distributed in multiple data blocks, the advantage of the retrieval efficiency of multi-threaded is more obvious.

Experiment 3: Use different methods to partition data blocks for the test data.

One of it is to construct an improved Merkle Tree with a depth of 9 as a full binary tree. The size of each data block is 6.3 M, and there is 256 blocks totally.

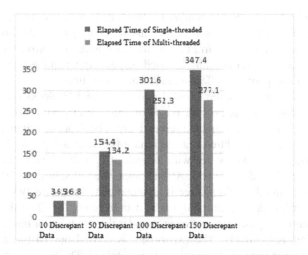

Fig. 8. Multi-threaded processing efficiency test result 1

The other one of it is to construct an improved Merkle Tree with a depth of 10. The size of each data block is 5 M, and there is 319 blocks totally. At this point, the left subtree of Merkle Tree is a full binary tree with a depth of 9, the right subtree is an incomplete binary tree with a depth of 7.

20 discrepant data are randomly added to two data blocks, one of which is located in the left subtree and the other in the right subtree, in order to test the efficiency of multi-threaded and single-threaded in this condition. The experiment is carried out 10 times. Finally, the average consumption time of 10 experiments is obtained. The test results are in figures as follows.

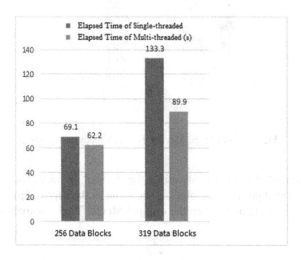

Fig. 9. Multi-threaded processing efficiency test result 2

It is seen in Fig. 9, when the number of data blocks is larger than 2^n, the depth of Merkle Tree's left subtree and right subtree are different greatly. At this point, if the few discrepant data are far apart from each other, it is able to improve the efficiency of the discrepancy comparison in multi-threaded. Especially when the balance of the left and right subtrees is poor, the effect is much more significant.

Experiment 4: The test data is divided into a total of 133 blocks according to that the size of each data block is 11.8 M. Based on this, the original Merkle Tree and the improved Merkle Tree are constructed respectively. The depth of original Merkle Tree is 8. At this point, the left subtree is a full binary tree with a depth of 7, and the right subtree is an incomplete binary tree with a depth of 7, with a total of 273 nodes. The depth of improved Merkle Tree is 8. At this point, the left subtree is full binary Tree with a depth of 7, and the right subtree is an incomplete binary tree with a depth of 3, with a total of 265 nodes. Place the different data in block 133. Test the construction time and the abnormal positioning efficiency under the condition of multiple threads of two Merkle Tree respectively. The experiment is carried out 10 times. Finally, the average consumption time of 10 experiments is obtained. The test results are in figures as follows.

As can be seen from Fig. 10, the Tree built with the improved Merkle Tree construction method has fewer nodes than the original Merkle Tree construction method, so the improved Merkle Tree wastes less construction time.

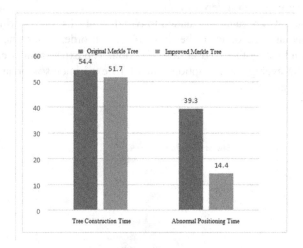

Fig. 10. Improved Merkle Tree efficiency test result

In the process of abnormal positioning, the number of node that need to be traversed to locate the abnormal data block is the same as the depth of the subtree within which this block is contained. When the original Merkle Tree construction method has a poor balance, its incomplete right subtree has the same depth as the full binary left subtree. Under the same condition, the right subtree constructed by improved Merkle Tree is the full binary tree with fewer depth. Therefore, when the abnormal data block is located in the incomplete subtree, the tree built with the improved Merkle Tree construction method requires to traverse fewer nodes, with shorter abnormal positioning time.

5 Conclusion

In this paper, we focus on the problem of data consistency test of power consumption information collection system unified interface platform. A comparison method of massive power consumption information collection test data based on improved Merkle Tree is proposed.

The method proposed in this paper establishes Merkle Tree for source data and test data respectively. First, verify the consistency by comparing Merkle Root. Then, compare the discrepancy in hash values of their corresponding nodes to find data blocks with discrepancy rapidly. Finally, precisely locate the discrepant data in the block.

Design experiments on the data source of a provincial power grid corporation. Test the efficiency difference between the comparison method of one to one, the method based on original Merkle Tree and the method based on improved Merkle Tree, as well as verify the effect of optimizing Merkle Tree in multi-threaded mode. The result proves that the efficiency of data comparison and execution is significantly improved by using the improved Merkle Tree. And it can further improve the execution efficiency after optimizing with multi-threaded.

The experimental results show that the method not only realizes the rapid comparison of mass data, but also realizes accurate locating of inconsistent data. The execution efficiency of this method is far higher than the traditional one to one comparison method, and can further optimize the efficiency of the discrepancy comparison by multithreading. It solves the problem that the simple hash algorithm can only be used to verify the consistency of the algorithm. When it is used in the power consumption information collection system, this method can effectively improve the efficiency of massive data consistency test and discrepancy localization.

References

1. Shen, L., Liu, X., Liu, Y., Dong, L.: Reliability compliance test for electrical energy data acquisition equipments. Electr. Measur. Instrum. **52**(1), 8–12 (2015). https://doi.org/10.3969/j.issn.1001-1390.2015.01.002
2. Hu, J., Zhu, E., Du, X., Du, S.: Application status and development trend of power consumption information collection system. Autom. Electr. Power Syst. **2**, 131–135 (2014). https://doi.org/10.7500/aeps20130617005

3. Lu, J., Li, Z., Zhu, Y., Xu, Z.: Communication bandwidth prediction for information gathering services in power distribution and utilization of smart grid. Power Syst. Technol. **40**(4), 1277–1282 (2016). https://doi.org/10.13335/j.1000-3673.pst.2016.04.044
4. Wang, X., Shi, Y., Zhang, J., Liang, B., Cheng, C.: Computation services and applications of electricity big data based on Hadoop. Power Syst. Technol. **39**(11), 3128–3133 (2015). https://doi.org/10.13335/j.1000-3673.pst.2015.11.017
5. Li, F., Chen, Y., Zhang, S., Hu, T.: Research on construction of unified database of large-scale power grid. Power Syst. Technol. **37**(2), 417–424 (2013)
6. Shi, W., Huang, L., Xu, J., Gu, Y., Jiang, H., Zhang, J.: Fault detection, identification and location based on measurement data from finite synchronized Phasor measurement. Power Syst. Technol. **39**(4), 1067–1074 (2015). https://doi.org/10.13335/j.1000-3673.pst.2015.04.029
7. Wang, Y., Cai, G.: Study on hash function based on random function. Comput. Eng. Des. **10**, 2679–2683 (2015)
8. Xu, P., Liu, S., Lan, J., Xiao, D.: An improved segment hash algorithm. Comput. Eng. **41**(1), 266–269 (2015). https://doi.org/10.3969/j.issn.1000-3428.2015.01.050
9. Hu, Q., Luo, Y., Yin, M., Qu, X.: Method of duplicate removal on alert logs based on attributes hashing. Comput. Sci. **43**(6A), 332–334 (2016). https://doi.org/10.11896/j.issn.1002-137X.2016.6A.079
10. Yan, W., Wu, W.: Data Structure (C Language Version). Tsinghua University Press, Beijing (2011)
11. Al-Harbi, R., Abdelaziz, I., Kalnis, P., Mamoulis, N., Ebrahim, Y., Sahli, M.: Accelerating SPARQL queries by exploiting hash-based locality and adaptive partitioning. VLDB J. **25**(3), 355–380 (2016). https://doi.org/10.1007/s00778-016-0420-y
12. Cao, Y., Liu, Y., Jia, X., Wang, D.: Image spam filtering with improved LSH algorithm. Appl. Res. Comput. **33**(6), 1693–1696 (2016). https://doi.org/10.3969/j.issn.1001-3695.2016.06.021
13. Zhang, H., Li, Y., Ma, F.: Merkle hash tree based security authentication scheme for WBSN. Mod. Electron. Tech. **40**(3), 65–70 (2017)
14. Xie, F.: Trusted cloud computing information security certificate method based on Merkle hash tree. Laser J. **37**(11), 122–127 (2016). https://doi.org/10.14016/j.cnki.jgzz.2016.11.122
15. Liu, F., Cai, Y., Wang, C., Yan, C.: Research of traverse and application of Merkle authentication tree. Comput. Eng. Appl. **48**(14), 98–101 (2012). https://doi.org/10.3778/j.issn.1002-8331.2012.14.021
16. Chen, L., Qiu, L.: A verifiable ciphertext retrieval scheme based on Merkle hash tree. Netinfo Secur. **4**, 1–8 (2017)
17. Li, H.: Application of MD5 hash encryption algorithm in SQL server 2005. Mod. Comput. **4**, 60–61 (2011). https://doi.org/10.3969/j.issn.1007-1423-B.2011.04.019
18. Ge, J., Li, J.: A new encryption algorithm for digital images based on MD5 value. Comput. Appl. Softw. **27**(6), 35–38 (2010). https://doi.org/10.3969/j.issn.1000-386X.2010.06.012

19. Xiao, B., Zheng, Y., Lonf, J., Guo, P.: QR code design of information security based on Rijndael encryption algorithm and SHA512 encryption algorithm. Comput. Syst. Appl. **24**(7), 149–154 (2015). https://doi.org/10.3969/j.issn.1003-3254.2015.07.027
20. Aizezi, Y., Wumaier, A.: Security payment scheme based on symmetric encryption and hash function in e-commerce. Microcomput. Appl. **33**(3), 46–48 (2017). https://doi.org/10.3969/j.issn.1007-757X.2017.03.014

Towards Realizing Sign Language to Emotional Speech Conversion by Deep Learning

Nan Song[1], Hongwu Yang[1,2(✉)], and Pengpeng Zhi[1]

[1] College of Physics and Electronic Engineering,
Northwest Normal University, Lanzhou 730070, China
[2] Engineering Research Center of Gansu Province for Intelligent Information
Technology and Application, Lanzhou 730070, China
yanghw@nwnu.edu.cn

Abstract. This paper proposes a framewrok for realizing sign language to emotional speech conversion by deep learning. We firstly adopt a deep belief network (DBN) model to extract the features of sign language and a deep neural network (DNN) to extract the features of facial expression. Then we train two support vector machines (SVM) to classify the sign language and facial expression for recognizing the text of sign language and emotional tags of facial expression. We also train a set of DNN-based emotional speech acoustic models by speaker adaptive training with an multi-speaker emotional speech corpus. Finally, we select the DNN-based emotional speech acoustic models with emotion tags to synthesize emotional speech from the text recognized from the sign language. Objective tests show that the recognition rate for static sign language is 92.8%. The recognition rate of facial expression achieves 94.6% on the extended Cohn-Kanade database (CK+) and 80.3% on the JAFFE database respectively. Subjective evaluation demonstrates that synthesized emotional speech can get 4.2 of the emotional mean opinion score. The pleasure-arousal-dominance (PAD) evaluation shows that the PAD values of facial expression are close to the PAD values of synthesized emotional speech.

Keywords: Sign language recognition · Facial expression recognition
Deep Neural Network · Emotional speech synthesis
Sign language to speech conversion

1 Introduction

Sign language is one of the most important communication methods among speech impaired and normal person. Since researches on sign language

The research leading to these results was partly funded by the National Natural Science Foundation of China (Grant No. 11664036, 61263036), High School Science and Technology Innovation Team Project of Gansu (2017C-03), Natural Science Foundation of Gansu (Grant No. 1506RJYA126).

recognition have been widely concerned [1], it have been a research hotspot in human-computer interaction. The wearing technologies such as the data gloves are used for sign language recognition [2] in the very beginning. In recent years, machine learning algorithms such as Hidden Markov model (HMM) [3], Back Propagation (BP) neural network [4] and Support Vector Machine (SVM) [5] are applied to sign language recognition for improving the recognition rate of sign language. At present, because deep learning becomes an important machine learning method, it also has been applied to sign language recognition [6] and has greatly improved the sign language recognition rate. Because emotion expression can make communication more accurate, facial expressions also play an important role in the communication between normal persons and speech impairments during their daily life. Therefore, facial expression recognition technologies also have a rapidly developing in the area of image processing. Several different methods such as SVM [7], Adaboost [8], Local Binary Pattern (LBP), Principal Components Analysis (PCA) [9] and Deep Learning [10] are applied to the facial expression recognition. At the same time, Deep Neural Network (DNN) - based speech synthesis methods are widely used in the field of speech synthesis [11,12], by which the text information can be converted into speech. However, most of the existing researches mainly focus on sign language recognition, facial expression recognition and emotional speech synthesis individually. Some researches adopt information fusion methods to integrate facial expressions, body language and voice information to realize the emotion recognition under the multi-modal fusion [13]. In the study, the sign language recognition and voice information are fused to command the robot' [14] to achieve the wheelchair navigation control of robot [15]. These studies show that multi-modal information fusion has gradually become a trend. Although sign language to speech conversion has studied in [16,17], the synthesized speech does not include changes in emotion so that ignore the speech expression of the deaf and mute people that make communication becomes easy for listeners to understand ambiguous.

To overcome the deficiency of current researches on sign language to emotional speech conversion, the paper proposed a framework that combines sign language recognition, facial expression recognition and emotional speech synthesis together to realize the conversion of sign language to emotional speech. We firstly use the static sign language recognition to obtain the text of sign language. At the same time, the expressed facial emotion information is obtained by the facial expression recognition. We also trained a set of DNN-based emotional speech acoustic models. The text and emotion information obtained from sign language recognition and facial expression recognition are finally converted into corresponding emotional speech with emotional speech synthesis.

2   Framework of Sign Language to Emotional Speech Conversion

The framework of the proposed sign language to emotional speech conversion consists of three parts including sign language and facial expression recognition, acoustic models of emotional speech training and emotional speech synthesis, as shown in Fig. 1. In the recognition step, the categories of sign language are obtained by sign language recognition and the emotion tags are obtained by facial expression recognition. In the acoustic model training step, we use an emotional speech corpus to train a set of emotional acoustic models with a DNN-based emotional speech synthesis framework. In the speech synthesis step, the text is obtained from the recognized categories of sign language by searching a defined sign language-text dictionary. Then we generate the context-dependent label from text of sign language by a text analyzer. Meanwhile, the emotional speech acoustic models are selected by using the emotion tags that is recognized from facial expression. Finally the context-dependent labels are applied on the emotional acoustic models to generate acoustic parameters for synthesizing emotional speech.

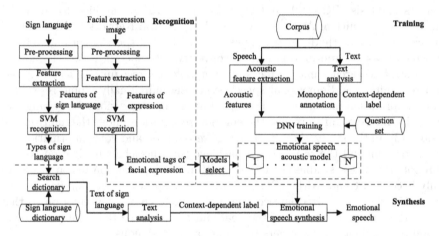

Fig. 1. The framework of sign language to emotional speech conversion for speech disorder.

3   Sign Language Recognition

Sign language recognition mainly has three steps including image pre-processing, feature extraction and SVM recognition. The image pre-processing step transforms the static sign language image into a grayscale image and converts the image format from 28×28 to 784×1 to forms a two-dimensional matrix for all images to construct a data cube. The x-axis of the data cube represents the number of different samples in a group, the y-axis of the data cube represents the dimension of a specific sample in a group, and the z-axis of the data cube represents the number of groups. The data cube is unified as a deep belief network (DBN) model readable data format. Sign language features are extracted using a 5-layer DBN model. The feature extraction consists of restricted boltzmann machine (RBM) adjustment and feedback fine-tuning. RBM is used to adjust the weights between two adjacent layers [18]. RBM has a connection between the hidden layer and the visible layer. There is no connection inside the layer. The relationship between the hidden layer and the visible layer can be expressed as an energy function as shown in Eq. 1,

$$E\left(v, h; \theta\right) = -\sum_{i=1}^{V}\sum_{j=1}^{H} w_{ij}v_i h_j - \sum_{i=1}^{V} b_i v_i - \sum_{j=1}^{H} a_j h_j \tag{1}$$

where v_i and h_i represent the state of the visible layer node and the hidden layer node respectively. The value is generally 0 or 1, $\theta = \{W, a, b\}$ is the model parameter, b_i and a_j represent the corresponding offsets, and w_{ij} represents the connection weight. The joint distribution of visible layer and hidden layer is shown in Eq. 2,

$$p\left(v, h\right) = e^{-E(v,h)} / \sum_{v,h} e^{-E(v,h)} \tag{2}$$

The conditional probabilities between the visible layer and the hidden layer are shown in Eqs. 3 and 4,

$$p\left(h_j = 1|v\right) = sigm\left(\sum_{i=1}^{V} w_{ij}v_i + a_j\right) \tag{3}$$

$$p\left(v_j = 1|h\right) = sigm\left(\sum_{j=1}^{H} w_{ij}h_j + b_i\right) \tag{4}$$

In the adjustment step, the weights of each layer are obtained with layer-by-layer training. The repeated three conversions between the visible layer and the hidden layer are performed to obtain the corresponding reconstruction targets respectively. The RBM parameters are adjusted by reducing the difference between the original object and the reconstructed object. The fine-tuning is to construct the deep confidence network by connecting all initialized RBMs in sequence in training order. The features of sign language can be obtained through feedback fine-tuning of the depth model. In the sign language recognition step shown in Fig. 2, the SVM is used to classify the sign language image with the extracted features to obtain the categories of sign language.

4   Facial Expression Recognition

Facial expression recognition step includes pre-processing, feature extraction and SVM recognition as shown in Fig. 3. In the pre-processing stage, we process some unimportant background information in the original image that may affect the result of feature extraction. First of all, the original input image is detected by a detector with 68 facial landmark points and is adjusted to the edge of the landmark. Then the image is trimmed with the complete facial expression information. Some non-specific information is deleted after the image is cut off

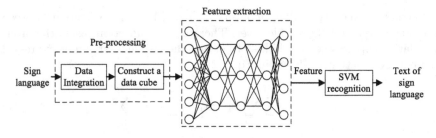

Fig. 2. Sign language recognition.

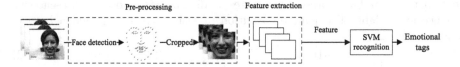

Fig. 3. Facial expression recognition.

to obtain a 96 × 96 image as input of the neural network model. In the feature extraction stage, we use a DNN model for feature extraction to extract 128-dimensional features from each of the imported emoticons. After feature extraction, a trained SVM classifier is used to classify the facial expression with the extracted features to get emotion tags corresponding to facial expressions.

5 Training Acoustic Models of Emotional Speech

The paper trained a set of DNN-based emotional acoustic models with the multi-speaker emotional speech corpus as shown in Fig. 4. The framework consists of three parts including data preparation, average voice model (AVM) training and speaker adaptation.

5.1 Data Preparation

In the data preparation phase, we use the WORLD vocoder [19] to extract the acoustic features from the multi-speaker emotional speech corpus. The acoustic features include the fundamental frequency (F0), the generalized Mel-generalized Cepstral (MGC), and the Band a periodical (BAP).

The paper adopt the initials and the finals as the unit of speech synthesis. A text analyzer is employed to obtain the initial, final, prosodic struct, word, and sentence information, which are used to form the context-dependent labels, through text normalization, grammar analysis, prosodic analysis, and phonological conversion by dictionary and grammar rule. The context-dependent label provides context information of the speech synthesis units including the unit layer, syllable layer, word layer, prosodic word layer, phrase layer, and sentence layer.

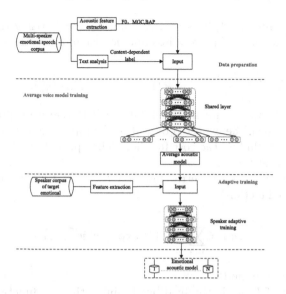

Fig. 4. Acoustic model training of emotional speech.

5.2 Training Average Voice Model

During the training AVM, the paper trained a set of DNN models as the emotional AVM by using the linguistic features (binary and digital) as input and acoustic features as output. The linguistic features were obtained from context-dependent label of the text corpus. The acoustic features were extracted from speech corpus including MGC, BAP, F0 and voice/unvoiced (V/UV). During training, the DNN models share various hidden layers between different emotional speakers to model its language parameters. Duration models and acoustic models were trained by a stochastic gradient descent (SGD) [20] of back propagation (BP) algorithm. Finally, a set of speaker-independent AVM were trained by the multi-speaker corpus.

During the training of the AVM model, the DNN structure uses a non-linear function to model the non-linear relationship between the linguistic features and acoustic features. Each hidden layer k uses the output $\mathrm{h}^k - 1$ of the previous layer to calculate the output vector h^k. $x = h^0$ is the input of the model. h^k can be defined as Eq. 5,

$$\mathrm{h}^k = \tanh(b^k + w^k h^{k-1}) \tag{5}$$

where, the b^k is offset vector and w^k is weight matrix. The mapping function uses $tanh$ in an node manner and can be replaced by other saturated non-linear functions such as Sigmod-shaped functions. The top level h^l and the supervised target output y are combined into a generally convex loss function $L(h^l, y)$.

The output layer uses the linear regression function as Eq. 6,

$$\mathrm{h}^l_i = \frac{e^{b^l_i} + w^l_i h^{l-1}}{\Sigma_j e^{b^l_j + w^l_j h^{l-1}}} \tag{6}$$

where w_i^l is the i line of w^l, h_i^l is positive, and $\Sigma_i h_i^l = 1$. Using the conditional log-likelihood function in Eq. 7 as a loss function to minimize the expectation on the (x, y) pair.

$$L(h^l, y) = -\log P\left(Y = y|x\right) = -\log\left(h_y^l\right) \tag{7}$$

The paper adopts a gradient training-based BP algorithm [21] to measure the difference between the expected output of the input x and y by back propagation derivative including the cost function. The optimal weights are searched to minimize the cost function.

Defining the cross-entropy error function $E\left(w\right)$ as Eq. 8 by using softmax [22] as an output function,

$$E\left(w\right) = \sum_{i=1}^{N} \log\left(1 + \exp\left(-y^n w^T x^n\right)\right) \tag{8}$$

where N is number of node.

The gradient of $E\left(w\right)$ is obtained by obtaining the cost function $E\left(w\right)$ relative to each weight $E\left(w\right)$ from w, which is defined in Eq. 9,

$$g_k = \frac{\partial}{\partial\left(w_k\right)} E\left(w\right) \tag{9}$$

According to [23], gradient descent can be written as Eq. 10,

$$w_{k+1} = w_k - \eta_k g_k \tag{10}$$

where η_k is the step size or learning rate.

Finally, the optimal parameters of the minimization cost function is determined by a corresponding stochastic gradient descent algorithm according to the Eqs. 8, 9 and 10.

The paper used a DNN structure including an input layer, a output layer and six hidden layer to train the AVM. The $tanh$ is used in the hidden layer and the linear activation function is used in the output layer. All speakers' training corpus share the hidden layer, so the hidden layer is a global linguistic feature mapping shared by all speakers. Each speaker has its own regression layer to model its own specific acoustic space. After multiple batches of SGD training, a set of optimal multi-speaker AVM model (average duration models and average acoustic feature models) is obtained.

5.3   Speaker Adaptation

In the speaker adaptation stage, a small corpus of the targeted emotions of the target speaker is used to extract acoustic features in the same way as AVM training, including F0, MGC, BAP and V/UV. Firstly, the speaker adaptation is performed by multi-speaker AVM model with the DNN models of the target emotional speaker to obtain a set of speaker-dependent adaptation models including duration models and acoustic models. The speaker-dependent model has the same DNN structure as the AVM, using six hidden layer structures, and the mapping function is the same as Eq. 5. At the same time, The maximum likelihood parameter generation (MLPG) algorithm [24] is used to generate the speech parameters from emotional acoustic models.

6 Sign Language to Emotional Speech Conversion

In order to obtain the text of sign language, we have designed a sign language dictionary based on the meaning of the sign language types defined in "Chinese Sign Language" [25], which gives the semantic text corresponding to each sign language. In the process of conversion from sign language to emotional speech, a sign language category is obtained by sign language recognition. Then the sign language dictionary is searched to obtain the text of recognized sign language. Finally, text analysis is performed on the text of sign language to obtain the context-dependent label that is used to generate linguistic features as the input of the models. A set of speech assessment methods phonetic alphabet (SAMPA) is designed for labeling initials and finals. A six level context-dependent label format is designed by taking into account the contextual features of unit, syllable, word [26], prosodic word [27], phrase and sentence. At the same time, the emotion tags are obtained by facial expression recognition to select the emotional acoustic model for synthesizing emotional speech. Sign language to emotional speech conversion process is shown in Fig. 5.

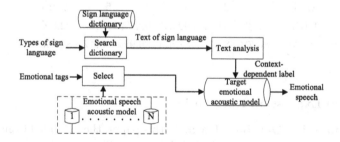

Fig. 5. Framework of sign language to emotional speech conversion.

7 Experimental Results

7.1 Sign Language Recognition

Sign Language Data. The sign language sample set constructed in the experiment mainly comes from the samples generated by two subjects. Each subjects plays 30 kinds of sign languages. The number of samples for each sign language is 1000. We train 30 deep learning models. The pre-defined 30 static sign languages are shown in Fig. 6.

Fig. 6. Examples of 30 kinds of static sign language.

Sign Language Recognition Rate. In order to verify the validity of the DBN model in sign language recognition, we randomly selected 4000 samples from the 30 kinds of sign language databases shown in Fig. 6 and used the DBN model and the PCA method to perform 5 cross-validation experiments. Each experiment has 3200 training samples and 800 testing samples respectively. Five experiments were numbered from 1 to 5. Finally, the recognition rates of SVM are obtained as shown in Table 1. From Table 1, it can be seen that in five cross-validation, the sign language recognition rate using the DBN model is better than that of the PCA method. This indicates that the features extracted through the DBN model can better reflect the essential features of sign language.

Table 1. Recognition rate of five cross validation.

Method	1	2	3	4	5	Average
DBN	91.3%	94.7%	92.5%	92.2%	93.1%	92.8%
PCA	86.7%	89.1%	87.5%	86.9%	88.4%	87.7%

7.2   Facial Expression Recognition

Facial Expression Database Data. In this paper, the extended Cohn-Kanade database (CK+) [28] and the JAFFE database [29] are used to train and test facial expression recognition. Each sequence image in the CK+ database begins with a neutral expression and ends with the emotional peak. The experimental database contains eight emotion categories. Contempt and neutral expression images are not used in the experiment. Images with obvious facial feature information are selected as a sample set. Six of the seven emoticons in the JAFFE database were tested without the use of neutral emoticons, each of which had an expression size of 256 × 256. Some examples of facial images are shown in Fig. 7.

Facial Expression Recognition Rate. We carried out 5 cross-validation experiments on the CK+ database to obtain the corresponding recognition rates for the 6 facial expressions. We conducted three cross-validation experiments

Fig. 7. Examples of facial expression database.

Table 2. Facial expression recognition rate under different database

Emotion	CK+	JAFFE (Original)	JAFFE (Transform)
Angry	96.7%	86.6%	96.6%
Disgust	93.1%	82.7%	93.0%
Fear	97.6%	75.0%	96.8%
Happy	98.3%	80.6%	87.1%
Sad	92.9%	83.8%	96.8%
Surprise	89.0%	73.3%	98.3%
Average	94.6%	80.3%	94.8%

on the JAFFE database and obtained the corresponding recognition rates for the six facial expressions. The facial expression recognition results are shown in Table 2.

It can be seen from Table 2 that the recognition rate on JAFFE in the original database is lower than that on CK+ database, mainly because the number of facial expression images in JAFFE original database is less than the number of facial expression images in CK+ database. In view of the above problems, the number of experimental images is increased by inverting the JAFFE database images in the experiment. The corresponding recognition rate of cross-validation experiments on six facial expressions after increasing 3 times of the database are shown in Table 2.

Image Visualization. In this paper, nn4.small2 neural network model [30] is used to extract the facial image features. Figure 8 shows the output of a cropped image after the first layer of the model. This figure shows the first 64 convolutions of all the filters of the input image. The network model definition is shown in Table 3.

7.3 Emotional Speech Synthesis

Speech Data. In our work, the emotional speech corpus contains 6 kinds of emotional speech recorded from 9 female speakers. Each speaker records 100 sentences for each emotion. The sample rate of the speech file is 16 kHz.

Fig. 8. An example of Convolution layer visualization.

Table 3. Definition of deep neural network

Type	Output size	#1 × 1	#3 × 3 reduce	#3 × 3	#5 × 5 reduce	#5 × 5	Pool proj
conv1 (7 × 7 × 3,2)	48 × 48 × 64						
max pool+norm	24 × 24 × 64						m 3 × 3,2
inception(2)	24 × 24 × 192		64	192			
norm+max pool	12 × 12 × 192						m 3 × 3,2
inception (3a)	12 × 12 × 256	64	96	128	16	32	m,32p
inception (3b)	12 × 12 × 320	64	96	128	32	64	ℓ_2,64p
inception (3c)	6 × 6 × 640		128	256,2	32	64,2	m 3 × 3,2
inception (4a)	6 × 6 × 640	256	96	192	32	64	ℓ_2,128p
inception (4e)	3 × 3 × 1024		160	256,2	64	128,2	m 3 × 3,2
inception (5a)	3 × 3 × 736	256	96	384	·		ℓ_2,96p
inception (5b)	3 × 3 × 736	256	96	384			m,96p
avg pool	736						
linear	128						
ℓ_2 normalization	128						

60-dimensional generalized Mel-Frequency Cepstral coefficients, 5-band non-periodic components and log fundamental frequency are extracted to compose a feature vectors. We use the WORLD vocoder to generate speech in steps of 5 ms.

Emotion Similarity Evaluation. We use the emotional DMOS (EMOS) test to evaluate the emotional expression of synthesized emotional speech. Ten subjects are played 100 original emotion speech files as a reference, and then 6 kinds of synthesized emotional speech files in sequence according to the emotion order. In the evaluation scoring process, it is conducted in accordance with the order in which the voices are played, and the assessors are asked to refer to the emotional expression experience in real life and score the emotional similarity scores for each synthesized sentence according to the 5-point system. The result is shown in Fig. 9.

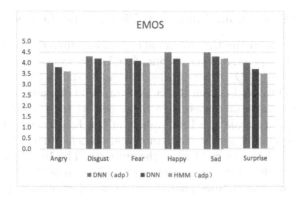

Fig. 9. The EMOS scores of synthesized emotional speech.

As can be seen from Fig. 9, the EMOS score of the emotional speech synthesized by the multi-emotional speaker adaptation method of DNN is higher than the EMOS score of the emotional speech synthesized by the DNN method and the HMM adaptation method. It also shows that the emotional speech synthesized by this method has better preference and natural emotion.

Objective Evaluation. This paper calculates the Root Mean Square Error (RMSE) between the original speech and the synthesized speech in terms of duration and fundamental frequency. The results are shown in Table 4. From the Table 4, we can see that the root mean square error of the DNN adaptive method is smaller than that of the DNN method and the HMM adaptive method. This means that the emotional speech synthesized by this method is closer to the original emotional speech. Therefore the synthesized emotional speech has better speech quality.

Table 4. The RMSE of duration and fundamental frequency between the synthesized emotional speech and original emotional speech.

Emotional type	F0/Hz			Dur/s		
	HMM (adp)	DNN	DNN (adp)	HMM (adp)	DNN	DNN (adp)
Angry	46.5	32.7	20.4	0.116	0.109	0.089
Disgust	41.7	34.5	28.7	0.141	0.135	0.092
Fear	48.5	38.7	22.7	0.131	0.128	0.086
Happy	52.1	45.1	21.8	0.104	0.103	0.102
Sad	38.5	35.4	22.1	0.230	0.228	0.202
Surprise	46.3	41.2	24.2	0.169	0.153	0.132

7.4   PAD Evaluation for Emotional Picture and Emotional Speech

In order to further evaluate the emotional expression similarity between the synthesized emotional speech and the original facial expression, we use the pleasure-arousal-dominance (PAD) three-dimensional emotional model to compare the difference of PAD values between the facial expression of pictures and emotional expression of the synthesized speech. In the paper, we use the abbreviated PAD emotion scale [31] to scale the PAD values of the facial expression of images and their corresponding synthesized emotional speech. First, all facial expression images were played at random to the subjects. The 10 subjects completed the PAD mood scale according to the emotional state of the images they felt when they observed the picture. Then the synthesized emotional speech is played randomly, which also requires the subjects to complete the PAD mood scale according to the emotional state that they feel when they listen to the emotional speech. Finally, the Euclidean distance of the PAD values between the facial expression picture and the emotional speech in the same emotional state is calculated. The results of the evaluation are shown in Table 5. From Table 5, we can see that the Euclidean distance of PAD value of facial expression and emotional speech is smaller in the same emotional state, indicating that the synthesized emotional speech can accurately reproduce the emotional state of facial expression.

Table 5. PAD evaluation for facial expression and synthesized emotional speech.

Emotional type	Picture results ([−1,1])			Speech results ([−1,1])			Euclidean distance
	P	A	D	P	A	D	
Angry	−0.84	0.64	0.82	−0.83	0.62	0.83	0.02
Disgust	−0.50	0.26	0.39	−0.48	0.21	0.39	0.05
Fear	−0.39	0.62	−0.63	−0.37	0.65	−0.65	0.04
Happy	0.67	0.78	0.37	0.68	0.81	0.45	0.09
Sad	−0.24	−0.40	−0.75	−0.26	−0.38	−0.73	0.03
Surprise	0.23	0.60	0.03	0.25	0.65	0.03	0.05

8   Conclusions

In this paper, we propose a method of sign language to emotion speech conversion that integrate the facial expression recognition. Firstly, the recognized sign languages and facial expression information are converted to the context-dependent label of the sign language and the corresponding emotional tags. At the same time, a DNN-based emotional speech synthesis is trained through an emotional corpus. Finally, emotional speech synthesis is done according to the emotional tags and the context-dependent labels of sign language text, so as to achieve the conversion from sign language and facial expression to emotional speech. The experimental results show that the average EMOS score of converted emotional speech is 4.2. At the same time, we use the PAD three-dimensional model to assess the expression of facial expression and synthesized emotional speech. The results show that the Euclidean distance of the PAD values are small between the facial expression of pictures and emotional expression of synthesized speech that demonstrate the converted emotional speech can express the emotional state of facial expression. Further work will use deep learning to optimize sign language recognition, facial expression recognition and emotional speech synthesis to improve the recognition rate and the quality of synthesized emotional speech.

References

1. Kalsh, E.A., Garewal, N.S.: Sign language recognition system. Int. J. Comput. Eng. Res. **3**(6), 15–21 (2013)
2. Assaleh, K., Shanableh, T., Zourob, M.: Low complexity classification system for glove-based arabic sign language recognition. In: Huang, T., Zeng, Z., Li, C., Leung, C.S. (eds.) ICONIP 2012. LNCS, vol. 7665, pp. 262–268. Springer, Heidelberg (2012). https://doi.org/10.1007/978-3-642-34487-9_32
3. Godoy, V., et al.: An HMM-based gesture recognition method trained on few samples. In: Proceedings of the 26th International Conference on Tools with Artificial Intelligence (ICTAI), pp. 640–646. IEEE (2014)
4. Yang, Z.Q., Sun, G.: Gesture recognition based on quantum-behaved particle swarm optimization of back propagation neural network. Comput. Appl. **34**, 137–140 (2014)
5. Ghosh, D.K., Ari, S.: Static hand gesture recognition using mixture of features and SVM Classifier. In: Proceedings of the Fifth International Conference on Communication Systems and Network Technologies (CSNT), pp. 1094–1099. IEEE (2015)
6. Oyedotun, O.K., Khashman, A.: Deep learning in vision-based static hand gesture recognition. Neural Comput. Appl. **28**(12), 3941–3951 (2017)
7. Hsieh, C.C., Hsih, M.H., Jiang, M.K.: Effective semantic features for facials recognition using SVM. Multimed. Tools Appl. **75**(11), 6663–6682 (2016)
8. Prabhakar, S., Sharma, J., Gupta, S.: Facial expression recognition in video using Adaboost and SVM. Pol. J. Nat. Sci. **3613**(1), 672–675 (2014)
9. Abdulrahman, M., et al.: Gabor wavelet transform based facial expression recognition using PCA and LBP. In: Signal Processing and Communications Applications Conference (SIU), pp. 2265–2268. IEEE (2014)
10. Zhao, X., Shi, X., Zhang, S.: Facial expression recognition via deep learning. IETE Tech. Rev. **32**(5), 347–355 (2015)
11. Hojo, N., Ijima, Y., Mizuno, H.: DNN-Based speech synthesis using speaker codes. IEICE Trans. Inf. Syst. **101**(2), 462–472 (2018)
12. Potard, B., Motlicek, P., Imseng, D.: Preliminary work on speaker adaptation for DNN-Based speech synthesis. Idiap (2015)
13. Caridakis, G., et al.: Multimodal emotion recognition from expressive faces, body gestures and speech. In: Boukis, C., Pnevmatikakis, A., Polymenakos, L. (eds.) AIAI 2007. ITIFIP, vol. 247, pp. 375–388. Springer, Boston, MA (2007). https://doi.org/10.1007/978-0-387-74161-1_41

14. Burger, B., Ferran, I., Lerasle, F.: Two-handed gesture recognition and fusion with speech to command a robot. Auton. Robots **32**(2), 129–147 (2012)
15. Sinyukov, D.A., et al.: Augmenting a voice and facial expression control of a robotic wheelchair with assistive navigation. In: Proceedings of the International Conference on Systems, Man and Cybernetics (SMC), pp. 1088–1094. IEEE (2014)
16. Yang, H., et al.: Towards realizing gesture-to-speech conversion with a HMM-based bilingual speech synthesis system. In: Proceedings of the International Conference on Orange Technologies (ICOT), pp. 97–100. IEEE (2014)
17. An, X., Yang, H., Gan, Z.: Towards realizing sign language-to-speech conversion by combining deep learning and statistical parametric speech synthesis. In: Che, W., et al. (eds.) ICYCSEE 2016. CCIS, vol. 623, pp. 678–690. Springer, Singapore (2016). https://doi.org/10.1007/978-981-10-2053-7_61
18. Feng, F., Li, R., Wang, X.: Deep correspondence restricted Boltzmann machine for cross-modal retrieval. Neurocomputing **154**, 50–60 (2015)
19. Morise, M., Yokomori, F., Ozawa, K.: WORLD: a vocoder-based high-quality speech synthesis system for real-time applications. IEICE Trans. Inf. Syst. **99**(7), 1877–1884 (2016)
20. Deng, L., Yu, D.: Deep learning: methods and applications. foundations and trends®in. Signal Process. **7**(3), 197–387 (2014)
21. Larochelle, H., Bengio, Y., Louradour, J.: Exploring strategies for training deep neural networks. J. Mach. Learn. Res. **1**(10), 1–40 (2009)
22. Wu, Z., et al.: A study of speaker adaptation for DNN-based speech synthesis. In: Sixteenth Annual Conference of the International Speech Communication Association, pp. 879–883. Interspeech (2015)
23. Fan, Y., et al.: Multi-speaker modeling and speaker adaptation for DNN-based TTS synthesis. In: International Conference on Acoustics, Speech and Signal Processing (ICASSP), pp. 4475–4479. IEEE (2015)
24. Hwang, H.T., et al.: A probabilistic interpretation for artificial neural network-based voice conversion. In: Signal and Information Processing Association Annual Summit and Conference (APSIPA), pp. 552–558. IEEE (2015)
25. China Association of the Deaf and Hard of Hearing: Chinese Sign Language. Huaxia Publishing House, Beijing (2003)
26. Yang, H., Oura, K., Wang, H.: Using speaker adaptive training to realize Mandarin-Tibetan cross-lingual speech synthesis. Multimed. Tools Appl. **74**(22), 9927–9942 (2015)
27. Yang, H., Zhu, L.: Predicting Chinese prosodic boundary based on syntactic features. J. Northwest Norm. Univ. (Nat. Sci. Ed.) **49**(1), 41–45 (2013)
28. Lucey, P., et al.: The extended Cohn-Kanade Dataset (CK+): a complete dataset for action unit and emotion-specified expression. In: Computer Vision and Pattern Recognition Workshops, pp. 94–101. IEEE (2010)
29. Lyons, M., et al.: Coding facial expressions with Gabor wavelets. In: Proceedings of the Third IEEE International Conference on Automatic Face and Gesture Recognition, pp. 200–205. IEEE (1998)
30. Amos, B., Ludwiczuk, B., Satyanarayanan, M.: Openface: a general-purpose face recognition library with mobile applications. CMU School of Computer Science (2016)
31. Xiaoming, L., Xiaolan, F., Guofeng, D.: Preliminary application of the abbreviated PAD emotion scale to Chinese undergraduates. Chin. Ment. Health J. **22**(5), 327–329 (2008)

Noise-Immune Localization for Mobile Targets in Tunnels via Low-Rank Matrix Decomposition

Hong Ji[1], Pengfei Xu[2(✉)], Jian Ling[1], Hu Xie[1], Junfeng Ding[1], and Qiejun Dai[1]

[1] Jiangsu Power Transmission and Transformation Company Limited, Nanjing 211102, People's Republic of China
[2] School of Computer, Nanjing University of Posts and Telecommunications, Nanjing 210003, People's Republic of China
mail.xupengfei@gmail.com

Abstract. Accurate and robust mobile targets localization is one of the important prerequisites for tunnel safety construction in complex environments. Currently, the mainstream localization methods for tunnel mobile targets are almost range-based, which usually suffer from low accuracy and instability due to the complex natural conditions and geographical environment. The reason for the low accuracy and instability is the existing methods mostly assume that the sampled partial Euclidean distance matrices are corrupted by Gaussian noise or/and outlier noise. However, in real applications, the noise is more likely to be unpredictable compound noise. Therefore, in this paper, we propose a noise-immune LoCalization algorithm via low-rank Matrix Decomposition (LoCMD) to address this challenge. Specifically, we adopt the Mixture of Gaussians to model the unpredictable compound noise and employ the popular Expectation Maximization technique to solve the constructed low-rank matrix decomposition model, and thus a complete and denoised Euclidean distance matrix can be obtained. Finally, the extensive experimental results show that the proposed LoCMD achieves better positioning performance than the existing algorithms in the complex environment.

Keywords: Noise-immune localization · Intelligent data processing
Matrix decomposition · Mixture of Gaussian · Tunnel

1 Introduction

In recent years, Wireless Sensor Networks (WSNs) have achieved rapid development and been widely used in various application domains, such as environmental monitoring, smart transportation [1]. Also, the large amount of real-time data generated between sensor nodes provides a powerful guarantee for real-time localization for mobile targets in tunnels, which will bring great convenience to the construction safety management of the tunnels and underground projects.

In a WSNs, due to the restrictions in hardware cost and power consumption, usually just a few nodes know their real coordinates and are called anchor nodes. All we need

© Springer Nature Singapore Pte Ltd. 2018
Q. Zhou et al. (Eds.): ICPCSEE 2018, CCIS 902, pp. 431–441, 2018.
https://doi.org/10.1007/978-981-13-2206-8_35

to do is using localization algorithms to gain the positions of all unknown-location nodes with anchor nodes real coordinates and inter-node measurements. In general, existing localization algorithms are classified into two categories: range-based and range-free [4]. Range-based algorithms based on Euclidean distance collected by using ranging methods, such as Received Signal Strength Indicator (RSSI) [5] and Time Difference of Arrival (TDOA) [6], can achieve more accurate positions, but the calculation and communication costs are relatively large, and they require certain hardware support. Range-free algorithms do not require additional equipment and consume less energy, but they can only achieve coarse-grained positioning. They are normally suitable for applications that do not require high accuracy. In the above two types of algorithms, we focus on the former, where typically few anchor nodes know their real locations. With inter-node Euclidean distances obtained from the collected data and real locations of anchor nodes, it is feasible to gauge the real locations of all unknown-location nodes.

Many previous range-based algorithms are based on certain assumptions that the data of Euclidean distance measurements between inter-nodes can be obtained efficiently and accurately. However, in practice, due to limited communication distance, energy capacity, environment factors or equipment failure, Euclidean distance measurements are likely to be noisy or even missing and this is particularly serious in tunnels. Moreover, to achieve energy savings and extend the working hours of the entire networks, it is possible to collect just part of Euclidean distances selected reasonably. Thus, the purpose of our localization algorithm is to calculate the real locations of all mobile targets with a Euclidean Distance Matrix (EDM), a part of entries is missing or noisy. To smooth unknown noise, Yang and Jian suggested that there are two kinds of noises in real localization systems, normal noise and outliers [7]. Xiao and Chen smoothed the Gaussian noises and outliers by using Frobenius-norm and L_1 norm regularization [3]. However, it is not the case in most real problems because in tunnels there are different types of noises, not just Gaussian noise and outliers. Existing algorithms based on matrix completion often just consider Gaussian noise and outliers and ignore other types of noises.

To address this challenge, this paper proposes LoCMD, a robust localization approach with unknown noise in Tunnel Wireless Sensor Networks (T-WSNs). We model the reconstruction of matrix in which part of elements are missing or noisy as a Low-rank Matrix Decomposition (LRMD) problem using the low rank characteristic of Euclidean Distance Matrix. The key idea is to smooth the unknown noise as a Mixture of Gaussians (MoG) [8]. We use an efficient algorithm to solve the LRMD with a MoG noise model based on Expectation Maximization (EM) method [11]. Thus, with a complete and denoised EDM, we can use Multi-Dimension Scaling (MDS) method to localize positions of mobile targets [12]. Furthermore, to improve positioning accuracy, we also propose a method of selecting anchor nodes when localizing in 3-Dimensional space.

The rest of this paper proceeds as follows. In Sect. 2, a robust localization method is introduced, including problem description, elaboration of matrix completion model under noise, model optimization and the localization method. Then, we conduct a series of simulation experiments in the Sect. 3 to test the performance of our algorithm. Finally, we make some summary in Sect. 4.

2 Noise-Immune Localization for Mobile Targets in Tunnels

2.1 Problem Description

In a tunnel, we suppose that some sensor nodes are deployed. Among them, a few nodes called anchor nodes are base stations which are arranged on both sides of the tunnel, and the remaining nodes are ordinary nodes which represent the mobile targets whose positions are unknown. The purpose of our localization algorithm is to obtain the real positions of the ordinary nodes. Usually, the distance measurements used by range-based algorithms between inter-nodes can be gained by two common technologies, such as RSSI and TDOA. However, limited by energy capacity and environmental impact in the tunnel, distance measurements are most likely to be inaccurate or missing.

Assuming n nodes distributed in 3-Dimensional space of a tunnel, the coordinates of these nodes can be expressed as $\{x_1, x_2, \ldots, x_n\} \in R^3$. The Euclidean Distance Matrix could be defined as $D \in R^{n \times n}$, where each element d_{ij} of matrix D is

$$d_{ij} = \left\| x_i - x_j \right\|^2, i, j \in \{1, 2, \ldots, n\}. \tag{1}$$

As mentioned above, the matrix D cannot be directly obtained but only incomplete matrix M. In addition, the existing elements of the matrix M may be inaccurate due to noise. Obviously, when we obtain a sampled matrix, the matrix can be decomposed into the final Euclidean Distance Matrix and the noise matrix. Thus, the relationship between matrix D and matrix M can be given by:

$$P_\Omega(M) = P_\Omega(D + N) \tag{2}$$

where N is an unknown noise matrix, and $P_\Omega(\cdot)$ is the orthogonal projection operator defined as:

$$[P_\Omega(M)]_{ij} = \begin{cases} m_{ij}, & if\,(i,j) \in \Omega \\ 0, & otherwise. \end{cases} \tag{3}$$

Based on the MDS-based algorithm [12], the relative coordinates of all unknown nodes can be exactly and simply estimated with at least 4 anchor nodes in 3-Dimensional space. Furthermore, the premise is that each element in the Euclidean distance matrix is known.

However, regrettably, as mentioned above, our matrix is partially missing or noisy. As a result, it is necessary to reconstruct EDM before using MDS-based algorithm to estimate the real coordinates of unknown nodes in T-WSNs. In this paper, the localization algorithm contains the following steps:

(1) Euclidean distance elements are collected with a constructed T-WSNs to build an initial Euclidean Distance Matrix.
(2) Completed and denoised matrix can be obtained from noisy and missing EDM via LRMD with MoG.
(3) Based on MDS-based algorithm, coordinates of all mobile targets can be acquired.

2.2   LRMD with MoG

To obtain the completed and denoised matrix, basing on the low rank characteristic of EDM, we introduce a Low-Rank Matrix Decomposition theory, which often has been used to predict unknown element of a low-rank matrix [13, 14]. At the same time, we use Mixture of Gaussians to smooth the unknown noise in tunnels.

In this paper, from Eq. (2), we model the problem of reconstruction matrix D as

$$M = UV^T + N. \tag{4}$$

Further, Eq. (4) can be written as

$$m_{ij} = u_i v_j^T + n_{ij} \tag{5}$$

where $m_{ij}(i, j = 1, 2, \ldots, n)$ represents each element of sampled matrix M, u_i and v_i represent the ith row vectors of the matrix U and V, respectively, and n_{ij} is called unknown noise in matrix N. To deal with unknown noise in a tunnel, we model the unknown noise with Mixture of Gaussians. Therefore, in this paper, we assume that the unknown noise n_{ij} satisfies the Mixture of Gaussians distribution, then each n_{ij} in Eq. (5) can be reformulated as:

$$p(n_{ij}) = \sum_{k=1}^{K} \pi_k p(n_{ij}|0, \sigma_k^2) \tag{6}$$

where $\pi_k \geq 0$ represents the mixing coefficient and $\sum_{k=1}^{K} \pi_k = 1$, $p(n_{ij}|0, \sigma_k^2)$ denotes the probability of the kth Gaussian distribution with mean 0 and variance σ^2 and K represents the number of Gaussian components. Based on Eq. (6), the matrix decomposition model of noise immunity can be converted into the following equivalent maximum likelihood function optimization model:

$$\max_{U,V,\Pi,\Sigma} L(U, V, \Pi, \Sigma) = \sum_{i,j \in \Omega} \log \sum_{k=1}^{K} \pi_k p(n_{ij}|u_i v_j^T, \sigma_k^2) \tag{7}$$

where $\Pi = \{\pi_1, \pi_2, \ldots, \pi_K\}$, $\Sigma = \{\sigma_1, \sigma_2, \ldots, \sigma_K\}$, $L(U, V, \Pi, \Sigma)$ denotes the log-likelihood function under the parameters U, V, Π, Σ.

2.3   Optimizing the Constructed Model Using EM Method

To estimate the parameters (U, V, Π, Σ) of the log-likelihood function in Eq. (7), EM method which was proposed by DEMPSTER, LAIRD and RUBIN can be used. In our case, the mean of each Gaussian component is made up of the variables U and V and all components share them. Set a set of implicit variables $z_{ijk} \in \{0, 1\}$ where $\sum_{k=1}^{K} z_{ijk} = 1$ and $k = 1, 2, \ldots, K$, indicating which Gaussian component the noise n_{ij} comes from. Through iterating between estimating the implicit variables by calculating their expected values (E step) and estimating the parameters U, V, Π, Σ of the log-likelihood (M step) [2].

E Step: The probability of mixture k for generating the noise of each element m_{ij} of matrix M can be calculated by:

$$E(z_{ijk}) = \gamma_{ijk} = \frac{\pi_k p(m_{ij}|u_i v_j^T, \sigma_k^2)}{\sum_{k=1}^K \pi_k p(m_{ij}|u_i v_j^T, \sigma_k^2)} \tag{8}$$

M Step: The parameters U, V, Π, Σ can be estimated by maximizing the following function as:

$$E_{ZP}(M, Z|U, V, \Pi, \Sigma) =$$
$$\sum_{i,j \in \Omega} \sum_{k=1}^K \gamma_{ijk} \left(\log \pi_k - \log \sqrt{2\pi}\sigma_k - \frac{\left(m_{ij} - u_i v_j^T\right)^2}{2\sigma_k^2} \right) \tag{9}$$

An easy way to solve this maximization problem is to alternatively update the parameters Π, Σ and the factorized matrices U, V as follows:

Update Π, Σ: For $k = 1, 2, \ldots, K$, Π and Σ can be given as follows:

$$N_k = \sum_{i,j} \gamma_{ijk}, \ \pi_k = \frac{N_k}{\sum N_k}, \ \sigma_k^2 = \frac{1}{N_k} \sum_{i,j} \gamma_{ijk} \left(m_{ij} - u_i v_j^T \right)^2. \tag{10}$$

Update U, V: The Eq. (9) related to U and V can be re-written as follows:

$$\sum_{i,j \in \Omega} \sum_{k=1}^K \gamma_{ijk} \left(-\frac{\left(m_{ij} - u_i v_j^T\right)^2}{2\pi\sigma_k^2} \right)$$
$$= -\sum_{i,j \in \Omega} \left(\sum_{k=1}^K \frac{\gamma_{ijk}}{2\pi\sigma_k^2} \right) \left(m_{ij} - u_i v_j^T \right)^2 = -\left\| W \odot (M - UV^T) \right\|_{L_2} \tag{11}$$

where

$$W \in R^{n \times n}, w_{ij} = \begin{cases} \sqrt{\sum_{k=1}^K \dfrac{\gamma_{ijk}}{2\pi\sigma_k^2}}, i, j \in \Omega \\ 0 \quad , i, j \notin \Omega. \end{cases} \tag{12}$$

Interestingly, this formula in last of Eq. (11) is equivalent to L_2 LRMD problem with a weight W which can be simply solved by many existing algorithms. The process is summarized in Algorithm 1.

Algorithm 1. Euclidean Distance Matrix completion based on EM method

Input: sampled Euclidean Distance Matrix M which is noisy and incomplete, index set Ω

Output: the completed matrix D

1. Initialize U, V, Π, Σ, K and threshold θ
2. Repeat
3.   **E Step:** compute γ_{ijk} by Eq. (8).
4.   **M Step:**
5.   **Update Π, Σ:** compute Π, Σ by Eq. (10).
6.   **Update U, V:** compute U, V by Eq. (11) and Eq. (12).
7.   **(Tune K):** Let n_i and n_j denote the ith and j^{th} Gaussian components, if $|\sigma_i{}^2 - \sigma_j{}^2|/(\sigma_i{}^2 + \sigma_j{}^2) < \theta$, then remove the jth Gaussian component from Π, Σ and let $\pi_i = \pi_i + \pi_j$, $\sigma_i{}^2 = (n_i\sigma_i{}^2 + n_j\sigma_j{}^2)/(n_i + n_j)$, $K = K-1$.
8. Until converge.
9. $D = UV^T$.

2.4   Localization Based on Completed Matrix for Mobile Targets

Based on the completed matrix which has been recovered from a small set of entries, MDS-based method can be used to determine the positions of mobile targets with precise distance measurements.

MDS is used to construct a suitable low-dimensional space that keeps the distance between the sample in this space and the similarity between samples in the high-dimensional space as consistent as possible. In localization, this similarity can be expressed in Euclidean distance. In this paper, the MDS-based method has two steps. First, the completed EDM will be used to obtained the relative coordinates of all nodes with MDS. Then, based on the real coordinates of anchor nodes, the relative coordinates can be converted into absolute ones. The key of this convert is to find a linear transformation including reflection, rotation and translation and fortunately the linear transformation is available. Assume that R_i and A_i $(i = 1, 2, .., n)$ represent respectively the relative coordinates and absolute coordinates of all nodes in a tunnel and nodes $1, 2, .., k$ are anchor nodes where $4 \leq k \leq n$. This approach of localization is specifically shown by the following Algorithm 2.

Algorithm 2 noise-immune Localization via Low-rank Matrix Decomposition

---

Input: Sampled Euclidean Distance Matrix M, index set Ω, the real coordinates of anchor nodes $R_i, i = 1,2, \dots, k$.

Output: Real coordinates of all mobile targets $A_i, i = k + 1, k + 2, \dots n$.

1.   Reconstruct the distance matrix D from Sampled Euclidean Distance Matrix M via algorithm 1

2.   W can be got by double center D, as:
$$W = -0.5 \times JDJ, \text{ where } J = I - 1 \cdot 1^T/n \text{ and } I \text{ is identity matrix.}$$

3.   Singular Value Decomposition on W:
$$[X, \Lambda, Y] = svd(W)$$

4.   Get the relative position coordinate matrix R of all nodes:
$$R = [R_1, R_2, \dots, R_n] = \sqrt{\Lambda_3} \cdot X_3^T$$
where $R_i \in R^{3 \times 1}$, $\Lambda_2 = \Lambda(1:3, 1:3)$, $X_2 = X(:, 1:3)$ in 3-Dimensinal space.

5.   The estimation error of anchor coordinates can be written as:
$$\alpha = \sum_{i=1}^{k} \|T_i - A_i\|^2.$$

6.   Repeat

7.   Select four nodes arbitrarily from k anchor nodes, let $e, f, g, h \in [1, k]$ be their index. The transform matrix Q can be obtained by:
$$Q = [A_f - A_e, A_g - A_e, A_h - A_e]/[R_f - R_e, R_g - R_e, R_h - R_e],$$

8.   Obtain the coordinates of mobile targets:
$$T_i = Q \cdot (R_i - R_e) + A_e.$$

9.   Compute α.

10. Until all anchor node combinations are selected.

11. Output real coordinates of all mobile targets computed based on four anchor nodes which make α value to be smallest.

---

3   Experimental Evaluation

To test the performance of the proposed algorithm, a network of 100 nodes is generated in $100 \, m \times 20 \, m \times 20 \, m$ area, in which all nodes are randomly distributed and 6 of them are anchor nodes. Assuming $X = [x_1, x_2, \dots, x_n] \in R^{3 \times 100}$ and $D \in R^{100 \times 100}$ represent the coordinate matrix of all nodes and the EDM respectively. The accuracy of the proposed algorithm is indicated by the reconstruction error and localization error, which are defined as follows:

(1) Reconstruction error:

$$e_r = \|D_r - D\|_F / \|D\|_F \tag{13}$$

where D_r represents the reconstructed EDM and $\|\cdot\|_F$ is the Frobenius norm.

(2) Localization error:

$$e_l = \|X_l - X\|_F / n \tag{14}$$

where X_l represents the matrix composed of coordinates obtained by the proposed algorithm.

To investigate the performance of the proposed algorithm in tunnels, a simulation is developed by calculating the LoCMD in three categories and each of them represents different measurement noise condition. Our algorithm is compared with the SVT-based localization algorithm in [9] and the SET-based localization algorithm in [10]. The three noise states are shown below:

(1) NN: No noise
(2) WG: With Gaussian noise
(3) WM: With mixture noises (Gaussian noise, outliers and random noise)

3.1 Localization under No Noise

In this condition, all measured distances between all nodes in the T-WSNs are assumed to be accurate. Figure 1 shows the reconstruction error and localization error of three algorithms including SVT-based algorithm, SET-based algorithm and LoCMD in different sampling rates.

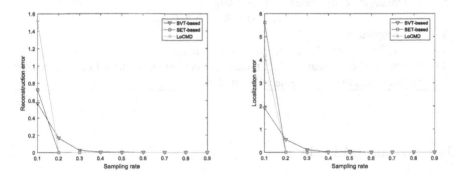

Fig. 1. Performances with no noise

It is obviously seen from the figure that with the increase of sampling rate, the reconstruction error and localization error rapidly decrease to close to zero. When the sampling rate reach 0.2, both LoCMD and SET-based algorithm reduce the two types of errors to a low level. At the same time, however, the reconstruction error of the SVT-based algorithm is about 0.18 due to the its inaccurate estimation of the rank of EDM. And Fig. 1 shows that reconstruction error and localization have the same trend.

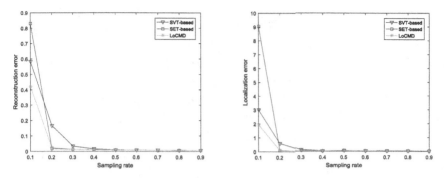

Fig. 2. Performances with Gaussian noise

3.2   Localization with Gaussian Noise

In the general real environment, normal noise usually follows the Gaussian distribution. So, to simulate the general environment, a Gaussian noise with 0 mean and variance 100 is added to the sample matrix. All the three algorithms perform well when the sampling rate is large. It means that all the three algorithms can adapt to general Gaussian noise environment and achieve accurate localization. But LoCMD and SET-based algorithm can obtain more accurate reconstruction of matrix and localization when the sampling rate is in a low level (Fig. 2).

3.3   Localization with Mixture Noises

In the above experiments, we discussed the performance of the three algorithms in unnoisy environment and a Gaussian noise environment and the performance of the three algorithms under normal condition is acceptable. However, in complex environments such as tunnels, unknown noise often does not satisfy strict Gaussian distribution. To better simulate the real environment, unknown noise is assumed to be a mixture of Gaussian noise, outliers, and random noise in this experiment. On the basis that Gaussian noise has been added, the entries of the matrix are corrupted with random noise between [0, 50] and 5% of the entries are added outliers with value of 5000.

As can be seen in Fig. 3, both SVT-based algorithm and SET-based algorithm cannot adapt to mixed noise, even the sampling rate has reached 0.9. In other words, these two algorithms suffered the reconstruction error and localization error in a real complex environment. And when facing the mixture noise, our algorithm, LoCMD has achieved good results if the sampling rate reached 0.3. In addition, with low reconstruction error, the localization error is also in a low level based on LoCMD. This means that LoCMD can have more accurate localization for mobile targets in tunnels.

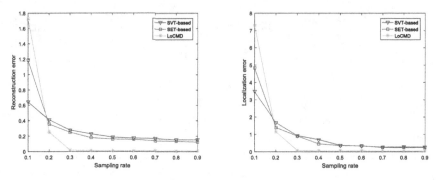

Fig. 3. Performances with mixture noises

4 Conclusion

In this paper, we propose a robust localization algorithm for mobile targets in tunnels via Low-rank Matrix Decomposition named LoCMD. In LoCMD, based on the low rank of Euclidean Distance Matrix, a Low-Rank Matrix Decomposition theory is introduced and the unknown noise is modeled by Mixture of Gaussians. Then, with the reconstructed matrix, an MDS-based method was used to locate the real coordinates of all mobile targets. Finally, simulation results show that the proposed algorithm can achieve more accurate localizations in complex noisy environment, comparing with traditional algorithms.

References

1. Patwari, N.: Locating the nodes: cooperative localization in wireless sensor networks. IEEE Signal Process. Mag. **22**(4), 54–69 (2005)
2. Meng, D., De La Torre, F.: Robust matrix factorization with unknown noise. In: IEEE International Conference on Computer Vision (ICCV). IEEE (2013)
3. Xiao, F.: Noise-tolerant localization from incomplete range measurements for wireless sensor networks. In: IEEE Conference on Computer Communications (INFOCOM). IEEE (2015)
4. Han, G.: Localization algorithms of wireless sensor networks: a survey. Telecommun. Syst. **52**(4), 2419–2436 (2013)
5. Girod, L.: Locating tiny sensors in time and space: a case study. In: IEEE International Conference on Computer Design: VLSI in Computers and Processors. IEEE (2002)
6. Girod, L., Estrin, D.: Robust range estimation using acoustic and multimodal sensing. In: IEEE/RSJ International Conference on Intelligent Robots and Systems, vol. 3. IEEE (2001)
7. Yang, Z.: Beyond triangle inequality: sifting noisy and outlier distance measurements for localization. ACM Trans. Sens. Netw. (TOSN) **9**(2), 26 (2013)
8. McLachlan, G.J., Basford, K.E.: Mixture models: Inference and applications to clustering, vol. 84. Marcel Dekker, New York City (1988)
9. Cai, J.F., Candès, E.J., Shen, Z.: A singular value thresholding algorithm for matrix completion. SIAM J. Optim. **20**(4), 1956–1982 (2010)

10. Dai, W., Milenkovic, O.: SET: an algorithm for consistent matrix completion. In: IEEE International Conference on Acoustics Speech and Signal Processing (ICASSP). IEEE (2010)

11. Dempster, A.P., Laird, N.M., Rubin, D.B.: "Maximum likelihood from incomplete data via the EM algorithm. J. Roy. Stat. Soc. Ser. B (Methodol.), 1–38 (1977)

12. Shang, Y.: Localization from mere connectivity. In: Proceedings of the 4th ACM International Symposium on Mobile Ad Hoc Networking & Computing. ACM (2003)

13. Wang, X.: Robust component-based localization in sparse networks. IEEE Trans. Parallel Distrib. Syst. **25**(5), 1317–1327 (2014)

14. Candes, E.J., Plan, Y.: Matrix completion with noise. Proc. IEEE **98**(6), 925–936 (2010)

Passenger Flow Forecast of Sanya Airport Based on ARIMA Model

Yuan-hui Li, Hai-yun Han, Xia Liu[✉], and Chao Li

Sanya Aviation and Tourism College, Sanya 572000, Hainan, China
576735855@qq.com, paolo_lx@qq.com

Abstract. By analyzing the passenger flow data of Sanya Airport collected from January 2008 to December 2016, the general trend and seasonal variation regularity of the passenger flow can be found. By constructing and testing the ARIMA forecast model, the results show that the ARIMA model has a good fitting effect on the passenger flow data, and its forecast error is small. Therefore, this model can be applied into the short-term forecast of airport passenger flow, and help to provide the corresponding decision-making basis for the airport operation management.

Keywords: Passenger flow · ARIMA model · Prediction

1 Introduction

Sanya Phoenix International Airport was officially launched on July 1st, 1994. It was initially designed to provide service to 1.5million passengers per year. However, the actual passengers quantity exceeded 5 million in 2007, and then achieved 10 million in 2011. In 2017, there are more than 38 million passengers, approaching the 20 million mark. Therefore, the airport has to serve far more passengers than its designed support capability, and the terminal building are in the state of overload operation. The contradiction between the rapid growth of passengers transport and the serious shortage of the development space has badly affected the image of the international tourism island and increased the safety risk of the airport operation.

The forecast of airport passenger flow is of great significance for the improvement of airport's future planning and management. Scientific and accurate forecast of passenger flow is the basis for determining the reasonable quantity of transportation facilities, ensuring the high efficiency of all the airport facilities, and providing the management with scientific decision-making basis.

2 Passenger Flow Data

This study selects the monthly passenger flow volume of Sanya Phoenix Airport from 2008 to 2017 as the research data, with 120 data samples in total, of which 108 sample-data collected from 2008 to 2016 are used as training sets to build the forecast model. 12 samples from the year of 2017 are used as test sets to test the performance of the forecast model. The specific data is shown in Table 1.

© Springer Nature Singapore Pte Ltd. 2018
Q. Zhou et al. (Eds.): ICPCSEE 2018, CCIS 902, pp. 442–454, 2018.
https://doi.org/10.1007/978-981-13-2206-8_36

Table 1. Monthly statistics of passenger flow of Sanya Airport from 2008 to 2017 (unit: 10,000 people)

Month	Year				
	2008	2009	2010	2011	2012
1	59.54	83.34	118.69	113.26	139.12
2	63.10	80.93	124.52	120.14	128.65
3	56.56	71.54	101.22	101.35	115.53
4	42.14	55.77	75.16	76.42	86.17
5	36.25	49.40	63.61	69.41	67.98
6	31.39	43.52	54.54	60.81	60.39
7	43.50	55.51	61.84	69.31	75.45
8	45.63	57.80	64.05	69.52	77.53
9	39.78	45.06	50.50	59.69	66.43
10	51.72	65.99	50.38	73.34	82.25
11	60.39	87.78	73.29	104.75	107.86
12	70.62	97.50	91.59	118.18	126.96
Month	Year				
	2013	2014	2015	2016	2017
1	138.45	162.96	170.05	177.00	194.46
2	151.21	165.34	173.31	179.19	187.38
3	132.48	144.81	168.27	159.79	183.13
4	90.44	103.59	121.18	130.04	150.76
5	81.37	97.03	112.10	120.02	142.50
6	73.56	87.30	101.28	112.47	135.84
7	86.92	104.98	115.62	125.87	142.88
8	93.28	113.99	123.96	137.79	153.11
9	82.24	97.96	106.35	125.10	135.77
10	96.85	117.73	126.95	138.22	152.59
11	120.47	141.84	139.78	153.48	171.24
12	139.41	156.70	160.35	177.14	189.33

Data source: http://tour.sanya.gov.cn/tongji_
yue.asp

3 Modelling

3.1 Choice of Forecasting Method

There are many ways to forecast passenger flow. According to incomplete statistics, there are about 300 methods in the world, of which 150 are comparatively mature, about 30 are commonly used, and more than 10 are widely used, but generally speaking, they are divided into two categories: one is linear and non-linear theory, the other is qualitative and quantitative forecast method. In general, frequently used forecast methods includes: time series model, grey prediction model, expert prediction model, exponential smoothing method, neural network, support vector machine, trend extrapolation method, regression analysis method, and so on [1].

Domestic scholars use different models to predict airport passenger flow and have made some achievements. For example, Wang Tingting (2017) uses the grey Markov model to predict the passenger throughput data of Guiyang Longdongbao Airport for 2006–2016 years. The results show that the grey Markov forecast model has a better prediction accuracy. The average annual error is 3.38% [2]. Liu Xia (2016) forecasted the passenger flow of Sanya Airport using the weighted combination of the Holt-Winter seasonal model, the ARMA model and the linear regression model. It was proved that the method can be used to predict airport passenger flow effectively with an average annual error of 3.98% [1]. Huang Bangju (2013) established a multiple linear regression model to predict passenger throughput at an airport in southwest China with an average annual error of 2.49% [3]. Qu Tuo (2012) combined a gray model and BP neural network to build a combined forecasting model. Predicting passenger throughput at Chengdu Airport, the average error is 2.74% [4].

From this perspective, we can see there is a clear difference in the prediction accuracy of different models. In terms of short-term forecasting, the ARIMA model is a more accurate method, and domestic scholars have applied to different areas of prediction. For example, Xue Dongmei (2010) used the ARIMA (3,1,2) model to forecast the total investment of society's fixed asset in Jilin Province over the next five years based on the total social fixed asset investment data of Jilin Province for the past 16 years [5]. Zhou Ye (2010) used the ARIMA (1,1,1) model according to the monthly data of China's air cargo traffic from January 2002 to December 2009, and forecasted the freight volume for the next six months [6]. Zhang Xiaofei (2006) used the smooth ARIMA (1,1,2) time series model based on sample data of China's GDP from 1978 to 2003 to make predictions on China's GDP for the next four years [7]. The fitting values are very close to the actual observed values (the error is less than 5%) and have high prediction accuracy, which indicates that the ARIMA(p,d,q) model is effective and feasible for short-term time series forecast.

The passenger flow of a local airport is affected by many different factors, such as the local industrial structure, holiday arrangements, airline arrangements, airline ticket pricing, and climate, which often have the characteristics of cyclical and non-linear changes. As a tourist city, Sanya has a significant difference between the peak season and low season. The monthly number of visitors fluctuates considerably (refer to Table 1), with obvious seasonal features. Since the monthly data of passenger flow feature seasonal changes, the ARIMA(sp,sd,sq)S(1) seasonal model can be used for forecasting.

In summary, the ARIMA(p,d,q) model and the ARIMA(sp,sd,sq)S(1) seasonal model are to be synthesized to build a composite seasonal model. This model will be put into the application of passenger flow analysis and forecast at Sanya Airport with a hope to provide theoretical support for the management and development of the local aviation passenger transportation market.

3.2   Brief Introduction of ARIMA Model

The ARIMA model, also known as the Auto Regressive Integrated Moving Average model, is abbreviated as the ARIMA(p,d,q) model, which was a time series prediction model proposed by the American scholar Box and British scholar Jenkins in 1970s. So it is also called Box-Jenkins model, referred to as B-J model.

The modeling idea is to treat the sequence changing over time as a random sequence and apply it to a corresponding mathematical model assisted with approximate the description. Through the analysis of the corresponding mathematical model, the inherent structure and complex characteristics of these dynamic data are more fundamentally understood. In this way the best prediction can be achieve in the sense of minimum variance.

The ARIMA model has the following structure:

$$\Phi(B)\nabla^d x_t = \Theta(B)\varepsilon_t$$

The stochastic interference ε_t satisfies the formula, $E(\varepsilon_t) = 0, Var(\varepsilon_t) = \sigma_\varepsilon^2, \forall s \neq t$, $E(\varepsilon_s \varepsilon_t) = 0$, and in the formula, $\nabla^d = (1-B)^d$ means that the d order differential operation is performed. $\Phi(B)$ is the auto regressive coefficient polynomial of the smooth and reversible ARMA(p,q) model, and $\Theta(B)$ is the corresponding moving smoothing coefficient polynomial and B is the delay operator [9].

The ARIMA model structure shows:

$$\nabla^d x_t = \frac{\Theta(B)}{\Phi(B)}\varepsilon_t$$

That is, the ARIMA(p,d,q) model is a combination of differential arithmetic and ARMA(p,q) models. Because the difference operation has powerful deterministic information extraction capability, it can be shown that the ARIMA(p,d,q) model can be fitted with time sequence as long as the non-stationary sequence is stable after difference operation of the appropriate order [8].

In the monthly and quarterly sequences, the data often contains seasonal changes. From this aspect, seasonal Model is just similar to continuous sequence. The difference is that the time unit of the continuous model is 1, and the time unit of the seasonal model is the corresponding period S. and the Seasonal model ARIMA(sp,sd,sq)S (1) can be expressed as:

$$\Phi(B^S)w_t = \Theta(B^S)u_t$$

and the operation obeys the following rules:

$$\Phi(B^S) = 1 - \Phi_1 B^S - \Phi_2 B^{2S} - \cdots - \Phi_P B^{PS}$$

$$\Theta(B^S) = 1 - \Theta_1 B^S - \Theta_2 B^{2S} - \cdots - \Theta_Q B^{QS}$$

$$w_t = \nabla_s^D \nabla^d z_t$$

Combining the ARIMA(p,d,q) model with ARIMA(sp,sd,sq)S(1) to get the General Multiplicative Seasonal Models, it is a multiplicative model. For the common sequences, the composite seasonal model can get satisfactory results [9]. Its model structure is:

$$\phi(B)\Phi(B^S)w_t = \theta(B)\Theta(B^S)u_t$$

ARIMA(p,d,q)(sp,sd,sq)S(1) is the general form of the ARIMA model. Where p is the order of the autoregressive model, d is the order of the difference, q is the order of the moving average, sp is the order of the auto regression of the seasonal model, sd is the order of the seasonal difference, and sq is the movement of the season model The average order, S is the seasonal period [10].

4   Basic Steps for Using ARIMA Model Prediction

4.1   Test Data Stability

If the sequence is stationary, go to the next step; If not, perform the difference operation to make the sequence stable.

According to the data in Table 1, sequence graph of passenger numbers made by SPSS 19 is shown in Fig. 1. We can see that Sanya Airport's passenger flow showed an overall upward trend and seasonal fluctuations (period of 12 months), so the data on the numbers of passengers has trends and seasonal regularity, and can be initially judged as

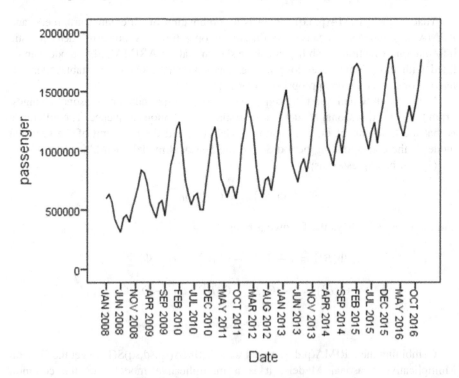

Fig. 1. Sequence of passenger numbers

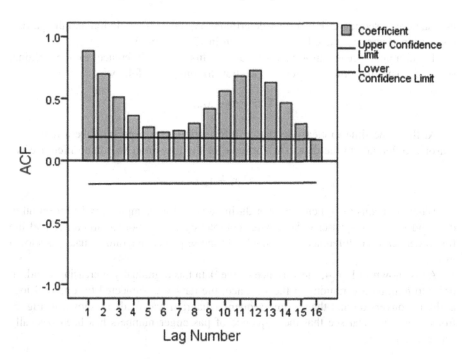

Fig. 2. Auto correlation of passenger numbers

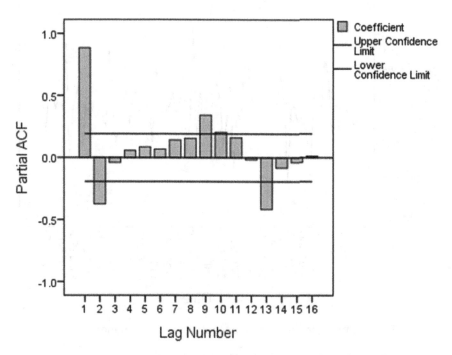

Fig. 3. Partial correlation of passenger numbers

non-stationary sequence. For further determination, the self-correlation and the partial-correlation charts are made below, as shown in Figs. 2 and 3.

To eliminate trend information, make a first-order difference to the original sequence. The first-order difference operation formula is as follows:

$$\nabla x_t = x_t - x_{t-1}$$

At the same time to clear the season information, you need to make a first-order seasonal difference to the data. The formula for the first-order seasonal difference is:

$$\nabla_k = x_t - x_{t-k}$$

Where k means periodicity number during a year, for example, k = 12 for monthly data and k = 4 for quarterly data. After completing the first-order difference and the first-order seasonal difference respectively, a time sequence diagram is made, as shown in Fig. 4.

As is shown in Fig. 4, the variance of the data has a gradually increasing trend. In order to eliminate variations in the variance, the data was subjected to a natural logarithmic conversion and than a time sequence diagram was made as shown in Fig. 5, from which, we can see that the sequence of passenger numbers has been basically stable at this time.

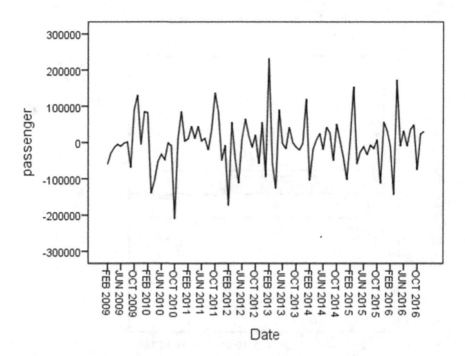

Transforms: difference(1), seasonal difference(1, period 12)

Fig. 4. Time sequence of first-order difference of passenger numbers

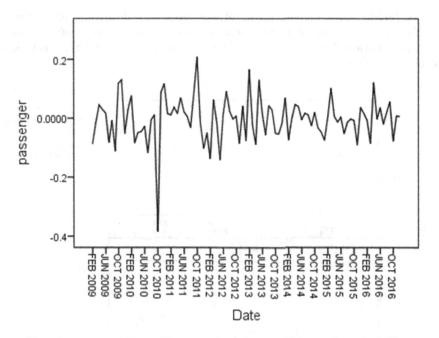

Transforms: natural log, difference(1), seasonal difference(1, period 12)

Fig. 5. First-order differential time sequence diagram after logarithmic conversion

Table 2. Various model fitting effects

(p,q)	(sp.sq)	R-Squared	N-BIC	(p,q)	(sp.sq)	R-Squared	N-BIC
(1,0)	(1,0)	0.969	22.270	(0,1)	(1,0)	0.969	22.270
	(0,1)	0.970	22.266		(0,1)	0.970	22.266
	(1,1)	0.970	22.309		(1,1)	0.970	22.309
	(2,0)	0.970	22.301		(2,0)	0.970	22.301
	(0,2)	0.970	22.304		(0,2)	0.970	22.304
	(2,1)	0.970	22.357		(2,1)	0.970	22.357
	(1,2)	0.970	22.363		(1,2)	0.970	22.363
	(2,2)	0.971	22.409		(2,2)	0.971	22.409
(p,q)	(sp.sq)	R-Squared	N-BIC	(p,q)	(sp.sq)	R-Squared	N-BIC
(1,1)	(1,0)	0.971	22.271	(2,0)	(1,0)	0.970	22.321
	(0,1)	**0.971**	**22.265**		(0,1)	0.970	22.319
	(1,1)	0.972	22.313		(1,1)	0.970	22.363
	(2,0)	0.972	22.306		(2,0)	0.970	22.356
	(0,2)	0.972	22.310		(0,2)	0.970	22.358
	(2,1)	0.972	22.361		(2,1)	0.971	22.411
	(1,2)	0.971	22.383		(1,2)	0.970	22.417
	(2,2)	0.972	22.418		(2,2)	0.971	22.463

To further verify the stationarity of sequences after differential operations, the graphs and tables of the auto-correlation function and partial-correlation function are made, as shown in Table 2, Figs. 6, and Fig. 7. It can be seen that the data is basically stable at this time.

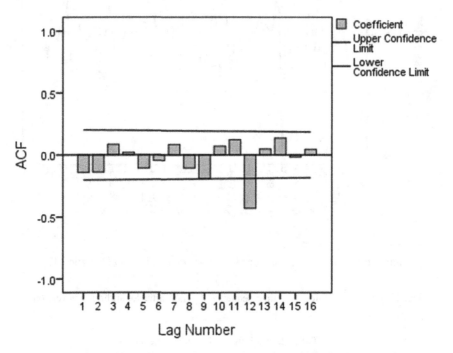

Fig. 6. Auto-correlation diagram after the first-order difference

4.2   Ordering ARIMA Models

From Table 2 and Figs. 6 and 7, it can be seen that the number of autocorrelation coefficients significantly not 0 is 1, the corresponding lag period is 12; the number of partial correlation coefficients not significantly 0 is 1, corresponding to The lag period is 12. At this point the sequence is basically stationary, except that the data has an effect on the season. After experiment, the second-order seasonal difference was performed on the sequence, and it was found that the seasonality of the sequence was not significantly improved. Taking into account that Sanya is dominated by the tourism industry, the differences between the peak seasons are obvious, and the seasonal effect of the passenger number sequence cannot be completely eliminated. Therefore, only the first-order seasonal differential can be used. Therefore, it is desirable that $d = 1$, $sd = 1$, and $S = 12$.

Observe Figs. 6 and 7, both of which are trailing first-order truncations, and there are first-order seasonal self-correlation and first-order seasonal partial-correlation. It can be estimated that the possible values of (p,q) and (sp,sq) are $(1,0)$, $(0,1)$, $(1,1)$, $(2,0)$, $(0,2)$, $(2,1)$, $(1,2)$, $(2,2)$ and so on.

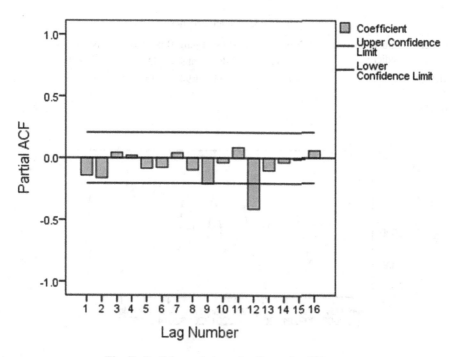

Fig. 7. Partial-correlation after first-order difference

Now each model is fitted separately, and then the judgement is made according to the value of the coefficient R-Squared and the standardized BIC. The larger the R-Squared value of the model and the smaller the BIC value, the better the fitting effect.

The models were built with the help of SPSS software and than each possible (p,q) (sp,sq) combination is compared, the results are as shown in Table 2 (for the sake of brevity, the data with the smaller R-Squared value and the larger BIC value are omitted.). Although the R-Squared value of the ARIMA(1,1,1)(0,1,1) model is slightly smaller than other models, its BIC value is the smallest. Therefore, ARIMA(1,1,1) (0,1,1)S(12) is finally selected as a fitting model.

4.3   Analysis and Forecast Results

The fitting effect of the selected model is tested and the results are shown in Table 3:

As can be seen from the table, this model can explain 97.1% of the information in the original sequence, the N-BIC value is the smallest, and the value of the Ljung-Box statistic is significant, indicating that the effect of ARIMA(1,1,1)(0,1,1)S (12) model fitting time series data is ideal. As can be seen from Fig. 8, the fitting effect of the model is very good, and the fitting value almost coincides with the observation value.

Based on the model ARIMA(1,1,1)(0,1,1)S(12), SPSS 19 statistical software was used to forecast the passenger flow volume at Sanya Airport from January to December, 2017. The results are shown in Table 4. As shown in Table 4, the Absolute Percentage of Error (APE) of the predicted value is mostly less than 5%, and the annual

Table 3. Model statistics

Model	Number of predictors	Model fit statistics		Ljung-Box Q(18)			Number of Outliers
		R-Squared	N-BIC	Statistics	DF	Sig.	
ARIMA (1,1,1) (0,1,1)	0	0.971	22.265	11.943	15	0.683	0

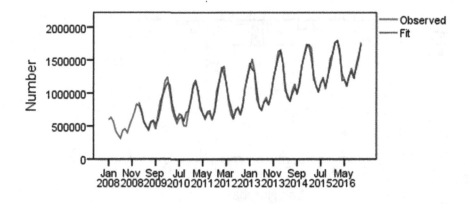

Date

Fig. 8. Observed and fit

Table 4. Comparison between predicted value and actual value of passenger flow in 2017 (Unit: 10,000)

Month	Predicted value	Actual value	APE
Jan.	194.06	194.46	0.21%
Feb.	196.44	187.38	4.84%
Mar.	181.27	183.13	1.02%
Apr.	143.19	150.76	5.02%
May	134.18	142.50	5.84%
June	125.01	135.84	7.97%
July	139.64	142.88	2.27%
Aug.	149.70	153.11	2.23%
Sep.	134.63	135.77	0.84%
Oct.	151.67	152.59	0.60%
Nov.	168.15	171.24	1.80%
Dec.	188.79	189.33	0.28%
In Total	1906.73	1938.99	1.66%

APE is 1.66%. So the prediction accuracy is very high. Therefore, this ARIMA model can be well applied to forecast passenger flow at Sanya Airport.

The ARIMA(1,1,1)(0,1,1)S(12) model is used to forecast the monthly passenger flow volume of Sanya Airport in 2018 and 2019. The results are shown in Table 5.

Table 5. Monthly forecast of passenger volume in two years (Unit: 10,000)

Month	Year	
	2018	2019
Jan.	205.50	217.59
Feb.	203.63	214.76
Mar.	193.95	204.96
Apr.	159.43	172.15
May	150.37	163.17
June	142.53	155.79
July	153.22	165.78
Aug.	163.63	176.43
Sep.	147.83	160.73
Oct.	164.13	176.57
Nov.	180.43	193.46
Dec.	200.93	213.78
In Total	2065.58	2215.18

5 Conclusion

According to the statistic analysis of the passenger flow data of Sanya Airport from January 2008 to December 2016, it can be found that the passenger flow volume has a clear seasonal regularity and tendency. Since Sanya is a tropical tourist city, the passenger flow volume here has a significant difference between the peak season and off-season, which makes the civil aviation market busy during the peak season and the off-season much of the aviation transportation capacity sits idle. If we can forecast the passenger flow in and out of the airport and arrange the transportation capacity in advance, we can bring obvious economic and social benefits.

By building and testing ARIMA model of airport passenger flow, the results show that the passenger flow forecast at Sanya Airport, based on ARIMA(1,1,1)(0,1,1)S(12) Model, is feasible and reliable. It has higher prediction accuracy and less error, therefore, it can provide decision-making basis for airport operation and management.

Acknowledgments. This research was financially supported by Hainan Provincial Natural Science Foundation of China (618QN258) and A cooperative science project between colleges and local government in Sanya (2014YD52). Thanks to associate professor Xia Liu, correspondent of this paper.

References

1. Xia, L.: Sanya airport passenger flow forecast based on combination forecast method. Comput. Syst. Appl. **25**(8), 23–28 (2016)
2. Tingting, W.: Prediction of passenger throughput in Longdongbao airport based on Grey Markov model. Technol. Econ. Areas Commun. **19**(5), 48–51 (2017)
3. Bangju, H.: The prediction for civil airport passenger through put based on multiple linear regression analysis. Math. Pract. Theor. **43**(4), 172–178 (2013)
4. Tuo, Q.: Application of combination model in airport passenger throughput prediction. Comput. Simul. **29**(4), 108–111 (2012)
5. Dongmei, X.: Application of the ARIMA model in time series analysis. J. Jilin Inst. Chem. Technol. **27**(3), 108–111 (2010)
6. Ye, Z.: Study on forecasting of the aeronautic cargo capacity in china based on ARIMA model. J. Nanchang Hangkong Univ. **12**(9), 22–27 (2010)
7. Xiaofei, Z.: Research and application of short-term sequence forecasting model based on ARIMA model. New Stat. **10**, 7–9 (2006)
8. Haolin, Y.: Application and Prediction With R, 1st edn. Publishing House of Electronics Industry, Beijing (2016)
9. Wentong, Z., Wei, D.: Advanced course in SPSS statistical analysis, 2nd edn. Higher Education Press, Beijing (2013)
10. Yifan, X.: Essentials and Examples of SPSS Statistical Analysis, 1st edn. Electronic Industry Press, Beijing (2010)

Comparison of LVQ and BP Neural Network in the Diagnosis of Diabetes and Retinopathy

Jiarui Si, Yan Zhang, Shuaijun Hu, Li Sun, Shu Li, Hongxi Yang,
Xiaopei Li, and Yaogang Wang[(✉)]

Tianjin Medical University, Tianjin 300070, China
wyg@tmu.edu.cn

Abstract. Diabetes Mellitus is a chronic, non-infectious disease that affects people's health, which has a rapid onset trend with the continuous improvement of China's economy and material life. At present, there is no practical solution to this condition. Diabetic retinopathy (DR) is the most important manifestation of diabetic microangiopathy, which can be divided into proliferative lesion and non-proliferative lesion. The traditional artificial diagnosis method has strong subjectivity and low accuracy. Because disease diagnosis can be regarded as a two-classification pattern recognition problem, the application of neural network method can provide the possibility for the application of artificial intelligence (AI) in disease-assisted diagnosis and treatment. Furthermore, it has significant meanings to find which kind of neural network has a better efficiency. In this paper, learning vector quantization (LVQ) neural network and back propagation (BP) neural network were used to diagnose diabetes and diabetic retinopathy with MATLAB and their recognition rate were compared. The datasets of diabetes and diabetic retinopathy were available in the UCI database. The experiment results were analyzed to evaluate the efficiency of each neural network classifier. The results demonstrate that both LVQ neural network and BP neural network can classify the two datasets effectively. However, compared with the LVQ neural network, the average accuracy rate and sensitivity of the BP neural network is higher.

Keywords: Diabetes · Proliferative diabetic retinopathy
Non-Proliferative Diabetic Retinopathy · Artificial Intelligence
LVQ neural network · BP neural networks · UCI database

1   Introduction

Deep learning is the best machine learning algorithms at present, which is neural networks consisting of multiple hidden layers. [1, 2] This article uses two kinds of neural network methods to diagnose diabetes and retinopathy and compares their efficiency [3].

As the incidence of diabetes is increasing year by year in a population society, the degree of attention of this disease in the population is also getting higher and higher. In laboratory tests, blood glucose concentration is the sole criterion for diagnosing diabetes [4, 5]. Diabetic retinopathy, as the most important manifestation of diabetes in micromodule, cannot be overlooked [6]. Clinically based on the presence or absence of

© Springer Nature Singapore Pte Ltd. 2018
Q. Zhou et al. (Eds.): ICPCSEE 2018, CCIS 902, pp. 455–466, 2018.
https://doi.org/10.1007/978-981-13-2206-8_37

retinal neovascularization, diabetic retinopathy without retinal neovascularization is called non-proliferative diabetic retinopathy (NPDR) (or simple or background type), and there will be retinal neonatal Angiogenic diabetic retinopathy known as proliferative diabetic retinopathy (PDR) [7]. There had been no good method for classifying the images of diabetic retinal proliferative and non-proliferative lesions automatically in the past, so exploring a intelligent method is an imaginable field [8–11].

LVQ neural network, which is a competitive layer Neural network can automatically learn to classify the input vector pattern, the classification by the competitive layer depends only on the distance between the input vectors [12]. When two input vectors are very close, the competition layer may classify them as one. There is no such mechanism in the design of the competition layer that strictly determines whether any two input vectors belong to the same class or belong to different classes. In order to specify the target classification result, the network can complete the accurate classification of the input vector pattern through supervised learning. The BP neural network, according to its nonlinear function fitting algorithm, can be divided into three steps: BP neural network construction, BP neural network training and BP neural network prediction [13]. Since the nonlinear function has two input parameters and one output parameter, the structure of the BP neural network is 2-5-1, that is, the input layer has 2 nodes, the hidden layer has 5 nodes, and the output layer has 1 node.

2 Materials and Methods

In the following paragraphs, we describe experiment datasets and neural network classifiers [14, 15].

2.1 Datasets

The datasets came from the UCI machine learning repository. The diabetes dataset was collected by the National Institute of Diabetes, Digestive and Kidney Diseases in the United States in 1990 [16]. There are 768 examples, all of which are women older than 21, divided into ten data attributes that belong to the classification task. Attribute information includes: (1) The quality of the data, (2) Classification variables, where the number 1 represents the diagnosis of diabetes and 0 represents the definitive diagnosis without diabetes, (3) The number of pregnancies, (4) Oral glucose tolerance test for 2-h oral glucose tolerance test, (5) Diastolic pressure, 6) Triceps Skinfold thickness, (7) 2 h serum insulin concentration, (8) Body mass index, (9) Diabetes lineage function, (10) Age.

Similarly, the diabetic retinopathy dataset was provided by the department of computer graphics and image processing faculty of informatics in the Debrecen University. It contained features extracted from the Messidor image set to predict whether the image contained signs of diabetic retinopathy. This data was collected in November 2014 and was a multivariate dataset with 1,151 instances. It has 20 attributes that belong to the classification task. All features represent detected lesions, descriptive features of the anatomy, or image-level descriptors. Method-based image analysis and feature extraction and our classification techniques were applied to BalintAntal (which is a system-based diabetic retinopathy screening system) and AndrasHajdu (a knowledge-based system). Attribute information includes: (1) The binary result of the quality assessment. 0 = quality is not good, 1 = quality is enough, (2) Pre-filtered binary results, where 1 indicates a severe retinal abnormality and 0 indicates a lack, (3)–(8) MA test results. Each eigenvalue represents the number of MAs found with confidence at α = 0.5–1. (MA is microalbuminuria), (9)–(16) For exudates, contains the same information as (2)–(7). However, exudate is composed of a set of dots and is not composed of pixels constituting the lesion, so the features are normalized by dividing the number of lesions in the diameter of the ROI to compensate for different image sizes, 17) The Euclidean distance between the macula and the center of the optic disc can provide important information about the condition of some patients. This feature can also be normalized with the diameter of the ROI, (18) the diameter of the optic disc, (19) Binary results based on AM/FM classification, (20) Classification tags: 1 = feature containing DR (cumulative tags for Messidor classifications 1, 2, and 3), 0 = features that do not include DR.

Of these 1151 instances of diabetic retinopathy, 805 were used as training set, 173 were used as validation set and 173 were used as test set. After studying the properties of all the data, we found that the second property of the dataset is the vascular changes we observed. However, it performs a binary transformation of all data, where 1 represents a severe retinal abnormality and 0 represents a lack of retinopathy. That means, in these binary data, 1 represents proliferative retinopathy (PDR) and 0 represents non-proliferative retinopathy (NPDR). But because in LVQ, you need to use the ind2vec function, and the ind2vec function does not recognize 0, so I changed all 0 of this property in the database to 2, and then the dataset was trained and tested. To illustrate the reliability of this LVQ method for classification data, 768 diabetes data were also classified by LVQ, of which 538 were training set, 115 were validation set and 115 were test set. The last attribute of this dataset is the basis for our diagnosis of diabetes and non-diabetes, with 1 representing a definitive diagnosis of diabetes and 0 representing without diabetes. Since this method also requires the ind2vec function, I also changed the 0 data in the categorical variable to 2, and then trained and tested.

The component of two basic data are shown in Table 1, the number of training, validation and test set distribution of two datasets are shown in Table 2.

Table 1. Basic data component of diabetes dataset and retinopathy dataset.

	Diabetes	Retinopathy
Benign/negative	500	94
Malignant/positive	268	1057
Total	768	1151

2.2 Classifiers

Learning Vector Quantization (LVQ) neural network is an input forward neural network used to train the competitive layer of supervised learning methods [17]. Its algorithm is evolved from the Kohonen competition algorithm. LVQ neural networks

Table 2. Training, validation and test set distribution of diabetes dataset and retinopathy dataset

	Diabetes	Retinopathy
Training sample size	538	805
Validation sample size	115	173
Test sample size	115	173
Total	768	1151

have a wide range of applications in the field of pattern recognition and optimization. The LVQ neural network consists of three layers, the input layer, the competitive layer, and the linear output layer. Full connection is adopted between the input layer and the competition layer. The number of neurons in the competition layer is always greater than the number of neurons in the linear output layer. Each competitive layer neuron is connected to only one linear output layer neuron and the connected weights are 1 constantly. However, each linear output layer neuron can be connected with multiple competitive layer neurons. The values of the competitive layer neurons and the linear output layer neurons can only be '1' or '0'. When an input pattern is sent to the network, the closest competitive layer neuron to the input pattern is activated, the neuron state is '1', and the other competitive layer neurons state is '0'. Therefore, the state of the neurons in the linear output layer connected to the activated neurons is also '1', while the states of other neurons in the linear output layer are all '0'.

In Fig. 1, P is the R-dimensional input mode; S1 is the number of neurons in the competition layer; IW-1,1 is the link between the input layer and the competition layer, and the weighting coefficient matrix; n1 is the input of the competitive layer neuron. a1 is the output of the competitive layer neuron; LW2,1 is the connection weight coefficient matrix between the competitive layer and the linear output layer; n2 is the input of the linear output layer neuron; a2 is the output of the linear output layer neuron. Compared with other pattern recognition and mapping methods, the advantage of LVQ neural network lies in its simple network structure. It can complete very complex classification processing only through the interaction of internal units, and it is also very easy to disperse various design in the design domain. The conditions converge to the conclusion. Moreover, it does not need to normalize and orthogonalize the input vector. It only needs to directly calculate the distance between the input vector and the competition layer, so as to realize pattern recognition, so it is simple and easy.

The mathematical theory of LVQ neural network and its algorithm implementation can be described as follow:

$$d_i = \sqrt{\sum_{j=1}^{R} \left(x_j - w_{ij}\right)^2} \quad i = 1, 2, \ldots, S^1 \tag{1}$$

$$w_{ij\_new} = w_{ij\_old} + \eta\left(x - w_{ij_{old}}\right) \, or \, w_{ij\_new} = w_{ij\_old} - \eta\left(x - w_{ij_{old}}\right) \tag{2}$$

BP neural network training uses neural function input and output data to train the neural network so that the trained network can predict the output of the nonlinear function [18, 19]. The algorithm flow is shown in Fig. 2.

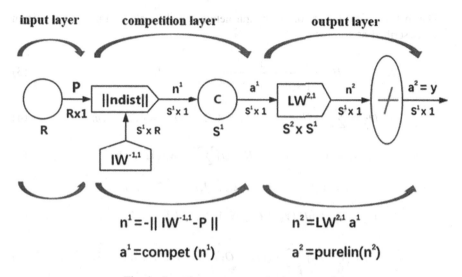

Fig. 1. Learning vector quantization network.

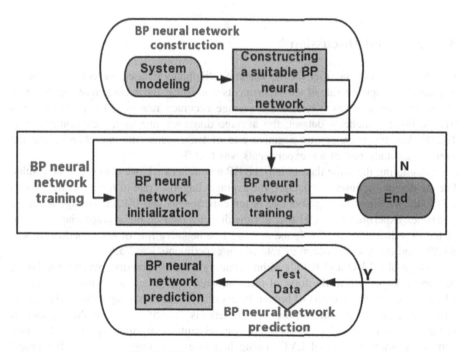

Fig. 2. BP neural network algorithm flow

The mathematical theory of BP neural network and its algorithm implementation can be described as follow:

$$H_j = \frac{1}{1 + e^{-\left(\sum_{i=1}^{n} w_{ij}x_i - a_j\right)}} \quad j = 1, 2, \ldots, l \tag{3}$$

$$O_k = \sum_{j=1}^{l} H_j w_{jk} - b_k \quad j = 1, 2, \ldots, l, k = 1, 2, \ldots, m \tag{4}$$

$$w_{ij} = w_{ij} + \eta H_j \left(1 - H_j\right)x(i) \sum_{k=1}^{m} w_{jk}(Y_k - O_k) \tag{5}$$

$$i = 1.2, \ldots, n; j = 1, 2, \ldots, l, k = 1, 2, \ldots, m$$

$$w_{ij} = w_{jk} + \eta H_j (Y_k - O_k) j = 1, 2, \ldots, l; k = 1, 2, \ldots, m \tag{6}$$

$$a_j = a_j + \eta H_j \left(1 - H_j\right) \sum_{k=1}^{m} w_{jk}(Y_k - O_k) \quad j = 1, 2, \ldots, l, k = 1, 2, \ldots, m \tag{7}$$

$$b_k = b_k + Y_k - O_k \quad k = 1, 2, \ldots, m \tag{8}$$

3 Result and Discussion

Based on the LVQ classifier for the diabetic retinopathy dataset, we got some results. The average diagnosis rate of ten experiments was 100%, the average true negative rate of ten experiments was 0%, and the average accuracy rate of ten experiments was 91.3%. For the diabetes dataset, the average diagnosis rate of ten experiments was 13.5%. And the average true negative rate of ten experiments was 94.8%, and the average accuracy rate of ten experiments was 67.6%.

Substituting the same dataset into the BP neural network, we also got some results. For the diabetic retinopathy dataset, the average diagnosis rate of ten experiments was 100%, the average true negative rate of ten experiments was 0%,the average accuracy rate of ten experiments was 91.8%. For the diabetes dataset, the average diagnosis rate of ten experiments was 57.6%, the average true negative rate of ten experiments was 85.2%, and the average accuracy rate of ten experiments was 75.6%.

Both in the LVQ and BP algorithm, training follows a random weighting/biasing rule and is expressed in the form of mean squared error. Neurons of hidden layers of LVQ are 20 and the neurons of hidden layers of BP is 25, both algorithms add test set with 'dividerand' function, whose default ratio is 7: 1.5: 1.5. Both the number of hidden layers of BP and LVQ are 2. The neural number of input layers of BP neural network is same as that of LVQ during handling each kind of dataset, the neural number of BP neural networks output layers is 1 while that of LVQ is 2.

3.1   LVQ Neural Network

In the LVQ neural network, a 'Plotconfusion' calculation was performed on the diabetic retinopathy dataset and diabetes dataset firstly. 'Plotconfusion' training produces a classification confusion matrix. In the confusion matrix diagram, the rows correspond to the forecasted output class and the columns show the actual target class. The diagonal unit shows the number and examples of the training network's correct estimation of the observation category, that means, it shows the percentage of real and predicted category matches. The Fig. 3(a) shows the confusion matrices for testing data of the retinopathy dataset and Fig. 3(b) shows the confusion matrices for testing data of the diabetes dataset.

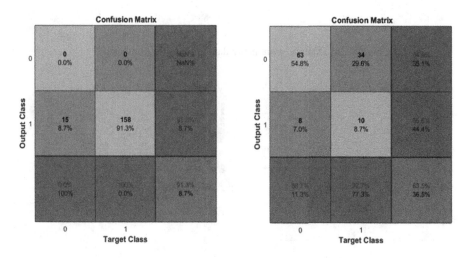

Fig. 3. (a) Diabetic retinopathy datasets (b) Diabetes dataset

Subsequently, Plotroc operations are performed in the LVQ network to draw ROC curves. A perfect test would show points in the upper-left corner. The results of the Plotroc operation of the diabetic retinopathy dataset are shown in Fig. 4(a), The results of the Plotroc operation of diabetic dataset are shown in Fig. 4(b).

In the diabetic retinal dataset, some network training results are not very significant after applying the LVQ algorithm. In order to verify the reliability of this algorithm, I applied the same method to diabetes dataset. For example, in the Plotroc function, the data of diabetic retinopathy did not cause obvious changes in the chart, and after applying the diabetes data, it was found that the two categorical variables had significant differences in the training ROC, and thus caused global ROC production difference. Furthermore, this indicates that LVQ neural network is reliable for the application of disease diagnosis.

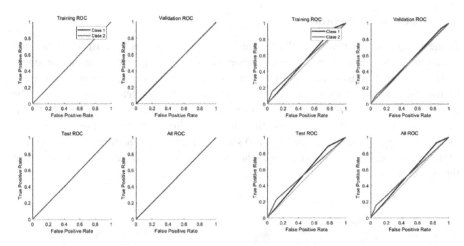

Fig. 4. (a) Diabetic retinopathy datasets (b) Diabetic datasets

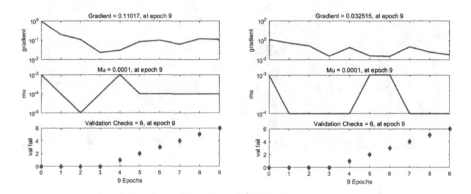

Fig. 5. (a) Diabetic retinopathy (b) Diabetic

3.2 BP Neural Network

We also performed classification on each dataset using BP neural network [20, 21]. Firstly, in the BP network, Plottrainstate training is performed. Plottrainstate is the plot training status value. Through this training, the result obtained is a graph of the gradient between the training data and the mean square error. The results of diabetic retinopathy are shown in Fig. 5(a), and the results of diabetes data are shown in Fig. 5(b).

Subsequently, Plotregression was used to fit the input dataset for the output dataset. The result of diabetic retinopathy was shown in Fig. 6(a). The result of diabetes data was shown in Fig. 6(b).

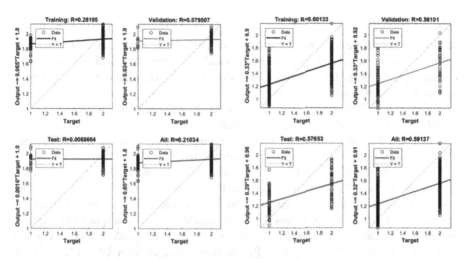

Fig. 6. (a) Diabetic Retinopathy Datasets (b) Diabetic Datasets

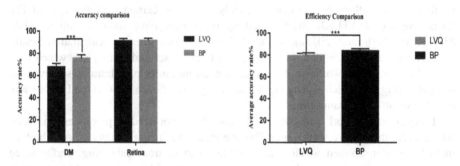

Fig. 7. (a) Accuracy comparison of the two neural networks based on two datasets (***P < 0.05) (b) Efficiency comparison of BP network and LVQ network(***P < 0.05)

3.3   Comparison of Accuracy Rate

The average diagnostic accuracy of ten experiments of the two neural networks based on diabetes data and diabetic retinopathy data are shown in Fig. 7(a). The average diagnostic accuracy of LVQ network based on diabetes data was 67.6%, while the average diagnostic accuracy of BP network based on diabetes data was 75.6%; The average diagnostic accuracy of LVQ network based on diabetic retinopathy data was 91.3%, while the average diagnostic accuracy of ten experiments of BP network based on diabetic retinopathy data was 91.8%.

In order to compare the efficiency of LVQ network and BP network, the average diagnostic accuracy of diabetic retinopathy data and diabetes data of BP network and LVQ network were calculated, and the average diagnostic accuracy of ten experiments of each network was showed in Fig. 7(b). The average diagnostic accuracy of two diseases of LVQ network was 79.4%, The average diagnostic accuracy of two diseases of BP network was 83.7%.

3.4   Discussions

We can speculate the reason why diabetes' diagnosis rate lower than that of diabetic retinopathy may be related to the attribute numbers of the data. Diabetic retinopathy dataset include more data attributes than diabetes dataset, which may be the main reason for the difference in accuracy.

Compared with BP neural network, the accuracy and sensitivity of LVQ network is slightly lower, this may due to the influence of the neuron number of hidden layer, although which is not optimal for either BP network or LVQ network, the performance of BP network was better than LVQ network relatively.

The neural network method has been applied to benign and malignant breast cancer classification. The dataset was about nuclear fiber images of breast cancer lesion tissue which was collected at the University of Wisconsin Medical School. Both BP and LVQ have achieved a good effect. In this paper, the neural network was applied to diagnose diabetes and diabetic retinopathy. In the work of some predecessors, LVQ had achieved a better performance than BP on breast dataset from UCI. [22] Without consideration the factor of data preference, The reason why the efficiency of BP was better than LVQ may relied on :(1) We performed the task 10 times to get a more stable result while the others performing once only may have certain contingency; (2) The former version MATLAB(2012b) used by other may have some deviation, while MATLAB(2017b and 2018a) we used have a more accurate calculation result. (3) Predecessor only divided the dataset into training set and testing set, which may cause the over fitting, while we divided the dataset into three part: training set, evaluate set and testing set, which enhancing the accuracy rate of each network and got a more credible result at the same time.

LVQ neural network and BP neural network do not need to preprocess the data, which make them more simple and effective than other neural networks. And both of them have been applied to all walks of life, such as troubleshooting, performance assessment, risk prediction and so on. This provides new ideas for using AI technology to realize the diagnosis of diseases.

4   Conclusion

Early diagnosis for diabetic retinopathy has significant meanings. The non-proliferative lesion belongs to one, two, and three stages generally, and the proliferative lesion is common in more than four stages. That is to say, judging the stage of the disease, is deciding what kind of measures should be taken to prevent from more serious situations. On the other hand, Assistant diagnosis for diabetes can improve the screening rate in populations.

We found that both the LVQ and BP had better effect in diagnosis of diabetic retinopathy than diagnosis of diabetes. Compared to the LVQ neural network, the average accuracy rate about two disease of BP is higher, has significant differences ($P < 0.05$). In general, the BP neural network should be preferred in intelligent diagnosis of diabetes and its complications. We will try to train more kinds of neural networks to find the best one to help clinical diagnosis in the future.

Acknowledgment. This research was supported by Natural Science Foundation of China under Grant No. 91746205 and China Postdoctoral Science Foundation under Grant No. 2017M 621092.

References

1. Yang, Y., Sun, J., Li, H., Xu, Z.: Deep ADMM-Net for compressive sensing MRI. In: Advances in Neural Information Processing Systems (2016)
2. Szegedy, C., Vanhoucke, V., Ioffe, S., Shlens, J., Wojna, Z.: Rethinking the inception architecture for computer vision. In: 2016 IWWW Conference on Computer Vision and Pattern Recognition, pp. 2818–2826 (2016)
3. Ryan, E.A., Holland, J., Stroulia, E., et al.: Improved A1C levels in type 1 diabetes with smartphone app use. Can. J. Diabetes **41**(1), 33–40 (2016)
4. Bell, K.J., Smart, C.E., Steil, G.M., et al.: Impact of fat, protein, and glycemic index on postprandial glucose control in type 1 diabetes: implications for intensive diabetes management in the continuous glucose monitoring era. Diabetes Care **38**, 1008–1015 (2015)
5. Wong, J.C., Foster, N.C., Maahs, D.M., et al.: Real-time continuous glucose monitoring among participants in the T1D exchange clinic registry. Diabetes Care **37**, 2702–2709 (2014)
6. Ferrara, N.: Vascular endothelial growth factor and age-related macular degeneration: from basic science to therapy. Nat. Med. **16**, 1107–1111 (2010)
7. Cho, B.H., Yu, H., Kim, K.W., et al.: Application of irregular and unbalanced data to predict diabetic nephropathy using visualization and feature selection methods. Artif. Intell. Med. **42** (1), 37–53 (2008)
8. Adegbola, R.A.: Childhood pneumonia as a global health priority and the strategic interest of the Bill & Melinda Gates Foundation. Clin. Infect. Dis. **54**(Suppl2), S89–S92 (2012)
9. Klupa, T., ek-Klupa, T., Malecki, M., et al.: Clinical usefulness of a bolus calculator in maintaining normoglycaemia in active professional patients with type 1 diabetes treated with continuous subcutaneous insulin infusion. J. Int. Med. Res. **36**, 1112–1116 (2008)
10. Gross, T.M., Kayne, D., King, A., et al.: A bolus calculator is an effective means of controlling postprandial glycemia in patients on insulin pump therapy. Diabetes Technol. Ther. **5**, 365–369 (2003)
11. Cernadas, E., Amorim, D.: Do we need hundreds of classifiers to solve real world classification problems? J. Mach. Learn. Res. **15**(1), 3133–3181 (2014)
12. Zhan, C., Lu, X., Hou, M., et al.: A LVQ-based neural network anti-spam email approach. ACM SIGOPS Oper. Syst. Rev. **39**(1), 34–39 (2005)
13. Liu, S.Q.: Research and application on MATLAB BP neural network. Comput. Eng. Des. **11**, 025 (2003)
14. Van, B.V., Lisboa, P.: White box radial basis function classifiers with component selection for clinical prediction models. Artif. Intell. Med. **60**(1), 53–64 (2014)
15. Kermany, D.S., Goldbaum, M., et al.: Identifying medical diagnoses and treatable diseases by image-based deep learning. Cell **172**, 1122–1131 (2018)
16. Antal, B., Hajdu, A.: An ensemble-based system for automatic screening of diabetic retinopathy. Elsevier Science Publishers B. V. (2014)
17. Ma, X., Liu, W., Li, Y., et al.: LVQ neural network based target differentiation method for mobile robot. In: Proceedings of the 12th International Conference on Advanced Robotics, ICAR 2005, pp. 680–685 (2005)

18. Sadeghi, B.H.M.: A BP-neural network predictor model for plastic injection molding process. J. Mater. Process. Technol. **103**(3), 411–416 (2000)
19. Yi, J., Wang, Q., Zhao, D., et al.: BP neural network prediction-based variable-period sampling approach for networked control systems. Appl. Math. Comput. **185**(2), 976–988 (2007)
20. Melin, P., Amezcua, J., Valdez, F., et al.: A new neural network model based on the LVQ algorithm for multi-class classification of arrhythmias. Inf. Sci. **279**, 483–497 (2014)
21. Demuth, H., Beale, M.: Neural network toolbox - for use with MATLAB. Matlab Users Guide Math Works **21**(15), 1225–1233 (1995)
22. Huo, S.: Research on Diagnosis of Breast Cancer Based on Machine Learning (2017)

A Heuristic Indoor Path Planning Method Based on Hierarchical Indoor Modelling

Jingwen Li[1], Liqiang Zhang[1], Qian Zhao[2], Huiqiang Wang[1], Hongwu Lv[1], and Guangsheng Feng[1(✉)]

[1] Harbin Engineering University, Harbin 150001, China
fengguangsheng@hrbeu.edu.cn
[2] Harbin University of Commerce, Harbin 150028, China

Abstract. Indoor path planning technology has become more and more extensive, e.g., indoor navigation and real-time scene construction. There are many difficulties in the existing indoor path planning researches, including the slower searching speed and the higher storage space cost, which is inability to meet user requirements in a complex indoor environment. This paper proposes a path planning method based on indoor maps. First, the indoor environment is hierarchically modelled, and then the corresponding path searching algorithm is developed to obtain the optimal path. Experimental results show that using the path planning method designed in this paper can effectively solve the above problems.

Keywords: Indoor path planning · Indoor map · Jump point search

1 Introduction

Path planning technology is one of the most widely-recognized technologies in the academia and business community today, which is widely used in almost every field [1]. In terms of cutting-edge technology, path planning technology can be used for robot autonomous collision-free motion, unmanned drones, and autonomous drone flight. In the military field, it can be used for missile autonomous positioning and anti-radar systems. In academia, indoor path planning is one of the key directions of current researches [2]. For example, in large shopping malls, navigation systems can use indoor path planning techniques to generate navigation routes. The core of path planning technology is path search algorithm [3], which is also the focus of this paper.

At present, the researches on path planning mainly focus on the outdoor environment [4–6], while only a few of researches pay attention on that in the indoor environment. The existing path search algorithm is mainly divided into two categories, one is deterministic and the other is heuristic. The earliest deterministic search algorithm is proposed by Dijkstra [7,8], which is mainly used to solve the problem of the shortest path with non-negative weights. As an accurate algorithm, it still has a high application value when the numbers of nodes

© Springer Nature Singapore Pte Ltd. 2018
Q. Zhou et al. (Eds.): ICPCSEE 2018, CCIS 902, pp. 467–476, 2018.
https://doi.org/10.1007/978-981-13-2206-8_38

and edges are small. According to Dijkstra's algorithm, Bellman and Ford proposed the Bellman-Ford shortest path algorithm that can solve the edge with negative weight [9]. The Floyd algorithm is developed to calculate the shortest path between any two nodes in the graph. The time-space cost of the existing deterministic algorithms is too high to apply to models with more nodes and edges.

In order to overcome the limitations of the deterministic algorithm, the heuristic search algorithm A* was proposed, which greatly improve the time complexity of traditional algorithms [10]. For the existing heuristic search algorithms, the search process is improved by constructing a heuristic function, where the search algorithm can be guaranteed toward the objective in an asymptotically optimal manner. In theory, if the heuristic function is developed well enough, the result can be obtained more shortly. The A* algorithm needs to maintain the OPEN priority queue and the CLOSE priority queue during the search process, in which the OPEN priority queue stores the nodes to be expanded. Since there are many nodes that need to be expanded, the problem of large memory consumption is caused inevitably. In the same time, when the rectangular grid method is applied, the searched path may be longer because the A* algorithm can only search in eight directions. In contrast to the A* algorithm, jump point search can significantly reduce the number of nodes to be expanded by setting the pruning rule of each node to be expanded, while the obvious disadvantage is that the search time is too long [11,12].

In order to overcome the shortcomings of current methods, this paper first proposes a three-level indoor environment modelling framework based on indoor maps. Then, in order to overcome the problem of high space cost and long path results, a data-map-based Jump Point Search (DMBJPS) algorithm is proposed, which converts a raster map into a data map containing direction and distance information. Finally, a Triangle-node based A* (TNBA*) algorithm is designed. The main contribution of this work is summarized as follows.

- A three-level indoor environment modelling framework is proposed to divide the indoor environment into three levels, in which each level was modelled by a corresponding environment modelling.
- Based on the data map, the jumping point search algorithm is proposed, which can speed up the subsequent node search process in skip point search by calculating the key nodes during the map preprocessing.
- An intrinsic cost function is designed by measuring the distance between the inbound and outbound edges. Compared with the existing algorithms, this function is more in line with the actual walking path and can obtain more reasonable results of the triangle sequence.

2   System Model

The indoor environment is more complex than the outdoor, which leads to the structural composition of the indoor environment incapable of being divided by a uniform standard as it is outdoors. The concept of layers in the indoor environment is universal, so the paths between layers and inner layers are all considered.

Taking into account the characteristics of the indoor environment, a three-level indoor model framework is proposed, which divides the indoor environment into multiple levels.

- The level between adjacent floors is defined as the first level. At this level, the optimal path is easy to find because there are only a few fixed paths.
- Each independent floor is defined as the second level. There are different areas in each layer, there is a lack of fixed routes and obstacles between areas. To overcome this difficulty, grid partitioning techniques can be applied to second-level modelling.
- The indoor level is defined as the third level. Because of the large number of obstacles in the room and the irregular shape and distribution, the third level can be modelled using triangulation techniques.

Notice that, the indoor level model should accurately represent the positional relationship between obstacles and accessible roads. Since the indoor space is not large, excessive triangle nodes are not generated after modelling, which can avoid undesirable time consumption.

3   Path Searching Based on the 3-Level Indoor Model

According to the discussion in the previous section, because there is only a fixed number of passable paths between layers, the optimal path can be discovered by brute force techniques. The purpose of the second level modelling is to find the optimal path between different areas in the same layer, and it is necessary to develop high speed path searching algorithm in consideration the complex indoor environments. In the third level, i.e., searching the optimal path inner one room, accurate obstacle avoidance search is needed due to the complexity and ambiguity of indoor scenes. Therefore, the path searching in the second and the third level is the focus of this paper.

3.1   Second-Level Path Searching

Since the data in the indoor map is static, the online path search can be accelerated by offline preprocessing of the map, based on which a data map containing key hop information can be formed. The data map contains direction and distance information from each node to key points and obstacles, by which the neighbours of the node to be expanded in a certain direction can be obtained rapidly. The details of the second-level path searching algorithm is described as follows.

After processing the map, a data map is generated, in which each point has its own direction and distance information to the expanding node in the corresponding direction. This information can serve as the basis for the traditional

---

Input: The initial indoor map
Output: The preprocessing data map

1 **for** *each point in the map which can be reached from four directions* **do**
2   **if** *the point has forced neighbours* **then**
3   this point is recorded as the forced point;
4   **end**
5 **end**
6 **for** *each point of all entry areas in the map* **do**
7   setting all directions as the searching directions apart from that closing the obstacles.
8 **end**
9 /*Forced jump points and passable area entry points are the basis of the entire data map, so these two types of points are defined as key points*/
10 /*Traverse the data map from left to right*/
11 **for** *each row obtained by traversing the data map* **do**
12   calculate the nearest key point in the current line according to the west key point information;
13   **if** *the nearest key point is not the western key point and is obstacle point* **then**
14   this distance is setted to be negative;
15   **end**
16   record the pint and distance information into the data map;
17 **end**
18 /*Traverse the map diagonally*/
19 **for** *each point in the diagonal obtained by traversing the data map* **do**
20   calculate the distance to the nearest key point or obstacle that the point can reach in the diagonal direction, where the distance to the obstacle is marked as a negative value;
21 **end**
22 generate the data map;
23 **return** the data map;

Algorithm 1. the preprocessing of the data map

---

jump search algorithm, which can speed up the search speed of traditional algorithms. Correspondingly, we design a new path search algorithm based on the generated data map, which can be described as follows.

Notice that, it is necessary to search the candidate paths according to the parent point and its direction information when calculating the successor point set V of *curNode*. In other words, if the search direction is the basic direction, the search process should start from *curNode* and search the candidate path in all five directions in the same half-plane as the basic direction. If the search direction is diagonal, then the search process should start from *curNode* and search along the diagonal direction and its two decomposition directions.

Input: the start point s and the data map
Output: the path

1 add s into the table *open*;
2 **while** *table open is not empty and the path has not been found* **do**
3   get the point *curNode* whose value of f is the minimal;
4   /*Notice that, $f = g + h$, g is the distance between *curNode* and the start point s, and h is the estimated Manhattan distance from *curNode* to the terminal */
5   add *curNode* into the table *close*, get the father point *fnode*, and delete *curNode* from table *open*;
6   obtain the successor nodes of *curNode*, denoted by set $\mathbb{V}$ according to the direction from *curNode* to *fnode*;
7   **for** *each point sucNode* $\in \mathbb{V}$ **do**
8   **if** *sucNode is terminal* **then**
9   **return** path;
10   **end**
11   **if** *sucNode* $\in$ *close* **then**
12   continue; /*this point has been expanded*/
13   **end**
14   **if** *sucNode* $\in$ *open* **then**
15   recalculate the value of g;
16   **if** *the new value of g is less than the old one* **then**
17   reassign the value of parent point, g and f;
18   **end**
19   **end**
20   **if** *sucNode* $\notin$ *open* **then**
21   mark the parent point of *sucNode* as s;
22   reassign the values of g, h and f;
23   **end**
24   **if** *the path is found or the table open is empty* **then**
25   **return** path;
26   **end**
27   **end**
28 **end**

Algorithm 2. obtaining the precedent point

3.2 Indoor Path Searching Based on TNBN*

Taking into account the complexity of the indoor environment caused by a large number of irregular shaped obstacles, the technology of triangulation is applied to the construction of environmental models. Based on this, this paper proposes an indoor path search algorithm based on triangulation. The algorithm improves the grid A* algorithm and designs a heuristic function based on the vertical distance to perform heuristic search on the triangular grid, which includes "adjacent point checking" and "path generating and smoothing".

On this basis, a serial of triangles adjacent to each other can be found, and one path can be formed by connecting the incoming and outgoing points.

Input: the data map and the moving object in the data map
Output: all the triangles including they can be expanded or not
1 **for** *each moving object* **do**
2   calculate the radius of the convex hull of a moving object;
3   calculate the number of constrained edges of a triangle node, denoted by *count*;
4   **if** *count=3* **then**
5 | this triangle is not expanded since all the edges are constrained;
6   **end**
7   **if** *count=2* **then**
8   **if** *this point is not the terminal* **then**
9 | this triangle is not expanded since there is only one direction that the moving object can entering
10   **end**
11   **end**
12   **if** *count=1* **then**
13   /*there are only two directions that one is entering this triangle crossing the one edge, denoted by *ca*, and the other one is leaving it crossing another edge, denoted by *cb**, where the intersection of *ca* and *cb* is *c*./
14   calculate the nearest obstacle point, denoted by *o*, and $d = |oc|$;
15   **if** $d \geq 2r$ **then**
16 | this point can be expanded since there is no collision during the object moving;
17   **end**
18   **if** $d < 2r$ **then**
19 | this point can not be expanded;
20   **end**
21   **end**
22   **if** *count=0* **then**
23   /*this case is similar to *count* = 1*, and check this point according to the case that *count* = 1/
24   **if** *one of the edges satisfies the requirements of obstacle avoidance* **then**
25 | this point can be expanded;
26   **end**
27   **end**
28 **end**

Algorithm 3. adjacent point checking

4 Experiments and Analysis

In this experiments, three metrics are critical for evaluate a method of path searching, namely, path length, time cost and space cost.

To test the performance of the proposed method DMBJPS, the indoor environment is modelled based on the grid method. In this experiment, the obstacles, start point and end point are generated randomly, and the test has been repeated 100 times. The comparison of path lengths among different map sizes are shown in Fig. 1. It can be found that the path lengths of all three methods are

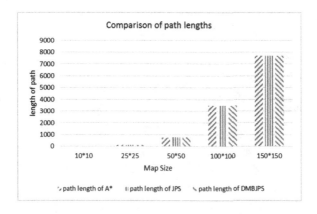

Fig. 1. The comparison of path lengths.

similar, and the DMBJPS can almost achieve the similar performance with the A* algorithm.

Figures 2(a) and (b) shows the comparison of path searching time and space cost among different map sizes. As showing in Fig. 2(a), the performance of DMBJPS has a significant improvement in contrast to the A*, and has a less searching time than the JPS algorithm. More importantly, the map size has a less impact on searching time of the DMBJPS than the others. The reason is that the DMBJPS can make a full use of the distance information to find the successors of the expanded nodes quickly. Figure 2(b) shows the performance of different methods in space cost. Similar to the path searching time, the DMBJPS has achieved a obvious improvement in space cost.

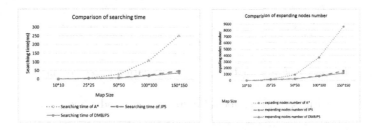

(a) The path searching time.   (b) The path searching time.

Fig. 2. the performance of different methods in time and space costs.

Figure 3 shows the comparison of path lengths between the proposed TNBA* and the existing centroid search algorithms. It can be found that the TNBA* algorithm can find shorter-distance path than the centroid search algorithm.

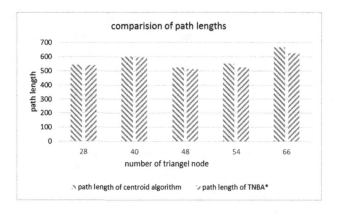

Fig. 3. The number of users over each radio link.

Figure 4(a) and (b) show the comparison of path with smoothing or not. The result of the TNBA* algorithm is a sequence of triangular nodes, and the path actually presented to the users are generally smoothed, which is more aesthetically achieve a better navigation effect, and the path length is shortened to a certain extent.

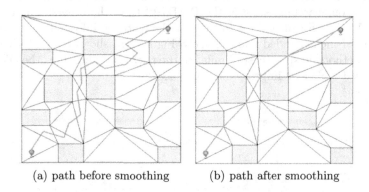

(a) path before smoothing   (b) path after smoothing

Fig. 4. The number of users over each radio link.

Through the above experiment, we can conclude that:

- The path length obtained by the TNBA* algorithm is shorter than that of the existing centroid algorithm.
- The TNBA* algorithm achieves a similar performance in terms of search speed with the traditional algorithms.
- The memory occupancy of the TNBA* algorithm is similar to the traditional algorithms.

5   Conclusions

This paper proposes a jump point search algorithm based on the data map, based on which can accelerate the traditional grid-based path search algorithm. It can be widely used in indoor environments existing a number of obstacles with regular shapes. More further, this paper also proposes an A* search algorithm based on triangular nodes. Compared with the existing algorithms, it obtains a shorter path length in consideration of the path smoothing, which can be widely applied to the indoor environments existing a large number of obstacles with irregular shapes.

Acknowledgements. This work is supported by the Natural Science Foundation of Heilongjiang Province in China (No. F2016028, F2016009, F2015029, F2015045), the Youth Innovation Talent Project of Harbin University of Commerce in China (No. 2016Q N052), the Support Program for Young Academic Key Teacher of Higher Education of Heilongjiang Province (No. 1254G030), the Young Reserve Talents Research Foundation of Harbin Science and Technology Bureau (2015RQQXJ082), and the Fundamental Research Fund for the Central Universities in China (No. HEUCFM180604).

References

1. Cho, K., Suh, J., Tomlin, C.J., Oh, S.: Cost-aware path planning under co-safe temporal logic specifications. IEEE Robot. Autom. Lett. **2**(4), 2308–2315 (2017)
2. Xiong, Q., Zhu, Q., Zlatanova, S., Zhiqiang, D., Zhang, Y., Zeng, L.: Multi-level indoor path planning method. Int. Arch. Photogramm. Remote Sens. Spat. Inf. Sci. **40**(4), 19 (2015)
3. Peng, B., Lü, Z., Cheng, T.C.E.: A tabu search/path relinking algorithm to solve the job shop scheduling problem. Comput. Oper. Res. **53**, 154–164 (2015)
4. Valencia, R., Andrade-Cetto, J.: Path planning in belief space with pose SLAM. Mapping, Planning and Exploration with Pose SLAM. STAR, vol. 119, pp. 53–87. Springer, Cham (2018). https://doi.org/10.1007/978-3-319-60603-3_4
5. Yu, L., Long, Z., Xi, N., Jia, Y., Ding, C.: Local path planning based on ridge regression extreme learning machines for an outdoor robot. In: 2015 IEEE International Conference on Robotics and Biomimetics (ROBIO), pp. 745–750. IEEE (2015)
6. Hossain, M.A., Ferdous, I.: Autonomous robot path planning in dynamic environment using a new optimization technique inspired by bacterial foraging technique. Robot. Auton. Syst. **64**, 137–141 (2015)
7. Dijkstra, E.W.: A note on two problems in connexion with graphs. Numer. math. **1**(1), 269–271 (1959)
8. Singh, Y., Sharma, S., Sutton, R., Hatton, D.C.: Optimal path planning of an unmanned surface vehicle in a real-time marine environment using dijkstra algorithm (2017)
9. Goldberg, A., Radzik, T.: A heuristic improvement of the Bellman-Ford algorithm. Technical report, STANFORD UNIV CA DEPT OF COMPUTER SCIENCE (1993)
10. Hagelbäck, J.: Hybrid pathfinding in starcraft. IEEE Trans. Comput. Intell. AI Games **8**(4), 319–324 (2016)

11. Harabor, D.D., Grastien, A., et al.: Online graph pruning for pathfinding on grid maps. In: AAAI (2011)
12. Zhou, K., Lingli, Y., Long, Z., Mo, S.: Local path planning of driverless car navigation based on jump point search method under urban environment. Futur. Internet **9**(3), 51 (2017)

Predicting Statutes Based on Causes of Action and Content of Statutes

Zhongyue Li[1,2], Chuhan Zhuang[1,2], Jidong Ge[1,2(✉)], Chuanyi Li[1,2],
Ting Lei[1,2], Peitang Ling[1,2], Mengting He[1,2], and Bin Luo[1,2]

[1] State Key Laboratory for Novel Software Technology, Nanjing University,
Nanjing 210093, China
gjdnju@163.com
[2] Software Institute, Nanjing University, Nanjing 210093, China

Abstract. In recent years, the Supreme People's Court of China has vigorously promoted the information construction of the people's court. A large number of information are stored in judgment documents, these documents contain the basic information of cases and statutes cited in cases. As far as we know, there are very few studies on the recommendation of statutes at present. We study the method by which judges choose the statutes in practical work, linking details of cases with the contents of statutes and choosing appropriate statutes to support the trial. Through data research on a large number of judgment documents, we propose a method based on the causes of action and contents of recommendation statutes, and use the Random Forest algorithm for training. In our approach, we take key words of statutes as the basic characteristics, which are used for finding more key words of documents to set up a key words dictionary for each cause of action. We use the dictionary to vector documents and train a classifier for each cause of action. We build a data set for this task and carry out evaluation experiments on the data set. The experimental results show that our method can perform well.

Keywords: Criminal judgment documents · Syntactic parsing
Text similarity measure · Word2Vec

1 Introduction

In recent years, the Supreme People's Court of China has vigorously promoted the information construction of the people's court. China Judgments Online website was built in 2013, and by January 2018 it has accumulated more than 45 million documents. Through the study of these data, we can put forward more intelligent information means to assist the judge to handle cases, and realize the modernization of the trial system.

A judgment document is written by a judge about the trial process and the result after the completion of a case. Judgment documents faithfully record cases. Therefore, these documents contain a lot of key information. The judgment documents are written in a fixed framework and can be divided into fixed logical segments. The detailed description of the claims of the parties and the facts of the cases appeared in the same

© Springer Nature Singapore Pte Ltd. 2018
Q. Zhou et al. (Eds.): ICPCSEE 2018, CCIS 902, pp. 477–492, 2018.
https://doi.org/10.1007/978-981-13-2206-8_39

logical segment which we call Case Segment. In addition, in documents, the judges clearly list the statutes cited in cases, as a supplement to the reasoning of documents. In our study, the subject is the recommendation of statutes. We define statutes recommendation as follow: when a new case arises, the statutes suitable for judging the case are recommended by the details of the case.

The statutes recommendation can be applied to many aspects of the legal domain. For judges, these legal statutes are selected by judges in the process of hearing cases after reading a lot of relevant laws and regulations. Judges need to combine the claims of parties and facts of the cases with statutes. The application of statutes recommendation can help judges to liberate from the complicated work of selecting, and let judges pay more attention to the details of the case. In addition, the recommendation can also help judges evaluate whether the statutes they have selected are appropriate. For the parties, when they meet a case, they usually have to seek the help of a lawyer. With the statutes recommendation they can get some relevant laws directly through the details of the case and understand the case better.

When establish a case, the judge determines the nature of the case. The nature of a case is decided by the law protecting the object that is infringed in the case. Then, the case is assigned to a judge who specializes in the nature of the corresponding case. The judges start to trial the case after determining the cause of action (CoA). A CoA is the abstract of a case, and it also reflects the nature of the case. The cause of action is determined correctly or not, directly affects the quality of handling case. A correct CoA can help the judges hold the core of the lawsuit, use the policy and the law correctly to judge cases. A correct CoA help with narrowing the scope of statutes.

Through the understanding of the trial process, if we want to recommend statutes, we must first determine the CoA to narrow the scope of statutes, and select the recommended statutes on the basis of the details of the case. Common statutes, which are often quoted in general cases and recommending these statutes can improve the efficiency of the judges' handling of these cases.

We use the documents that are downloaded from China Judgments Online. Through information extraction, we extract the information we need and convert the documents into XML format. In Sect. 3, we propose a method of statutes recommending based on CoAs and the content of statutes. With the key words of recommended statues as the basic feature, a keyword dictionary is obtained. After vectoring the document by the key dictionary, we treat the statues as classes and use Random forest to train classifiers. We have carried out experiments on two CoAs: divorce cases and traffic accident cases. We collect 300,000 judgment documents as the data source and train a classifier for each CoA. We conduct evaluation experiments, and experimental results show that our proposed method is effective.

To summarize, we make the following contributions:

(1) We propose a method for statutes recommending. A key words dictionary is designed to establish the connection between statues and documents, and can reduce the dimension of input data. Our model is more simple, light and efficient.
(2) We propose a method to format document. We divide document into seven segments and convert the judgment documents into XML format. It can help us to understand documents and get most related information of document.

2   Related Work

Recommender Systems (RSs) are software tools and techniques providing users with suggestions or items. The term first appeared in [1–3] in the 1990s. Approaches to recommender system involve mainly collaborative filtering, content-based filtering, and hybrid methods. Among the three approaches, collaborative filtering and content-based filtering are more common. Amazon proposed "Amazon.com Recommendations: Item-to-Item Collaborative Filtering" in 2003 [4]. This described item-based collaborative filtering Algorithm, which was widely used in business systems. RSs in legal domain becomes as a popular research topic in recent years. Moens proposed an XML retrieval model to retrieve legislations in constructed and unconstructed forms [5]. Kim et al. established a model to recommend statutes based on law ontology [6]. Rosso et al. proposed a passage retrieval system based on JIRS to extract conflicting sentences from contract text [7]. Liu et al. designed a three-phase prediction (TPP) algorithm to predict associated statutes for lay-people [8]. Though previous works have been done on statutes recommendation, such as Kim's work and Liu's, those work mainly focus on providing recommendations for lay users rather than professional users.

Text mining is the process of deriving unknown, understandable, usable information from large sum of text data. It involves information retrieval, natural language processing, machine learning etc. Because text data usually lacks structure [9], such data can hardly be managed in direct ways. Most studies apply the Vector Space Model (VSM) [10] to transform text to multi-dimensional vector so that it can be further processed. In VSM, documents and queries are represented in vectors with several dimensions, each of which correspond to a separate term. TF-IDF (Term Frequency–Inverse Document Frequency) [11, 12] is one of the method to compute the term weight. It is used to evaluate the importance of a word to a certain corpus. In our study, we build a statute keyword collection to evaluate how important a word is in legal statues. Word2vec was proposed by Mikolov et al. [13, 14]. The model is used to reconstruct the representation of words into form of vectors. In this model, words with similar context are located close to each other in the vector space. The word2vec model use two architectures: continuous bag-of-words (CBOW) and continuous skip-gram, with the former being faster while the latter being slower but better for infrequent words. The model works well on Chinese as well. Bai et al. utilized word2vec in sentiment computing [15].

Classification: Many classical algorithms are used for classification, such as k-nearest neighbor (k-NN), Naïve Bayes (NB), SVM, random forest etc. Among the classification algorithms, SVM and random forest are used in many research. In the TPP algorithm, Liu et al. trained a multi-labeled SVM classifier to classify a case into several statutes [8]. Duan et al. applied SVM-RFE to classify cancer [16]. We trained a model based on SVM and random forest at the final stage, which makes classifications and gives recommendations of the top relevant statues.

3   Approach

3.1   Overview

Each case corresponds to a judgment document as a written summary of the case. We call the paragraphs which describe details of the case as Case Segment. In Case Segment, the judge summaries the basic situation of the case, including the cause, the process and the result, as well as the parties' claims and the main evidence. A judgment document also includes a number of legal statutes used by the judge to hear the case. Our goal is to recommend appropriate statutes for a case based on the Case Segment.

A law itself is in accordance with nature of case and CoAs, e.g. in the field of civil law, divorce laws are specially used to hear divorce cases. A statute is often accompanied by the nature of case and CoAs it relates to. We recommend statutes based on CoA because it can narrow the recommended scope of statues availably.

In this paper, we introduce a method of statutes recommending based on CoAs and the content of statutes, and convert recommendation problem into classification problem by treating statutes as classes. As shown in the Fig. 1, our method is carried out for a certain CoA. In document of XML format, cited statuses are extracted and labeled. After statistics, we filter statues which have very low frequency to recommend. We use the TF*IDF weighting method to extract key words from recommend statues and train a Word2Vector model using documents. From the Word2Vector model, we get a keyword dictionary and use it to do the document vectorization using keyword's frequency in the document. In the process of vectorization, we remove stop words which we introduce in Sect. 3.4. Then we use Random forest to train a classifier for the CoA.

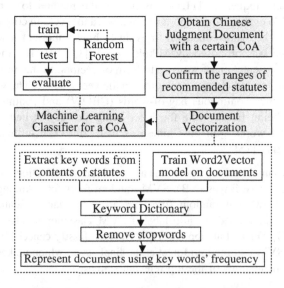

Fig. 1. Overview of our proposed method

We do the same preprocessing on the new cases, i.e., use the keyword dictionary of the CoA of the case to do the vectorization. According to the CoA of the case, the corresponding statutes classifier is selected for recommendation.

3.2 Corpus Preprocessing

This section introduces the preprocessing operation of the training set. The purpose is to get the information related to the training data, and to eliminate the effect of noise data.

In our study, the documents we use are written in text formats with natural language, and we cannot get information directly from them. Therefore, we covert them into XML format. Through information extraction, we extract and label key information that hidden in contents of the documents. Finally, we output data that computers can understand and use. The structure of judgment documents is fixed and has obvious stylized characteristics [17]. The judgment documents are written in a fixed order and can be divided into seven segments, each of which contains a number of natural paragraphs. These segments are Header segment, Participants segment, Judicial Records segment, Case segment, Judicial Analysis segment, Judgment Result segment and End segment. We define an extraction rule for each segment and use different information extraction methods to extract and label information according to the characteristics of information. Finally, we convert the judgment documents into XML format.

In this paper, we only focus on the CoA, the basic situation of the case and the cited statutes. Due to the lack of strict writing regulations in judgment documents and the failure of some judges in writing, some undesirable documents have emerged.

3.3 Determining Range of Recommended Statutes

In our study, we only focus on common statutes used frequently, because the statutes which have low frequency have little data to train and can interfere training. We choose the statute that is cited more than 200 times in 20,000 documents. A statute is composed of two parts: the name and the number. For example, "《中华人民共和国婚姻法》第一条", the name of this statute is "《中华人民共和国婚姻法》", the number is "第一条".

Due to the wrong writing of judges, there will be errors in the name of statutes in some documents, which affect the recommendation effect. Therefore, we need to standardize the writing. In this paper, we stipulate that a standard statute should consist of correct name and number of strips, and do not contain punctuation marks and Arabic numerals, and we implement a method of standardization of statutes. In any steps related to statutes, statutes data must be standardized before it can be used as a valid data.

According to the definition of common statutes, we determine the range of recommended statutes by the following steps. We select 40,000 XML format documents and divide them into two averagely. For each copy of documents, the frequency of all cited statutes that appear in the corpus is counted and statutes with a frequency greater than 200 both in two copies are selected. For each case, we use $L = \{l_1, l_2, \ldots, l_m\}$ to denote the recommended statutes set. A standardized statutes content library is set up for statutes in the recommended scope so that we can get the content of a statute with the name and number.

3.4   Establishing Keyword Dictionary

Any text-based system requires some representation of documents, and the appropriate representation depends on the kind of task to be performed [18]. Before training, we represent the Case segment as the format that the computer can understand. We use representative words to do vectorization of the text and establish a keyword dictionary and words in this dictionary which are the contact point between the statutes and the Case segment. Figure 2 shows the process. We train a Word2Vector model based on corpus to calculate the similarity of words and extract key words from the content of recommended statutes as basic features. Through the Word2Vector model, we can get similar words with high similarity with key words that we extract from statutes. We combine the similar words and the key words of statutes into a keyword dictionary.

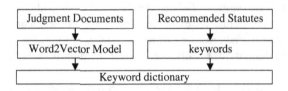

Fig. 2. The process of establishing a keyword dictionary

Word Segment: Word segment is a basic step in Chinese language processing. Chinese scholars have made great progress in Chinese word segment after their efforts. Nowadays, there are many excellent word segment tools, such as JIEBA, ICTCLAS, SCWS, LTP, NLPIR and so on. Among these word segment tools, JIEBA is one of the most efficient systems that is easy to use and supports python. In this paper, we choose the exact pattern to do word segment, which tries to cut the sentence most accurately, and is suitable for the application of text analysis.

Stopwords Removal: In the sequence of words after segment, there are a lot of common words and no information words which are called stopwords. In order to make our results more accurate, these words need to be removed. There are many mature and complete stopword list that can be used, and we choose a stopword list with a length of 2788 as the basic list. In addition, we remove the following words in view of the characteristics of legal field. (1) Removal of legal words with high frequency. These words appear frequently in the statutes and documents. We don't think these words can be used as the characteristics statutes. (2) Removing time, places, person names, proper nouns. (3) Removing words with a very high frequency but a little practical significance. These words include modal auxiliaries, adverbs, prepositions and conjunctions. (4) Removing words or punctuation marks with a length of 1. In Chinese, a word usually consists of at least two Chinese characters. In our research, we stipulate that the length of a word is at least 2. (5) Since the study of this paper is the Chinese statutes recommendation, we have removed English words and only reserved the Chinese words.

Word2Vector Model: For each CoA, we are training a Word2Vector model to compute word similarity. We collect 200,000 judgment documents and use the Case Segment as the corpus for each CoA to do this job. In the process of training, we exclude words with a frequency less than 5.

Extraction of Statutes: We extraction key words from contents of recommended statutes for each CoA. For each statute in range of recommended, we obtain the content of it from the standard statutes content library according to the name and number of the statute and select some words that can represent the content. We evaluate the importance of a word in of a statute with the TF-IDF method. TF-IDF method is based on a theory: a word that appears in many documents does not have a good distinction, and words that appear in fewer files should be given a larger weight. The TF-IDF weight is computed as the product of two terms: TF and IDF. TF (Term Frequency) indicates the frequency of words appearing in a document. The second term is IDF (Inverse Document Frequency). The main idea of IDF is that the less document that contains a word, the greater the distinction of the word has, and the word has greater IDF value. When calculating TF-IDF value, we do not consider the stopwords. In this paper, we use the keyword extraction interface based on TF-IDF in JIEBA to extract every key word. When we use the interface, two parameters are required: Sentence and topK. We define Sentence as the content of a statue, and topK is the max number of key words that each statute can have. Here is the formula for calculating TF-IDF of a word w in a statute s:

$$TF\text{-}IDF(w,s) = TF(w,s) * IDF(w,s)$$

$$TF(w,s) = \frac{frequency\ of\ w\ in\ s}{number\ of\ words\ in\ s}$$

$$IDF(w) = \log \frac{number\ of\ statutes}{number\ of\ statutes\ with\ w\ in\ it}$$

Establish Keyword Dictionary: We get 30 similar words with high similarity with key words that we extract from statutes and combine the similar words and the key words of statutes into a keyword dictionary. We use $K = \{k_1, k_2, \ldots, k_n\}$ to represent the dictionary.

3.5 Vectorization of Documents

In this paper, we use the Case segment of judgment documents as training data. The core content of a judgment document is composed of the claims of litigants, the evidence, the facts, the cited statutes and the judgment result. As shown in Fig. 3, the judge confirms the evidence according to evidence and deduces the fact from facts, the fact is connected with the cited statutes, and the judgment result is made based on the litigants' claims and the cited statutes. Among these terms, the claims of litigants, the evidence and the facts all belong to the Case segment of the case. The Case segment in a judgment document is the most direct basis for recommendation. Therefore, in this paper, we choose to use the Case segment as training data.

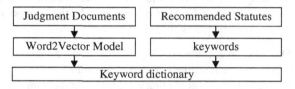

Fig. 3. The core structure of judgment documents

We use the keyword to do the vectorization of Case segment. Our idea comes from the bag-of-words model. Many machine learning algorithms require the input to be represented as a fixed-length feature vector. When it comes to texts, one of the most common fixed-length features is bag-of-words [19]. According to bag-of-words model, the keyword dictionary is denoted as $K = \{k_1, k_2, \ldots, k_n\}$. For a document, we calculate the frequency of each keyword in it. For k_j, the frequency of it in a document d_i is denoted as W_{ij}. For a set of documents $D = \{d_1, d_2, \ldots, d_n\}$, the vectorization result is $W = \{w_1, w_2, \ldots, w_n\}$.

3.6   Training and Recommending

In this paper, we choose Random Forest algorithm to train the classifier for each CoA. Random forests method [20] is a successful ensemble learning method. There are two sources of randomness in the algorithm, random inputs and random features, which make random forests an accurate classifier in different domains [21].

For a CoA, we introduce how to use the keyword dictionary to do the vectorization of documents. In order to carry out the training, we need establish training data of cited statutes data according to the recommended range of statutes. We denote the recommended range of statutes as $L = \{l_1, l_2, \ldots, l_m\}$, and each statute is treated as a class. For a statute l_j and a document d_i, if l_j is cited in d_i, then y_{ij} is 1, otherwise y_{ij} is 0. Then we get the classification vector y_i of d_i. For a set of documents $D = \{d_1, d_2, \ldots, d_n\}$, we get $Y = \{y_1, y_2, \ldots, y_n\}$ to represent the cited situates by each document. Figure 4 shows the relationship between L, d_i and y_i.

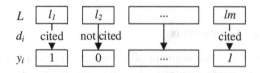

Fig. 4. Relationship between L, d_i and y_i

After the above operations, for a CoA, the recommendation range is $L = \{l_1, l_2, \ldots, l_m\}$, documents set is $D = \{d_1, d_2, \ldots, d_n\}$, we can get the vector of cases $W = \{w_1, w_2, \ldots, w_n\}$, the vector of cited statutes reference $Y = \{y_1, y_2, \ldots, y_n\}$. Statutes recommendation is a multi-label classification problem, so we use Random Forests to train the classifier for each CoA. When a new case occurs, we quantify the basic situation of the case of the case using the keyword dictionary of its CoA. Then we use the vectorization result as an input and use the classifier of its CoA to predict the statutes. The results given by the classifier are 0/1 sequences $R = \{r_1, r_2, \ldots, r_n\}$, we need to transform the sequence into a set of statutes. For r_i, if the value is 1, the l_i is recommended. Figure 5 shows the process of recommending.

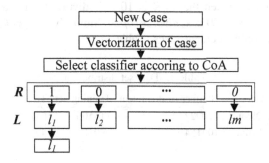

Fig. 5. Process of recommending

4 Evaluation

4.1 Preparing Data

In order to evaluate our method, we choose divorce disputes cases from first-instance civil procedure and traffic accident cases from first-instance criminal procedure to carry out experiments. We collect the judgment documents of these two CoAs. In order to avoid discrepancy in the statutes due to the difference between the closing time of the case, the area of the case and the court level, we chose documents from different levels of courts throughout the country, and the closing time of the cases are from 2013 to 2015, including 150,000 documents of divorce disputes cases and 150,000 documents of traffic accident cases. We convert these documents into XML format and establish a XML corpus with the steps mentioned in Sect. 3.2.

After corpus preprocessing, we have removed documents that lack of CoA, Case segment and statutes. However, there are still some of the following documents that are not suitable for training: (1) Documents with less content. There is too little description of the case in the kind of documents. These descriptions are not enough to support the reasoning, so we exclude document less than 20K. (2) After statistics on frequency of statutes cited in documents, we find that some documents only cite the certain statute with the highest frequency. We filter out these documents to prevent the impact of these documents.

We establish a keyword dictionary for each CoA in the method mentioned in Sect. 3.3 following five steps: (1) Using the corpus of each CoA to train Word2Vector model. (2) Determining the recommended range of statutes. Statutes with a frequency more than 200 in both two 20,000 documents are reserved. (3) Setting up a standard statutes content library for each CoA. We can obtain the contents of a statute with the name and number of it. (4) Standardizing name of statutes. We remove the punctuation, convert the Arabic numerals in the number of statutes into Chinese, and complement the statute name of the abbreviation to the full name. (5) Using JIEBA to extract keywords for each statute. The maximum number of keywords per statute is *topK*, and we take *topK* = 10 and *topK* = 20 respectively. Then we combine these key words with Word2Vector model to build the keyword dictionary. For divorce disputes case, the dictionary obtained when the *topK* is 10 is k_divorceTop10, and the dictionary obtained when the *topK* is 20 is *k_divorceTop20*. Similarly, we get *k_trafficTop20* and *k_trafficTop20* for traffic accident cases (Table 1).

Table 1. Data set distribution

CoA	Number of document	Training set	Test set
Divorce	120922	114506	6416
Traffic	104154	99183	4971

4.2   Experiment Metrics

In this paper, we use three metrics to evaluate the result. They are Precision@topK, Recall@topK, F@topK. In metrics, topK is the maximum number of keywords that can be obtained per statute using the JIEBA word segmentation tool.

Precision@topK is a measure of the correctness of the recommended statutes. Precision@topK is higher, indicating that in the recommended statutes, the more the appropriate number of statutes. Precision@topK is defined as:

$$Precision@topK = \frac{1}{N}\sum\frac{R}{T}$$

where N denoted the total number of documents. R denotes the number of statutes which are predicted right. T denotes the total number of statues which are recommended.

Recall@topK is a measure of the recall of recommended results for the true cited statutes in a case. The higher the Recall@topK, the more statutes cited actually are recommended. The formula is as follows: where S is the actual number of references in the instrument. Recall@topK is defined as:

$$Recall@topK = \frac{1}{N}\sum\frac{R}{S}$$

where S denotes the actual number of statutes cited in a document.

When we use Precision@topK and Recall@topK, there may appear contradictory situation, then we can use F@topK in order to evaluate comprehensively. The F@topK value is larger, the better the recommended effect is. F@topK is defined as:

$$F@topk = \frac{2 * (Precision@topK) * (Recall@topK)}{Precision@topK + Recall@topK}.$$

4.3   Experiment and Results

In order to evaluate the method proposed in this paper, we design and conduct the following experiment.

Different topK Comparison of Random Forest: We use Random Forest for divorce disputes cases and choose different *topK* values to experiment with. We set the *topK* value of 10, 20, respectively, corresponding to two keyword dictionaries k_divorceTop10 and k_divorceTop20. We use 114506 judgment documents of divorce disputes cases as a training set and 6416 judgment documents as the test set. The test results are shown in Table 2. As can be seen from the table, topK has little effect on the overall recommendation.

Table 2. Different topK experiment result

CoA	$topK = 10$	$topK = 20$
Precision@topK	92.26	92.17
Recall@topK	59.38	59.35
F@topK	69.01	68.95

We observe the contents of the statutes in the recommended range and the contents of the keyword dictionary k_divorceTop10, k_divorceTop20. We found that the content of a statute is generally not very long, therefore, *topK* is 10 can basically cover the keywords of the statute, and *topK* 20 will introduce many disturbing words that are not helpful to the recommendation. It results in a large number of interference terms in k_divorceTop20, thus reducing the recommendation effect.

Different CoAs Comparison of Random Forest: In order to evaluate the effect of Random Forest method on different CoAs, we conduct a recommendation experiment on divorce disputes cases and traffic accident cases. In this experiment, *topK* value is 20. We use k_divorceTop20 as the keyword dictionary for divorce disputes cases and use k_trafficTop20 for traffic accident cases. We train a classifier for each CoA using the method of Random Forest mentioned in Sect. 3.4. The experimental results are shown in Table 3. We can see that the Random Forest method has good performance on both cases. Among them, the result of traffic accident cases is higher than the divorce disputes case.

We observe the frequency of citations of each statute and find that there is a big difference between the frequency in divorce disputes and traffic accident. Statutes are more evenly cited in traffic accident cases. The frequency of statutes in divorce disputes cases vary widely. The statutes with higher frequency are trained better and are more

Table 3. Different CoAs comparison experiment result with Random Forest

CoA	Traffic accident	Divorce disputes
Precision@topK	92.26	81.33
Recall@topK	59.38	83.51
F@topK	69.01	80.45

likely to be recommended, while less frequent ones are less effective than the former. The recommended effect is not so good for the statutes that appears less in the documents and more statutes with low frequency cannot be correctly recommended in divorce disputes cases.

Comparison Experiment of Random Forest and SVM: In order to evaluate the recommended effect of random forest on this problem, we choose SVM model for comparison experiment. We use the divorce dispute cases for experiments. For each statute in the recommended range of divorce dispute, we use the method in Sect. 3.4 to set up a keyword dictionary with *topK* = 20 and train a single SVM probability model. In SVM method, we recommend the K most probable statutes for users. The test results are shown in Fig. 6 below. It can be found out that with the increase of the recommended number K, the precision gradually decreases, while the recall increases gradually and the value of f decreases continuously. When the K value is 2, two statutes with the highest probability are recommended, so the accuracy is higher and the recall rate is lower. When K is 10, the recommended range is expanded to 10, while most of the documents cite only 3 or 4 statues, so the accuracy is lower and the recall rate is very high. In addition, we find that the training time of SVM is much longer than the training time of Random Forest. On the whole, the performance of the SVM probability model is not as good as that of the Random Forest model. Because the sample number is very unbalanced, Random Forest model can deal with the problem very well. When topK is taken as 20 and K is 2, comparison between Random Forest and SVM is in Table 4.

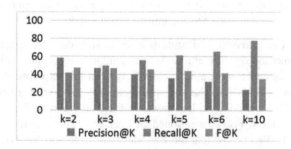

Fig. 6. Different K values test results with SVM

Different CoAs Comparison of SVM: In order to test the effect of different cases using SVM, we use divorce dispute and traffic accident to train their SVM models for each statute respectively, and the *topK* value is 20. The results are shown in Figs. 7, 8

Table 4. *topK* = 20 with Random Forest and *K* = 2 with SVM

	SVM	*topK* = 10
Precision	58.34	92.26
Recall	41.51	59.38
F	47.2	69.01

and 9. It can be seen that when *K* takes different values, the traffic accident cases perform better then divorce dispute cases, and precision is far higher than the divorce dispute cases'. As mentioned earlier, the statutes of traffic are more evenly cited, so the average training effect is better and the precision value is higher. However, SVM models for traffic accident and divorce dispute are not as good as Random Forest models.

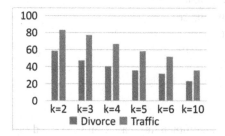

Fig. 7. Precision@K of SVM under two CoAs

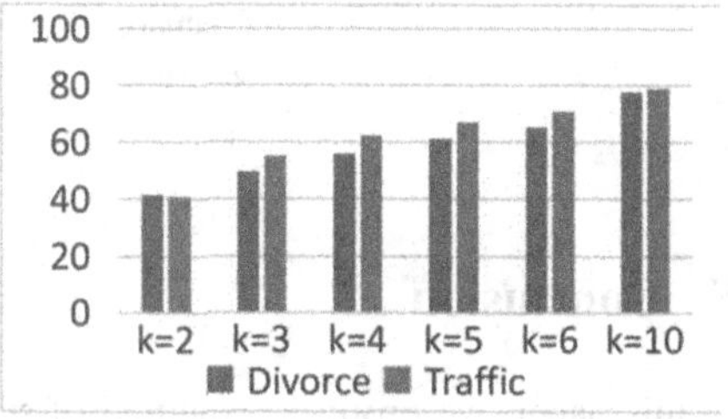

Fig. 8. Recall@K of SVM under two CoAs

Different Basic Characteristics Comparative of Random Forest: In the method proposed in this paper, the keyword dictionary is based on the CoA and the content of statutes. Firstly, key words are extracted from the content of statutes, and the key words are used as the basic features to expand a keyword dictionary. Therefore, we design an experiment using keywords extract from documents instead of statutes as the basic features to evaluate our method.

First of all, we select randomly 20,000 judgment documents from the training set of divorce dispute cases, and use TF-IDF weighting method for each document to extract 50 keywords into the keyword dictionary K_case. There are about 40,000 words in K_case. After using K_case to quantify the Case segment, Random Forest is used for training. Table 5 shows the comparison result. As can be seen from the table, the model based on documents contents has a higher precision than the model based on statutes content, but the recall rate is lower than the latter.

This is because the difference between the frequency of statute cited in divorce disputes is particularly large, and TF-IDF is a weighting method based on word frequency. As a result, when many keywords are directly extracted from documents, many relevant word of statutes with low reference frequency are omitted. These statutes are trained less effectively, resulting in a lower recall rate. When we use the model based on statutes content, the keyword dictionary is set up from each statute and statutes with the low frequency can also get features to be trained. Therefore, the model based on statutes content has a better performance.

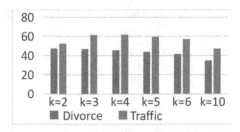

Fig. 9. F@K of SVM under two CoAs

Table 5. Different basic characteristics

Characteristics	K_case	$topK = 10$
Precision	93.39	92.26
Recall	54.16	59.38
f1	65.20	69.01

5 Conclusion

In this paper, we propose a method of statutes recommending based on CoAs and the content of statutes. Firstly, we preprocess documents by transforming judgment documents into XML documents. For each cause of action, we choose some statutes that are cited more frequently as recommended statutes after the document statistics of the cause of action, and use the documents of the cause of action as corpus for training the Word2Vec model, so that we can count the word similarity. Secondly, we extract keywords from contents of statutes via TF-IDF weighting method and get similar words of the keywords using the Word2Vec model, and then set up a keyword dictionary which is the connection point between statutes and Case segments of documents. Thirdly, we use the keyword dictionary to do vectorization of Case segments of documents and use the vector as input data. Finally, we choose Random Forest algorithm to train classifier for each CoA. Our experiment results show that our approach can recommend statutes more accurately.

In our work, there are still some disadvantages. The keyword dictionary is based on contents of statutes via TF-IDF. However, TF-IDF is related to frequency of words and ignores semantics, so words in the keyword dictionary is disordered and structures of sentences are not preserved. Thus, in the future work, we will consider how to preserve the information of sentences' structures. In addition, we find that the keyword dictionary contains many meaningless words that have bad impact on our training and cannot be predicted. We try to filter these words manually later and we get better training result. Thus, we will study how to get a better keywords dictionary with less manual work.

Acknowledgements. This work was supported by the National Key R&D Program of China (2016YFC0800803).

References

1. Hill, W.C., Stead, L., Rosenstein, M., Furnas, G.W.: Recommending and evaluating choices in a virtual community of use. In: Proceedings of Human Factors in Computing Systems (CHI 1995), pp. 194–201. ACM (1995)
2. Resnick, P., Iacovou, N., Suchak, M., et al.: An open architecture for collaborative filtering of netnews. In: Proceedings of the Conference on Computer Supported Cooperative Work (CSCW 1994), pp. 175–186. ACM (1994)
3. Shardanand, U., Maes, P.: Social information filtering: algorithms for automating "Word of Mouth". In: Proceedings of Human Factors in Computing Systems (CHI 1995), pp. 210–217. ACM (1995)
4. Linden, G., Smith, B., York, J.: Amazon.com recommendations: item-to-item collaborative filtering. IEEE Internet Comput. **7**(1), 76–80 (2003)
5. Moens, M.-F.: Combining structured and unstructured information in a retrieval model for accessing legislation. In: International Conference on Artificial Intelligence and Law, pp. 141–144. ACM (2005)
6. Kim, W., Lee, Y., Kim, D., et al.: Ontology-based model of law retrieval system for R&D projects. In: Proceedings of the 18th Annual International Conference on Electronic Commerce: e-Commerce in Smart Connected World, p. 26 (2016)
7. Rosso, P., Correa, S., Buscaldi, D.: Passage retrieval in legal texts. J. Log. Algebraic Program. **80**, 139–153 (2011)
8. Liu, Y.-H., Chen, Y.-L., Ho, W.-L.: Predicting associated statutes for legal problems. Inf. Process. Manag. **51**, 194–211 (2015)
9. Croft, W.B., Metzler, D., Strohman, T.: Search Engines - Information Retrieval in Practice. Pearson Education, London (2009)
10. Salton, G., Wong, A., Yang, C.-S.: A vector space model for automatic indexing. Commun. ACM **18**(11), 613–620 (1975)
11. Salton, G., Buckley, C.: Term-weighting approaches in automatic text retrieval. Inf. Process. Manag. **24**(5), 513–523 (1988)
12. Salton, G., Allan, J., Buckley, C.: Automatic structuring and retrieval of large text files. Commun. ACM **37**(2), 97–108 (1994)
13. Mikolov, T., Sutskever, I., Chen, K., et al.: Distributed representations of words and phrases and their compositionality. In: Proceedings of Annual Conference on Neural Information Processing Systems, pp. 3111–3119 (2013)
14. Mikolov, T., Chen, K., Corrado, G., et al.: Efficient estimation of word representations in vector space. CoRR abs/1301.3781 (2013)
15. Bai, X., Chen, F., Zhan, S.: A study on sentiment computing and classification of Sina Weibo with Word2Vec. In: Proceedings of IEEE International Congress on Big Data, pp. 538–363. IEEE (2014)
16. Duan, K.-B., Rajapakse, J.C., Nguyen, M.N.: One-versus-one and one-versus-all multiclass SVM-RFE for gene selection in cancer classification. In: Marchiori, E., Moore, J.H., Rajapakse, J.C. (eds.) EvoBIO 2007. LNCS, vol. 4447, pp. 47–56. Springer, Heidelberg (2007). https://doi.org/10.1007/978-3-540-71783-6_5
17. Tian, L.: Individuation with standardization—to thinking of the judicial writing. Hebei Law Sci. **26**(7), 160–164 (2008)
18. Lewis, D.D.: Text representation for intelligent text retrieval: a classification-oriented view. In: Text-Based Intelligent Systems, pp. 179–197 (1992)

19. Le, Q., Mikolov, T.: Distributed representations of sentences and documents. In: Proceedings of the 31st International Conference on Machine Learning (ICML 2014), pp. 1188–1196 (2014). JMLR
20. Breiman, L.: Random forests. Mach. Learn. **45**(1), 5–32 (2001)
21. Huang, X., Pan, W., Grindle, S., et al.: A comparative study of discriminating human heart failure etiology using gene expression profiles. Bioinformatics **6**, 205 (2005)

Adaptive Anomaly Detection Strategy Based on Reinforcement Learning

Youchang Xu, Ningjiang Chen$^{(\boxtimes)}$, Hanlin Zhang, and Birui Liang

Guangxi University, Nanning, China
chnj@gxu.edu.cn.com

Abstract. In complex and changeable cloud environment, the monitoring and anomaly detection of cloud platform become very important. Good anomaly detection can help cloud platform managers to make quick adjustments to ensure a good user experience. Although many anomaly detection models have been put forward by researchers in recent years, the application of these anomaly detection models to a given service still faces the challenge of parameter adjustment, which is time-consuming and exhausting, and still fails. In order to solve the problem of parameter adjusting, in this paper, an adaptive anomaly detection framework is proposed, the process of parameter adjustment is transformed into a general Markov decision process by means of reinforcement learning, which realized the automation of parameter adjustment, reducing the workload of operator and the effective detection rate of the anomaly detection model is improved, we compared it on three typical KPI (Key Performance Indicator) curves with artificial adjustment mode and other optimization strategies, in the end, we verified the effectiveness of the strategy used in this paper.

Keywords: Anomaly detection · The decision-making process of Markov
Automatic parameter adjustment

1 Introduction

With the rapid development of cloud computing, in order to ensure the stability, safety and reliability of the cloud platform, operator selected a large of monitoring indicators from different levels of cloud environment, forming KPI curve to represent the running status of cloud system. The anomaly detection of the basic performance indicators, such as CPU and memory utilization, can effectively reduce the failure rate of the virtual machine allocation and reallocation in the cloud platform [1].

Although many anomaly detectors [2, 3] have been put forward in these years, we still face great challenges when detector is applied to a given service. The first is **Definition Challenge**: it is difficult to precisely define anomalies in reality [4], for most anomalies, the operator can't give with quantitative form, and they are more liked to use the observed KPI curve to describe the anomaly, which makes the error always in the communication. The second is **the Challenge of Maintain Precision**: because of the complexity of users' habits in cloud computing, the characteristics of monitoring data are constantly changing, which will cause the current anomaly detection model to

© Springer Nature Singapore Pte Ltd. 2018
Q. Zhou et al. (Eds.): ICPCSEE 2018, CCIS 902, pp. 493–504, 2018.
https://doi.org/10.1007/978-981-13-2206-8_40

fail to a certain extent, so developer need to adjust the anomaly detection model according to the new data characteristics.

These two challenges have created huge human costs, in order to solve those challenges, in this paper, the automatic adjustment of parameters is realized based on the reinforcement learning algorithm. First of all, we defined the adjustment action set of different anomaly detection algorithm with unified form, solved the problem of adjusting precision and adjusting speed during parameter adjustment. Secondly, we optimized the algorithm of Q-learning by setting the dynamic reward function and the comparison value of state, which reduced the number of iterations for finding the optimal parameter. Finally, we carried out the comparison experiment with the way of manual and other optimization strategies on the recall, precision, F-Score, and the effectiveness of the strategy is verified. The main contributions of this article are as follows:

(1) Automatic identification of data characteristics of monitoring data using differential technology;
(2) The anomaly detection model is automatically adjusted by using reinforcement learning technology, and the iterative process is optimized.

The remaining part of this paper is organized as follows: the second chapter introduced the automatic adjustment model proposed in this paper, the third chapter introduced the comparative experiments on three typical KPI data. The fourth chapter introduced the related work of this paper, and the fifth chapter summarizes the work of this paper and prospects for the future.

2   Adaptive Detection Method

The challenge of anomaly detection in cloud environment is two iterative processes, the first is the communication process between the developers and the operator who maintain the KPI curve. The second is the process of developers adjusting the anomaly detection model based on the change of data characteristics. The two iterative processes consume a lot of labor cost, so we hope to build an intelligent model to reduce labor cost. This paper builds an adaptive anomaly detection framework based on reinforcement learning technology, as shown in Fig. 1. The framework is divided into 4 main modules, the resource module is the basic service, the data processing module is divided into data collection and data characteristic recognition module, which separately collected and identified the monitoring data. The anomaly detection module contains a variety of anomaly detection algorithms, which are used to detect different anomaly scenarios. The automatic adjustment module is used to automatically adjust the anomaly detection model.

2.1   Data Characteristics Recognition Module

The data characteristics recognition module determines whether the current anomaly detection model needs to be re-adjusted, in cloud environment, the KPI data type are classified as periodic, unstable and stable, we used the difference technique to distinguish the global fluctuation trend of the monitoring data, and recognize whether the

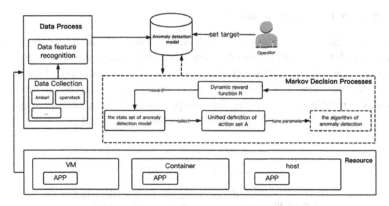

Fig. 1. Adaptive anomaly detection framework

data is periodic, we used the wavelet transform to distinguish the relationship between the global fluctuation of the monitoring data and the local fluctuation, so as to recognize whether the data is stable or unstable. We first differentiate the current data, if the global fluctuation is very large before the difference processing, but the global fluctuation is very small after the difference processing, it is periodic data, otherwise it is non-periodic data. In the non-periodic data, we comparing the global variance and local variance of the current data, and if the global variance approximates the local variance, it is stable data. If the global variance is far greater than the local variance, it's unstable data, the specific process is described in Table 1.

Table 1. Data characteristics recognition algorithm

Algorithm 1: Data characteristics recognition
Input: Monitoring dataset DS = $\{p_1, p_2, ..., p_i, ..., p_n\}$, where p_i is the monitored value.
Output: Monitoring dataset type
1: Calculate global variance: $V_g(DS)$
2: Differential processing: diff(DS), Calculate the global variance after differential $V_g'(diff(DS))$
3: IF $V_g(DS) \gg V_g'(diff(DS))$ //*Comparison of global variance*
4:      Output Periodic data
5: ELSE
6: Wavelet transforms the DS to obtain local fluctuations: Mallat(DS)
7: Calculate local variance: $V_l(Mallat(DS))$
8: IF $V_g(DS) \gg V_l(Mallat(DS))$ // *Comparison of global and local variance*
9:      Output Unsteady data
10: ELSE
11:      Output Stable data
12:    END IF
13: END IF

2.2   Anomaly Detection Module

The anomaly detection module contains a variety of anomaly detection algorithms for detection of different types of anomaly scenes. In order to make operator directly adjust the anomaly detection model, we convert the anomaly detection model measurement index to the indicators that operators are concerned and good at. In the anomaly detection, the most concerned index for the operator is the recall and precision of the anomaly detection. The calculation formulas are shown as 1 and 2:

$$recall = \frac{\# \, of \, ture \, anomalous \, points \, detected}{\# \, of \, ture \, anomalous \, points} \tag{1}$$

$$precision = \frac{\# \, of \, ture \, anomalous \, points \, detected}{\# \, of \, anomalous \, points \, detected} \tag{2}$$

F-Score is a comprehensive index of recall and precision, which can comprehensively measure the detection effect of anomaly detection model. Therefore, we use the F-Score as a single measure of the status of anomaly detection model, see Eq. 3.

$$F\text{-}Score = \frac{2 \cdot recall \cdot precision}{recall + precision} \tag{3}$$

According to the actual needs, the operator set the range of the specified F-Score to the anomaly detection model, such as F-Score $\geq$ x. At this time, the F-Score become the target of the automatic parameter adjustment module. Therefore, the conversion of operation indicators to parameter adjustment indicators has been completed.

2.3   Automatic Adjustment Module

In the automatic adjustment module, based on reinforcement learning technology, the adjustment problem of anomaly detection model is incorporated into the Markov decision process framework. We use the tuple (S, A, P, R, γ) to describe the anomaly detection model adjustment process as: under the action A of parameter adjustment, the anomaly detection model obtains a new state S and an immediate reward R. If the new state is already the target state, the iterative process end. otherwise, the action will continue to be selected to act on the new state to obtain the reward value until the state of the anomaly detection model is the target state, which is the implementation of a strategy, after multiple strategies have been executed, the strategy with the largest accumulated return is selected as the optimal strategy. The parameters under the optimal strategy are the best parameters of the anomaly detection model. In the iterative process of the optimal parameter calculation, the factors that affect the convergence speed are the selection of the parameter adjustment action and the setting of the reward function.

The Selection of Action Set: The anomaly detection model contains a variety of anomaly detection algorithms, and the range of parameter values of different anomaly detection algorithms is different. If the amplitude of the parameters is set to a fixed value, may be too high or too low for other algorithms when they are applicable to some algorithms. If the adjustment amplitude is too high, it may cause the optimal parameter always be missed during the adjustment process, and the iterative process can't be stopped. If the adjustment amplitude is too low, the adjustment speed will be slow, and the anomaly detection model can't be corrected in time. The anomaly detection algorithm is divided into two types: one's parameter has a specific range of values, for example, the Holt-Winters algorithm has three parameters α, β, γ, range from $(0, 1]$. The other's parameter has no specific range of values, for example, the parameter of the ARIMA algorithm have no specific range and can only be adjusted or estimated based on the existing data. In the process of parameter adjustment in this paper, considering the rate and precise range of parameter adjustment, the strategy we adopt is shown in Eq. 4, which is based on [2, 9] and our experience of parameter adjustment.

$$p(m) = \begin{cases} \frac{|k|}{10} \cdot n, m \in [-k, k], & k > 0, n = 1, 2, 3 \dots & \textit{if range is known} \\ \frac{1}{10} \cdot m \cdot n, & m \neq 0, n = 1, 2, 3 \dots & \textit{if range is unknown} \\ n, & m = 0, n = 1, 2, 3 \dots & \textit{if range is unknown} \end{cases} \quad (4)$$

In formula 4, m is the initial parameter, and n is the number of parameters that the algorithm needs to adjust. There are two factors that affect the rate of parameter adjustment. One is the magnitude of adjustment, the reason is as described above. The other is the number of parameters, the set of parameter adjustment actions is composed of Descartes products with each parameter up, down, and non-adjustment. Excessive action set elements make it take more time to select actions, therefore, we appropriately increase the adjustment range according to the number of parameters n.

Iterative Process Optimization: The Q-Learning [5] algorithm is the main method to solve the problem of model free reinforcement learning. Its basic idea is to record the utility value of the state in each action by establishing a function table. The utility value represents the validity and value of the action selected under the current state, and also as the basis for the next strategy to select the action, and updates the action state value of the current state through the action state value of the next state, as shown in Fig. 2 (the data in the diagram is used for demonstration):

(a)	a1	a2	a3	a4
s0	0	0	0	0
s1	0	0	0	0
s2	0	0	0	0
s3	0	0	0	0
sT	0	0	0	0

(b)	a1	a2	a3	a4
s0	0	0.1	0	0
s1	0	0	0	0
s2	0	0	0	0
s3	0	0	0	0
sT	0	0	0	0

(c)	a1	a2	a3	a4
s0	0	0.1	0	0
s1	-0.3	0	0	0
s2	0	0	0	0.7
s3	0	0.2	0	0
sT	0.5	0	0	0

Fig. 2. $Q(s, a)$ table

	a1	a2	a3	a4
s0	0.2	0.1	-0.4	0.3
s1	-0.3	0.6	-0.5	0.2
s2	0.3	-0.4	-0.6	0.7
s3	0.75	0.2	-0.6	0.3
sT	0.5	-0.4	0.4	0.2

Fig. 3. Convergent function table $Q(s, a)$

The initial value of function table $Q(s, a)$ is (a) in Fig. 2. In one strategy, s_0 is selected randomly from the action of non-negative value in the initial state, in Fig. 2(b) a_2 is selected so that the state becomes s_1, and the utility value is 0.1 by formula 5, where r is the immediate reward given by the reward function, $Q(s_{t+1}, a_{t+1})$ is the utility value of the next state, 0 in Fig. 2, and the update function table as shown in Fig. 2(c) at the end of a strategy.

$$Q(s_t, a_t) = r + \gamma(\max(Q(s_{t+1}, a_{t+1}))) \tag{5}$$

In the process of action selection, the Q-Learning algorithm is selected according to the non-negative value of the corresponding $Q(s, a)$ function table, such as the next policy, Fig. 2(c) as the basis, the optional action at s_0 is $\{a_1, a_2, a_3, a_4\}$, and the optional action at s_1 is $\{a_2, a_3, a_4\}$. The execution of each strategy will update the $Q(s, a)$ function table until the $Q(s, a)$ table converges as Fig. 3, at this time, the best strategy is to choose the maximum cumulative return value, for example, the optimal strategy in Fig. 3 is a sequence $\tau = \{a_4, a_2, a_4, a_1, a_1\}$.

In the original Q-Learning algorithm, the setting of the reward function is static. It gives rewards or penalties depending on whether the current action makes the model state better than the initial state or whether it is superior to the previous state. But at this time, there will be a state of s_1 in Fig. 3. When the function table is not yet convergent, there are always multiple actions to choose from in this state. Most actions do not bring optimization to the current model, which makes many invalid iteration steps in finding optimal strategies. We want to reduce the steps that can't make the model state transition to better in the overall adjustment process, so that the function table can converge faster, so we set the dynamic return function as formula 6:

$$R = \frac{F_t}{F_T} \cdot (F_t - F_{max}) \tag{6}$$

F_t is the F-Score value obtained from the anomaly detection model under the current parameter adjustment action, that is, the current state value. F_T is the parameter adjustment index from operator, $\frac{F_t}{F_T}$ make the current state value closer to the target value, and the bigger the reward value is, F_{max} is the maximum state value set during the execution of a policy. $F_t - F_{max}$ makes the action be rewarded only if the anomaly detection model gets better state value under the regulation action, otherwise it will be punished, which is beneficial to a strategy to reach the optimal state faster. Based on the above strategy, we get the pseudo code for obtaining the best policy based on Q-Learning algorithm, as shown in Table 2:

Through Algorithm 2, we compare the action selection process of static reward function, as shown in Fig. 4:

When the state s_1 is updated to s_2 in Fig. 4(a), according to the static reward function, action a_4 makes the state of the model better than the previous state, so we get the reward. However, because the state s_1 is punished, the whole model is not optimized. According to the dynamic reward function presented in this paper, we will get the punishment, as shown in Fig. 4(b). In the next policy execution, there are 4 actions that can be attempted in the state S2 (a) table, while there are only 3 of them in (b). Therefore, (b) table will arrive at the next state faster, and this advantage will be more obvious in the accumulation of multiple strategies. Comparing the convergent function table $Q(s, a)$ in Fig. 3. Under the proposed strategy, the function table $Q(s, a)$ will converge as shown in Fig. 5.

Compared to the static reward function, the dynamic reward function is stricter in the selection of the best strategy, so the utility value had more negative value, and the

Table 2. Optimal strategy seeking algorithm

Algorithm 2:Optimal strategy seeking algorithm based on Q-Learning
Input: Initialization parameters (x, y)

1:*Action Set* $A = \{x, x + p(x), x - p(x)\} \times \{y, y + p(y), y - p(y)\}$
2:InitializatiQ$(s, a), \forall s \in S, a \in A(s)$, Given parameter a, γ
3:Repeat:
4:Given initial state s, Choose action a according to ε greed strategy
5:Computational anomaly detection model score F_{max}
6:    Repeat(episode):
7:    Select action a according to $Q(s_t, a_t)$ in the state s
8:$Q(s_t, a_t) \leftarrow Q(s_t, a_t) + a[R_t + \gamma max_a Q(s_{t+1}, a_{t+1}) - Q(s_t, a_t)]$
9:    Calculation of the current anomaly model score F_t
10:        $s_t \leftarrow s_{t+1}; a_t \leftarrow a_{t+1}$
11:        $R_t = \frac{F_t}{F_T} \cdot (F_t - F_{max})$ // Calculation of reward
12:        IF $R_t > 0$
13:        $F_{max} = F_t$ //update the current best model
14:    until s is the terminated state
15: until all $Q(s, a)$ convergence
16:output: $\pi(s) = argmax_a Q(s, a)$

---

	a1	a2	a3	a4
s0	0	0.1	0	0
s1	-0.3	0	0	0
s2	0	0	0	0.1
s3	0	0.2	0	0
sT	0.5	0	0	0

(a)

	a1	a2	a3	a4
s0	0	0.1	0	0
s1	-0.3	0	0	0
s2	0	0	0	-0.2
s3	0	0.1	0	0
sT	0.3	0	0	0

(b)

Fig. 4. $Q(s, a)$ table comparison

	a1	a2	a3	a4
s0	-0.2	0.7	-0.7	-0.3
s1	0.3	-0.2	-0.1	-0.5
s2	-0.2	0.5	-0.8	-0.6
s3	0	-0.1	-0.3	-0.4
sT	-0.2	0.5	-0.1	-0.8

Fig. 5. The convergence function table $Q(s, a)$ under the dynamic reward function

adjustment action corresponding to the negative value will not be selected again, so the function table $Q(s, a)$ will converge faster. It is known from the (1) that an adjustment action is a non-adjustment action, so each line of the function table has at least one non-negative value, and it does not appear in a state without the optional action of a regulation action. In experiment, we verified by experiments that the convergence speed of function table $Q(s, a)$ is faster under dynamic reward function.

3 Experiment

3.1 Experimental Design

In order to verify the effectiveness of the proposed strategy in this paper, we conduct relevant experiments on the open desensitization data set [6] in the data center of Baidu. The physical environment of this experiment is 6 Sugon servers with 24 cores Intel(R) Xeon(R) CPU E52620, 64 GB memory and the operating system is CentOS 7.2. The software environment is openai/gym, and the programming environment is Anaconda 3.6. The comparative object is decision trees, K-means, and the strategy of parameter selection and estimation in [7]. In the anomaly detection model, the Holt-Winters, ARIMA and wavelet algorithms are selected as the detection algorithms for periodic data, stable data and unstable data. In order to verify the effectiveness of the strategy in the anomaly detection, we observe and analyze the change process of the recall and precision of the anomaly detection results and compare the F-Score values of the different anomaly detection models. In order to verify the optimization of the strategy in the iterative process, the number of iterations per adjustment process is compared with the original Q-learning algorithm.

3.2 Experimental Result

Verification of Anomaly Detection Effect: The recall represents the ratio of true anomalies detected to total true anomalies. The precision represents the ratio of the detection of the true anomaly to the total anomaly point. F-Score is used as an indicator to comprehensively measure the recall and precision. In our strategy, we set up the $F\text{-}Score \geq 0.60$, $F\text{-}Score \geq 0.70$, and $F\text{-}Score \geq 0.80$. We calculate recall, precision and the overall F-score value of the decision tree algorithm, the clustering algorithm, the parameter selection estimation method and the reinforcement learning method, and get the following experimental results in the 30 day. The following experimental results are obtained.

From Figs. 6 and 7, it can be seen that in the process of anomaly detection, the recall and the precision of the manual adjustment method are further reduced when each time data characteristic changed. And other strategies can maintain good anomaly detection after adjustment. On the recovery of anomaly detection model adjustment, from the fourth day, the tenth day, the eighteenth day, the twenty-third day and the twenty-eighth day of Fig. 7, our strategy (RL) and parameter selections method adjust the recovery fastest in the face of changes in data characteristics. Although both the decision tree strategy and the k-means method have high reliance on the characteristics of the new data, the decision tree strategy utilizes the markup update of the expert system, so that it is faster than the adjustment of the anomaly detection effect by the k-means. In the overall detection effect, the overall trend of the decision tree strategy shown in Figs. 6 and 7 is relatively stable, but the overall average is low. The overall fluctuations of RL, k-means, and parameter selection methods are large, but the overall average is high. We further compare the effects of different strategies in the overall anomaly detection process by using the precision comparison chart and the F-Score comparison chart.

From Fig. 8, we can see that under the same recall rate, the precision of RL strategy is the highest, followed by parameter selection and estimation methods. Decision trees and k-means have their own advantages under different recall rates. According to formulas 1 and 2 for calculating the recall rate and precision, the same recall rate indicates the same number of true anomaly points detected under the same detection data. The higher the precision, the fewer the number of points marked as anomaly. Therefore, the fewer anomaly points detected as false, the fewer false alarms occur. From the F-Score value of the overall anomaly detection process in Fig. 9, we can see that RL has the highest comprehensive score in terms of recall and precision. Therefore, the effectiveness of the proposed strategy in the adjustment precision of the anomaly detection model is verified.

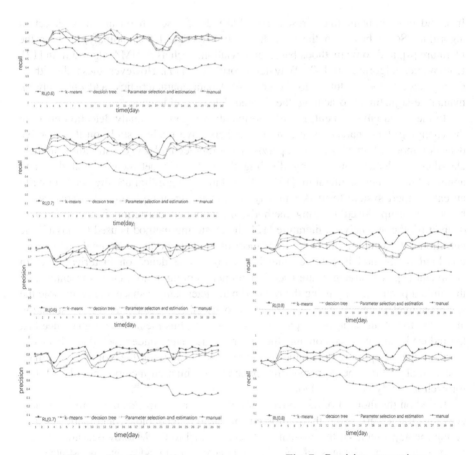

Fig. 6. Recall comparison   **Fig. 7.** Precision comparison

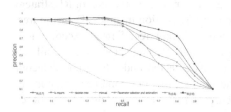

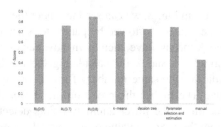

Fig. 8. Recall and precision change diagram   **Fig. 9.** Comparison of F-Score values

4 Related Work

In cloud environment, many researchers have done research on anomaly detection algorithm. Some based on the data distribution to detect the anomaly, such as the literature [8], and some methods based on deviation, such as ARIMA algorithm in [12], Holt-Winters algorithm in [10], Wavelet algorithm in [11]. However, these algorithms do not have a good solution to the change of data characteristics, and only rely on manual re-adjustment to achieve the desired detection efficiency.

In the cloud environment, in order to maintain a good anomaly detection effect in the changing data characteristics, the researchers have made a study on the adaptive detection model. Some based on supervised learning technology, constructing anomaly classifier to identify outliers by dividing the difference between normal data and abnormal data, such as literature [13], but this kind of algorithm usually needs to build an extra expert system to mark anomaly data, resulting in high operation cost. Some based on unsupervised learning methods, the anomaly clustering is constructed to distinguish the anomaly. In literature [12], the clustering method is used to classify the detection data, which is far away from most of the data points as anomaly. However, this kind of method is complex and has a high dependence on historical data. In document [7], two strategies are used in parameter configuration. One is to enumerate the limited parameters by using the reduced parameter sample space and enumerate the spare parameters in advance. The other is to use the targeted parameter estimation algorithm to get the appropriate parameters. However, this method can't guarantee that the reduced sample space contains the optimal parameters under each data characteristics in the pre-proposed parameter sample space, and for the complex anomaly detection algorithm, the corresponding parameter estimation method should be tested for each anomaly detection algorithm.

Based on the thought of the above work, this paper constructs an adaptive anomaly detection framework using the reinforcement learning technology, automatically triggering the adjustment of the anomaly detection model to the Markov decision process by perceiving the changes of the data characteristics,in addition, the strategy of selecting parameter adjustment action for different anomaly detection algorithms and the optimization algorithm based on the optimal parameters based on Q-Learning algorithm are proposed, which realizes the automatic adjustment of the anomaly detection model in the face of the change of data characteristics, and ensures a good anomaly detection effect in the cloud environment.

5 Conclusion

Anomaly detection is an important technology in the cloud platform. However, because of the complexity of the data changes in the cloud environment, the anomaly detection model needs to be constantly adjusted. In this paper, we introduce an adaptive detection method based on reinforcement learning, which automatically triggers the transformation of the anomaly detection model to the Markov decision process by perceiving the changes in the characteristics of the monitoring data, and we put forward the selection strategy of parameter adjustment action and the optimization algorithm for obtaining the best parameters, and realize the automatic adjustment of the anomaly detection model when the data characteristics is changed. In the future work, we will further optimize the iterative process of the parameters of the Markov decision process, reduce the time of the parameter selection process, and improve the adaptability and sensitivity of the model in the anomaly detection process.

References

1. Agrawal, B., Wiktorski, T., Rong, C.: Adaptive anomaly detection in cloud using robust and scalable principal component analysis. In: International Symposium on Parallel and Distributed Computing. IEEE (2017)
2. Yang, Y.M., Yu, H., Sun, Z.: Aircraft failure rate forecasting method based on Holt-Winters seasonal model. In: International Conference on Cloud Computing and Big Data Analysis. IEEE (2017)
3. Tian, H., Ding, M.: Diffusion wavelet-based anomaly detection in networks. In: International Conference on Parallel and Distributed Computing, Applications and Technologies, pp. 382–386. IEEE (2017)
4. Yan, H., Breslau, L., Ge, Z., Massey, D., Pei, D., Yates, J.: G-RCA: a generic root cause analysis platform for service quality management in large IP networks. In: Proceedings of the 6th International Conference, Co-NEXT 2010, pp. 5:1–5:12. ACM, New York (2010)
5. Aksaray, D., et al.: Q-Learning for robust satisfaction of signal temporal logic specifications. In: Decision and Control. IEEE (2016)
6. Data Set Homepage. https://github.com/baidu/Curve. Accessed 19 June 2018
7. Liu, D., et al.: Opprentice: towards practical and automatic anomaly detection through machine learning. In: Internet Measurement Conference, pp. 211–224. ACM (2015)
8. Tang, X.M., Yuan, R.X., Chen, J.: Outlier detection in energy disaggregation using subspace learning and Gaussian mixture model. Int. J. Control Autom. **8**, 161–170 (2015)
9. Pena, E.H.M., Assis, M.V.O.D., Proenca, M.L.: Anomaly detection using forecasting methods ARIMA and HWDS. In: Chilean Computer Science Society, pp. 63–66. IEEE (2017)
10. Ghanbari, M., Kinsner, W., Ferens, K.: Anomaly detection in a smart grid using wavelet transform, variance fractal dimension and an artificial neural network. In: Electrical Power and Energy Conference, pp. 1–6. IEEE (2016)

11. Zeb, K., et al.: Anomaly detection using wavelet-based estimation of LRD in packet and byte count of control traffic. In: International Conference on Information and Communication Systems. IEEE (2016)

12. Wazid, M., Das, A.K.: An efficient hybrid anomaly detection scheme using k-means clustering for wireless sensor networks. Wirel. Pers. Commun. **90**(4), 1971–2000 (2016)

13. Erfani, S.M., et al.: High-dimensional and large-scale anomaly detection using a linear one-class SVM with deep learning. Pattern Recognit. **58**(C), 121–134 (2016)

Research on Country Fragility Assessment of Climate Change

Yanwei Qi[1], Fang Zhang[2($\boxtimes$)], and Zhizhong Wang[1]

[1] School of Computer Science and Telecommunication Engineering,
Jiangsu University, Zhenjiang 212013, China
[2] School of Electrical and Information Engineering,
Jiangsu University, Zhenjiang 212013, China
1347411167@qq.com

Abstract. Climate change has attracted people attention in recent years. IPCC's assessment reports have informed people that climate change also has an influence on the regional instability. Droughts, sea level rise and resource shortages can further aggravate instable societies and fragile states. We establish a fragility evaluation system (FES) to evaluate the fragility of a state, using weighted mean method to calculate the fragility index. In the proposed system, we use pressure-state-response (PSR) model to measure the effect of climate change which mainly contributes to the degree of national vulnerability. Consequently, we divide states into five levels according to the fragility score. Back-Propagation neural network model is employed to predict the fragility index of Central Africa which is based on proposed system. The results can well demonstrate the effectiveness and rationality of our model.

Keywords: Climate change · Regional instability
Pressure-state-response model

1 Introduction

In recent years, the climate around the world has undergone significant changes, such as: global warming, melting glaciers and so on. IPCC (Intergovernmental Panel on Climate Change) third assessment report [1–5] made a brand- new summary of climate change in the 20th century. It forecasts the trend of climate change in the 21st century and analyzes the impact of global change on natural and human social systems. A fragile state is significantly susceptible to crisis in one or more of its sub-systems. A fragile state may fail to meet the needs of people, which will may eventually lead to the split of a county if we don't take any action. Climate change has a lot of impact on all aspects of human life whether now or the future, then, the influence may be possible to lead to the weakening and breakdown of social and governmental structures. As a result, a county may be reduced to a fragile state. So, formulating suitable evaluation system is gradually become an urgent task for many countries especially fragile states.

Since climate change plays a vital role in determining a country's fragility, it's significant to develop a suitable model to determine a country's fragility and simultaneously measures the impact of climate change. Besides, the approaches of data

© Springer Nature Singapore Pte Ltd. 2018
Q. Zhou et al. (Eds.): ICPCSEE 2018, CCIS 902, pp. 505–515, 2018.
https://doi.org/10.1007/978-981-13-2206-8_41

processing become more and more important, for example, MapReduce [6] is a popular programming paradigm for big data analysis. A series of efficient algorithms for determining the currency of dynamic datasets are proposed to deal with the velocity of big data [7]. To identify how climate change increases fragility, we propose an evaluation system based on PSR model [8, 9]. Besides, we use our model to determine the fragility of a country which is one of the top 10 most fragile states [10] and show how the state may be less fragile without the influence of climate changes.

2   Establishment of Vulnerability Evaluation Index System

A national system should include many elements such as economy, society, nature, policies and resources. There are complex relationships existing among these elements. In order to accurately determine the degree of national vulnerability and to measure the impact of climate change, we establish the Fragility Evaluation System (FES). In the proposed system, we use system theory to explain the relationship between climate change and national system, so the pressure-state-response" model (PSR) is selected to assess the degree of national vulnerability. The dynamic process of the system under the circumstance of climate change meet the pattern of PSR. In this pattern, external factors pose pressure on the system and generate stimulus inputs which can cause changes of system state. As a result, the changes response to the input in some way and eventually manifests itself as the country's fragility. In our system, the pressure is defined as the risk of a national system when the climate changes and one cannot take a timely response to the risk, namely exposure. State is defined as the direct expression of pressure and effective response, namely sensitivity. Responses are defined as a set of measures and responses made to reduce the effects of stress, namely adaptive capacity.

We establish the evaluation system with four levels: target-domain-presentation-factor. In this system, the next level indicator group reflects the situation of the upper level indicator group. The target layer called state fragility index is a comprehensive assessment of the country's vulnerability score under the circumstance of the climate change. The country's vulnerability is determined by the evaluation scores. The domain layer consists of exposure, sensitivity and adaptive capacity. In the PSR, stress, state, and response correspond to exposure, sensitivity and adaptive capacity respectively. The exposure is a pressure indicator which reflects the negative impact of climate change on people's survival status. Sensitivity is a state indicator which is easily influenced by the exposure. Besides, it reflects the situation of a country's society, nature and economy under the influence of climate change. Adaptive capacity is a responsive indicator that reflects a series of responses to exposure and sensitivity. It includes the construction of infrastructure, the improvement of science and technology and government policy to deal with the negative effects of climate change. The presentation layer and the factor layer are used to extend and specify the target layer. The interior elements of the presentation layer and domain layer are designed to further interpret the meaning of the domain layer.

From the Fig. 1, we can see that progressive relationships exist in our proposed indicator system. The efficient metric is set up in the following subsection in the proposed system.

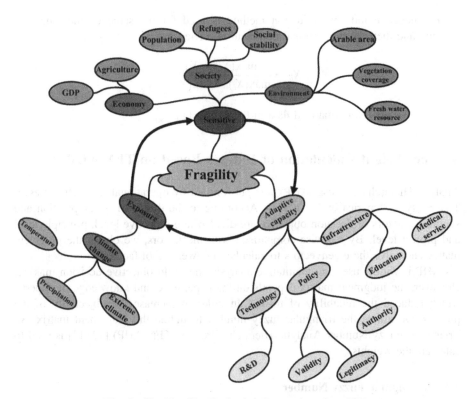

Fig. 1. The Fragility Evaluation System based on PSR

3 Data Pre-processing

Suppose that the original indicator data matrix consists of m evaluation objects and n evaluation indexes in an evaluation system:

$$X = (x_{ij})_{m \times n}$$

where $(x_{ij})_{m \times n}$ represents the observation of the jth index of the ith object. In order to reduce and eliminate the negative effect of the extreme data on the calculation, we need to normalize the data in the matrix.

For positive indexes:

$$X_{ij} = \frac{X_{ij} - \min(X_i)}{\max(X_i) - \min(X_i)}$$

For negative indexes, valuation method is used for the same chemotactic processing, and then the dimensional treatment is carried out.

$$X_{ij} = \frac{\max(X_i) - X_{ij}}{\max(X_i) - \min(X_i)}$$

Then we get the normalized data in the matrix $(x_{ij})_{m \times n}$.

4 The Weight Calculation of Indexes Based on TFN-AHP

Analytic Hierarchy Process (AHP) was put forward by operational research professor T.L. Saaty in America in 1970s [11]. According to the theory of decomposition and synthesis, factors of decision objective are divided into objective level, principle level and project level. By pairwise comparison with the factors, we obtain the judgment matrix and solve the eigenvectors to calculate the weights of factors.

AHP is widely used in quantitative analysis and multi-objective decision making. However, the judgment matrix depends on the experience and knowledge of experts, which reduced the reliability of results. In order to increase the objectivity of the process, we utilize the triangular fuzzy number to define the judgement matrix. So Triangular Fuzzy Number-Analytic Hierarchy Process (TFN-AHP) [12, 13] is used to calculate the weights.

4.1 Triangular Fuzzy Number

Triangular Fuzzy Number is proposed by mathematician Van Laarhoven in Netherlands in 1983, using triangle geometry to describe uncertainty. Triangular fuzzy number is denoted as $\tilde{M} = (l, m, u)$. l, m, u and their subordinate function is presented in Fig. 1.

In Fig. 2, $l \leq m \leq u$, u and l represent the upper and lower bound of $\tilde{M}$ respectively. The fuzzy extent is determined by the difference between u and l. The larger $|u - l|$ is, the larger the fuzzy extent is.

There are some basic rules in two triangular fuzzy numbers $\tilde{M}_1 = (l_1, m_1, u_1)$ and $\tilde{M}_2 = (l_2, m_2, u_2)$ as follows.

Addition: $\tilde{M}_1 \oplus \tilde{M}_2 = (l_1 + l_2, m_1 + m_2, u_1 + u_2)$
Multiplication: $\tilde{M}_1 \otimes \tilde{M}_2 = (l_1 l_2, m_1 m_2, u_1 u_2)$
Reciprocal: $\tilde{M}^{-1} = \left(\frac{1}{u}, \frac{1}{m}, \frac{1}{l}\right)$

4.2 Weight Calculation Based on TFN-AHP

Firstly, we establish the hierarchical structure including the main indexes. Then, we define the judgement matrix by giving score to measure the relative importance. The score is not an exact value but a fuzzy interval. We use these intervals to construct fuzzy matrix $R = (r_{ij})_{n \times n}$, $r_{ij} = r_{ji}^{-1}$, $i, j = 1, 2, \cdots, n$.

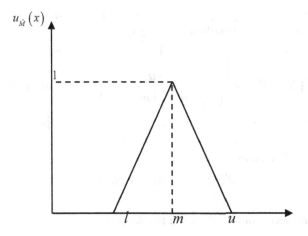

Fig. 2. The relationship among l, m and u

$$R = \begin{bmatrix} 1 & r_{12} & \cdots & r_{1n} \\ r_{21} & 1 & \cdots & r_{2n} \\ \cdots & \cdots & 1 & \cdots \\ r_{n1} & r_{n2} & \cdots & 1 \end{bmatrix}$$

$r_{ij} = (l_{ij}, m_{ij}, u_{ij})$ is a triangular fuzzy element which represents the relative importance between r_i and r_j. Besides, m_{ij} can measure relative importance in AHP.

Suppose the relative importance between r_i and r_j is r_{ij}, $r_{ij} = r_{ji}^{-1} = \left(\frac{1}{u_{ji}}, \frac{1}{m_{ji}}, \frac{1}{l_{ji}} \right)$.

For fuzzy judgement matrix $R = (r_{ij})_{n \times n} i, j \in \{1, 2, \cdots, n\}$, if the median matrix $\vec{R} = (m_{ij})_{n \times n}$ is a consistent judgement matrix, the $R = (r_{ij})_{n \times n}$ is a consistent fuzzy judgement matrix.

According to the fuzzy judgement matrix, the weight of each index corresponding to the higher level is

$$
\begin{aligned}
S_i &= \sum_{j=1}^{n} M_{E_i}^J \otimes \left(\sum_{i=1}^{n} \sum_{j=1}^{n} M_{E_i}^J \right)^{-1} \\
&= \left(\sum_{j=1}^{n} l_{ij}, \sum_{j=1}^{n} m_{ij}, \sum_{j=1}^{n} u_{ij} \right) \otimes \left[\sum_{i=1}^{n} \sum_{j=1}^{n} l_{ij}, \sum_{i=1}^{n} \sum_{j=1}^{n} m_{ij}, \sum_{i=1}^{n} \sum_{j=1}^{n} u_{ij} \right]^{-1} \\
&= \left(\frac{\sum_{j=1}^{n} l_{ij}}{\sum_{i=1}^{n} \sum_{j=1}^{n} u_{ij}}, \frac{\sum_{j=1}^{n} m_{ij}}{\sum_{i=1}^{n} \sum_{j=1}^{n} m_{ij}}, \frac{\sum_{j=1}^{n} u_{ij}}{\sum_{i=1}^{n} \sum_{j=1}^{n} l_{ij}} \right)
\end{aligned}
$$

Where $M_{E_i}^J$ means the importance of i relative to j in the fuzzy judgement matrix.

Let $S_1 = (l_1, m_1, u_1)$, $S_2 = (l_2, m_2, u_2)$. We use the following Piecewise function $V(S_1 \geq S_2)$ to measure the possibility of $S_1 \geq S_2$:

$$V(S_1 \geq S_2) = \begin{cases} 1 & m_1 \geq m_2 \\ \dfrac{l_2 - u_1}{(m_1 - u_1) - (m_2 - u_2)} & m_1 < m_2, l_2 \leq u_1 \\ 0 & 其它 \end{cases}$$

The $d'(r_i)$ is defined as

$$d'(r_i) = \min_{j=1,2,\cdots,n; i \neq j} V(S_i > S_j).$$

The weights vector of all indexes in this level is

$$w' = (d'(r_1), d'(r_2), \cdots, d'(r_n))^T.$$

After calculating sums and normalizing, we get the weights of indexes,

$$w = (d(r_1), d(r_2), \cdots, d(r_n))^T.$$

Finally, we obtain the weights of all indicators shown as Table 1.

Table 1. The weight of indicators

Target	Domain	Presentation	Factor	Weight	
State fragility index	Exposure (0.291)	Climate change (1)	Temperature	0.4	+
			Precipitation	0.4	+
			Extreme climate	0.2	+
	Sensitivity (0.317)	Economy (0.25)	GDP	0.7	−
			Agriculture	0.3	−
		Society (0.5)	Population	0.1314	+
			Refugees	0.7631	+
			Social stability	0.1055	+
		Environment (0.25)	Arable area	0.3108	−
			Vegetation coverage	0.1958	−
			Fresh water resource	0.4934	−
	Adaptive capacity (0.392)	Infrastructure (0.5396)	Education	0.5	−
			Medical service	0.5	−
		Policy (0.2970)	Authority	0.4	−
			Legitimacy	0.3	−
			Validity	0.3	−
		Technology (0.1634)	R&D	1	−

4.3 Fragility Grades

We use the proposed system to make a comprehensive evaluation of the country and get its fragile score which reflects the current fragile degree of the country. Research institutes around the world have given different standards on the evaluation of a country's current state. The World Bank's Country Policy and Institutional Assessment System (CPIA) [5] quantifies approximately 76 countries yearly. The CPIA is a positive indicator with a range of 0 point to 5 points, in which the lower score means the higher degree of the vulnerability. Table 2 shows the CPIA classification criteria.

Table 2. CPIA assessment criteria

Score	0–2.5	2.6–3.0	3.1–3.2	3.3–5
Grade	Serious	Moderate	Slight	Stable

According to our evaluation system, in order to measure the impact of climate change, we take the social environment, the natural environment and the economic environment into consideration. Then, we give the FES assessment criteria as Table 3.

Table 3. FES assessment criteria

Score	0–2	2–4	4–6	6–8	8–10
Grade	Stable	Slight	Moderate	Serious	Extreme

5 The Fragility of CAR

In this section, Central African Republic (CAR) is selected as example to analyze the fragility, which is one of the top 10 most fragile states. CAR is a landlocked country within the interior of the African continent, which covers a land area of about 620,000 km^2 and had an estimated population of around 4.6 million as of 2016. CAR is one of the least developed countries in the world announced by the United Nations. The economy is dominated by agriculture and the industry is weak. Figure 3 shows Topographical outline of CAR.

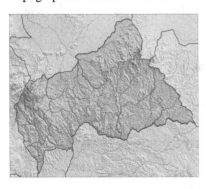

Fig. 3. The topographic map

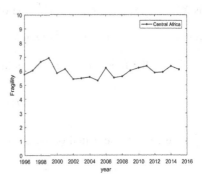

Fig. 4. The fragility of CAR over twenty years

We use the model to evaluate the trend of changes in the vulnerability index of CAR over the 20 years from 1995 to 2015. We can conclude that the average fragility of CAR is 6.12 from Fig. 4, which belong to serious fragility according to Table 3.

5.1   Back-Propagation Neural Network

Artificial Neural Network is an active research field in international academic circles. It has been widely used in predictive analysis, image processing, control and optimization. Back-Propagation neural network is the most basic model in this area.

BP neural network is a multi-layer feedforward network trained by error inverse propagation algorithm. Its topological structure consists of three kinds of layers including input layers, hidden layers and output layers. Every layer includes a number of neurons. Figure 5 shows the basic topological structure of BP neural network model.

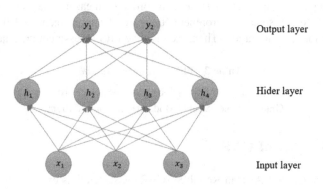

Fig. 5. BP neural network model

Supporting that the tth layer in network has t neurons, the next layer has t_1 neurons. The weight of link vectors from the tth layer's neurons to the next layer's neurons is $w_i = (w_{i_1}, w_{i_2}, \cdots, w_{i_t})$ and $W_t = (w_1, w_2, \cdots, w_i, \cdots, w_{t_1})^T$. The input of the tth layer is $X_t = (x_{t_1}, x_{t_2}, \cdots, x_{t_i}, \cdots, x_{t_t})^T$, the output is $Y_t = (y_{t_1}, y_{t_2}, \cdots, y_{t_i}, \cdots, x_{t_{t_1}})^T$, there is relationship $y_{t_i} = w_i \cdot X_t = \sum_{j=1}^{t} w_{ij} x_{tj}$.

Using the following equation as the active function:

$$f(v) = \frac{1}{1 + e^{-\partial v}}$$

We can get the actual output of sample X_t in the k neuron of t layer is

$$y_{t_k} = f\left(\sum_{j=1}^{t} w_{ij} x_{tj} - \theta_k\right)$$

After applying steepest descent method to accelerate convergence, we can get the weights of every hidden layer and the relationship between the output and input. The BP neural network model [14–16] is completely trained.

When the trained network receives input values, based on the trained algorithm, we can predict the output values.

5.2   BP Neural Network Model for Prediction

Let the weights of Temperature, Precipitation and Extreme climate be input, since they are selected to measure climate change in proposed FES. Let the weights of GDP, Agriculture, Population, Refugees, Social stability, Arable area, vegetation coverage and Fresh water resource be output, which are sensitive to climate change.

In order to mitigate the effects of climate change, we assume that the annual precipitation is evenly distributed, the annual mean precipitation variance is zero, the temperature tends to be stable, the annual variance in temperature is zero and the annual average number of extreme weather is zero. These assumptions are used to judge the changes of vulnerability without these indicator' effects. We utilize the original data to train the network, then, use the weakened climatic indicators to obtain weights for the other eight indicators. Finally, we use proposed FES to predict the degree of national vulnerability in two decades, as shown in Fig. 6.

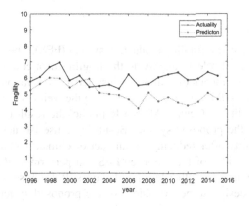

Fig. 6. The actuality and prediction of CAR fragility (Color figure online)

The blue line shows the change in vulnerability of countries over the past two decades without changing the weight of the indicator, and the red line shows the change in vulnerability after drastically reducing the impact of climate factors. From the figure, we can see that these two lines have basically the same degree of vulnerability during 1996–2003, and the degree of vulnerability of the two straight lines gradually separates over time. Obviously, without the influence of climate, the vulnerability of the country will gradually be lower than that of the climatic factor. From this we can see that the degree of vulnerability of a country to climate change is slow and important.

Figure 7 shows the score of the sensitivity layer from 1996 to 2016, the blue line represents the score of the sensitivity before weakened climate index, and the red line represents the score of the sensitivity after weakened the climate index. Obviously, the sensitivity index directly affects Exposure's score at the upper level, which has a great impact on the weight of climate change.

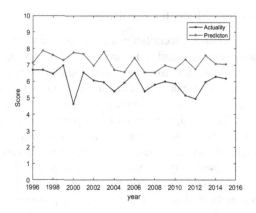

Fig. 7. The score of sensitivity layer (Color figure online)

6 Conclusion

In this article, we propose a fragility evaluation system (FES). The aim is to well reflect the climate plays a vital role in aggravate the fragility of a country. The FES which stresses the important impact of climate is inspired by the rationale of PRS model. TFN-AHP based on AHP is adopted to determining the weight of indexes. We use FES to calculate the fragility of Central African Republic, the result can well demonstrate the effectiveness of the proposed system. Besides, we use BP network to predict the change in vulnerability after reducing the impact of climate factors. The results well demonstrate the basic idea of the proposed FES can perform well on Central African Republic, it may be useful in other less fragile states. We will study this issue and use the idea of FES to design some extensions of the proposed system to fit less fragile states in the future.

References

1. Nguyen, T.T.X., Bonetti, J., Rogers, K., Woodroffe, C.D.: Indicator-based assessment of climate-change impacts on coasts: a review of concepts, methodological approaches and vulnerability indices. In: Ocean & Coastal Management, February 2016, Wollongong, pp. 18–43 (2016)
2. Scandurra, G., Romano, A.A., Ronghi, M., Carfora, A.: On the vulnerability of small island developing states: a dynamic analysis. In: Ecological Indicators, Rome, pp. 382–392 (2018)

3. Montaud, J.M., Pecastaing, N., Tankari, M.: Potential socio-economic implications of future climate change and variability for Nigerien agriculture: a countrywide dynamic CGE-microsimulation analysis. In: Economic Modelling, Bayonne, pp. 128–142 (2017)

4. Liang, Y.L., Yan, X.D., Huang, L., Lu, H., Jin, S.F.: Prediction and uncertainty of climate change in China during 21st century under RCPS. J. Trop. Meteorol. 102–110 (2018)

5. Chao, Q.C., et al.: Scientific assessment and political decision-making of climate change. Yuejiang Acad. J. 28–45 (2018)

6. Song, Y., Wang, H., Li, J., Gao, H.: MapReduce for big data analysis: benefits, limitations and extensions. In: Che, W., et al. (eds.) ICYCSEE 2016. CCIS, vol. 623, pp. 453–457. Springer, Singapore (2016). https://doi.org/10.1007/978-981-10-2053-7_40

7. Ding, X., Wang, H., Gao, Y., Li, J., Gao, H.: Efficient currency determination algorithms for dynamic data. Tsinghua Sci. Technol. 21(5) (2017)

8. Yu, C.L., Zhu, Y.B., Dang, X.X.: Human settlements ecological security evaluation based on PSR model and the development strategy of Jinghe river basin. In: Pioneer Theory, May 2017, Xi'an, pp. 1–3 (2017)

9. Liu, G., Yuan, X.M., Huang, J., Wang, H.M.: Evaluation of urban resilience to flood based on PSR framework—a case of area in Suzhou-Wuxi-Changzhou. In: Resource Development & Market, April 2018, Nanjing, pp. 593–598 (2018)

10. Fragile States Index Homepage. http://fundforpeace.org/fsi/data. Accessed 11 May 2018

11. May, P., Ehrlich, H.-C., Steinke, T.: ZIB structure prediction pipeline: composing a complex biological workflow through web services. In: Nagel, W.E., Walter, W.V., Lehner, W. (eds.) Euro-Par 2006. LNCS, vol. 4128, pp. 1148–1158. Springer, Heidelberg (2006). https://doi.org/10.1007/11823285_121

12. Zhou, B.Z., Yang, S.Q., Zhang, K.: Application of triangular fuzzy number - analytic hierarchy process in prediction of outburst hazard. In: Coal Technology, April 2018, Xuzhou, pp. 170–172 (2018)

13. Jiang, Z., Feng, X., Feng, X., Shi, J.: An AHP-TFN model based approach to evaluating the partner selection for aviation subcontract production. In: IEEE International Conference on Information and Financial Engineering, Zhenzhou, pp. 311–315 (2010)

14. Mo, M., Zhao, L.Z., Gong, Y.W., Wu, Y.: Research and application of BP neural network based on genetic algorithm optimization. In: Modern Electronics Technique, May 2018, Nanjing, pp. 41–44 (2018)

15. Ehteshami, M., Farahani, N.D., Tavassoli, S.: Simulation of nitrate contamination in groundwater using artificial neural networks. Model. Earth Syst. Environ. 1–10 (2016)

16. Kong, X.Q., Chen, Y., Liu, B.Y., Zhu, P.Z.: Application of network based principle component analysis in prediction of water demand. In: Water Resources and Power, April 2018, Huizhou, pp. 26–28 (2018)

Data Analysis and Quality Management Research on the Integration of Micro-Lecture-Oriented Design Theory Courses with Maker Practice

Tiejun Zhu[✉]

Art School, Anhui Polytechnic University, Wuhu, China
ztj@ahpu.edu.cn

Abstract. Chinese design education field and students majoring in design tend to attach more importance to practice rather than theory. Due to the deep-rooted routine, the teaching effect of design theory courses in Chinese colleges and universities has not been effectively improved for a long time. What's worse, it is also difficult to effectively match and integrate universal practice resources with specialized design theory courses. In such case, it is particularly crucial to create autonomous, targeted and unique practice resources for design theory courses and establish relevant platforms. Therefore, this paper, by empirically taking *Advertising Copywriting* as a pilot, introduces Micro-lecture advocated by China's Ministry of Education as a teaching method to highlight the contents of design theory in a segmented manner and achieve the precise teaching of core knowledge. Furthermore, course-based Maker practice task is creatively designed and course-based Maker platform is established in order to cultivate students' Maker concept and mindset so that they can carry out vivid Maker action in Maker practice. In this way, the current situation where the teaching quality of design theory courses is ineffective will be fundamentally changed through comprehensive reform with contextualized, diversified and practice-oriented curriculum teaching content, form and structure. In the midst of carrying on the beneficial attempt, this program will serve as a reference for the all-round promotion of teaching reform and teaching quality control and management with specific and detailed examples and analysis.

Keywords: Micro-lecture · Maker practice · Design theory · Data analysis
Quality management · Case study

1 Introduction

Within the Chinese curriculum system of design major, although the proportion of theoretical courses has risen under the background of international education exchange, its teaching model and teaching methods still follow the traditional manner. In spite of the introduction of multimedia information technology as a supplement, the effect is not that obvious. The main reasons are as follows: firstly, Chinese design education field attaches more importance to design practice over a long period of time so that the theoretical teaching links are so weak; secondly, Chinese students majoring in design

© Springer Nature Singapore Pte Ltd. 2018
Q. Zhou et al. (Eds.): ICPCSEE 2018, CCIS 902, pp. 516–529, 2018.
https://doi.org/10.1007/978-981-13-2206-8_42

tend to be more utilitarian in terms of learning purpose. The majority of them have the intention of making money by taking part-time jobs; thirdly, the social application of design major is so broad that students can easily get job opportunities in spare time, which leads to the distraction and occupation of students' energy in learning, especially in theoretical courses; fourthly, Chinese contemporary design field lacks highly concise and deeply accumulated design theory so that far-reaching influence has not been shaped.

Considering the above reasons and the present situation that traditional teaching model fails to have much breakthrough, it is extremely difficult to stimulate students majoring in design to have interest in theoretical courses in a short time. Accordingly, it is impossible to fundamentally change the situation by merely adding the proportion of theoretical courses and purely innovating teaching methods. Instead, multimedia information technology needs to be utilized to change the content model and key knowledge point of design theory as well as to greatly improve design theory free from traditionally tedious and dogmatic aspect. In addition, innovative and upgraded teaching methods are badly needed to make design theory courses more accessible to contemporary young students with more acceptable forms. Meanwhile, various school resources should be employed to provide students with experimentation task and practice platform matched and integrated with theoretical knowledge. Based on the above concepts, this paper aims to discuss how to master the focus of design theory with short and condensed Micro-lecture by combining the Maker education advocated by the field of Chinese higher education to create "Micro-lecture plus Maker", an innovative teaching practice platform for design theory courses.

2   Research Background

The number of Chinese colleges and universities has been on the rise in recent years. Based on the statistics in the website of China's Ministry of Education, "there are altogether 2,914 colleges and universities in the whole China by the end of May 31, 2017" [1]. According to incomplete statistics, in almost 3,000 colleges and universities, at least more than 1,500 of them have set up design major. "Nearly one thousand vocational colleges offer design major" [2] and the number of applicants for design major has been rising year after year. Not only comprehensive and arts colleges and universities, but also specialized ones of various industries have provided design major. In this sense, art design major has been a very popular one in Chinese higher education. In the meantime, with the popularity of deign education, internationalized school has been on the rise and the internationalization of Chinese colleges and universities has accelerated dramatically in recent years. Through the communication and cooperation with foreign universities, the setting of development program and curriculum system of many design majors has borrowed experience and practices from western developed countries which put much emphasis on theoretical courses. As a result, the proportion of the credit of design theory courses has been increasing in total credits. Apart from traditional courses on illustrating art history, some colleges and universities have "set up certain cutting-edge and popular art theory courses" [3]. Despite the fact that art theory courses has been gradually emphasized and its number and credit are guaranteed, traditional teaching content, methods and resources are still the greatest bottlenecks which restrict the teaching effect of design theory courses based on the researcher years of experience on teaching design theory courses.

In recent years, Micro-lecture, as a prevalent teaching method promoted by Chinese education departments at all levels, serves as a "Micro-environment" for courses and a new teaching resource by integrating important teaching points, difficult knowledge points, teaching reflection, assignments and exercises as well as communication between teachers and students. The application of design theory courses Micro-lecture can not only solve design major students' lack of energy in learning with relatively short theoretical teaching time, but also highlight the focus of teaching contents to achieve the precise teaching and deep understanding of core knowledge. Furthermore, the contextual presentation of multimedia also effectively changes students' low interests in classroom learning.

As "mass entrepreneurship and innovation" acts as a strong call from Chinese government, Maker education is thereupon comprehensively promoted for "Maker education serves as a significant means to carry out innovation and entrepreneurship education" [4]. The core of Maker is to "innovate, design and share", which is what students majoring in design uphold and pursue as well as the truth of conducting innovation. Another concept of Maker education, based on interest and hobby, is actually quality-oriented education which places much emphasis on investigation, cooperation and spanned development. Therefore, it is of great significance to combine advanced Maker concept and culture, design Maker activities integrated with diversity and utilize present scientific, flexible and innovative teaching methods to concretely alter the teaching of design theory courses from "teaching theory with theory" to "immediately guiding practice with theory".

Based on the researcher years of experiences on teaching design theory courses, this research particularly chooses *Advertising Copywriting* (a basic disciplinary course as well as a theoretical course set up by School of Arts, Anhui Polytechnic University for Visual Communication Design major) as a pilot. The reason can be concluded as follows. On the one hand, this course belongs to a basic disciplinary and theoretical course for Visual Communication Design major. In addition, although advertising copy plays a significant role as a part of advertisement, the basic theoretical principles and requirements of advertising copywriting are relatively systematic and many key points need paying attention to in terms of writing methods. In this sense, *Advertising Copywriting* belongs to a typically strong theoretical design theory course which students fail to focus on for a long time in classroom. On the other hand, advertising copywriting needs practical training in theoretical learning. While traditional courses, merely with theoretical teaching and simple classroom exercise, fail to meet this requirement. Under this circumstance, this paper creatively chooses *Advertising Copywriting* as a pilot by means of Micro-lecture lasting for 16 weeks with 48 lessons altogether. Meanwhile, a sally port is, by means of integrating with curriculum Maker practice to achieve teaching application, created to obviously improve teaching effect of design theory courses. What's more, the specific practice and correlation effect of innovative reform can be empirically analyzed and demonstrated through concrete operation, full record and immediate reflection. As a result, the above can not only serves as a reference for other colleges and universities in China, but also is conducive to the promotion and application of experience.

3 The Targeted Management of Micro-Lecture-Oriented Design Theory Course

It is theoretically stated that the rudiment of the earliest Micro-lecture is derived from the fragmented video devised by a teacher from Peking University in China in the 1980s. However, there are two other versions more widely acknowledged in the field of education. "One is the 60-s Course put forward by Professor LeRoy A. McGrew from the University of Northern Iowa in the United States in 1993. The other is The One Minute Lecture that is proposed by T.P. Kee from the Napier University in Britain in 1995, short for OML" [5]. In either event, the micro in forms and the concision in contents are highly stressed. For a long time, a majority of students majoring in design have found design theory courses less attractive. Now, students' interests in such courses will be developed with the help of the two core elements aforementioned. As such, Micro-lecture is meant to be a boost in the educational reform of theoretical design courses. This paper will target at the management of the contents that are inherited in theoretical design courses by introducing Micro-lecture teaching resources and generating a Micro-environment for communication. Steps that emphasize key points, teaching structures and exercises will also be included in the Micro-design.

3.1 Micro-environment

The design and creation of Micro-environment serve as the premise and foundation to live up to Micro-management and full penetration. Micro-environment, created on the basis of Micro-lecture, requires an extraordinary integration of contents and other elements including micro key points, Micro-structure and Micro-exercise. More importantly, it asks for an interaction between targeted contents and fixed time for small-class teaching. According to the research years of experience in teaching design theory courses and long-term observations and experimental analysis in the past years, Micro-environment is designed to display the ratio of nine students and teaching period in every twenty minutes. The reasons are as follows:

There are nine students for each group. Firstly, the total number of students in a class is thirty-six, so all students can be divided into four groups, with nine students in each group. Each lesson lasts ninety minutes and the total teaching time for four groups of students is eighty minutes. If the time for shifting and operation is considered, the ratio also works. Secondly, nine students can also be divided into three groups where students in each group can fully communicate with each other. Thirdly, one teacher only has to instruct nine students in this case so that both teachers and students can get benefits thanks to a small number of students in class. On the one hand, teachers can concentrate more on students' learning process. On the other hand, students will take the initiative to be more focused on study for fear of the high attention from teachers when there are not too many students. In this sense, teaching will be more effective.

On the contrary, according to the past experience, teaching result does not seem to be as productive when there are more than ten students in a class.

Teaching period for each group of students is set for twenty minutes. It is set according to the researcher observation and experimental analysis based on personal experience in teaching such theoretical courses. It is found that students can be highly concentrated in only twenty minutes and they only become greatly less focused when a class exceeds twenty minutes. It is interesting that, no matter what teaching methods and teaching contents are chosen, student's attention will not increase until the class is over. If the internal index about the attention of students taking such courses is planned to be from 0–100, one teaching period for the course of Advertising Copywriting will be set for forty-five minutes. The variation of student's concentration is shown in the Fig. 1 below as teaching time changes.

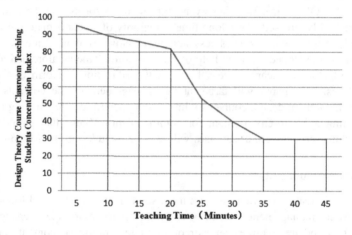

Fig. 1. The variation trend about student's concentration index in design theory courses

3.2   Micro Key Point

The course Advertising Copywriting covers a number of key points and techniques that cater to different types of advertising copywriting. Pure theoretical teaching and emphasis in classes alone appear to be pale and hollow, only to make it difficult for students to memorize, digest and internalize. "Micro-video, serves as the core resource of Micro-lecture" [6]. After considering the important teaching points of Advertising Copywriting, the researcher collect and create twelve Micro-video clips in the form of WMV so that videos can be played online by making the most of the advantages of Micro-video. Teachers can play these Micro-videos in class, upload them on class chatting room in social media like Wechat or QQ while students free from barriers caused by time and space can watch or download these videos online through their smart phones and other mobile terminals such as PAD and laptop. The main contents are shown in Fig. 2:

Take video No. 5 named after *Encounter Advertising Slogan* as an example. Advertising slogan is the crucial part for advertising copywriting. It can either

No.	Micro-video Name	Teaching Key Points
1	The development origin of advertising copywriting	The origin and development of advertising copywriting concept
2	Quality and thinking mode of advertising copywriting writer	The advertising copywriter writer's professional role, quality ability, innovative spirit and the thought process
3	Advertising copywriting principles and strategies	First premise, fundamental basis and strategic thinking of advertising copywriting
4	Advertising copywriting and creative idea	The core content, creative proposition and creative thought of advertising copywriting creative idea
5	Encountering advertising slogan	The writing technique of advertising slogan and its value utility
6	Writing technique of advertising copywriting headline	Writing's creative approach of advertising copywriting headline
7	Writing technique of advertising copywriting body and accompanying text	Writing's creative approach of advertising copywriting body and accompanying text
8	Emotional and rational appeal means of advertising copywriter	The appropriate use of emotional and rational appeal techniques in advertising copywriting writing
9	Appeal object of dialogue	How to use advertising copywriting to maximally approach and impress on appeal object
10	Graphic advertising copywriter	Features and writing skills of graphic advertising copywriter
11	Broadcast, television and network advertising copywriter	Features and writing skills of broadcast, television and network advertising copywriter
12	Commercial and public service advertising copywriter	Features and writing skills of Commercial and public service advertising copywriter

Fig. 2. List of the micro key points of Micro-video for the course

determine the future of a company or exert a direct effect on the long-term value of a brand. Normal class teaching usually elaborates on some main points such as the definition, features, functions and writing principles and techniques of advertising copywriting in general. However, too much key elements involved may bore and demotivate students. In this sense, Micro-video instead focuses on the question whether students have fully grasped the techniques of advertising copywriting. Micro-video, added with vivid examples for illustration, also helps present micro key points. As a result, students will deeply understand the information transmitting function of advertising slogan and the profound and lasting value and significance of its societal effect.

3.3 Micro-structure

Micro-structure that is compact and optimized can effectively grab and lead student's attention the whole time. In general, students are able to stay focused on study in 20 min. However, this does not secure an excellent teaching effect. The brilliant design of Micro-structure aims to enable students to study with whole concentration so that the positive effect that a twenty-minute Micro-lecture generate can greatly surpass that of a ninety-minute normal one. Thus, the twenty-minute Micro-lecture is designed to include four parts, namely the eight-minute Micro-lecture teaching, four-minute Micro-exercise, four-minute group communication for Micro-theme, and four-minute Micro-comment by teachers. Each part is sure to be short yet concise in time while main points are interlinked with each other. By doing so, neither a second nor a part will be a waste. Figure 3 shows the compact and optimized Micro-structure of Micro-lecture.

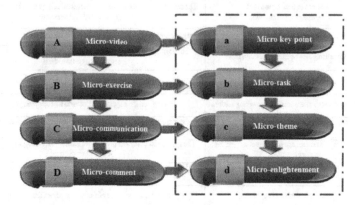

Fig. 3. The flat pattern of Micro-structure for Micro-lecture

3.4   Micro-exercise

Micro-exercise is closely related to the major contents of Micro-video. It is a Micro-task that is clear in requirements, explicit in tasks and distinct in goals. It also requires students to create their own Micro-design in class. Take a Micro-lecture where Micro-video No. 12 is played as an example. In this Micro-video about *Commercial and Public Service Advertising Copywriting*, a four-minute Micro-exercise will later be designed to ask students to come up with a slogan for public service advertising based on a picture in which three traffic lights in a crossroad all turn green.

Every Micro-exercise is presented as a *Micro-lecture Learning Task List* (Fig. 4) which is different from traditional ways like teaching materials presentation or paper distribution. Students are also more likely to accept and embrace the new ways of teaching. In this scenario, teachers will upload exercises on class chatting room via QQ when Micro-lecture is coming to an end, which enabling students to go check them later online. Besides, a full coverage of WIFI network in school makes it possible for students to bring their own laptops in class to fill in their own studying lists. Room for resource links and feedbacks will be spared on every studying list as well so that students can study by themselves at any time after class. If there are any questions or suggestions, students are open to share their ideas on the certain column. In this way, teacher can be more ready to solve students' customized questions and doubts after class.

3.5   Micro-communication

Micro-communication is the essential part to realize the Micro-management of the design theory course and make sure the teaching results, so its time is eight minutes, making up two fifths of the total time of the Micro-lecture. In the real practice, the researcher divides the Micro-communication into two parts, namely, the Micro-theme

Micro-lecture Learning Task List of
"Commercial and Public Service Advertising Copywriting"

I Learning Key Points
Features and writing skills of Commercial and public service advertising copywriter.
II Resource Link
1. Textbook P215-218 2. https://www.douban.com/group/topic/34870549/ 3. http://www.docin.com/p-1272689961.html 4. http://www.shichangbu.com/article-30004-1.html 5. http://www.tooopen.com/copy/view/38234.html
III Learning Method
1. Combining with micro-video, master the key knowledge. 2. Real-time practice, group discussion, thematic communication
IV Learning Task
Showing picture: Three traffic lights in a crossroad all turn green! According to this picture to associate and quickly design a public service advertising slogan.
V Follow-up study preview (optional)
VI Confusion & Suggestion
(Tip: here is filled by students)

Fig. 4. Content of learning task list for Micro-lecture

communication among students and the Micro-comment communication among teachers. Each stage is limited in four minutes.

In the process of Micro-theme communication among students, nine students will be divided into three groups and they are distributed around teachers. The Micro-topic of the communication depends on the content of the Micro-practice. Students can make a speech in their group according to the actual performance of the micro task. Every student has one minute to state his own actual performance from the perspective of the planning thoughts, design methods and creativity contents. Then, the remaining one minute is used to conclude the inspiration and harvest of the Micro-exercise and Micro-communication in the whole group and recommend one student to make a statement in the next step. Throughout the process, teacher stands in the middle of three groups and controls the time and tempo, making sure that every student has comparatively equal time to make a speech.

As for the teachers' Micro-comment, representative of each group uses several words to conclude and highlight the bright sides of the actual performance of the task and the enlightenment students received through intercommunication. In this sense, students can not only train their summing-up ability, but also can improve their impromptu expression ability. In order to make every student have access to this kind of exercises and broaden their scope of communication, teachers keep a record of the grouping situation and students with representative speech each time. With this method, on the one hand, it requires regrouping the members in the next micro communication. On the other hand, it also needs adjusting the representative students. The concluding remark of a group is restricted in one and a half minutes. In the next two and a half minutes, teachers will comment on each group's performance and give feedbacks about their communication in the process of Micro-practice and Micro-theme. Meanwhile, the teacher will focus on the key and difficult points of the course and the matters needing attention in the learning process.

Through operating and implementing each step of the Micro-lecture, the teaching of design theory course has realized its elaborate design and management. It can be said that the brain of every student is operating effectively in the twenty minutes' Micro-lecture. They need practice and make a speech in every step after the micro video, so the students will watch the teaching video wholeheartedly and response quickly. Such teaching plan and teaching mode can realize the full penetration and the fine grasp of the design theory in a better way. In addition, "Micro-lecture provides teaching supplement material of high quality for teachers" [7]. In the Micro-lecture, the teachers on the one hand play a role in coordinating and connecting, on the other hand, they are supervisors and instructors. On the whole, teachers shift their roles from leaders in conventional teaching to learning collaborators. Students in turn become the dominant role in the teaching process. In this way, a good teaching result is obviously achieved and the learning efficiency of the students improves multiply.

4 Comprehensive Innovation Integrated with Maker Practice Teaching

"According to *Horizon Report: 2015 Higher Education Edition* published by New Media Consortium, Maker education will bring a profound education reform among teachers and students in colleges and universities" [8]. The combined application of innovation of Micro-lecture and new teaching method can greatly increase students' learning interests in design theory course. However, the determining factor, practical teaching, affecting the learning efficiency of students cannot be guaranteed completely, especially for the course like *Advertising Copywriting*. If the actual effect of the advertising copy is not tested by practice, it is hard to tell whether the advertising copy is actually good or bad. Therefore, introducing the Maker education resource creatively, conducting Maker practice and launching genuine Maker project can meet the realistic demand of the students, realize the ideal course teaching in students' minds and thus guarantee the teaching results in high degree. Besides, the actual performance of the Maker project is integrated into the important component in the Evaluation System of Basic College Course. In this way, it completely abandons the evaluation method made up of final exam and course paper in traditional Chinese design theory course. The creative part of the Maker project is that it highlights the assessment combining theory with real practice. In the meantime, it implements process evaluation integrated with actual performance of *Micro-lecture Learning Task List*, speech and communication in the group and the whole Maker practice, and thus form an effective evaluation mode geared to international standards.

Maker practice is arranged by utilizing another half period of curriculum, namely, it is accomplished within twenty four teaching hours in five times. The first four times will occupy twenty teaching hours with five teaching hours for each, and the last four teaching hours will be applied to summarize the Maker practice project

4.1 Maker Environment

The School of Arts in Anhui Polytechnic University, which offers this curriculum, has established a creative industry research institute supported by Wuhu National Advertising Industrial Park. The institute with innovative ideas combines the strengths of universities with entity enterprise. In this sense, it can make a dynamic integration between talent cultivation and innovation entrepreneurship. Besides, it can effectively promote the transition between achievement transformation and social service and it creates a better Maker environment for students majoring in design. Due to the fact that the enterprises entering the creative industry research institute are emerging companies, which badly need advertising planning and advertising campaign. Thus, the Maker practice project meets the end of enterprises and offers the opportunities for students to make practice, so the first course is to let students feel the atmosphere of the corporation, appreciate the conception of Maker and fit the Maker environment. All the students will be divided into six groups with six people consisting of one group. Each group gets access to the enterprises settling in the creative industry research institute. These enterprises will introduce the corporate business and corporate culture to students, making students know more about the corporation and lay the foundation to choose the subject related to advertising copy Maker practice. In terms of course teaching, such action breaks into the closed "first classroom", closely connects the abundant "second classroom" and steps into the lively "third classroom".

4.2 Maker Movement

"Maker education, the combination of Maker movement and education, is the project-based learning" [9]. Therefore, the key point of the next Maker practice curriculum is to decide, assign and complete the subjects of Maker practice, which is based on the mini Maker movement and task in the *Advertising Copywriting*. Take the Maker subject of Anhui Yuexi Cultural Development Co., Ltd. as an example, the company is now collaborating with General Post Office of Anhui Province and creating a display for the design and construction achievement under the theme of post office of colleges and universities. Therefore, experimental topic of Maker is the planning and designing of advertising copy targeted at youth courier stations. In this experiment, six students are divided into three groups, two students are in each group, and three groups will carry out the Maker race. On the one hand, the environment, atmosphere and pressure of business competition will be simulated through contest. On the other hand, tireless efforts and hard feelings and experience at the start-up phase will be delivered through the close cooperation and interdependence between every two members. Thus, this experiment, by nature, can also be deemed as an entrepreneurship simulation training. When fully devoted to the experimental topic of Maker, learners will regard the success and failure of the company and the project as those of their own businesses in the mode of actual competition. Hence, apart from the deep research and accurate analysis on enterprises, projects, markets and consumers, they will also conduct a series of Maker activities and training including direction selection, planning making, risk assessment, dynamic adjustment as well as marketing strategy, which are under the guidance of teachers and business associates. In this sense, their practical experience about ad copy can be accumulated and their entrepreneurial quality and ability will be effectively improved.

4.3   Maker Culture

Teaching modes of Maker education manifested in *Advertising Copywriting* have led students into a brand-new Makerspace. While engaging in Maker practice, they will be exposed to rich and deep Maker culture. The principle of Maker-oriented teaching is to implant the Maker DNA into students so that they will be equipped with the vitality to innovate, the capability to imagine as well as the self-confidence and courage to start their own business. By effectively carrying out the curriculum Maker movement, learners will be stimulated with a unique interest, passion and excitement of copy-writing application and market experience from the bottom of their hearts. More importantly, they can truly understand the spirits of DIY featuring hands-on science, sharing spirit characterized by team work, perseverance during the process of creation as well as the stimulation after success, which are the truth and essence of the Maker culture based on free and creative thinking. Influenced, edified and penetrated by this culture, students will acquire the basic quality and capacity of Makers, even only through a micro curriculum Maker teaching.

4.4   Maker Learning

The setting of Maker environment, the infiltration of Maker culture and the penetration of Maker movement can cultivate students' Maker spirit and provide them with Maker experience. For the curriculum Maker practice teaching, what's crucial is to "promote high-level Maker learning for learners" [10]. Although the content taught in the course is ad copywriting, at the very beginning of Maker practice, the Maker learning boundaries and areas have long been beyond the mere advertising copywriting. In fact, it has been involved with broad knowledge, such as management, marketing, psychology, economics and sociology, which is exactly the charm and the most obvious efficacy of Maker practice. Coherent systemic learning and accompanying practical application will make students understand more deeply real skills about the advertising copywriting that can never be learned in books, which is the essence of Maker learning. The Context and Trend of Advertising Copywriting Curriculum Maker learning are shown in Fig. 5:

Therefore, the curriculum Maker learning created by *Advertising Copywriting* is not only knowledgeable and coherent, but also practical and creative. "What distinguishes it from the common education and teaching is that the former pays attention to developing learners' creativity while imparting knowledge" [11]. It can be considered that Maker learning has completely out of the traditional learning model, becoming a

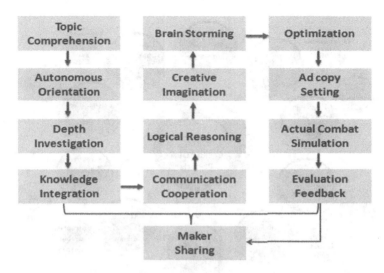

Fig. 5. Flow chart of curriculum Maker learning

new type of learning experience which mixed knowledge, practice, thinking, creating, enjoyment as well as fun.

4.5 Maker Platform

After all, the curriculum Maker practice is just the first step, and its final step is to shift from Maker to innovation and entrepreneurship. Thus, top priority should be given to the construction of a scientific and systematic Maker platform characterized by abundant resources as well as cooperation and sharing, among which "collaborate and sharing is an important characteristic of Makerspace" [12]. As the full innovation and reform pilot unit in Anhui Province, Anhui Polytechnic University, which has opened this course, not only specifically sets up an innovative college and start-up college, but also establishes a combination of a three-dimensional innovation and entrepreneurship practice platform as well as a scientific innovation and entrepreneurship ecosystem. Based on this, Maker education and practice of *Advertising Copywriting* has formed a comprehensive multi-level open platform system under the mixed influence of the course, profession, college, university, alliance, government, and enterprise which are closely connected. Figure 6 shows its structural framework and functional mechanism.

It's true that twenty four teaching hours of Maker practice and experience is far from enough. However, without the supervision and requirement from teachers, students will consciously devote their after-school time to Maker practice projects and to adapting themselves to the Maker platform system, which is exactly the real breakthrough and the highest level of the curriculum reform and Maker practice.

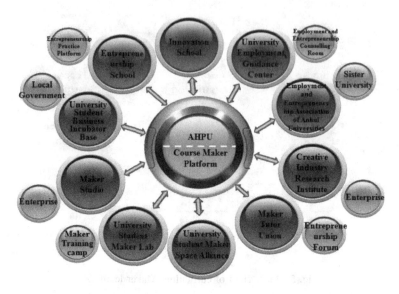

Fig. 6. Platform framework of the course Maker

5   Conclusion

Micro-lecture in the *Advertising Copywriting* and other design theory courses present contents featuring precise setting, exquisite analysis and subtle "Micro-governance", which, for one thing, demonstrates the contents in a segmented yet concentrated manner, without lack of contact and with emphasized focus; for another thing, makes the structure process of teaching course more closely and improves the teaching efficiency significantly. For individual students, though their overall study time has been reduced, what's cut down is the time when their attention greatly decreases while what remains is the golden time when their learning efficiency is at the highest level. Therefore, the introduction of Micro-lecture mode has played a significant role, either from the directness of the knowledge point of the course and the block-level combination of the content structure, or from the accuracy and efficiency of student learning.

The organization and implementation of the Maker practice is the most obvious characteristic and the highlight of the course teaching reform. Through the curriculum Maker learning and the special training of Maker, students will truly feel the Maker atmosphere and culture. In addition, they will understand the importance of studying design theory and the significance and value of employing the theory to guide practice and how to transfer it into practice. The Maker platform jointly constructed by course-university, university-local government, university-enterprise and university-university enables students to integrate into the Makerspace in advance and provides them access to varieties of support and guidance concerning innovation and start-up. With this, they will acquire Maker quality and Maker spirit that are indispensable at the present moment and in the future.

Acknowledgements. This work is supported by Anhui Province key education and teaching projects of universities and colleges "School running characteristics exploration and innovation of international engineer institute of Anhui Polytechnic University" (2016jyxm0091); 2017 National Social Science Fund pre-research Project of Anhui Polytechnic University "Collaborative innovation assimilation and alienation research of intangible cultural heritage protection of the Yangtze River Delta" (2017yyrw01); Key projects of higher education research in Anhui Polytechnic University "Study on Teachers' education and development of Anhui Polytechnic University under the background of internationalization of Higher Education" (2017gjzd002); "International compound students training mode of Local colleges and universities in global MOOCs era" (2014jyxm43).

References

1. Name List of National Higher Universities and Colleges. http://www.moe.edu.cn/srcsite/A03/moe_634/201706/t20170614_306900.html. Accessed 1 Nov 2017
2. Zhang, Z.: The necessity of increasing theory teaching time of vocational school design majors. Sci. Educ. Pap. Collects **6**, 34–53 (2016)
3. Zhao, X.X.: The importance research on design theory course undergraduate teaching. Art Educ. Res. **8**, 155–156 (2015)
4. Hao, S.S.: Research on the curriculum system of maker education based on the makerspace. Mod. Technol. Educ. **8**, 109–114 (2017)
5. Guang, Z.G.: Micro-lecture. Inf. Technol. Educ. **6**, 14 (2016)
6. Hu, T.S.: China Micro-lecture development stages. J. Distance Educ. **4**, 36–42 (2013)
7. Wang, M.: Micro-video: evolution, location and application. Electron. Educ. J. **4**, 88–94 (2017)
8. Johnson, L.: The NMC horizon report: 2015 higher education edition. New Med. Educ. **1**, 1–2 (2015)
9. Zheng, J.C.: Maker education and curriculum construction. Inf. Technol. Educ. **6**, 153–154 (2016)
10. Ji, Y.: Studying of Creative Learning Surrounding: Virtual Makerspace and Creation, Master's Paper, Yunnan University (2016)
11. Xie, C.R.: Online Teaching Design and Assessment. Beijing Normal University Press (2015)
12. Liu, W.W.: A study of academic makerspace construction mechanism: implications from the practices in colleges and universities in HEMI. Educ. Dev. Stud. **11**, 78–84 (2017)

A Cloud-Based Evaluation System for Science-and-Engineering Students

Qian Huang[1,2], Feng Ye[1,3(✉)], Yong Chen[3], and Peiling Xu[4]

[1] College of Computer and Information, Hohai University,
Nanjing 211100, China
yefeng1022@hhu.edu.cn
[2] Key Laboratory of Symbolic Computation and Knowledge Engineering
of Ministryof Education, Jilin University, Changchun 130012, China
[3] Postdoctoral Centre, Nanjing Longyuan Micro-Electronic Company,
Nanjing 211106, China
[4] Suzhou College of Information Technology, Suzhou 215200, China

Abstract. With the burgeoning of IT industry, more and more companies and universities concentrate on the scientific evaluation of science-and-engineering students. Existing evaluation strategies typically lie on grades or scores of the courses taken by students, which have obvious drawbacks nowadays and cannot lead to a proper improvement of education management. This paper proposes an overall student evaluation system architecture that includes three levels, i.e. data collection infrastructure which comprises student data collection and student data transmission, data center which is composed of four sub-levels, and unified portal which defines unified applications and classes of visiting terminals. In the proposed architecture, the main contribution lie in the upper most sub-level of data center, i.e. evaluation services. Four categories of student evaluation services, i.e. moral trait, civic literacy, knowledge level and comprehensive ability, are defined. Furthermore, in order to have a satisfactory feedback in the practical teaching process, the most important comprehensive ability for science-and-engineering students is fractionized into four sub-categories and visualized from several different aspects for a good feedback in practical teaching process.

Keywords: Student evaluation · Comprehensive ability · Visualization

1 Introduction

With the rapid development of modern IT technologies, more and more information systems have been armed with cloud computing, internet of things and big data [1]. For example, the State Grid Jiangsu Electric Power Company has gradually introduced the informatization strategy and its information system is becoming more and more integrated with business [2]. At the same time, more and more companies and universities realize that the capability of science-and-engineering students is closely related to the future of the whole industry. Therefore, more attentions are paid to the evaluation of science-and-engineering students, especially in a cloud-based system. Taking into account the fact that more and more data are generated and transmitted over 3G/4G networks and WIFI, and that many information systems are still PC-oriented, this paper

© Springer Nature Singapore Pte Ltd. 2018
Q. Zhou et al. (Eds.): ICPCSEE 2018, CCIS 902, pp. 530–538, 2018.
https://doi.org/10.1007/978-981-13-2206-8_43

presents a cloud-based evaluation system for science-and-engineering students with hybrid network architecture [3].

The rest of this manuscript is organized as follows. Section 2 firstly gives the overall design of the proposed student evaluation system, and then introduces the general data flow and the simple student evaluation services. Section 3 performs comprehensive ability analysis for science-and-engineering students and discusses various visualization aspects. Finally, the concluding remarks are drawn in Sect. 4.

2 Overall Design

The overall student evaluation system architecture is depicted in Fig. 1.

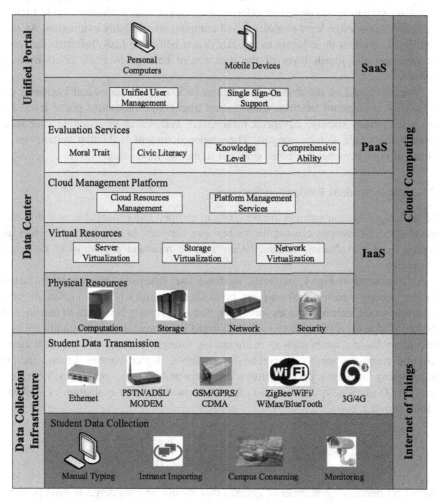

Fig. 1. Overall design of the proposed student evaluation system

As shown in Fig. 1, the student evaluation system includes three levels, i.e. data collection infrastructure, data center and unified portal. Among the three levels, the data collection infrastructure level lies on the Internet-of-Things technology, whereas the data center and united portal levels lie on the Cloud Computing technology.

2.1   Student Data Flow

Firstly, student data are collected through manually typing, Intranet importing, campus consuming or monitoring.

Secondly, the collected data are transmitted via various networks and finally stored at the data center either built by the school or rented from third-party providers.

Thirdly, at the data center, hardware and software resources are virtualized and managed to offer better services to students and teachers. For example, a typical student evaluation system is suggested to offer at least moral trait evaluation, civic literacy evaluation, knowledge level evaluation and comprehensive ability evaluation. As seen from Fig. 1, the first three layers of the data center belong to IaaS (Infrastructure-as-a-Service); and the fourth layer of the data center belongs to PaaS (Platform-as-a-Service).

Fourthly, based on the student data center, various applications and interfaces can be built for interactions between students and teachers at the unified portal level. For the contemporary society, advanced information systems require unified user management and single sign-on support. The proposed student evaluation system can be utilized at both personal computer side and mobile terminal side.

2.2   Simple Student Evaluation Services

The student evaluation services are designed in an open and expandable way that each individual in the campus can submit his/her own services. In case his/her services are adopted by the evaluation system, he/she can be refunded according to the usage information.

As illustrated in Fig. 1, currently we have four categories of student evaluation services, among which moral trait and civic literacy generally have similar or same evaluation scores unless one does something that is obviously beneficial or detrimental. The knowledge level is typically evaluated based on examination scores and credits of compulsory basic courses such as mathematics, foreign language and degree courses. Take the computer science discipline for example, degree courses general include programming languages, data structures, computer architecture, operating systems, etc. Equation (1) gives a typical definition for the knowledge level evaluation of student s in time period t:

$$K(s,t) = \frac{\sum_{c=1}^{C(t)} score(c,s,t) \cdot credit(c,t)}{\sum_{c=1}^{C(t)} credit(c,t)}, \tag{1}$$

where $C(t)$ denotes the number of courses that should be considered in time period t, and c is the course index varies in the range $[1, C]$. $score(c,s,t)$ indicates the final score of student s in course c offered in time period t, whereas $credit(c, t)$ indicates the credits of course c offered in time period t.

In the next section, the comprehensive ability level will be discussed in detail.

3   Comprehensive Ability Analysis

The requirement of comprehensive ability differs according to the country, society, organization and even the discipline. For example, IT companies usually expect engineering talents to promote the continuous upgrading and development of software/hardware products. In contrary, universities and colleges require their students to have a firm and comprehensive grasp of basic theories and profound and systematic specialized knowledge in the discipline concerned. How to evaluate students' comprehensive ability is a challenging problem today.

In this section, we mainly focus on the comprehensive ability analysis for science-and-engineering students.

3.1   Related Work

Ren et al. [4] proposed a teaching reform approach based on CDIO capability maturity for undergraduate students. Zheng et al. [5] proposed a software capability evaluation method for computer majored students. In [6], a multilevel fuzzy comprehensive evaluation method was proposed to evaluate the practical ability of students. In [7], a fuzzy logic feature mapping clustering algorithm was presented to evaluate theoretical foundation and application skills of undergraduates. In [8, 9], ability evaluation systems were proposed for vocational school students.

Existing works such as [4–7] do not implement visualization tools, whereas [8, 9] present only simple metrics. Therefore, a reasonable and good performance comprehensive ability analysis strategy is still needed.

3.2   Proposed Comprehensive Ability Analysis

For science-and-engineering students, the authors believe that the evaluation of comprehensive ability should consider at least four aspects, i.e. active learning ability, sense of cooperation, presentation skills and innovation practices.

Active Learning Ability. In [10], the authors addressed an important problem in deep learning, i.e. how to train a good classifier with minimum labels. Science-and-engineering students need to address many similar problems as many scientific and engineering problems are required to be solved with limited resources.

To evaluate the active learning ability, the Hale exponential [11] is utilized based on the knowledge level evaluation in Sect. 2.2. Specifically, the Hale score of student s in time period t is defined as:

$$H(s,t) = e^{\alpha \left(5 + \frac{K(s,t) - \overline{K(s,t)}}{STD(K,t)} \right)},$$

$$(2)$$

where $K(s,t)$ is defined in Eq. (1), whereas $\overline{K(s,t)}$ and $STD(K,t)$ are the expectation and standard deviation of $K(s,t)$. In addition, α is a predefined coefficient.

Based on the definition of the Hale exponential, the active learning ability of student s can be defined as:

$$AL(s,t_1,t_2) = H(s,t_2) - H(s,t_1), \tag{3}$$

where t_1 and t_2 are different time periods. For example, t_1 could be an autumn semester and t_2 could be the subsequent spring semester.

Among all the students considered, the one with the largest $AL(s,t_1,t_2)$ value is considered to be the one with strongest active learning ability between the specific periods. An example is given in Table 1, where the parameter α is set to 0.1.

Table 1. Example of active learning ability evaluation

s	Student name	$K(s,t_1)$	$K(s,t_2)$	$H(s,t_1)$	$H(s,t_2)$	Progress	Rank
1	Michael	88	90	1.83	1.89	0.06	2
2	Jane	82	72	1.70	1.46	−0.24	6
3	Jack	83	82	1.72	1.69	−0.03	3
4	Pony	81	80	1.68	1.64	−0.04	4
5	Robin	79	73	1.63	1.48	−0.15	5
6	John	65	85	1.37	1.76	0.40	1

Sense of Cooperation. The sense of cooperation of student s during a specific time period t can be defined as a weighted sum of cooperative project score $Proj(s,t)$ and cooperative competition score $Comp(s,t)$:

$$SC(s,t) = \rho \cdot Proj(s,t) + \theta \cdot Comp(s,t), \tag{4}$$

where ρ and θ are empirical weights.

Take the second part in Eq. (4) as an example. For state-level competition awards, $\theta = 5$; for provincial level competition awards, $\theta = 3$; otherwise $\theta = 1$. As for $Comp(s,t)$, the value can also vary due to the specific contribution in the cooperation. For example, if s is the first one in the certificate of award, $Comp(s,t) = 1$; if s is the second one in the certificate of award, $Comp(s,t) = 0.8$; otherwise $Comp(s,t) = 0.3$.

Presentation Skills. This comprises written presentation skills and oral presentation skills. The basic presentation score is zero for all students. For written presentation skills, each Science citation indexed paper contributes 5 and each engineering indexed paper contributes 2. For oral presentation skills, an oral certificate or speech contest certificate contributes 2 and 5 respectively. Equation (5) gives a formal definition:

$$PS(s,t) = \sigma \cdot (5 \cdot Num\_Paper_{SCI}(s,t) + 3 \cdot Num\_Paper_{EI}(s,t)) \\ + \tau \cdot (5 \cdot Num\_Cert_{SpeechContest}(s,t) + 3 \cdot Num\_Cert_{Oral}(s,t)), \tag{5}$$

where s and t are the same in previous equations. σ and τ are predefined weights for written presentation and oral presentation respectively.

Innovation Practices. Innovation practices are evaluated according to important publications, distinguished competitions and issued patents. Taking the computer science major as an example, journals and conferences recommended by the China Computer Federation are assigned weights in the final score; and distinguished competitions include only widely acknowledged ones such as the "Challenge Cup" [12] and the "China Software Cup" [13]. In this paper, the innovation practices of student s in time period t is evaluated as:

$$
\begin{aligned}
SC(s,t) = & \sum_{p=1}^{P_{CCF}} PWeight(p,s,t) \cdot ARate(p,s,t) \\
& + \sum_{c=1}^{C_{Comp}} CWeight(c,s,t) \cdot WRate(c,s,t),
\end{aligned}
\tag{6}
$$

where $PWeight(p,s,t)$ and $CWeight(p,s,t)$ denote the level of paper p or competition c related to student s in time period t. For example, for A-class papers and Competitions, $PWeight$ and $CWeight$ can be defined as 5. $ARate(p,s,t)$ and $WRate(c,s,t)$ refer to how much the student s contributes to the achievement. For example, if the student s is the second author of a paper p published in time period t, $ARate(p,s,t)$ is defined as 80%. P_{CCF} and C_{Comp} are the number of contributions related to student s in time period t in terms of publications and competitions, respectively.

3.3 Visualization

According to the definitions in Sect. 3.2, a radar chart can be employed to depict the comprehensive ability of a specific student. An example is given in Fig. 2 for student John, who has the greatest active learning ability according to Table 1. It can be seen that John has made great progress in examinations this semester. However, he did not take part in any innovation activity and his presentation needs to be improved.

The example radar chart in Fig. 2 is drawn by the R language, and the corresponding sample code is given below:

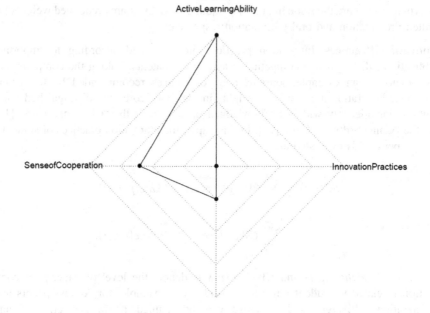

Fig. 2. Example radar chart of comprehensive ability

```
maxmin <- data.frame(
  ActiveLearningAbility = c(20,0),
  SenseofCooperation = c(20,0),
  PresentatonSkills = c(20,0),
  InnovationPractices = c(20,0)
)

dat.A<- data.frame(
  ActiveLearningAbility = 20,
  SenseofCooperation = 12,
  PresentatonSkills = 5,
  InnovationPractices = 0
)

dat.A1<-rbind(maxmin,dat.A)
library(fmsb)
radarchart(dat.A2, axistype=0, seg=4,centerzero = TRUE)
```

4 Concluding Remarks

In this manuscript, a cloud-based student evaluation system is presented for science-and-engineering students. Firstly, the overall design is given and the typical student data flow and simple student evaluation services are discussed. Then, the key contribution of this paper, i.e. the comprehensive ability analysis, is described from four aspects: active learning ability, sense of cooperation, presentation skills and innovation practices. As for active learning ability, the Hale exponential is employed. As for sense of cooperation, cooperative project scores and cooperative competition scores are taken into account. As for presentation skills, both written presentations and oral presentations are evaluated. As for innovation practices, only important publications and distinguished competitions are considered. A sample radar chart is also given to illustrate the overall performance of a specific student so as to make a quick judgment. In addition, other typical charts such as the jitter plot can also be provided for comparison among students.

Acknowledgments. This work is partly supported by the National Natural Science Foundation of China under Grant No. 61300122, the 2017 Jiangsu Province Postdoctoral Research Funding Project under Grant No. 1701020C, the 2017 Six Talent Peaks Endorsement Project of Jiangsu under Grant No. XYDXX-078, the Fundamental Research Funds for the Central Universities under Grant No. 2017B42214 and 2013B01814, and the National Key Technology Research and Development Program of the Ministry of Science and Technology of China under Grant No. 2016YFC0400910 and 2017ZX07104001.

References

1. Li, B., Liu, B., Lin, W., Zhang, Y.: Performance analysis of clustering algorithm under two kinds of big data architecture. J. High Speed Netw. **23**(1), 49–57 (2017)
2. Wang, C., Wang, Q., Huang, Q., Ye, F.: A classification method based on improved bia model for operation and maintenance of information system in large electric power enterprise. In: Barolli, L., Zhang, M., Wang, X. (eds.) EIDWT 2017. LNDECT, vol. 6. Springer, Cham (2017). https://doi.org/10.1007/978-3-319-59463-7_64
3. Huang, Q., Wang, X., Du, X., Ye, F.: A hybrid architecture for video transmission. In: Proceedings of the 2nd Asia-Pacific Engineering and Technology Conference, pp. 217–223. DEStech, Shanghai (2017)
4. Ren, S., Hu, Z., Wu, B.: Teaching reform and practice on fostering engineering capability for the undergraduate student. In: Zhu, M. (ed.) ICCIC 2011. CCIS, vol. 235, pp. 246–251. Springer, Heidelberg (2011). https://doi.org/10.1007/978-3-642-24022-5_41
5. Zheng, L.J., Ren, Y.C.: A software capability evaluation method for computer majored students. Comput. Technol. Dev. **6**, 233–240 (2013). (In Chinese with English Abstract)
6. Ren, X.L.: The study and exploration of practice and application computer personnel training mode. Microcomput. Appl. **30**(12), 23–25 (2014). (In Chinese with English Abstract)
7. Su, Z.H.: Research on the evaluation of undergraduates' theoretical foundation and application skills based on FLFM clustering algorithm. Master Thesis, Northeast Normal University, Changchun, China (2014). (In Chinese with English Abstract)

8. Xing, H.H.: The study of evaluation system of ability of students major in computer application in higher vocational schools. Master Thesis, Shandong Normal University, Jinan, China (2016). (In Chinese with English Abstract)

9. Tang, X.T.: The design and implementation of student ability evaluation system under the environment of J2EE and Android. Master Thesis, Guangxi University, Nanning, China (2015). (In Chinese with English Abstract)

10. Zhou, Z., Shin, J., Zhang, L., Gurudu, S., Gotway, M., Liang, J.: Fine-tuning convolutional neural networks for biomedical image analysis: actively and incrementally. In: Proceedings of the 2017 International Conference on Computer Vision and Pattern Recognition, pp. 4761–4772. IEEE, Hawaii (2017)

11. Jiang, Q.Y., Xie, J.X., Ye, J.: Chapter 3: Mathematical Model., 4th edn. Higher Education Press, Beijing, China (2011)

12. http://www.tiaozhanbei.net/

13. http://www.cnsoftbei.com/do/showsp.php?fid=6&id=36

From Small Scale Guerrilla Warfare to a Wide Range of Army Operations the Development Direction of Software Production and Education

Lei Xu$^{(\boxtimes)}$, Huipeng Chen, Hongwei Liu, Yanhang Zhang, and Qing Wang

School of Computer Science and Technology, Harbin Institute of Technology, Harbin 150001, China
xulei82@hit.edu.cn

Abstract. With the great improvement of computer computing power, the requirement of the software can be greatly improved in terms of performance, scale and reliability. From the programming of one or several people, it has rapidly transitioned to the collective programming of hundreds or even thousands of people. The increase in quantity has brought qualitative changes, and the era of programming that relies on individual ability no longer exists. It's unavoidable that large-scale group army will be fought. Group army is different from guerrilla warfare, which will require more discipline and cooperation of teams. Our software education also needs to cater this change.

Keywords: Group army operations · Guerrilla warfare · Strategies of fighting

1 Introduction

Early morning, December 7, 1941. Japan's Imperial Navy's aircraft carrier planes and miniature submarines suddenly attacked the U.S. Navy's Pacific Fleet at Pearl Harbor in Hawaii and the U.S. Army and Navy airports on the island of Oahu. This is the most famous Pearl Island incident in World War II history [1].

On November 26, 1941, a fleet of Japanese navies led by six aircraft carrier planes left Japan for Pearl Harbor under the command of vice admiral ChuichiNagumo. The fleet maintained a complete radio silence on the way. In addition to the six aircraft carriers, the Japanese fleet includes two battleships, three cruisers, nine destroyers and three submarines. In addition there are eight tankers and two destroyers in the North Pacific for waiting.

We are not going to explore the purpose of Japanese attacking on Pearl Harbor. We are just concerned about how such a battle can be won?

In February 1941, Japanese admiral Yamamoto Isoroku made the "Z plan", and its success depended on two unreliable hypotheses: First, the us Pacific fleet would be anchored in Pearl Harbor when the attack occurs; second, a large fleet of aircraft carriers would not be found when crossing half of the Pacific Ocean. Only the gambler would take the risk, and Yamamoto was the master of gambling. He often said to his

© Springer Nature Singapore Pte Ltd. 2018
Q. Zhou et al. (Eds.): ICPCSEE 2018, CCIS 902, pp. 539–544, 2018.
https://doi.org/10.1007/978-981-13-2206-8_44

staff that the gambler's thinking often played a role in his thinking: half was calculated and half luck. He was determined to make a big bet: "win or lose. If we fail at Pearl Harbor, the battle will not be fought." If the raid succeeds, it will have a strategic advantage over a year or so [2].

It takes about 12 days for a large aircraft carrier to cross the Pacific Ocean. During this period, in order to maintain the success of the attack, the entire fleet must ensure radio silent and cannot be interconnected. How to achieve the coordinated participation of large fleets in combat? The fleet was in the sea for a long time. How to supply? And so on numerous issues need to be solved. These issues must be formulated before the start of the campaign (that is, before the departure of the entire fleet), and then only by the strict implementation of the plan by all the members participating in the operation. With a slight oversight, the history will be rewritten.

The attack on Pearl Harbor is a successful war example in the history of the war. It fully reflects the characteristics of large-scale army operations: careful planning, strict enforcement and close cooperation with each other. These characteristics are precisely the characteristics of our software development today. These characteristics require that all of us who participate in software development have the characteristics of military personnel: disciplinary ability, executive ability, coordination ability, and then adaptability. And how should our education train students to have above abilities?

2 Software Development of Large-Scale Group Army Operations

The original computer software was developed by relatively underlying binary or assembly language, requiring very specialized software developers. And because of the computer's computing power and resource constraints, the function of software was not very powerful. The software can be completed by one or a few people, and its development was more concerned about the developer's technical capabilities and expertise. With the rapid development of computer technology, computer computing ability and resources have been greatly enhanced. Great scale of industrial production made the prices of computer get lower and lower. In particular, the emergence of computer networks has led to the widespread use of computers in various fields to accomplish various tasks, and the demand for computer software has increased dramatically. The popularization of computers has made the world's connections become very extensive, which has led to changes in the business models of various industries. The changes in business models have also driven the demand for computer software to increase. With the advent of various tools, the development of computer software has become easier to grasp, and a person with a high school education can do software development after a simple training.

Although today's software development does not require too much technical and professional knowledge, another problem arises as the software scale expands. A Windows2000 development [3] requires 250 project managers, 1, 700 developers, and 3, 200 testers. Together, it requires a large team of 5,150 people. And the 5,150 May come from many countries: the United States, Britain, the Czech republic, Iraq, Russia, India, China, Japan,... from different cultures: Buddhism, Christianity, Islam,

Confucianism,... have different languages: English, Russian, Chinese, Japanese,... These people may also live and work in different countries. How to integrate such a composition into a unified whole is a system engineering problem and organization work and planning work are very complex, but Microsoft did. Can we do it? We have to solve such a system engineering problem, what kind of quality talents do we need, and what kind of basic knowledge must these talents have? This is a problem that lies in our educational world.

Today's software development not only includes the management of the company's developers, but also involves commissioning the part of the code to third-party companies or individual for developing. How to design the interface, how to achieve integration, and how to verify the quality of the software module are also problems that cannot be ignored.

Guerrilla warfare was once the main tactics used by the Chinese Communist Party in the early days of the Anti-Japanese War. The guerrilla warfare needs to break up the whole into parts and fight with the enemy in a flexible and mobile manner. It requires a very strong single combat capability. One can become a guerrilla group, A few people can also become a guerrilla group. The strategies are that I will back when the enemy advance, I will disturb the enemy when they stop, I will attack the enemy when they are tired, and I will pursue the enemy when they retreat. These strategies are effectively against the enemy in the premise of preserving our military strength, but it is not possible to eliminate the enemy on a large scale. By the time of the War of Liberation, large-scale operations of the Army Group were carried out in the Liaoshen, Pingjin, Huaihai, and Dujiang battles. Each campaign required the participation of millions of people, and joint operations of multiple regiments and multiple arms. And our group operations are successful. In the early stages of software development, guerrilla fighters are effective. However, when the software scale becomes very large, guerrilla warfare is not able to accomplish such a task. It must be a group fighting, and the army must have strategies of fighting together.

To put it simply, a group battle requires the following processes:

(1) Make up your mind: the process of determination requires the following steps: to understand the task; to determine the situation; to develop programs; to evaluate the process of selection.
(2) Plan the organization: formulate the battle plan; give the battle order; organize and coordinate the battle; organize the command system; organize various safeguards; check the army combat readiness;
(3) Combat control: monitor and supervise the correction; make new plans in a timely manner.
(4) Summary of combat experience: after the battle, the first thing the army should do is to sum up experience. This is not in the realm of combat command. But the experience and lessons of the last combat command will inevitably produce incalculable reflections on future combat commands. After all, no army is born to be a warrior. No commander was born to be a military commander. Military commanders can master operational command and be ever victorious even win and win only if they constantly learn from experience. This is the way of the military, the true meaning of the millennium.

It is known from the knowledge of software engineering and development process that the current software development needs the following stages:

(1) Needs analysis: to understand the needs of users; to determine the user's real needs and priorities; to develop technical solutions; to evaluate the process of selection;
(2) Design stage: to determine development plan; to determine exchange plan; to determine software architecture; to determine organization mode;
(3) Coding and testing stage: monitoring and supervision; make new plans in time to meet changing needs.
(4) Project Summary: Whether it is a successful project or a failed project, checking the entire project implementation process is an effective feedback, which can effectively adjust the software development process, refine the development experience and train the developers.

From the above comparison, the process of software development is similar to that of the large-scale group army. However, even if each campaign is well designed, it cannot guarantee the absolute victory of the battle. Therefore, Yamamoto said: half rely on computing, half luck. In the same way, even if the software development process is carefully designed, it is not guaranteed to be absolutely successful and requires a bit of luck. What we call "luck" here can be attributed to the basic quality of the personnel and the cooperation of the environment. Environmental cooperation means that there are not too many unexpected situations: there is no natural factors such as earthquakes and tsunamis; there is no objective factor such as the interruption of funds or the cancellation of certain causes. The quality of personnel is a major factor in the success of the project. Careful design naturally requires a high level of personnel which may depend on the quality of one or two individuals. It is relatively easy to solve. When implementing, it is the key to the success of the project that all developers are able to accurately perform.

3 Software Education Needs to Be Focused on Large-Scale Army Operations

As the software goes deep into all fields, as the development tools become more and more abundant, as software development levels become higher and higher, many software developments no longer need to understand the knowledge of the computer itself, which makes the entry into the field of software development become lower and lower. Students of all majors will learn some knowledge about software development in college, such as programming language. And these non-computational students, because of their knowledge of certain areas of expertise, have an advantage of programming in their field than those of pure computers. So, as more and more people are programming, what are the advantages of a software college graduate? What are the reasons for the students' survival? This is a question that anyone with software education must think about.

Although the lack of knowledge of a large number of fields of specialization at Software Academy graduates, the basic courses required for mass production have been added to our education system and the setting of these courses has established foundation of large-scale operations for students. Training and internships, in turn, provide students with a platform to apply these basic knowledge to make students have practical experience. This is very important.

However, although we have designed good courses and practical opportunities, there are still some problems in the implementation process:

(1) Many of our teachers did not have the experience of a large group army. Some teachers have some experience, but it is guerrilla warfare experience. A lot of problems cannot be given a good answer if the experience of guerrilla warfare has been extended to the teaching of the operations of the Army. Teacher isn't so confident, and the class effect is not very ideal. We have to get teachers to engage in large-scale group warfare to exercise experience, even if it is failed experience.

(2) Teachers do not know enough about the importance of management courses. Large-scale group army operations not only need technology, the most important thing is the grasp of the war situation that is the battle's design planning and the correct implementation of plan. All this require the support of correct management philosophy. If our teachers think management is not a mainstream course at Software Academy, how can we convince students to master the concepts and skills that plays an important role in a large-scale battle?

(3) Students think that management classes are extremely simple, just look at it once and learn it. This is a very childish idea. If you can become a militarist by a war, there will be no losing battle. Although there are many art of war, although the art of war strategy is no longer a secret to anybody, not everyone who learns the "the art of war" can be invincible. This is not a simple problem such as $1 + 2 = 3$. It requires a global grasp and a systematic thought process. Students' naive understand precisely explains that we have not yet articulated the essence of such a course and also need to enhance a lot.

So far, our teachers and students of Software Institute think that technology is still our main course and management classes are worthless. Such a way of thinking will make the living space for students graduating from Software Academy very narrow, cannot even survive.

4 Conclusion

Our software engineering education is to teach students the ability to fight on a large scale. It requires the cooperation of theory and practice to achieve the teaching effect. If students have highly quality talents, he or she may become a commander in future software development. Even if he can not be a commander but an ordinary soldier. He also needs to know how to become a good soldier and a disciplined soldier. A soldier who can only fight alone will not be able to play a good part in the operations of a large-scale army. The reason why the "talent" in the TV drama "Soldier Assault" was eliminated by "Old A" and cannot become a special forces soldier is that he didn't understand that the modern war needs a collective force rather than just individual power.

Acknowledgements. This research is supported by 2016 key projects of Online Education Research Fund (General Education) of the online education research center of the Ministry of education.

References

1. Churchill, W.: World war II memoir. Southern publishing house (2003)
2. [Japan] Real pine.: 365 Days Before Pearl Harbor. Shanghai Translation Publishing House China (2008)
3. [US] Michael corsomaro: Microsoft's Secret. Peking University Press, Beijing (1996)
4. Yang, J., Xu, D., Liu, H., Qi, W.: Research on the Teaching Reform of C Language Course. Science Education Article Collects (2014)
5. Li, N., Wang, W.H., Wang, Y.: Teaching innovation beyond "Path-depended" on C language course. J. Anyang Inst. Technol. **2**, 033 (2012)
6. Shi, H., Zhang, J.M.: Discusses several difficulties' teaching method in C language course. Comput. Knowl. Technol. (2012)
7. Jin-fang, T.: Exploration on teaching method of C language course. High. Educ. Forum **3**, 030 (2007)
8. Kay, J.: Problem-based learning for foundation computer science courses. Comput. Sci. Educ. **10**, 109–128 (2000)
9. Lee, T.S., Yi, S.H.: The effects of tetris game on the cognitive abilities of students with mental retardation. J. Korea Game Soc. **12**, 77–85 (2012)
10. Chen, X., Lin, C.: Tetris game system design based on AT89S52 single chip microcomputer. In: IITA 2009, Third International Symposium on Intelligent Information Technology Application, pp. 256–259. IEEE (2009)

Gathering Ideas by Exploring Bursting into Sparks Through the Cross–To Discuss Interdisciplinary Role in Cultivating Students' Innovation

Lei Xu, Lili Zhang[✉], Yanhang Zhang, Hongwei Liu, and Yu Wang

School of Computer Science and Technology, Harbin Institute of Technology,
Harbin 150001, China
zhanglili@hit.edu.cn

Abstract. With the continuous development of science and technology, inter-section of different disciplines has become a new growth point of science and technology innovation. This paper starts with the interdisciplinary effect of promoting science and technology innovation. Taking the cultivation of students majoring in computer as an example, this paper demonstrates in detail the importance of intersection of different disciplines in cultivating students' innovation ability and expounds how to effectively exert the advantages of intersection of different disciplines in practice and innovation teaching in order to inspire students' innovative thinking, guide students' innovative activities and form a virtuous circle of development.

Keywords: Intersection of different disciplines · Innovative training
Computer science

1 Introduction

Intersection of different disciplines is the academic activity of two or more subjects cooperating with each other in the same target. In the 21st century, the widest and deepest development in various disciplines has prompted subjects' cross and penetration as new growth points for scientific discoveries, where most scientific breakthroughs are most likely to occur. As a result, revolutionary changes will take place. And glorious fields of academic and scientific research are often the stage and growth points of talent cultivation. With the continuous emergence of intersection of different disciplines in the field of scientific research, the cultivation of interdisciplinary talents has gained more and more widespread attention. Under the new situation that a single training model cannot meet the rapid development of science and technology and the demand of the society for the composite talents, It is especially important to cultivate multi-disciplinary talents for the sake of penetration and support among multi-discipline, different disciplines or different professions, building a talent training new mode of multi-disciplinary cross and penetration, cultivating innovative talents with multi-discipline and meeting the students' multi-level and various needs. Under such a trend, focusing on the characteristics of computer students' training, this paper

© Springer Nature Singapore Pte Ltd. 2018
Q. Zhou et al. (Eds.): ICPCSEE 2018, CCIS 902, pp. 545–551, 2018.
https://doi.org/10.1007/978-981-13-2206-8_45

emphasizes on roles and realization ways of intersection of different disciplines in training students' innovation so as to fully display the function of interdisciplinary training in students' cultivation, promote students to utilize interdisciplinary advantages and promote their own development.

2 Intersection of Different Disciplines is the Driving Force and Source of Scientific and Technological Innovation

Scientific and technological innovation is an important part of China's innovation system, and the cultivation of innovative ability is the basis and guarantee of innovation. For social groups, innovative ability pays more attention to the provision and guarantee of basic conditions. For individuals, innovative ability emphasizes that you should have new theories, new concepts or capabilities of inventing new technologies and new product. If the improvement of innovation ability is implemented in the cultivation of students, students should have broad basic knowledge, solid professional knowledge, innovative thinking, strong abilities of identifying problems, asking questions, solving problems and practicing. Nowadays, with the wide and deep development of disciplines, it has become more and more distant to be able to use single science to solve problems independently. It has become an important way to comprehensively solve the problems through multiple disciplines. Therefore, the understanding and application of interdisciplinary research have become key to enhance innovation ability.

Nowadays, international universities are increasingly emphasize interdisciplinary training. A grassroots movement from Stanford University, for example, has succeeded in creating a new project called bio-x to explore the intersection of biology, computer science, medicine and engineering. This project is located in Clark Center, opened in 2003. Clarke Center is located between the school and the medical center, designed to accelerate and facilitate interdisciplinary research. Each lab is equipped with at least two scientists from different fields, and it is not closed: Equipped with a small wheel, the walls can be moved (or removed). In China, the interdisciplinary development has also received the attention of higher education. Peking University established the Frontier Interdisciplinary Research Institute in 2006 to carry out innovative research related to advanced science and high technology such as life, information and materials. Jiaotong University also set up the Bio-X Center earlier as a interdisciplinary research platform related to life science for scientists. Our school has also established a basic interdisciplinary college to expand interdisciplinary research and applications in physics, chemistry, materials, electrical and mechanical terms, life and other disciplines. All of these illustrate interdisciplinary importance in science and technology innovation system. And for the people in the technological innovation environment, such environment and conditions are also needed to cultivate people's innovative ability.

3 Intersection of Different Disciplines is the Only Way for Cultivating Students' Innovation

Under the new situation of interdisciplinary development, the integration and collision of more disciplines will not only bring more scientific discoveries, but also provide a more favorable platform for the cultivation of talents. On such a platform, it is undoubtedly important to inspire students' innovative thinking, enhance the innovative ability to solve complex problems, and more effectively realize the responsibility of "Teach and educate people, serve the community".

3.1 The Characteristics of Cultivating Students' Innovation Ability

The cultivation of the students in the innovative ideas and practice is closely connected with today's technology development and technology innovation. More and more disciplines in developing cross and penetrate each other, which formed a new research direction and area. And the cultivation of students' innovation ability tend to it. Therefore, with the deep intersection of different disciplines, the innovative cultivation of students embodies notable features of "wide professionalism and heavy orientation". That is to say, while cultivating students' professional foundation, we should pay attention to cultivating a broad range of knowledge in order to be able to comprehend by analogy. At the same time, we should closely follow the development of relevant professions and social needs, and properly adjust the training programs so that they can be in line with practical applications.

Take computer science as an example, the specialty itself originated from mathematics, physics, electronics and other disciplines, which has a nature of cross. It has gradually moved from a professional technology to a common technology. This requires that computer professionals not only have to master skilled computer applied technologies, but also should be familiar with industry expertise. The integration of computer technology and other professions is the development trend of computer technology. In this regard, undergraduate teaching process should not only meet the needs of professional development, but also take fusion of knowledge points of different majors into account. Let students have a solid foundation of knowledge and a wide range of knowledge, as well as enhance the practical ability, innovation and entrepreneurship ability and sustainable development. At the same time, as a specialized field with strong applicability and extremely fast development and update speed, the computer specialty is different from traditional professions such as mechanical, electrical, energy and materials. The replacement period of new and old technologies is very short and other professional applications and social needs are the guide of the replacement of these technologies. Therefore, on the other hand, this determines that students should master the professional knowledge and keep abreast of the technological developments of related disciplines.

3.2   Intersection of Different Disciplines is the Need of Cultivating Students' Innovation

There are complex and comprehensive problems in economy, society and science and technology. It is not a subject that can be used alone. Therefore, the demand of all trades and professions for technical and managerial talents is no longer single, but presents a multiple and multi-layered demand situation. Students who have a comprehensive understanding of the relevant disciplines and are able to comprehend by analogy can quickly adapt to social needs. Therefore, the cultivation of compound talents has become an important aspect that we must pay attention to for cultivating students. However, it is also a hot issue to fully conduct students who have a clear division of disciplines under the traditional education mechanism. Intersection of different disciplines is undoubtedly an effective way to solve this problem.

Multidisciplinary cross innovation projects led by our faculty have won several awards. The innovative project, "Research on the Radiation Characteristics of Carbon nanotube antenna in the THz Band", combines students from the School of Electronics and Information Engineering, Materials and Aerospace. Students who participated in this project not only consolidated their professional knowledge, but also expanded their knowledge and vision in related fields. According to the characteristics of students' innovation cultivation, intersection of different disciplines has become a necessity to cultivate students. Just like the urgent needs of science and technology innovation for intersection of different disciplines, students' innovative ideas and practices are also to some extent technological innovation. Talents training mode of multidisciplinary cross emphasizes the infiltration and integration of disciplines and meets the diversified learning needs of students, which is essentially a comprehensive, systematic and holistic change. In order to meet the needs of integrated development of disciplines and the new requirements of the society for personnel training, colleges actively changed their educational philosophies and explored new ways of cultivating talents. Mechanisms for training talents with different disciplines also came into being, which achieved multidisciplinary infiltration and cross training and improved the overall quality of students. It not only quickly grasp the skills required for jobs, but also can use "multidisciplinary" vision to find new jobs and create employment opportunities. Especially for the increasing proportion of undergraduates' continuing their studies, in-depth understanding of other disciplines also makes the direction of the subject more selective.

4   To Make Interdisciplinary Play Effective Role in Innovation Training

Because of the important role of intersecting of different disciplines in the promotion of innovation ability, how to grasp and cultivate interdisciplinary training in the cultivation of students is also an important issue we face. The following aspects are the main ways of intersecting of different disciplines.

4.1 Take Advantage of the Situation and Cultivate Interest

Interest is the best teacher. Confucius once said that "those who know it are better than those who are good at it, those who are good at it are not as good as those who enjoy it." This remark of Confucius reveals us the shortcut to cultivation of innovation. Many students have a strong interest in majors other than the major. If they can have an applicable intersection with the major in the field of technology, actively guiding the students to develop interest will make a reasonable cross with the major and will often do more with less. For the computer profession, it has become a common tool discipline and multidisciplinary contact is one of the most important features. We must also actively explore and guide students' interest in the cross field. Therefore, it is necessary to carry out regular and effective professional education for students. In teaching, attention should be given to guiding and stimulating students' interests in order to help them to explore unknown things by using things that they already know by means of giving top priority and comprehending by analogy, which will be of great benefit to students in similar study situations in the future. This allows students for life. At the same time, it is effective to highlight the main role of students, turn the original passive acceptance of knowledge into the active exploration of knowledge, stimulate students' interest in learning, change pressure as a driving force and mobilize the enthusiasm and initiative of students' learning, in order to promote the development of creativity.

4.2 Make a Distinction Between the Important and the Lesser One and Make a Reasonable and Modest Arrangement

The cultivation of students' innovation, especially the innovation of undergraduate students, should also emphasize the main body and characteristics. It is necessary to carry out cross-disciplines on the basis of students' professional development and to distinguish between primary and secondary one. We can neither kill students' interests or hobbies, but guarantee effective implementation of the professional main body to some extent. For computer science, as a general basic discipline, natural attributes have made it necessary to cross and integrate with other majors to realize the application value. Students tend to establish more research directions in the intersection of disciplines, which will play an important role in their own future development. The basis for the development needs a solid foundation of software and hardware in order to better integrate with other professions. This requires teachers to have a reasonable grasp of the degree of cultivation of their students' innovation. It is suitable to ensure the backbone and cross properly.

4.3 Highlight Practice that Combined with Practice

Influences of social demand orientation on technology development are obvious. In the innovation cultivation of students, the interdisciplinary subjects should also be combined with the practice. The refinement of innovative ideas, and the development of innovative practices should be carried out around the actual needs. In this way, it can not only achieve the application of learning, but also can clearly define the direction and accuracy of the intersection of disciplines. For computer majors, specialized courses are characterized by flexibility, practicality and comprehensive design. Therefore, in the process of cultivation of innovation, we must combine the curriculum and the actual situation of students to design. Inspiring students to find problems, analyze problems and solve problems, so that students gradually develop the habit of independently accessing to knowledge and creatively using of knowledge. Students should integrate the needs of practical technology application and pay attention to the latest scientific and technological knowledge of the development of the subject and related disciplines.

5   Successor

It is impossible for a professional teacher to be proficient in all subjects. But it requires teachers to have a wide range of knowledge, be proficient in or familiar with the subjects and disciplines of his major research. Therefore, teachers should often learn from those who in similar disciplines to study the entry points of penetration between disciplines. In particular, teachers should be able to ask key questions and discuss specific measures of problems with other teachers in similar disciplines. At the same time, the normalization of the brainstorming among teachers is necessary so as to put forward ideas of the subject of practical value. We will encourage teachers and other professional teachers in the school to jointly guide the teaching activities of science and technology innovation, and promote the large-scale and regularized development of cultivation of talents of multidisciplinary infiltration.

Acknowledgements. This research is supported by 2016 key projects of Online Education Research Fund (General Education) of the online education research center of the Ministry of education.

References

1. Wang, J., Shuai, C.M., Li, Y.Q.: A preliminary study on the mode of personnel training in interdisciplinary - an example of "Li Siguang" program. In: Proceedings of the International Conference on Environmental Systems Science and Engineering (ICESSE 2011), Dalian, Liaoning, pp. 650–665 (2011)
2. Gao, D.F., Li, T.F.: Review and summary and prospect on construction of national experimental teaching demonstration centers. Exp. Technol. Manag **34**(12), 1–5 (2017)
3. Chris, J., Anson, C., Steve, H.: Learning elsewhere: tales from an extracurricular game development competition. In: Proceedings of the 18th ACM Conference on Innovation and Technology in Computer Science Education (ITiCSE 2013), Canterbury, England, pp. 70–75 (2013)
4. Gao, Y., Liu, J.H., Chen, L.L.: Exploration of multi disciplinary and interdisciplinary talents training based on electronic information engineering. In: Proceedings of the Fourth International Conference on Information, Communication and Education Application (ICEA 2013), Hong Kong, pp. 248–252 (2013)
5. Sun, D.L.: Research on the Diversification Theory and Practice of Personnel Training Mode in Interdisciplinary, 1st edn. Economic Management Press, Beijing (2012)

6. Li, J.M.: Cross Boundary and Integration Based Research on the Training of University Talents, 1st edn. Suzhou University Press, Suzhou (2016)
7. Zhang, L.P., Yao, W.: Research on the Training of Engineering Science and Technology Talents Based on Industrial Innovation, 1st edn. Zhejiang University Press, Hangzhou (2013)
8. Mo, Z.W.: The Practice and Exploration of the Training of Freshman Students, 1st edn. University of Electronic Science and Technology Press, Chengdu (2008)

A High Precision and Realtime Physics-Based Hand Interaction for Virtual Instrument Experiment

Xu Han[1], Ning Zhou[2(✉)], Xinyan Gao[1], and Anping He[3]

[1] School of Software, Dalian University of Technology, Dalian 116620, China
[2] School of Electronic and Information Engineering, Lanzhou Jiaotong University,
Lanzhou 730070, China
TomZhoun@gmail.com
[3] School of Information Science and Engineering, Lanzhou University,
Lanzhou 730000, China

Abstract. Virtual hand interaction can greatly improve the operation significance of virtual instrument experiment. In order to apply virtual hand to virtual instrument interaction, as well as enhance the mechanical feedback effects, a physics-based real-time hand interaction method with high precision is proposed. We present a high precision restricted motion control algorithm by classifying the conventional instrument motion modes and considering the constraints and environmental force. We improve the tendon model of the existing tendon algorithm as well as the dynamics equation, to give a true mechanical feedback of physical interactions. Using Leap Motion and Modelica for hand data acquisition and equation calculation, we have realized an example of interaction, which proves the real-time and interactive authenticity of the method.

Keywords: Physics-based simulation · Tendon
Hand-computer interaction

1 Introduction

At present, virtual experiment is a hot research topic due to its experimental safety and resource saving. However, the problem existing in current virtual experiment is that: without the actual operation but only keyboard and mouse, if it is safe and scientific to operate the instrument in real life. Therefore, to make virtual experiment more real with better teaching and operational significance, human hand interaction is necessary.

Project supported by the National Nature Science Foundation of China (61650207), and the Chinese NSF NO. 61402121, the Fundamental Research Funds for the Central Universities of Lanzhou University, No. 861914, and Guangxi Science and Technology Project (AB17129012).

Q. Zhou et al. (Eds.): ICPCSEE 2018, CCIS 902, pp. 552–563, 2018.
https://doi.org/10.1007/978-981-13-2206-8_46

The particularity of virtual instrument interaction is that the operating parts are restricted objects. Compared with free objects, restricted objects need to consider environmental constraints, such as translation direction, rotation direction and environmental force. Existing motion control algorithms mostly interact with free object [1–4] without considering above constraints, which result to low precision in instrument interaction.

As for human interaction, the effects of physical interaction algorithm are not obvious enough. Interactive algorithms are typically divided into physics-based and geometry-based. The former can improve the interaction authenticity and avoid unnatural object movement, with large computation and real-time difficulties. The latter is computationally small but easy to generate unrealistic interactions. However in realization, there is no obvious difference between two algorithms, which means that the mechanical results obtained by large computational quantity do not get well feedback [1,3,5].

In order to solve above problems and apply virtual hand to virtual instrument interaction, this paper presents a physics-based real-time hand interaction method with high precision. We divide conventional instrument interactions into three categories, which are pressing, sliding and rotating, and put forward specific motion control algorithms, in order to achieve restricted object motion control. Next, we use tendon simulation to feedback the mechanical result of physical interaction. By modifying tendon model and dynamics equation, we modify current tendon control method to fit skin model and animation pipeline. Finally, we realize a simple interaction example by using Leap Motion for hand data acquisition and Modelica for physical calculation, which confirms the real-time and high precision of the method.

The main contribution in this paper is:

- A physics-based real-time restricted motion control method with high precision.
- A tendon model and simulation method for interaction mechanics feedback.
- A real-time virtual instrument interaction example by using Leap Motion and Modelica.

2   Related Work

Our work includes physics-based motion control and tendon simulation, so we will concentrate on these two parts.

For physics-based motion control, one way is used by motion controller [2–4]. Pollard and Zordan [2] matches the current scene with pre-recorded data and simulates using a mixture of PID and PD controllers. Wang [3] and Zhao [4] use synthetic motion controller to simulate object movement and hand grasping. The problem of using motion controller is that it requires enough good parameters to match the motion trajectory. Another way is to use physics-based optimization methods [1,5] or pure mechanical solutions [6,7]. Liu [1,5] distinguishes gesture types by initial state and object motion prediction, then solves an optimization method. Brubake [6] uses mechanics analysis to find out the contact force between human body and environment. However, the above method does not consider the restricted objects motion.

In addition, neural networks and machine learning can also be used. Ye [8] calculates the contact force based on object motion and uses random sampling to synthesize interaction results. Pham [9] uses artificial neural networks to calculate additional forces generated by hand. Blana [10] matches hand gestures and internal forces, then use the machine to verify. However, above methods cannot achieve real-time at present.

Next, about tendon simulation, we focus on tendon models and control algorithms. About tendon model, commonly use simple linear model [11,12], or complex one like network model [13,14] .These models describe tendons as a kinematic path and the constraints handing is simple. At present, the most realistic method is to use a slim physical structure called strand [15–17], which often use NURBS, Euler-on-Lagrange or other deformable models. This approach makes it easy to add motion constraints, with disadvantages of large computation and sensitive to parameter.

For tendon control method, most current methods are to calculate tendon force, and then control tendon through different constraint equations or optimization methods. For example, Sueda [15,17] solves tendon motion using linear least squares method. Geijtenbeekp [18] uses a CMA strategy to optimize muscle position, then control skeletal movement. These methods have good simulation results, but due to complex constraint handling, it is difficult to achieve real-time.

3   Restricted Motion Control

This paper focuses on interactions of common electronic instruments such as oscilloscopes, rheostats and so on. The basic operation modes can be grouped into three categories: press, slide and rotate, corresponding to buttons, sliders and knobs, which are all restricted objects. Based on the consideration of restricted motion direction and environmental external force, this section gives specific physics-based methods.

3.1   Motion Control Representation

We decompose the total joint torque(τ) supplied to drive the hand and object as follow

$$\tau = \tau_{hand,object} + \tau_{hand} \tag{1}$$

τ_{hand} is torque of current joint movement, and $\tau_{hand,object}$ is total torque required by object.

For τ_{hand} in Eq. 1, in this paper, we simplify finger model as a single rod model [19] with variable length. Cause during the interaction, joint angle changes little.

The key point in this section is how to find $\tau_{hand,object}$. According to the Newton-Euler's thoughts, object motion is decomposed into translation and rotation, and restricted movement are the combination of two states, as shown in Eq. 2.

$$\tau_{hand,object} = A^T \tau_R + J^T f_p \tag{2}$$

where τ_R is the torque to make objects rotate, f_p is the force to make objects translate, J and A are the Jacobian matrix. In addition, this paper sets when there is static friction between hand and object, put it as a component of f_p. Once the forces and torques in Eq. 2. are obtained, we can use them to better control the movement of the hand and objects, with the help of kinematics.

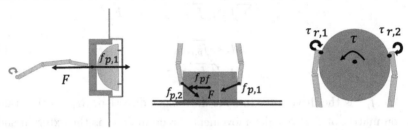

Fig. 1. Force illustration of three interactive motions. (left) button (middle) slider (right) knobs.

3.2 Button Motion Analysis

The button in this paper is a hard button, which means the resistance is provided as a threshold force. Button movement direction is limited and cannot rotate. So it is not necessary to consider rotation torque and environmental external force. Button only accepts force in restricted direction, and force in vertical direction will be offset by its mechanical structure.

Force analysis of button is shown in Fig. 1 (left), where $\overrightarrow{f_{p,i}}$ is the contact force which finger i provides, $\overrightarrow{F}$ represent the threshold force. Describe it as an optimization problem and get the contact force of each finger.

$$min_{\overrightarrow{f_{p,1}},...,\overrightarrow{f_{p,k}}} \| \sum_{i=1}^{k} M_i \overrightarrow{f_{p,i}} - \overrightarrow{F} \|^2$$

$$s.t. < \overrightarrow{f_{p,i}}, \overrightarrow{n_i} > < \theta \tag{3}$$

$$\tau_{hand,i} + J_i^T \overrightarrow{f_{p,i}} \leq \tau_{max,i}$$

where M_i is the transformation matrix of $\overrightarrow{f_{p,i}}$ to button movement direction, $\overrightarrow{n_i}$ is the normal of contact surface, θ is the friction angle of button cap surface, $\tau_{hand,i}$ is the joint torque in nature state, $\tau_{max,i}$ is the max torque which joint can provide. Constraint 1 means the contact force is in the friction cone. Constraint 2 means total torque of finger do not exceed the maximum torque that the finger can maintain.

3.3   Slider Motion Analysis

The physical structure of slide button includes slide rail and slider, and the extrusion between rail and slider will generate sliding friction as resistance. The slider movement direction is limited and cannot rotate, which is the same as the button. But extra environmental forces exist, and for slider it is the extra sliding friction at slider bottom when finger presses.

As shown in Fig. 1 (middle), $\overrightarrow{f_{p,i}}$ is the finger contact force. $\sum \overrightarrow{f_{pf}}$ is total of external sliding friction. $\overrightarrow{F}$ represent the total of acceleration terms and original friction. Corresponding optimization problem as follows:

$$min_{\overrightarrow{f_{p,1}},...,\overrightarrow{f_{p,k}}} \| \sum_{i=1}^{k} (M_{1,i}\overrightarrow{f_{p,i}} - \overrightarrow{f_{pf,i}}) - \overrightarrow{F} \|^2$$

$$s.t. < \overrightarrow{f_{p,i}}, \overrightarrow{n_i} >< \theta \qquad (4)$$

$$\tau_{hand,i} + J_i^T \overrightarrow{f_{p,i}} \le \tau_{max,i}$$

where $M_{1,i}\overrightarrow{f_{p,i}}$ is the drive force in slide movement direction, $M_{1,i}$ is the transformation matrix of $\overrightarrow{f_{p,i}}$ to slider movement direction. $\overrightarrow{f_{pf,i}}$ is the external slide friction, with coulomb friction we get $\overrightarrow{f_{pf,i}} = \mu M_{2,i}\overrightarrow{f_{p,1}}$. μ is the friction coefficient, $M_{2,i}\overrightarrow{f_{p,1}}$ is the pressure provided by hand, $M_{2,i}$ is the transformation matrix of $\overrightarrow{f_{p,1}}$ to vertical direction of slider movement.

3.4   Knobs Motion Analysis

The torque resistance of knobs comes from bearings or other rotatable parts. Rotation axis is limited and cannot translate, and forces in the other axes will be offset by the mechanical structure. So hand only provides torque, regardless of translation and extra force. Besides that, the rotational torque is provided by hand contact force and torsional torque which is produced by hand rotation itself, and they have the same force arm. Considering the static friction constraint, numerically we all consider as contact force.

As shown in Fig. 1 (right), $\overrightarrow{\tau_{r,i}}$ is the rotation torque that hand provides, $\overrightarrow{\tau}$ is the total of acceleration terms and resistance torque. Corresponding optimization problem as follows:

$$min_{\overrightarrow{f_{r,1}},...,\overrightarrow{f_{r,k}}} \| \sum_{i=1}^{k} M_{1,i}(r\overrightarrow{n_i} \times \overrightarrow{f_{r,i}}) - \overrightarrow{\tau} \|^2$$

$$s.t. < \overrightarrow{f_{r,i}}, \overrightarrow{n_i} >< \theta \qquad (5)$$

$$\tau_{hand,i} + A_i^T \overrightarrow{\tau_{r,i}} \le \tau_{max,i}$$

where $\overrightarrow{n_i}$ is the direction from the center mass of knobs to the contact point, r is the radius of knobs, $r\overrightarrow{n_i} \times \overrightarrow{f_{r,i}}$ is finger rotation torque, M_i is the transformation matrix of $\overrightarrow{\tau_{r,i}}$ to rotation axis, $\overrightarrow{f_{r,i}}$ is the corresponding contact force of $\overrightarrow{\tau_{r,i}}$.

3.5 Restricted Motion Solution

The direction of finger contact force is unknown, which lead to constraint matrix M_i in Eqs. 3, 4 and 5 is unknown. Considering the restricted motion of objects, we create a local coordinate system for objects. For translation-limited objects like buttons and sliders, set the movement direction to the x-axis, and for rotation-constrained objects such as knobs, set the rotation axis to the x-axis. Let the other two lines perpendicular to each other and the x-axis be the y-axis and the z-axis. The positive directions of the three axes are expressed as unit vectors $\overrightarrow{a}, \overrightarrow{b}, \overrightarrow{c}$ in the world coordinates. Then the finger contact force and the resistance force/torque can be expressed as

$$\overrightarrow{f_{p/r,i}} = f'_{p/r,i_x}\overrightarrow{a} + f'_{p/r,i_y}\overrightarrow{b} + f'_{p/r,i_z}\overrightarrow{c}$$
$$\overrightarrow{F} = F'_x\overrightarrow{a} \tag{6}$$
$$\overrightarrow{\tau} = \tau'_x\overrightarrow{a}$$

Use Eq. 6 to simplify Eqs. 3, 4 and 5 from vector operation to scalar operation with constraints unchanged, the optimization problems are amended as follows

Button :

$$\min_{\overrightarrow{f'_{p,i}}} |\sum_{i=1}^{k} f'_{p,i_x} - F'_x| + \alpha \sum_{i=1}^{k}(f'_{p,i_y} - R_1)^2 + \beta \sum_{i=1}^{k}(f'_{p,i_z} - R_2)^2 \tag{7}$$

Slider :

$$\min_{\overrightarrow{f'_{p,i}}} |\sum_{i=1}^{k}(f'_{p,i_x} - \mu f'_{p,i_z}) - F'_x| + \alpha \sum_{i=1}^{k}(f'_{p,i_y} - R_1)^2 \tag{8}$$

Knobs :

$$\min_{\overrightarrow{f'_{r,i}}} |r \sum_{i=1}^{k}(n'_{iy}f'_{r,i_z} - n'_{iz}f'_{r,i_y}) - \tau'_x| + \alpha \sum_{i=1}^{k}(f'_{r,i_x} - R_1)^2 \tag{9}$$

where weight α and β are 0.1, 0.1, R_1 and R_2 are offset parameter to let forces in the other directions consider the angle of finger. After getting the direction component, the final finger contact force is obtained by Eq. 6.

4 Tendon Simulation

This section will describe tendon model and tendon control method, which are used to feedback the mechanics result in Sect. 3. The method in this section refers to Suede [17], we present a simplified tendon model and constraints, as well as modify the dynamics equation to fit animation pipeline.

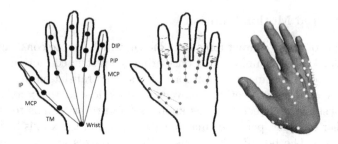

Fig. 2. (left) Hand skeleton diagram (middle) tendon diagram dotted line indicates tendon and point indicates tendon point. Color indicates the finger joints it depends: red-MCP, orange-TM, blue-wrist (right)skinned hand model, white points are tendon points. (Color figure online)

4.1   Tendon Model

The hand skeleton structure used in this paper is shown in Fig. 2 (left). It includes 16 joints with total degree of freedom (DoF) of $q \in R^{33}$. Wrist joint has 6 DoF with free translation and rotation. Joint TM and MCP have 3 DoF with relative rotation. PIP and DIP's DoF is 1, with one-axis rotation.

Tendon structure is shown in Fig. 2 (middle). Each tendon is controlled by multiple tendon points. All tendon points are joints in animation pipeline, which depend on hand joints to transform. For each tendon point DoF $s_i \in R^3$, including xyz coordinates transformation, so the total degree of freedom $S \in R^{81}$. Skinned hand model is shown in Fig. 2 (right).

4.2   Tendon Control

The reference coordinate of the dynamics equation is the parent coordinate of tendon point, and a reference frame is needed. We define it is when the palm is fully open and facing down quietly. Based on this, the dynamics equation is

$$\begin{pmatrix} M & \varnothing^T \\ \varnothing & 0 \end{pmatrix} \begin{pmatrix} v_{next} \\ \lambda \end{pmatrix} = \begin{pmatrix} h\Delta f \\ -\mu\varnothing \end{pmatrix} \tag{10}$$

Equation 10 is the adaptation of Sueda [17]'s dynamics equation in animation pipeline. Where M is the mass matrix. $\varnothing$ is the position-level equality constraint matrix. $\varnothing$ is the partial navigation of spatial variables. v_{next} is the velocity in next frame with 6 DoF including current point and reference point on fingers' bone, λ is the Lagrange multiplier, Δf is the difference of tendon force between current and reference frame, which can get from Eq. 1. h is the time step. μ is stabilizer weight, which equals $1/h$. $\mu\varnothing$ is used to put the point back to constraint manifold.

About constraints, including fixed constraint, surface constraint and sliding constraint. We put fix constraints to make origins and ends of tendons unmoved, surface constraint for the others to be unembedded in fingers' bones. Sliding constraint will be replaced by a semi-ellipsoid for real-time judgement to avoid unnatural flexing. The long axis is the connection of the two tendon endpoints for thumb and adjacents ones for the other finger. And short axis is established by the maximum angle of protrusion. When tendon point deviates from the semi-ellipsoid, map it to the edge.

5 Experiment and Result

This paper uses an oscilloscope as an operation example. The basic interaction process is shown in Fig. 3. At the beginning of each frame, the hand joint information is obtained to get the possible object motion state of the next frame. Then according to object motion mode in Sect. 3 to classify the interactive category and calculate joint torque as well as transform to tendon force. After that, do tendon simulation and transform tendon points, then finish current frame.

5.1 Experimental Configuration

The experimental environment is 32 GB CPU RAM and 4 GB GPU RAM, as well as 8 GB RAM and 1 GB GPU RAM. Experiment tools are Leap Motion and Modelica [20,21] for hand data acquisition and equation solved. In addition, virtual hand in this example satisfies the feature of "virtual synchronization priority". When the feedback force cannot be synchronized with the reality, virtual hand will prioritize the synchronization with objects in the virtual scene.

5.2 Motion Control Result

For three motion modes, we give the computing time as shown in Fig. 4. From Fig. 4 (left) we can find with average 5% CPU utilization, the worst calculation

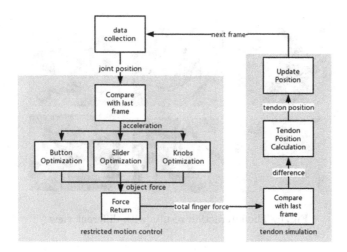

Fig. 3. Basic process of the example.

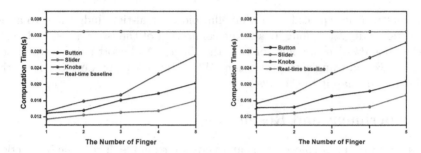

Fig. 4. Computation time of different hand numbers (left) 32 GB CPU with 5% utilization rate (right) 8 GB CPU with 19% utilization rate.

time is 0.027 s, which means 37 frames per second. Due to the high configuration, the same experiment was conducted under 8 GB CPU RAM and 1 GB GPU RAM. The result is shown in Fig. 4 (right). With average 19% CPU utilization, the worst-case calculation time is 0.03 s, which is 33 frames per second. All of them can meet real-time requirements.

The interaction results of three motion modes are shown in Fig. 5, which shows the correct interaction with restricted object. During interactions, there is a situation that virtual and reality are out of synchronization. Figure 6 (Left) shows when the hand maximum force can meet conditions of synchronous movement, they will be synchronized. Figure 6 (middle) shows when the maximum hand force can only meet the threshold force, the object and virtual hand can move synchronously with a lower speed than realistic hand. Figure 6 (right) shows that when the maximum hand force cannot meet threshold force, neither the object nor the hand will move. For this situation, due to the "virtual synchronization priority" feature, it is normal,which also shows the high authenticity.

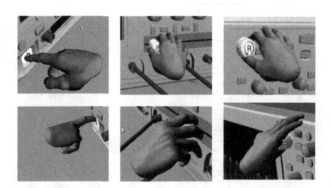

Fig. 5. The interaction results with different view.

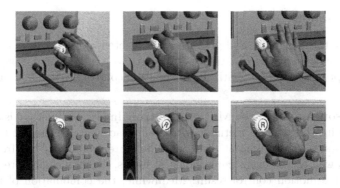

Fig. 6. Movements in different interactive conditions at the same time. (left) synchronous movement (middle) synchronous movement in different speed (right) unmoved.

5.3 Tendon Simulation Result

Tendon simulation is used to describe the force feedback of physical interactions. According to Sect. 4 building hand model, result is shown in Fig. 7 (up), we can find when interaction tendon is more prominent than nature states due to tendon force. Figure 7 (bottom) illustrates how the tendon changes by increasing the contact force from zero to maximum. We can find tendon will protrude rapidly until the peak.

About computational efficiency, the tendon algorithm uses 0.06 ms for a single tendon point and 0.3 ms in 8 GB CPU RAM and 1 GB GPU RAM, which can meet real-time requirements.

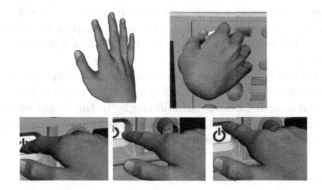

Fig. 7. (up) tendon performance in natural and interactive states (right) Tendon performance in rapidly increasing tendon force.

5.4   Evaluation

To evaluate the specificity and precision of restricted motion control method, as well as the efficiency and real-time performance of tendon simulation, we compared the proposed method with the free motion control method and Sueda [17].

Comparison Against Free Motion Control. In fact, our method is an extension of the free interaction method. It takes into account the contact force components outside the constraint direction which are ignored by the free motion control, as well as the extra environmental forces, which means more precision than the free motion control result. Meanwhile, the combination of the three interaction modes can also satisfy the regular conventional instrument operation and restricted movement.

Comparison Against Sueda [17]. Our method can be used for real time, and their methods apply to animated scenes. We use a new tendon model, then describe and control it directly at the animation pipeline. Because of the simpler tendon model and no need to map from strand points to animation pipeline, our method is faster.

6   Conclusion

In this paper, we introduce a physics-based real-time hand interaction method with high precision, which is used to interact with the restricted objects in the virtual instrument experiment and enhance the mechanical feedback effect. We classify conventional instrument motion mode and establish the physical motion control algorithm, as well as improve existing tendon simulation method in tendon model and the dynamics equation to fit the animation pipeline. Finally we realize a simple interaction example by means of Leap Motion and Modelica, which shows real-time interaction with high precision, as well as better mechanical feedback effect.

Our method can be well applied to circuit simulation experiments, but there are still improvements. Recent motion modes for restricted objects are enough for common instrument motions but still single. In the future we will do more composition and expansion, and extend from a single experimental apparatus to more circuit experiments to improve the application range.

References

1. Liu, C.K.: Synthesis of interactive hand manipulation. In: ACM Siggraph/eurographics Symposium on Computer Animation, pp. 163–171 (2008)
2. Pollard, N.S., Zordan, V.B.: Physically based grasping control from example. In: ACM Siggraph/eurographics Symposium on Computer Animation, SCA 2005, Los Angeles, July, pp. 311–318 (2005)

3. Wang, Y., et al.: Video-based hand manipulation capture through composite motion control. ACM Trans. Graph. **32**(4), 43 (2013)
4. Zhao, W., Zhang, J., Min, J., Chai, J.: Robust realtime physics-based motion control for human grasping. ACM Trans. Graph. **32**(6), 207 (2013)
5. Liu, C.K.: Dextrous manipulation from a grasping pose. ACM Transactions on Graphics (TOG). vol. 28. No. 3. ACM, p. 59 (2009)
6. Brubaker, M.A., Sigal, L., Fleet, D.J.: Estimating contact dynamics. In: IEEE International Conference on Computer Vision, pp. 2389–2396 (2010)
7. Li, Y., Fu, J.L., Pollard, N.S.: Data-driven grasp synthesis using shape matching and task-based pruning. IEEE Trans. Vis. Comput. Graph. **13**(4), 732 (2007)
8. Ye, Y., Liu, C.K.: Synthesis of detailed hand manipulations using contact sampling. ACM Trans. Graph. **31**(4), 1–10 (2012)
9. Pham, T.H., Kheddar, A., Qammaz, A., Argyros, A.A.: Towards force sensing from vision: observing hand-object interactions to infer manipulation forces. In: Computer Vision and Pattern Recognition, pp. 2810–2819 (2015)
10. Blana, D., Chadwick, E.K., Bogert, A.J.V.D., Murray, W.M.: Real-time simulation of hand motion for prosthesis control. Comput. Method Biomech. Biomed. Eng. **20**(5), 540 (2017)
11. Carvalho, A., Suleman, A.: Multibody simulation of the musculoskeletal system of the human hand. Multibody Syst. Dyn. **29**(3), 271–288 (2013)
12. Johnson, E.R., Morris, K., Murphey, T.D.: A Variational Approach to Strand-Based Modeling of the Human Hand. DBLP. Springer, Heidelberg (2009)
13. Dogadov, A., Alamir, M., Serviere, C., Quaine, F.: The biomechanical model of the long finger extensor mechanism and its parametric identification. J. Biomech. **58**, 232–236 (2017)
14. Valero-Cuevas, F.J., Yi, J.W., Brown, D., Mcnamara, R.V., Paul, C., Lipson, H.: The tendon network of the fingers performs anatomical computation at a macroscopic scale. IEEE Trans. Bio-Med. Eng. **54**(2), 1161–6 (2007)
15. Sachdeva, P., Sueda, S., Bradley, S., Fain, M., Pai, D.K.: Biomechanical simulation and control of hands and tendinous systems. ACM Trans. Graph. **34**(4), 42 (2015)
16. Spillmann, J., Teschner, M.: CORDE: Cosserat rod elements for the dynamic simulation of one-dimensional elastic objects. Eurographics Association (2007)
17. Sueda, S., Kaufman, A., Pai, D.K.: Musculotendon simulation for hand animation. In: ACM SIGGRAPH, p. 83 (2008)
18. Geijtenbeek, T., Panne, M.V.D., Stappen, A.F.V.D.: Flexible muscle-based locomotion for bipedal creatures. ACM (2013)
19. Baillieul, J.: Introduction to robotics mechanics and control. IEEE Trans. Autom. Control **32**(5) (2003)
20. Åkesson, J., Årzén, K.E., Gäfvert, M., Bergdahl, T., Tummescheit, H.: Modeling and optimization with optimica and jmodelica.orglanguages and tools for solving large-scale dynamic optimization problems. Comput. Chem. Eng. **34**(11), 1737–1749 (2010)
21. Fritzson, P., Engelson, V.: Modelica a unified object-oriented language for system modeling and simulation. Lect. Notes Comput. Sci. **1445**(1445), 67–90 (1998)

Application of Project Management in Undergraduates' Innovation Experiment Teaching

Qing Wang[✉], Huipeng Chen, Hongwei Liu, Lei Xu,
and Yanhang Zhang

School of Computer Science and Technology, Harbin Institute of Technology,
Harbin 150001, China
sunnywang@hit.edu.cn

Abstract. In view of unique characteristics of the experiment course of low power embedded system design and implementation, the project management theory is introduced into undergraduates' innovation experiment teaching. Under the framework of speculative knowledge, the key elements of the curriculum are presented, and the project schedule of course is emphatically analyzed.

Keywords: Project management · Project schedule management
Innovation experiment teaching

1 Introduction

As the most advanced management mode in twenty-first Century, Project Management (PM) has been widely applied in aerospace, computer, electronic communications and other technology fields. In the field of computer, project management can effectively control research risk, improve management efficiency and ensure the realization of the final goal for large-scale software projects [1]. At present, the method of project management is generally applied in the development of IT company and in the process of scientific research by graduate students in Colleges and universities, but it is less applied in the experiment and innovation of undergraduate. The scenario results in the lack of project management knowledge and difficult in the use of project management tools for engineering undergraduates. Taking the innovation experimental course "Low Power Embedded System Design and Implementation" as a practical project, this paper introduces the concept and method of project management into undergraduate innovation experimental teaching, and arranges curriculum contents with project management theory, which has a reference function for future experimental teaching.

2 Analysis of Project Elements in Innovation Experiment

From the beginning to the completion, there will often be conflicts or contradictions between goals pursued during different stages through the whole project, and the resolution of these problems may work in concert with whether the project is successful

© Springer Nature Singapore Pte Ltd. 2018
Q. Zhou et al. (Eds.): ICPCSEE 2018, CCIS 902, pp. 564–572, 2018.
https://doi.org/10.1007/978-981-13-2206-8_47

or aborted in final. Project management is the practice of initiating, planning, executing, controlling, and closing the work of a team to achieve specific goals and meet specific success criteria at the specified time, and the primary challenge of project management is to achieve all of the project goals within the given constraints [2]. The factors affecting the project success in many aspects, whenever, the scope, time, cost and quality are the most important key factors, and the latter three are called TQC.

Strictly speaking, it is impossible to ensure that each target gets the best performance at the same time in a project. In the actual issue, project managers must make a compromise solution to keep a balance among the variety of factors, so that it has the least impact on the objectives of the project. In general, quality, cost and time are mutually restrictive, as shown in Fig. 1. Usually there is no specific precedence among the three elements, and the priority is determined by the characteristics of the project itself. Therefore, the first step of project management is to clarify the detailed analysis of the project.

Fig. 1. Project management trade-offs

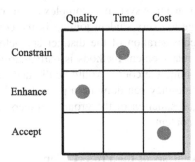

Fig. 2. Project priority matrix analysis

2.1 Curriculum Characteristics

The experimental course "Low Power Embedded System Design and Implementation" is an elective high quality innovation experimental course, which aims at cultivation of innovative thinking and development of manipulative ability. In the course, students are required to design and accomplish a low power consumption MP3 player taking use of MSP430F5529 MCU of TI company as the control unit, and to achieve the curriculum goal, they have to complete the whole process including circuit schematic design, PCB layout design, component welding and system debugging by themselves.

As an elective experimental course dominated by students' independent innovation, this course is open to all undergraduates in the university, so that there are some unique characteristics as follows.

The Complex Composition of Students

Since undergraduates of all the university can join in the course, students of different specialties and grades will appear in the class, and this kind of personnel composition is in conformity with the actual project. As it is difficult to completely integrate all aspects of teaching subjects, such as majors, grades, existing knowledge structure, regular school hours, curriculum expectations and so on, it is necessary to use some means of project management to achieve the balance of personnel distribution.

The Shortage of Teaching Hours

The teaching hours of the course is 48 school hours, according to the class arrangements of 6 h per week, this course must be completed in two months. Unlike traditional courses whose experiment articles are in accordance with the order of increasing difficulty, there is less class hours used to familiarize with the development environment and previous knowledge in the innovation experimental course, so students should complete the problems with less difficulty in their spare time.

The Difficulty in Curriculum Goals

To achieve the ultimate goal of this course which is an electronic product which can be used to play songs, each group must complete every details in developing progress including all kinds of knowledge of software and hardware, it is may be the first time to develop an entire system as complex as this one to some students, therefore, it is a great challenge to every student joining in the course.

In consideration of the distinct characteristics of this course, the introduction of project management methods is particularly necessary. On the first day of class, students joining the course would be divided into several groups. Each group is a project group, and they can draw up a project implementation plan of their own according to the actual situation of the group members, and effectively complete the final experimental content.

2.2   Analysis of the Project Scope

Usually, scope describes what you expect to deliver when the project is complete, and it should define the results to be achieved in specific, tangible, and measurable terms. The final deliverable of this course is a MP3 player that can be used, and the acceptance criteria of the project is to accomplish the function of playing songs of MP4 format. Project scope is the keystone interlocking all elements of a project plan, and we will make the definition of the scope using the following checklist as Table 1 depicts.

Table 1. Project scope checklist of the course

	Content	Analysis of the course
1	Project objective	Develop a MP3 player before the semester ends
2	Deliverable	A MP3 player which can play songs of MP4 format
3	Milestones	Software programming, circuit schematic design, PCB layout design and MP3 player
4	Technical requirements	C language programming, EDA design, electronic component welding

2.3   Analysis of the Project TQC

After identifying the scope of the course, we will define the TQC of the course.

The final deliverable of the course is a MP3 player, but not everyone could achieve the top target. Considering the actual situation, we pay more attention to whether students complete the deliverable by themselves from the quality aspect.

During the course, the laboratory will provide the corresponding electronic devices and tools, so students need consider less about the cost. Whenever, in order to reduce experimental difficulty and guide workload, we specified the CPU chip as MSP430F5529 and decoder as VS1300, and introduced a regular development and debugging environment called IAR for students. Take such a solution, restricts the freedom of students' independent innovation to a certain extent, but it can ensure that each group can implement the final product within a certain cost range.

As mentioned before, the class teaching time is 2 months, but it is far from enough for students to finish the project, so we extend the deadline to the end of semester. Students will gain their corresponding grades as long as delivering achievements before the last day of the semester. However, they will not get any marks if they submit nothing before the deadline. Therefore, in order to get the score of the course, the most important task to the students is to complete a product within the specified time.

2.4   Analysis of the Project Priorities

One of the primary jobs of a project manager is to manage the trade-offs among time, cost and quality. To do so, they must define and understand the nature of the priorities of the project. We will analyse the priorities of the course taking use of priority matrix which was found useful in practice.

Figure 2 illustrates the priority matrix of the project in the course.

Let's take a look at the constrain row. The project must meet the completion date, specifications and scope of the project, or budget. In the course, students should hand in their works before deadline, so the time limit cannot be compromised. Given the scope of the project, the enhance row presents the criterion should be optimized. Since students are expected to develop learning skills, enhancing the quality means adding value to the course, they should take advantage of every opportunity to enhance the performance. In doing so, going over budget is acceptable though not desirable.

3   Project Schedule Management

In PMBOK2017, the process of project management are grouped in ten knowledge areas, namely the project integration management, project scope management, project schedule management, project cost management, project quality management, project human resource management, project communications management, project risk management, project procurement management and stakeholder management [3]. Figure 3 illustrates the relationships of the ten areas [4]. In terms of the innovation course, students do not need to take into account human resources management, risk management, procurement management and stakeholder management. Since time

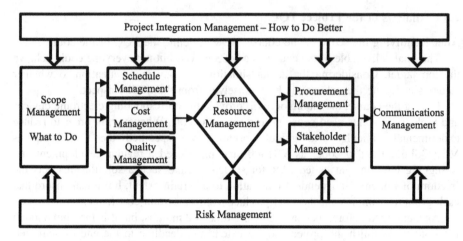

Fig. 3. Ten knowledge areas of project management

element is vital to students as described in Sect. 2, we will mainly discuss time management in this section.

Project time management is also known as schedule management, which includes the processes required to manage the timely completion of the project. The processes are defining activities, sequencing activities, estimating activity resources, estimating activity durations, developing schedule and controlling schedule.

3.1   Define Activities

Defining activities is the first step of schedule management, it is a process of identifying and documenting the specific actions to be performed to produce the project deliverables. Through the decomposed work packages, project members and manager could estimate, schedule, execute, monitor, and control the whole project more accurately.

The MP3 player we need to finish is a complete system combining software function and hardware architecture. Its development process can be divided into three parts: skills preparation, system design and development, and debugging and integration.

The main chips used in MP3 players are MSP430F5529 microcontroller and VS1300 decoder. The related knowledge and development environment of these two devices are the key preparation to the overall system. Therefore, students should first familiar with the development and debugging steps of the two chips. These necessary preparations include C language programming, IAR integrated environment development and debugging, and the use of SCH software. In addition, SD card, LCD display screen, touchscreen, JTAG debug port and other components are used at the periphery of the system, so students must arrange corresponding spare time to learn about relevant specifications and development methods.

The final deliverable is a MP3 player, which requires students to design hardware component as well as software program, and finally to combine the two parts into a whole product. To achieve the final goal, they must draw corresponding circuit schematic diagram, weld PCB board, program function programs, and integrate the overall system, most of the students may be unfamiliar to the circuit design and implementation content, so they need more time to complete the hardware part.

After the completion of the design and realization work, students must coordinate each component to complete the functions of the system, and check the acceptance of the system in the later debugging step. The whole project can be divided into 11 stages as listed in Table 2.

Table 2. Project scope checklist of the course

Activity name	Activity definition	Activity description	Milestones or not?
A	Knowledge preparation	Be familiar with usage method of MSP430F5529 and VS1300 chips; master usage of IDE	
B	Electronic devices chosen	Choose the actual chips used as I/O component and have an intimate knowledge of them	
C	Structure design of system	Design the whole structure of the system; define specific function of each module	
D	Circuit schematic design	Design circuit schematic diagram of the system taking advantage of CAD software	
E	PCB layout design	Design PCB layout board and hand over to factory to manufacture the actual one	
F	Component welding	Solder the chips to PCB board including MPU, decoder, the resistance, capacitance and so on	√
G	Hardware debugging	Check whether every electronic devices are soldered well and work normally	√
H	Software design	Design software function modules according to Activity 2	
I	Programming	Implement the modules using C or C++ language within IAR environment	
J	Software function debugging	Test the software function and adjust the incorrect functions	√
K	System integration debugging	Check whether the MP3 player can play songs or not and make adjustments	√

3.2   Sequence Activities

After the project activity is defined, the next step in schedule management is to sort or determine the dependency between them, and this step is called sequencing activities. Usually, the network diagram is applied to represent the dependency relationship between activities, and the general format is Activity-On-Arrow (AOA) or Arrow Diagramming Method (ADM) [5]. Figure 4 reveals the network diagram of the course in the format of ADM, arrows of the graph represent activities, and nodes represent the connections and interdependence between activities, and the activity name is in correspondence with Table 2. In order to facilitate the subsequent scheduling, we divide the overall development process into three stages named preparation phase, develop phase, and debug phase as Fig. 4 shows.

In Fig. 4, the develop phase is divided into two branches, and this can be a breakthrough in shortening the time duration. Activity D, E, F, and G are all designed to complete the hardware component, while activity H, I, and J are all designed to accomplish software function, so we can divide students in the same group into two teams to take charge with hardware and software respectively.

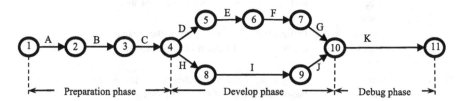

Fig. 4. Network diagram

3.3   Estimate Activity Durations

In previous sections, we have confirmed that students could earn credits so long as they submit products before the semester ends, that means that we can identify the whole project activity duration as 18 weeks, and we will estimate respective duration of each stage.

The simplest way is to allocate time on average in accordance with three stages as Fig. 5 displays. However, the difficulty of each link is different, so that the time needed is different from each. In consideration of characteristics of each stage, we can cut down 3 weeks from preparation phase to develop phase as Fig. 6 exhibits.

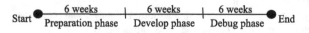

Fig. 5. Distribute time on average

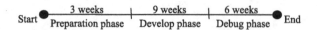

Fig. 6. Distribute more time in develop phase

Activity name	Activity definition	Start time (Week)	End time (Week)	Duration (Week)	Phase 1			Phase 2									Phase 3					
					1	2	3	4	5	6	7	8	9	10	11	12	13	14	15	16	17	18
A	Skills preparation	0	0.5	0.5																		
B	Devices chosen	0.5	1	0.5																		
C	System structure design	1	3	2																		
D	Circuit schematic design	3	7	4																		
E	PCB layout design	7	9	2																		
F	Component welding	9	11	2																		
G	Hardware debugging	11	12	1																		
H	Software design	3	4	1																		
I	Programming	4	8	4																		
J	Software debugging	8	12	4																		
K	System integration debugging	12	18	6																		

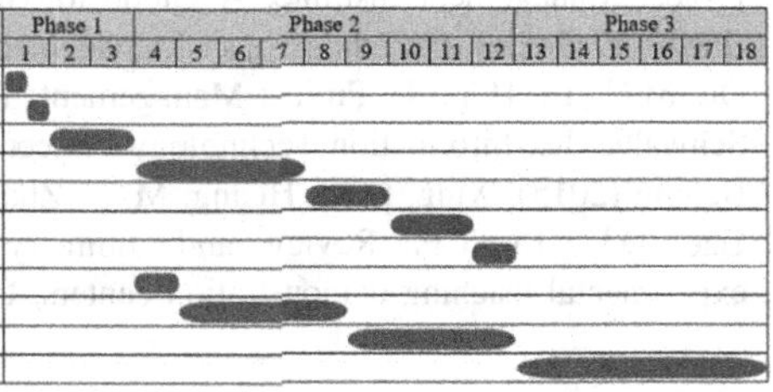

Fig. 7. Gantt chart of the course

3.4 Develop Schedules

The tools and techniques for schedule planning are mainly include Gantt chart, critical path method, critical chain scheduling, and program evaluation and review technique.

Figure 7 demonstrates detail schedules in the format of Gantt chart, from which we can figure out clearly durations of each activity and sequence between each other. As shown in Fig. 7, the hardware development task including D, E, F, G, and software development task including H, I, J are carried out at the same time, and this arrangement could reduce 9 weeks of develop time.

4 Conclusion

Experimental teaching plays the vital role in colleges and universities personnel training's system, it is an important means to cultivate students' original thoughts and creative ability [6]. Project management is a scientific management method which proved practical in actual project, and in the basic of the corresponding theoretical knowledge, teachers could arrange the content of class hours more reasonably, and students could complete the course content more efficiently. In addition, the introduction of project management could help students get familiar with project management knowledge and project management tools, to cultivate students' scientific spirit, to shorten the learning time at the beginning of careers. In summary, it is of great significance to the improvement of the experimental teaching mode and the improvement of the quality of the experimental teaching in the future.

Acknowledgements. This research is supported by 2016 key projects of Online Education Research Fund (General Education) of the online education research center of the Ministry of education.

References

1. Oak, V.J., Laghate, K.: Analysis of project management issues in information technology industry: an overview of literature. Int. J. Syst. Assur. Eng. Manag. 7(4), 418–426 (2016)
2. Phillips, J.: PMP Project Management Professional Study Guide. McGraw-Hill Professional, New York (2003)

3. Project Management Institute: A Guide to The Project Management Body of Knowledge (PMBOK Guide), 6th edn. Project Management Institute, Inc., Newtown Square (2017)
4. Qu, M.C., Li, H.Y.: IT Project Management. Tsinghua University Press, Beijing (2016)
5. Schwalbe, K.: Information Technology Project Management, 7th edn. China Machine Press, Beijing (2015). Xing, C.X., Huang, M.X., Zhang, Y. (translated)
6. Gao, D.F., Li, T.F.: Review and summary and prospect on construction of national experimental teaching demonstration centers. Exp. Technol. Manag. 34(12), 1–5 (2017)

Exploration and Research on the Training Mode of New Engineering Talents Under the Background of Big Data

Bing Zhao(✉), Jie Yang, Dongxiang Ma, and Jie Zhu

Department of lectronic Engineering, Heilongjiang University,
Harbin, Heilongjiang, China
zb0624@163.com

Abstract. The new economic and technological revolution drives the rapid development of the new industrial innovation model with interdisciplinary cross-integration characteristics, and shows great demand for a large number of new engineering talents with innovative entrepreneurial ability and multidisciplinary cross-border integration ability. Under the background of big data, colleges and universities at all levels need to combine big data technology with the training of new engineering talents and speed up the construction and reform of new engineering majors. This paper mainly discusses how to make use of big data technology to reform the training of talents. This paper puts forward some ideas and approaches for cultivating new engineering talents with high comprehensive quality who have the ability of innovation and pioneering and the ability of interdisciplinary cross-border integration.

Keywords: Big data · New engineering · Talents training

1 Introduction

In 2012, the United Nations issued a white paper big data for development, challenges and opportunities which pointed out that the age of big data had come. It would have a profound impact on all fields of society [1]. In August 2015, there was a circular- circular of the state council for the implementation of the platform for action for the development of big data which made a clear mark on thinking of the development of big data [2]. It indicated that big data had become an important strategic resource for the development of our country. Big data had brought new vitality to science and education and at the same time had put forward new challenges to traditional education [3–5].

At present, the world is experiencing a new round of technological and industrial revolutions that drive the development of a new economy and the formation of new industries. The new industry innovation model requires that engineers and technicians have a cross-sectoral and interdisciplinary knowledge and capability reserve. The new economic technology with the characteristics of "interdisciplinary integration" will spawn a number of emerging disciplines with cross-boundary features [6]. The rapid development of the new economy and new

© Springer Nature Singapore Pte Ltd. 2018
Q. Zhou et al. (Eds.): ICPCSEE 2018, CCIS 902, pp. 573–580, 2018.
https://doi.org/10.1007/978-981-13-2206-8_48

industries has posed new challenges to the personnel training model for higher education, the need for institutions of higher learning from a strategic height to re-layout of new engineering construction, adhering to the inclusive, open and inclusive concept, to explore diversified, personalized and innovative personnel training mode, cultivate innovative and entrepreneurial capacity and the interdisciplinary integration ability of high comprehensive quality personnel [7].

China has the largest engineering education in the world and the "Excellent Engineer Education and Training Program" was launched in 2010 [8]. In June 2016, we joined the International Engineering and Education Organization "Washington Protocol" and became an official member [9]. In November 2016, the State Council printed and distributed the "Plan for the Strategic Emerging Industries of the Thirteenth Five-year Plan" [10], proposing that emerging industries are the key development direction and main driving force for the future economic and social development and will drive the demand for new type of engineering talents. In February 2017, more than 100 representatives from many famous universities in China held a seminar about comprehensive university engineering education in Fudan university, discussed the new era of engineering personnel training, the new economy of engineering education needs and challenges, the comprehensive university of new engineering research and practice and other issues, and discussed the connotation characteristics of the new engineering, new engineering construction and development path selection, reached a Fudan consensus. The Ministry of Higher Education promulgated the "Notice of Higher Education Department of the Ministry of Education on Carrying out New Engineering Research and Practice" and launched the "New Engineering Research and Practice" project [11]. Colleges and universities throughout the country have launched discussions and studies on new engineering subjects, introducing big data technology into the training of engineering talents, giving full play to the role of big data in colleges and universities, and using it to guide educational and teaching reform. It is the need of current education reform and the direction of development and progress of education to promote colleges and universities to cultivate talents with high comprehensive quality who have the ability of innovation and pioneering and the ability of interdisciplinary integration.

2   Problems in the Training Mode of Engineering Talents

At present, in the big data technology background, there are some shortcomings in the traditional training mode of engineering talents. The aim of cultivating applied talents in colleges and universities is not clear enough, neglecting the cultivation of students' practical working ability. The curriculum system has not been optimized according to the needs of the society and the changes of the times. The poor teaching methods result in the low teaching quality and the low level of teaching, which leads to the inability to adapt to the new industrial needs of the new economic development [12]. The purpose of new engineering training mode reform is to cultivate the society oriented. Therefore, the evaluation mechanism of complex talents adapting to the rapid development of new economy and new industry should focus on process evaluation and social feedback [13].

3 Reform of the Training Mode of New Engineering Talents Under the Background of Big Data

3.1 Establishing the Goal of Cultivating New Engineering Talents

In the future, not only in the field of education, but also in all fields, a large amount of data should be analyzed and processed, and useful information should be extracted. Because of the large amount of data and the diversity of data structure, the engineering education must adjust the training program and direction in the new situation in order to adapt to the social needs.

The Goal of Training New Engineering Talents Should Be Adapted to the Needs of the Society. It is necessary for colleges and universities to train engineering talents that meeting the needs of the society, fully understanding the social economic and industrial needs, clarifying the training objectives of comprehensive engineering talents, and speeding up the cultivation of talents in the key areas of emerging industries. To ensure the usefulness and pertinence of talent training. It is necessary to analyze a large number of data among universities, students and society in order to cultivate innovative talents guided by social demand. Colleges and universities must take the initiative to go deep into the social reality to understand the current situation of industry development, and aim at the forward demand of industry. Enterprises also need to communicate with colleges and universities, timely feedback of talent demand information. At the same time, colleges and universities will transform the information obtained into teaching knowledge and impart it to students, so that students can master the knowledge, skills and methods, which are closely combined with social reality and industry demand. At the same time, they will learn to pay attention to the innovation spirit, industry attitude, social values. So, that will help students to adapt to the needs of the actual social job, and realize the innovation in the work post as soon as possible.

The Goal of Training New Engineering Talents Should Exceed the Needs of Society. With the development of new technology, the emerging industries emerge in endlessly. Therefore, the cultivation of new engineering talents in colleges and universities should open their horizons. The colleges and universities should transcend the existing technological level and the present situation of the industry, and pay a forward-looking attention to the development of new and innovative technologies and the latest developments of the industry. They need to gradually change from passive response to social needs to actively support and lead the development of innovation and technology, cultivating new technology talents facing the future new technology and new industry. In order to achieve the talent training model beyond the current social needs. It requires the integration of science and education in colleges and universities, and the combination of production, teaching research and using. In order for students to maintain high industry acumen, it is necessary to transmit the latest scientific research results and development trends to students in a timely manner. At the same time, they can cultivate students' ability to judge independently, deal with problems independently and update their knowledge constantly. So that students can better accept the new knowledge and field of the future society, and become a new type of engineering talents with global vision, practical ability and can lead the development of technology and industry.

3.2   Establishing a Diversified Teaching Content and a Three-Dimensional Curriculum System

With the emergence and rapid development of big data, our curriculum system should be changed accordingly. We can set up the mode according to the idea of "mainstream orientation remains unchanged, multiple branches assist". We can continue to optimize and adjust the curriculum system through big data analysis of social development needs and the needs of big data era. We should not only inherit the continuity of traditional professional courses, but also open up the students visual field through the setting up of general education. At the same time, we should also carry out cross - boundary integration and cross - boundary integration of disciplines, and cultivate students global integration and cross - boundary integration ability.

General Education and Professional Education are Complementary. General education is an education involving knowledge and abilities in the field of life. It is non-professional, non-professional, and covers a wide range of aspects. It is mainly aimed at developing students' sense of responsibility, establishing a correct attitude towards life and mastering basic knowledge and skills. Professional education mainly trains the students' academic ability and application ability in the professional direction. Therefore, teachers should constantly learn and master big data's knowledge and technology, strengthen their own professional level and teaching ability, and study in depth the influence of big data on society. At the same time, they also need to make use of the mutual assistance of general education and professional education to add ethics, morality, industry responsibility, social feedback and other issues to the learning process of professional knowledge. They need to develop creativity, communication, and leadership from a broader perspective.

Cross-Cutting and Interdisciplinary Integration of Engineering Disciplines. Cross-combination of engineering subjects and cross-border integration of disciplines is an important way to train new engineering talents. The subject between the modes of the knowledge combination is infinite, the data is huge. Therefore it provides unlimited possibilities for the development of innovative talents. The knowledge structure of different disciplines, collision thinking mode will help students to develop innovative thinking. The innovation itself is a process of differentiation-integration-differentiation and keeping breaking boundaries and stretching out. Therefore, the intersection and integration of subjects is an important way for students to acquire new knowledge, and also an effective way to train innovative new engineering talents.

3.3 Pay Attention to the Unity of Teaching and Learning

The time of big data make the educational and teaching methods need to adapt to the changes of the times. It is no longer a teacher-themed "teaching" and student-centered "learning", but requires the scientific use of big data technology to analyze the student learning situation. It has changed the traditional teaching idea and optimizes the resources of education and teaching. In the training mode of new engineering talents, the teaching methods pay attention to the interaction and penetration of the two. The teaching method leads the learning method according to the learning method, which has pertinence and feasibility, so as to achieve the expected goal of the unity of teaching and learning.

Multi-mode Teaching Methods. At present, teachers in higher education generally adopt the teaching method of lecturing, which can make it easier for teachers to control the teaching process and enable students to acquire a large amount of systematic scientific knowledge. However, it is not easy to arouse students' initiative in learning, and thus cause students to passively accept knowledge. Therefore, in the teaching process, teachers can analyze students' learning habits through learning behavior and results, such as big data. Learning aptitude, motivation, knowledge acquisition and problems, etc. It is necessary for teachers to consider all kinds of elements of teaching methods and to optimize the combination of teaching methods.

Autonomous Deep Learning. As the main body of teaching activities, students need good learning habits and autonomous learning ability. Students need to move from shallow to deep learning and from accepting learning to discovering learning. They need to actively engage in interactive communication under the guidance of teachers. With the extensive application of big data's technology, more and more teaching resources are converted into electronic data form. The teaching reform in colleges and universities should also adapt to the trend of the times. Encourage students to use internet resources and other ways to find answers. Through this kind of measures to solve the problem, the learning process is transformed from passive to active, and the comprehensive analytical ability and innovative consciousness are also cultivated.

3.4   Scientific Evaluation

The reform of teaching methods requires innovation and scientific evaluation. The cultivation of new engineering talents pays attention to the promotion of comprehensive ability and the development of creativity. The continuous development of big data technology is the evaluation of the quality of process teaching in colleges and universities. It has provided advanced scientific evaluation methods. For example, using big data technology to collect and analyze the data of students' daily learning, then to quantify the performance in the daily classroom, and finally to give a certain weight. This can more comprehensively reflect students' learning results and teachers' teaching quality. At the same time, big data technology has been widely used. It is also possible to construct a teaching quality evaluation system with multi-subject participation. In addition to the social evaluation feedback mechanism, colleges and universities can aim to cultivate talents according to the degree of adaptation of graduates to the needs of the society. The contents and methods of teaching should be adjusted dynamically.

4   The Concrete Implementation of Reform

The reform of personnel training mode must adhere to the goal of training high-quality composite new engineering talents. Based on the data mining and data clustering, carried out cross-disciplinary engineering and cross-disciplinary integration and token the three-stage teaching method and accorded to assessment and social feedback evaluation to form new personnel training model based on big data, and the overall implementation technology path of training mode reform is shown in Fig. 1.

Talent training requires the implementation of all-round, in-depth cooperation platform for interaction, establishes effective communication and data exchange. Using data mining technology to obtain the latest demand data from companies and research institutes, and integrating the information into all aspects of university teaching activities, achieves social needs to combine with talent development, targeted and dynamic adjustment of personnel training objectives, so that colleges and students in the early stages of theoretical learning to fully understand the social needs.

Colleges and universities should be based on personnel training objectives, adjust the curriculum, the establishment of a new teaching system, the basic course of a large class on a unified network platform for general education, the establishment of humanities, social science and the basic knowledge of natural science integration, theory and the combination of the application of the curriculum system to achieve mutual penetration, multiple coexistence purposes.

Teaching is based on different stages of learning, using different training methods, mainly divided into three training sections, including "guide learning by need", "promote learning by practicing", "learn by learning", so that achieving the purpose of teaching to meet the needs of society, practical assistance learning, internship guiding target.

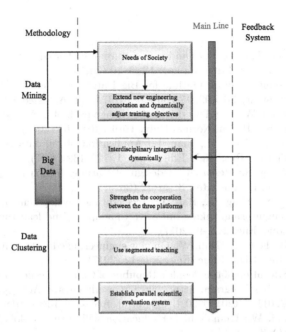

Fig. 1. The implementation technology path of training mode reform.

Universities should conduct follow-up surveys of graduates and communicate with enterprises and scientific research institutes regularly, using data clustering technology to analyze feedback information of employers, execute statistics and analysis for the social adaptability of output talents, this information as a basis for talents training objectives and teaching models adjusting dynamically.

5   Summary

The arrival of big data's time has not only changed our lives, It also brings the opportunity of development and the challenge of innovation for the cultivation of talents in colleges and universities. It is an important subject in current university education research that how to use big data technology to improve the teaching quality. Colleges and universities should make full use of big data related technology to improve the teaching quality of colleges and universities. The research on the development trend of new engineering talents can provide a constructive reference for further strengthening the effect of teaching quality and promoting the improvement of teaching level in colleges and universities under the new situation.

Acknowledgements. This work is supported by Heilongjiang University New Century Education and Teaching Reform Project (NO. 2017B15). Many thanks to the anonymous reviewers, whose insightful comments made this a better paper.

References

1. Zheng, H.: The characteristics and new culture of higher education research in the era of big data. High. Educ. Explor. **12**, 5–10 (2017)
2. The State Council: Action Outline for Big Data Development. http://www.gov.cn/zhengce/content/2015-09/05/content_10137.html. Accessed 1 Mar 2017
3. Cheng, Y., Zhao, W.: Big data drives the change path of higher education innovation. J. Educ. Sci. Hunan Normal Univ. **16**(6), 101–105 (2017)
4. Xu, X., Zhao, X., Xiao, H., Bian, L.: Application of big data in undergraduate teaching evaliation. Res. High. Educ. Eng. **1**, 127–132 (2017)
5. Chen, M., Wang, S.: Research on design of learning feedback supported by big data on exercise. E-Educ. Res. **3**, 35–42 (2018)
6. Wu, A., Hou, Y., Yang, Q., Hao, J.: Accelerating development and construction of emerging engineering, taking initiative to adapt to and lead the new economy. Res. High. Educ. Eng. **1**, 1–9 (2017)
7. Xu, L., Hu, B., Feng, H., Han, W.: On new engineering education in comprehensive universities. Res. High. Educ. Eng. **2**, 6–12 (2017)
8. Ministry of Education of the People's Republic of China: Excellent Engineer Education and Training Program. http://www.moe.edu.cn/s78/A08/gjs_left/moe_742/s5632/s3860/201109/t20110920_124884.html. Accessed 3 Feb 2018
9. Wan, Y., Chai, W.: China's higher education will truly go global.ZHONGGUO JIAOYU BAO, 3 June 2016
10. The State Council: "13th Five-Year" national strategic emerging industries development plan. http://www.gov.cn/zhengce/content/2016-12/19/content_5150090.html. Accessed 22 Dec 2017
11. Ministry of Education of the People's Republic of China: Ministry of Education, Ministry of Higher Education notice on the implementation of new engineering research and practice. http://www.moe.gov.cn/s78/A08/A08_gggs/A08_sjhj/201702/t20170223_297158.html. Accessed 22 Dec 2017
12. Cheng, X., Jiang, X.: A probe into the new strategy of cultivating applied talents in colleges and universities under big data. Inf. Syst. Eng. **8**, 174 (2017)
13. Zhang, Q., Zhu, H., Li, Z.: On the background, motivation and safeguard mechanism of innovation and entrepreneurship education. Res. High. Educ. Eng. **3**, 162–165 (2017)

An Empirical Study on the Influence Factors of Mobile Phone Dependence of College Students Based on SPSS and AMOS

Zhi-peng Ou and Xia Liu[(⊠)]

Sanya Aviation and Tourism College, Sanya 572000, Hainan, China
hainanozp@126.com, paolo_lx@qq.com

Abstract. Technology has an influence on people's lives, as the development and application of technology, people's lives, recreation, and the way of learned have a great change in recent decades. At present, as a mean of communication, mobile phones have entered many aspects of people's live, such as communication, entertainment, internet business, and learning. However people reverse the positions of the host and the guest, and depend on it. This thesis expounds the influence of mobile phones on college students' learning and life through investigating the students' dependence of mobile phones in Sanya Aviation & Tourism College, and drew a conclusion that college students have a depth of dependency on mobile phone. The results showed that some college students have a depth of de-pendency on mobile phones, they always use mobile phones as the main tools for internet and entertainment, at the same time, more and more college students are willing to communicate with friends on the Internet instead of communicating face to face; and we also found that female college students have a higher de-pendency than that of male college students; and the price of mobile phone, online time and spending are positively correlated with mobile phones dependency, which also can directly reflected the dependency on mobile phones; Finally, we found that the higher dependency on mobile phone, the greater influence of study and life. This shows that mobile phones have a depth of dependency on college students, and far from its basic function. And mobile phone has a main affect up-on college students' learning and life. Using mobile phone reasonable can be beneficial to the college students' life and study.

Keywords: SPSS · Amos · Mobile phone · College student · Empirical study

1 Introduction

Every information technology revolution has brought great changes of information dissemination, which has greatly affected upon people's lives, learning and work. Mobile phone was invented in the early 19th century, and as time passed by, has gradually been used into every aspect of people's lives. And it has become the most widely used media tool of all time. Information transmission has developed rapidly in the mobile phone era, and changed the way of young people's learning and life. Those changes have been reflected in the most advanced aspects especially higher education. The mobile phone is no longer a simple communication tool, but also plays many

© Springer Nature Singapore Pte Ltd. 2018
Q. Zhou et al. (Eds.): ICPCSEE 2018, CCIS 902, pp. 581–593, 2018.
https://doi.org/10.1007/978-981-13-2206-8_49

important roles, such as entertainment, information query and so on. At the same time, with the popularization and frequency of mobile phone, college students' lives have already changed, however, colleges have not found suitable countermeasures and effective adjustments. Many scholars have also raised some insights and ideas, Sun [1] proposed to make the best use of the circumstances and explore new ways of teaching, while Jia [2] and others believed that mobile learning will become an important part for college students learning; at the same time, Wang [3] provided that should guide college students used mobile phones correctly, and use its rational characteristics to guide students; Li [4] and Xia [5] proposed to change the shortcomings of mobile phones, and should implement effective intervention on the dependency of mobile phones. All of the above is the influencing factors, but no more in-depth analysis of the influencing factors. We collected and analyzed 1811 samples of college students' mobile phone usage in this survey at the Sanya Aviation & Tourism College, and made a statistical analysis. This thesis using various methods to simulate the degree and relevance of the influencing factors, such as difference analysis, testing of validity and reliability, correlations, regression equation analysis, and structural equation, which provided a reference for formulating corresponding measures and countermeasures. The data was been processed by SPSS 18.0 and Amos 21.0 [6–8].

2   The Results of Survey

Frequency statistics of classification of the variables in this study were conducted and the result were shown in Table 1. There were 1811 students engaged as the research samples, including 946 males and 865 females, basically equivalent in number. Among them, there were 1204 fresh students, 527 sophomores and 80 junior students, accounting for 66.5%, 29.1% and 4.4% respectively.

Among the survey samples, the majority students use iPhone and Huawei mobile phones, accounting for 28.9% and 19.2% respectively. Of course, there are many other brands, accounting for 37.8%. In the evening, most students play mobile phones or have a rest, accounting for 33.1% and 23.8% respectively, but the majority would not play their mobile phones until early in the morning without rest, accounting about 70.8%. Their main expenses of mobile phone resulted from surfing the net, accounting for 74.9% and their majority net expenses per month were used to pay regular net charges, accounting for 86.5%. Most students studied online via APP, accounting for 59%. In case of adopting the mobile phone management policy, most students would prefer to put their phone into the bag seldom or for each time, accounting for 31.6 & and 28.9% respectively while fairly a large number of students would put their phone into the bag regularly or absolutely not, accounting for 20.2% and 19.2 respectively. Considering the students' attitude towards the online teaching on Moji Cloud, the majority students indicated that they would actively cooperate, accounting for 73.3%. Nevertheless, the majority students believed they may persist in not playing mobile phone for half a day, accounting for 71.1%. Most students used learning APP regularly, accounting for 55.2%. In respect of the monthly expenses in mobile phone, the majority students spent RMB30-50 and RMB51-100, accounting for 45.2% and 32.2% respectively. Meanwhile, almost equivalent number of students would concern or not

Table 1. Table of frequency statistics (N = 1811)

		Frequency	Percentage (%)
Gender	Male	946	52.2
	Female	865	47.8
Grade	Freshmen	1204	66.5
	Sophomore	527	29.1
	Junior	80	4.4
Brand of mobile phone	Huawei	348	19.2
	Samsung	64	3.5
	iPhone	524	28.9
	Lenovo	14	0.8
	MI	177	9.8
	Others	684	37.8
Primary activities in the evening	Play mobile phone	600	33.1
	Rest	431	23.8
	Study	208	11.5
	Play outside	156	8.6
	Play games	162	8.9
	Others	254	14
Whether will you play the mobile phone until early in the morning without rest?	Yes	528	29.2
	No	1283	70.8
Main consumption of mobile phone per month	Call	414	22.9
	Send short message	40	2.2
	Surf the internet	1357	74.9
Main network consumption per month	Regular net charges (such as flow, wireless, etc.)	1567	86.5
	Open various memberships	58	3.2
	Charge the games	80	4.4
	Others	106	5.9
Main online learning approaches	Learning APP	1069	59
	Webpage	323	17.8
	Video	325	17.9
	Other	94	5.2
In case of mobile phone bag management policy, will you put your phone in before class?	Absolutely not	348	19.2
	Seldom	573	31.6
	Often	366	20.2
	Each time	524	28.9

(continued)

Table 1. (*continued*)

		Frequency	Percentage (%)
Attitude towards the network teaching by Moji Cloud	Very good and cooperate actively	1328	73.3
	Pretend to cooperate but actually look at other others	170	9.4
	Not to matter	313	17.3
How long will you persist in not playing mobile phone?	Half a day	1288	71.1
	2 h	296	16.3
	30 min	160	8.8
	Cannot leave for a moment	67	3.7
Frequency of using APP for learning	Very high	230	12.7
	High	335	18.5
	General	1000	55.2
	Low	190	10.5
	Not yet use	56	3.1
Mobile phone expenses per month	Less than RMB30	210	11.6
	RMB30-50	819	45.2
	RMB51-100	583	32.2
	More than RMB100	199	11
Whether you will concern your phone brand?	Yes	731	40.4
	No	877	48.4
	Unclear	203	11.2
Main approach of interaction with your friends	Communicate face to face at ordinary times	354	19.5
	Chat online	527	29.1
	Interact in the circle of friends	84	4.6
	All approaches	846	46.7

concern the phone brand, accounting for 40.4% and 48.4% respectively with the remaining 11.2% unclear. Besides, the majority students would prefer to interact with their friends by all approaches including face-to-face communication, online chatting and the circle of friends, etc., accounting for 46.7% (Table 2).

Among the survey samples for this time, they would mostly play mobile phone, have a rest or chat, accounting for 57.8%, 51.5% and 49.1% respectively during the break time. When they were playing their mobile phones, they would mostly watch videos, news, communicate with others, study and play games, accounting for 64.5%, 53.9%, 47.6%, 43% and 40% respectively. Whey they were browsing the websites, they would most search information, watch video or news, accounting for 65%, 56.9% and 54.4% respectively.

Table 2. Frequency statistics table of multiple choice questions

		Frequency	Percentage(%)
Main activities during break	Play mobile phone	1047	57.80%
	Rest	932	51.50%
	Chat	890	49.10%
	Other	264	14.60%
Main concern while playing mobile phone	Stars' gossip	398	22.00%
	Watch news	977	53.90%
	Study	779	43.00%
	Watch video	1168	64.50%
	As the communication tool	862	47.60%
	Play games	725	40.00%
	Other	132	7.30%
Main concern while browsing the website	Search information	1177	65.00%
	Watch news	986	54.40%
	Watch video	1031	56.90%
	Other	153	8.40%

3   Significance Test of Difference

The independent sample t was adopted to test the gender difference among main research variables, as shown in Table 3. The result indicates that the males' (M = 2.180, SD = 1.013; M = 2.590, SD = 1.200; M = 2.790, SD = 1.019; M = 2.821, SD = 0.913) expenses in surfing the internet, time spent in surfing the internet, phone price and dependency on mobile phone are significantly lower (t (1809) = −2.022, p < 0.05; t (1809) = −10.392, p < 0.001; t (1809) = −2.630, p < 0.01; t (1809) = −2.560, p < 0.05) than those of the females (M = 2.270, SD = 0.991; M = 3.160, SD = 1.158; M = 2.910, SD = 0.941; M = 2.921, SD = 0.762) [9]. Both the living expenses per month and the influence of mobile phone on the students' study and living are insignificant in terms of gender (t (1809) = −1.544. p > 0.05; t (1809) = 0.627. p > 0.05).

The one-way variance was adopted to analyze and test the grade difference of main research variables [10], as shown in Table 4, which indicates no significant grade difference among main research variables (F (2, 1808) = 0.809, p > 0.05; F (2, 1808) = 0.428, p > 0.05; F (2, 1808) = 1.536, p > 0.05; F (2,1808) = 0.585, p > 0.05; F (2,1808) = 2.089, p > 0.05; F (2,1808) = 0.425, p > 0.05).

Table 3. Analysis of gender difference of various research variables

	Male (N = 946)		Female (N = 865)		t	p
	M	SD	M	SD		
Living expenses per month	2.320	0.921	2.390	0.926	−1.544	0.123
Expenses in surfing the internet per month	2.180	1.013	2.270	0.991	−2.022	0.043
Time spent in surfing the internet	2.590	1.200	3.160	1.158	−10.392	0.000
Phone price	2.790	1.019	2.910	0.941	−2.630	0.009
Dependency on mobile phone	2.821	0.913	2.921	0.762	−2.560	0.011
Influence of mobile phone on study and living	4.228	0.736	4.207	0.685	0.627	0.530

Table 4. Analysis of grade difference of main research variables

	Freshmen (N = 1204)		Sophomore (N = 527)		Junior (N = 80)		F	p
	M	SD	M	SD	M	SD		
Living expenses per month	2.350	0.912	2.330	0.926	2.480	1.079	0.809	0.445
Expenses in surfing the internet per month	2.210	1.001	2.260	0.993	2.200	1.107	0.428	0.652
Time spent in surfing the internet	2.830	1.205	2.940	1.221	2.760	1.305	1.536	0.215
Phone price	2.860	0.976	2.820	0.987	2.780	1.079	0.585	0.557
Dependency on mobile phone	2.849	0.815	2.889	0.862	3.038	1.130	2.089	0.124
Influence of mobile phone on study and living	4.207	0.711	4.237	0.698	4.250	0.812	0.425	0.654

4   Reliability and Validity Test

KMO value is 0.902, higher than 0.8. The Bartlett degree of sphericity reaches the significance level of 0.001, indicating the factor analysis is applicable to these samples [11]. When the factors with their two characteristic roots higher than 1 were extracted, as the factor loading value of A15 and A16 items were too low, which greatly differed from the factor loading value of other items, these two items were then deleted.

Table 5. Exploratory factor analysis

Construct	Index	Factor loading	Characteristic root	Variance as explained
Dependency on mobile phone	A4	0.573	5.997	39.983%
	A5	0.733		
	A6	0.844		
	A7	0.812		
	A8	0.658		
	A10	0.795		
	A11	0.832		
	A12	0.832		
	A13	0.784		
	A16	0.634		
Influence of mobile phone on study and living	A1	0.856	3.111	20.737%
	A2	0.842		
	A3	0.827		
	A9	0.764		
	A14	0.629		

After deletion, when the factors with two characteristic roots higher than 1 were extracted, it explained the total square deviation of 60.72%; besides, the factor loading value of each item was higher than 0.55, indicating sound construct validity of the double factor structure.

On the basis of reliability test and confirmatory factor analysis of constructive study, destructive study and study behaviours, the result is shown in Table 4. The α coefficient of reliability is all above 0.8, indicating sound reliability of each sub-questionnaire. The result of confirmatory factor analysis indicates the fit index of each model: $\chi 2 = 767.774$, $df = 78$, $\chi 2/df = 9.843$, RMSEA $= 0.07 < 0.08$, GFI $= 0.95 > 0.9$, AGFI $= 0.923 > 0.9$, NFI $= 0.952 > 0.9$, RFI $= 0.936 > 0.9$, IFI $= 0.957 > 0.9$, TLI $= 0.942 > 0.9$, CFI $= 0.957 > 0.9$. All the fit indexes have fulfilled the fitting requirement except $\chi 2/df > 5$. Considering $\chi 2$ value may be greatly affected by the sample size, which is large in this research, it will be inappropriate to consider this index as the single judgment standard; therefore, this model is believed of fitting. Besides, the factor loading value of all items in respect of corresponding variable is all above 0.4, which is significant, indicating the convergent validity of each sub-questionnaire is fairly acceptable. The correlation coefficient of the dependency on mobile phone and the influence of mobile phone on study and living is only 0.18, which has reached the level of significance, indicating sound discrimination validity of the samples (Table 6).

Table 6. Reliability and validity test

Construct	Index	Factor loading	Significance	α
Dependency on mobile phone	A4	0.572	***	0.916
	A5	0.676	***	
	A6	0.742	***	
	A7	0.708	***	
	A8	0.558	***	
	A10	0.773	***	
	A11	0.811	***	
	A12	0.837	***	
	A13	0.791	***	
	A16	0.606	***	
Influence of mobile phone on study and living	A1	0.864	***	0.849
	A2	0.857	***	
	A3	0.73	***	
	A9	0.667	***	
	A14	0.42	***	

5   Correlation Analysis

The correlation analysis was adopted to analyse and study the relationship among all research variables [12], as indicated in Table 5. The result thereof indicates that the pairwise correlation between the main research variables is all positively significant ($p < 0.01$) with the correlation coefficient between $0.09 \sim 0.461$ except the pairwise correlation between the expenses in surfing the internet per month and the time spent in surfing the internet with the influence of mobile phone on study and living respectively, which is insignificant ($p > 0.05$) (Table 7).

Table 7. Correlation analysis of all main research variables

	Living expenses per month	Expenses in surfing the internet per month	Time spent in surfing the internet	Phone price	Dependency on mobile phone	Influence of mobile phone on study and living
Living expenses per month	1					
Expense in surfing the internet per month	0.257***	1				

(continued)

Table 7. (*continued*)

	Living expenses per month	Expenses in surfing the internet per month	Time spent in surfing the internet	Phone price	Dependency on mobile phone	Influence of mobile phone on study and living
Time spent in surfing the internet	0.198***	0.264***	1			
Phone price	0.461***	0.228***	0.169***	1		
Dependency on mobile phone	0.132***	0.215***	0.292***	0.097***	1	
Influence of mobile phone on study and living	0.09***	0.004	0.014	0.076**	0.177***	1

Note: * represents $p < 0.05$; ** represents $p < 0.01$; *** represents $p < 0.001$.

6 Regression Analysis

The regression analysis was adopted to further study the influence of living expenses per month [13], expenses in surfing the internet per month, time spent in surfing the internet and phone price on the dependency on mobile phone. The result of regression analysis is shown in Table 8. The result thereof indicates significant regression equation ($F = 54.637$, $R2 = 0.108$, $p < 0.001$); besides, VIF value is between $1.101 \sim 1.324$, smaller than 10, indicating no serious collinearity to the model. Moreover, both the expenses and time spent in surfing the internet are positively significant to predict the dependency on mobile phone ($\beta = 0.138$, $t = 5.797$, $p < 0.001$; $\beta = 0.246$, $t = 10.561$, $p < 0.001$) while the living expenses per month and the phone price cannot significantly predict the dependency on mobile phone ($\beta = 0.046$, $t = 1.803$, $p > 0.05$; $\beta = 0.003$, $t = 0.115$, $p > 0.05$).

Table 8. Regression of living expenses per month, expenses and time spent in surfing the internet and phone price to the dependency on mobile phone

	β	t	p	VIF	F	R2
Living expenses per month	0.046	1.803	0.072	1.324	F = 54.637	R2 = 0.108,
Expenses in surfing the internet	0.138	5.797	0.000	1.142	P < 0.001	P < 0.001
Time spent in surfing the internet	0.246	10.561	0.000	1.101		
Phone price	0.003	0.115	0.908	1.296		

Table 9. Regression of living expenses per month, expenses and time in surfing the internet, and phone price to the influence of mobile phone on study and living

	β	t	p	VIF	F	R2
Living expenses per month	0.076	2.818	0.005	1.324	F = 4.887, P < 0.01	R2 = 0.011, P < 0.01
Expenses in surfing the internet per month	−0.035	−1.4	0.162	1.142		
Time spent in surfing the internet	0.000	−0.014	0.989	1.101		
Phone price	0.049	1.851	0.064	1.296		

The regression analysis was adopted to further study the influence of the living expenses per month, the expenses in surfing the internet per month, the time spent in surfing the internet and the phone price on the influence of mobile phone on study and living. The result of such regression analysis is shown in Table 9. As indicated by the result thereof, the regression equation is significant (F = 4.887, R2 = 0.011, p < 0.01), VIF value is between $1.101 \sim 1.324$, less than 10, indicating no serious collinearity to the model. In addition, the living expenses per month is positively significant of predicting the influence of mobile phone on study and living ($\beta = 0.076$, t = 2.818, p < 0.01) while the expenses and time spent in surfing the internet and the phone price cannot significantly predict the influence of mobile phone on study and living ($\beta = -0.035$, t = −1.4, p > 0.05; $\beta = 0.000$, t = −0.014, p > 0.05; $\beta = 0.049$, t = 1.851, p > 0.05).

7 Structural Equation Model

The structural equation model was adopted to further study the relationship among main research variables [14]. As the variable of dependency on mobile phone has more items, this variable is categorized into five item parcellings to simplify the model. The final model as achieved is shown as in this Fig. 1 and its main fitting indexes include: $\chi2 = 705.636$, df = 67, RMSEA = 0.073 < 0.08, GFI = 0.949 > 0.9, AGFI = 0.920 > 0.9, NFI = 0.940 > 0.9, RFI = 0.919 > 0.9, IFI = 0.946 > 0.9, TLI = 0.926 > 0.9 and CFI = 0.946 > 0.9 respectively. All fitting indexes have fulfilled the fitting requirements except $\chi2$/df > 5. Considering $\chi2$ value may be greatly affected by the sample size, which is large in this research, it would be inappropriate to consider this index as the single judgment standard; therefore, this model shall be believed of fitting.

The route coefficient among main variables in the model is shown in Table 10. The expenses in surfing the internet per month, the time spent in surfing the internet and the living expenses per month can be positively significant of predicting the dependency on mobile phone to different extents ($\beta = 0.145$, t = 5.9, p < 0.001; $\beta = 0.282$, t = 11.41, p < 0.001; $\beta = 0.054$, t = 2.224, p < 0.05); the living expenses per month and the phone price can be positively significant of predicting the influence of mobile phone on study and living ($\beta = 0.076$, t = 2.691, p < 0.01; $\beta = 0.068$, t = 2.416, p < 0.05).

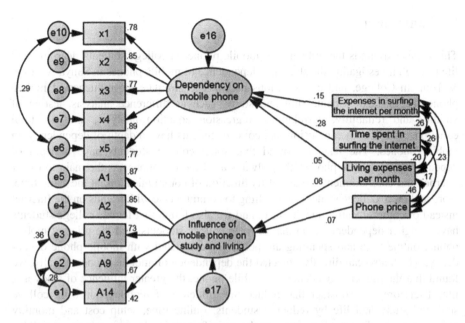

Fig. 1. The structural equation model of mobile phone dependency

Table 10. Route coefficient in structural equation model

Dependent Variable		Independent variable	Estimate	S.E.	C.R.	P
Dependency on mobile phone	←	Expense in surfing the internet per month	0.145	0.02	5.9	***
Dependency on mobile phone	←	Time spent in surfing the internet	0.282	0.017	11.41	***
Influence of mobile phone on study and living	←	Living expenses per month	0.076	0.012	2.691	0.007
Influence of mobile phone on study and living	←	Phone price	0.068	0.012	2.416	0.016
Dependency on mobile phone	←	Living expenses per month	0.054	0.021	2.224	0.026

Note: ***represents p < 0.001.

8   Conclusion

This thesis expounds the influence of mobile phone on college students' learning and life through investigating the students' dependence of mobile phones in Sanya Aviation & Tourism College, and using various methods to simulate the influence of mobile phones on college students' life and study, such as difference analysis, testing of validity and reliability, correlations, regression equation analysis, and structural equation. And draw a conclusion that college students have a depth of dependency on mobile phones. The results showed that some college students have a depth of dependency on mobile phones, they always used mobile phones as the main tools for internet and entertainment instead of its function of communication, at the same time, more and more college students are willing to communicate with friends on the Internet instead of communicating face to face; and we also found that female college students have a higher dependency than that of male college students; and the price of mobile phone, online time and spending are positively correlated with mobile phone dependency, which also can directly reflected the dependency on mobile phones; Finally, we found that the higher dependency on mobile phone, the greater influence of study and life. Therefore, we consider that reduce the influence of mobile phones on college students' study and life by reducing students' online time, setup cost and monthly living expenses. The results sorted out the main factors for the management of college students, and providing a reference for proposing new management and control model.

Acknowledgement. Thanks to associate professor Xia Liu, correspondent of this paper.

References

1. Sun, Y.: Thinking about influence of using mobile phone in classroom on undergraduate students and mobile phone aided teaching. J. High. Educ. **9**, 41–44 (2017)
2. Jia, M., Ye, J., Bian, W.: Probe into the Influence of smartphone on undergraduate students. Heilongjiang Sci. Technol. Inf. **13**, 140–141 (2017)
3. Wang, G.: Influence of application of mobile phone medium on psychological health education of undergraduate students. Educ. Teach. Forum **3**, 43–44 (2018)
4. Li, X., Tu, X., Han, X.: Investigation of influence of mobile phone on undergraduate students and the countermeasure analysis. Education **9**, 104–105 (2017)
5. Xia, Yu., Guo, T.: Influence of dependency on mobile phone on the study of undergraduate students and the intervening measures. Mod. Vocat. Educ. **4**, 62–63 (2017)
6. George, D., Mallery, P.: SPSS for windows step by step: a simple guide and reference. Comput. Soft. **100**, 357 (2003)
7. Preacher, K.J., Hayes, A.F.: SPSS and SAS procedures for estimating indirect effects in simple mediation models. Behav. Res. Methods Instrum. Comput. **36**(4), 717–731 (2004)
8. Igbaria, M., Zinatelli, N., Cragg, P., et al.: Personal computing acceptance factors in small firms: a structural equation model. MIS Q. **21**(3), 279–305 (1997)
9. Ying, H.G., Min, W.H., Bo, H.Y.: Significance test of difference for genetic absolute distance. Acta Ecol. Sin. **25**(10), 2534–2539 (2005)
10. Yang, Y., Zhang, H.: Significance test of difference between coefficients of variation. J. Northeast Agric. Univ. (1994)

11. Grigg, C., Wong, P., Albrecht, P., et al.: The IEEE reliability test system-1996. A report prepared by the reliability test system task force of the application of probability methods subcommittee. IEEE Trans. Power Syst. **14**(3), 1010–1020 (2002)
12. Andrew, G., Arora, R., Bilmes, J., et al.: Deep canonical correlation analysis. In: International Conference on Machine Learning, JMLR.org, p. III-1247 (2013)
13. Moore, T.: Regression analysis by example. Technometrics **43**(2), 236 (2013)
14. Wen, Z., Kit-Tai, H., Herbert, W.: Structural equation model testing: cutoff criteria for goodness of fit indices and chi-square test. Acta Psychologica Sinica **36**(2), 186–194 (2004)

The Reliability and Validity Analysis of Questionnaire Survey on the Mathematics Teaching Quality in Higher Vocational Colleges

Yuan-hui Li, Xia Liu[✉], and Hai-yun Han

Sanya Aviation and Tourism College, Sanya 572000, Hainan, China
576735855@qq.com, paolo_lx@qq.com

Abstract. Using the SPSS 19 software as the statistics analysis tool and taking the questionnaire survey data as the observation sample, this paper analyzes the reliability and validity of the questionnaire survey results of the mathematics course teaching situation in a higher vocational college in Sanya, and finds 3 potential factors affecting the quality of the classroom teaching: the teachers' professional competence, the students' interest in learning and their performance after class. This helps to better understand the connotation of classroom teaching quality assessment and establish a proper and effective system and measures correspondingly.

Keywords: Teaching quality · Reliability · Validity · Factor analysis

1 Introduction

Classroom teaching, as the most basic and important form of teaching organizations in higher vocational colleges, is an important link to achieve the goal of talent cultivation and guarantee the teaching quality. The teachers' teaching survey with students as respondents is an important part of teaching quality assessment. Perfecting and improving the student-centered classroom teaching quality evaluation system and establishing a set of scientific and all-around classroom teaching quality evaluation index system have great theoretical values and practical significance for diagnosing and improving the classroom teaching quality, promoting the development of mathematics faculty and strengthening the central position of teaching.

In this process, researchers focus on the scientificity and feasibility of the teaching quality evaluation, trying to find whether students' evaluation of the teaching quality is reliable and valid.

In this paper, SPSS was used to test the reliability and validity of survey results of the questionnaire on the teaching of mathematics in a vocational college in Sanya. This questionnaire was designed to test mathematics teachers' working attitude, classroom interaction, teacher-student relationships, students' interest in learning mathematics and other indexes. A total of 920 valid questionnaires were recovered. The reliability and validity of the questionnaire were tested.

Q. Zhou et al. (Eds.): ICPCSEE 2018, CCIS 902, pp. 594–604, 2018.
https://doi.org/10.1007/978-981-13-2206-8_50

2 Research Object and Survey Methods

2.1 Research Object

At present, various teaching quality assessment institutions and systems established by higher vocational colleges in China, generally fall into a two-level assessment model: college and department. The assessment system for teaching performance consists of colleges, departments, teaching and research teams (teachers) and students of all classes, which forms a four-in-one teaching quality assessment system to ensure the organized teaching [1].

In the section of department assessment, it is necessary to get the student's evaluation of the teachers' professional competence for a particular course and the students' interest in this course. This paper selects the mathematics, a public basic course, as the research object, and chooses the students who take courses of "Advanced Mathematics" and "Economic Applied Mathematics" as the target group for this survey.

2.2 Questionnaire Structure

In the design of a questionnaire on the teaching of mathematics in higher vocational colleges, the first thing is to construct an index system.

The quality of mathematics teaching includes two basic aspects: the quality of students' learning and the quality of teachers' teaching. From the perspective of student learning, the good mathematics learning quality includes good mathematics learning interest, excellent mathematics learning capacity, and the comprehensive ability to use mathematical methods to solve problems. From the teacher's teaching level analysis, the quality of good mathematics teaching requires teachers to have rich accumulation of mathematics, good classroom teaching organization skills, and skilled and flexible teaching skills [2].

The mathematics course evaluation system can be built from 2 aspects: students' evaluation of teachers' professional qualification and students' interest in learning mathematics, so that indexes in the system can achieve mutual demonstration, testing and explanation. The correlation between teachers' professional qualification and students' interest in learning mathematics can be discussed and analyzed by investigating the relationship between them, to provide reference for working out or revising the course syllabus and teaching plan in the future.

The questions in the questionnaire falls into 3 parts: identity description, teacher evaluation, and interest in mathematics.

In the part of identity description, qualified respondents are screened out by investigating students' class number. In order to evaluate teachers from various aspects of teachers and overall level, 11 question items are designed and for each question, the questionnaire used the Likert scale and adopted a positive scoring method, where 1 means "strongly dissatisfied", 2 means "not satisfied", and 3 means "neutral", 4 means "satisfactory" and 5 means "very satisfied" [3]. The part of interest in mathematics is designed to test students' attitudes towards mathematics-related contests, elective courses, extracurricular knowledge and after-school exercises.

On the basis of previous surveys, necessary adjustments were made to the questionnaire, and the questionnaire was used for this survey. The contents of the questionnaire are shown in Table 1:

Table 1. Questionnaire

Question NO.	Questions	Option menu
Q1	The teacher has enthusiasm for teaching and shows an earnest and devoted attitude towards teaching	strongly dissatisfied; not satisfied; neutral; satisfactory; very satisfied.
Q2	The teacher can combine theory with practice and highlight content upgrading	
Q3	The teacher has a clear mind and imparts knowledge precisely	
Q4	The teacher can stress the key point and clearly explain difficult points, in teaching	
Q5	In class, the teacher well communicates and interacts with students, encourages students to ask doubtful questions and gives guidance to find answers	
Q6	The teacher declares a reform on the examination and evaluation mode, at the very beginning of the course	
Q7	The teacher builds harmonious relations with students and provides guidance to students and well communicates with students after class	
Q8	The teacher can serve as a model of virtue for students, and show rigorous scholarship, noble morality, excellent teaching spirit and professionalism	
Q9	The teacher provides a lot of reference materials and related information to students	
Q10	The teacher checks students' homework, provides guidance and answers students' questions timely and conscientiously, and gives valuable comment on homework	
Q11	Please give your judgement on the overall teaching of your teacher	

(continued)

Table 1. (*continued*)

Question NO.	Questions	Option menu
Q12	If there is a mathematics contest, do you want to participate in or follow it with interest?	Yes, I do; No, I don't; I'm not sure
Q13	If there is a mathematics-related elective course, do you want to attend it?	Yes; No
Q14	What is your attitude towards mathematical knowledge (after-class)?	It's useful, so I'd like to know about it; The mathematics is not very useful, learning or not makes no difference; The mathematics is too difficult to learn and I am so weak at it
Q15	Whether the QR code in the exercise of "Applied Economic Mathematics" is helpful to your study?	very helpful; helpful; neutral; a little helpful; useless

2.3 Data Collection

Students were asked to scan the QR code and answer questions online, and data was automatically collected by the backstage. To ensure the representativeness and coverage of the survey, the number of respondents from each administrative class should not be less than 2/3 of the total number of each class. The survey targeted 1208 students and 920 questionnaires were obtained. The response rate was approximately 76.2%, exceeding expectations. Questionnaires with missing question items answered would fail to be submitted, so invalid questionnaires would be removed from the backstage. Therefore, the valid sample size was 920.

3 Statistical Processing Methods

3.1 Reliability Evaluation

The reliability was first introduced into psychological measurement by Pearman in 1904. The concept of reliability first appeared in the True Test Theory (CTT) developed in the first half of the 20th century. Its theoretical framework revolves around Reliability, Validity, and the Item Difficulty and Discrimination Index of Item Analysis [4].

Reliability refers to the degree of consistency between results of repeated questionnaire surveys. The consistency of survey results refers to the consistency between answers to questions in the same questionnaire [5]. If the consistency of questionnaire survey result is relatively high, it indicates that the positive correlation between one respondent's different survey results on the same questionnaire item in different questionnaire surveys is quite strong.

The reliability coefficient is an index for evaluating reliability. The mathematical formula (true variance/measured variance) is used to calculate the reliability coefficient.

If we use T to represent the true value, X to represent the measured value and E to represent the random error, their relationship equations are expressed as:

$$X = T + E, \; \sigma_X^2 = \sigma_T^2 + \sigma_E^2.$$

Therefore, the reliability coefficient can be expressed as:

$$R_X = \sigma_T^2 / \sigma_X^2 = 1 - \left(\sigma_E^2 / \sigma_X^2 \right).$$

According to different measurement methods, reliability can be divided into test-retest reliability, alternative-forms reliability and internal consistency reliability.

Test-retest reliability is the correlation coefficient between questionnaire results from the use of the same questionnaire and the same measurement method to investigate the same respondents twice at different times [6].

Alternate-forms reliability refers to the correlation coefficient between questionnaire results from the use of 2 equivalent questionnaires to investigate the same respondents at different times [6].

Internal Consistency Reliability is used to evaluate the consistency among various questions in a questionnaire [6]. Internal consistency reliability mainly reflects the reliability relationship between various questions within a questionnaire, that is, whether different question items focus on the same concept.

In practice, test-retest reliability and alternative-forms reliability are rarely applied, constrained by factors, such as the application condition and time limit.

Cronbach's α is an index commonly used in the academic circle to measure the internal consistency reliability. Its calculation formula is:

$$\alpha = \frac{k}{k-1} \left(1 - \frac{\sum S_i^2}{S_X^2} \right)$$

Where S_i^2 is the variance of all respondents' answers to Question (i), S_X^2 is the variance of all respondents' answers to all questions, k is the total number of questions in the questionnaire [7].

The value of α ranges between 0 and 1. A larger value of α indicates a higher correlation between question items in a questionnaire, that is, a better internal consistency reliability. In practice, it is generally believed that if $\alpha \geq 0.9$, it indicates an excellent level of internal consistency. If $\alpha \geq 0.7$, it indicates a good level of internal consistency; if $0.5 \leq \alpha < 0.7$, it indicates that the questionnaire is valuable and acceptable, but questionable and requires large modifications; if $\alpha < 0.5$, it indicates a poor level of internal consistency, so the questionnaire is unvalued and unacceptable.

Statisticians generally agree that Cronbach's α is the most stringent among all reliability coefficients. This coefficient is the most accurate reliability index for evaluating the internal consistency. The index is also used as a measure of reliability in this research. The questionnaire has 2 dimensions. One dimension is designed to investigate students' evaluation of teachers' professional qualification, while the other dimension is to investigate students' interest in learning mathematics. Statistical software SPSS19 was used to analyze data of the questionnaire, to get the questionnaire reliability coefficients shown in the table below.

Table 2 shows the calculation results of the reliability coefficients. It can be seen that the overall Cronbach's α of the questionnaire is 0.944, and the standardized item-based Cronbach's α is 0.950, indicating an excellent overall reliability of the questionnaire. The reliability of the questionnaire in the dimension "students' evaluation of teachers' professional qualification" is excellent. However, the reliability coefficient in the dimension "students' interest in learning mathematics" is 0.555, indicating a poor level. Therefore, the correlations between question items in this dimension should be strengthened by modifying contents or expressions of the existing question items or adding some new items.

Table 2. Reliability coefficients

Dimension	Number of samples	Number of items	Alpha	Standardized item-based Alpha
Students' evaluation of teachers' professional qualification	920	11	0.982	0.982
Students 'interest in learning mathematics	920	4	0.555	0.673
Overall evaluation of 15 question items	920	15	0.944	0.950

In addition, statistics about the reliability coefficients in the two dimensions is analyzed.

The mean value, variance and standard deviation of the reliability coefficients in the two dimensions are shown in Table 3. An analysis of the above data can obtain results in Table 4.

Table 3. Statistics about reliability

Dimension	Mean value	Variance	Standard deviation	Number of items
Dimension 1 (students' evaluation of teachers' professional qualification)	18.84	89.669	9.469	11
Dimension 2 (students' interest in learning mathematics	6.11	6.8788	2.622	4

Table 4 shows that the correlation coefficient of each question item in Dimension 1 is greater than 0.8 (most of the coefficients are greater than 0.9), indicating an excellent and good reliability.

In Dimension 2, the correlation coefficients of Item 12 and Item 15 are relatively small, indicating a weak correlation of the 2 questions with the overall Item. That is to

Table 4. Item-total statistics

Dimension	If this variable (question) is deleted	Scale mean if item deleted	Scale variance if item deleted	Corrected item-total correlation	Squared multiple correlation coefficient	Cronbach's Alpha if item deleted
1	Q1	17.20	75.086	0.898	0.829	0.980
	Q2	17.09	73.944	0.911	0.864	0.980
	Q3	17.10	73.553	0.929	0.893	0.980
	Q4	17.08	73.437	0.923	0.890	0.980
	Q5	17.15	73.819	0.916	0.853	0.980
	Q6	17.18	75.022	0.901	0.821	0.980
	Q7	17.11	73.324	0.915	0.850	0.980
	Q8	17.17	74.359	0.907	0.845	0.980
	Q9	17.06	73.655	0.906	0.843	0.980
	Q10	17.12	73.702	0.909	0.848	0.980
	Q11	17.17	76.735	0.824	0.685	0.982
2	Q12	4.32	4.865	0.342	0.182	0.487
	Q13	4.71	5.408	0.539	0.358	0.454
	Q14	4.66	4.914	0.426	0.274	0.445
	Q15	4.64	2.578	0.357	0.140	0.615

say, the 2 questions are poorly designed, because they cannot clearly reveal whether students have interest in learning mathematics (for example, whether students do not want to attend or learn mathematics owing to their poor foundation). If these 2 questions are deleted, the reliability of this dimension can be improved. However, these 2 items are important indexes for revealing students' interest in learning mathematics, so the items should not be deleted. The only choice is to redesign these 2 items or add new variables (question items). Therefore, the design of Dimension 2 has some defects, so the reliability of Dimension 2 should be re-tested after modifications.

3.2 Validity Evaluation

Validity refers to the degree of the difference in the true value reflected by the difference in the measured value of respondents, that is, the validity of the questionnaire. Validity is an index characterizing the gap between the measured value and true value of the research object. It characterizes the extent to which the research object is accurately described by a questionnaire, reflects, and reflects whether the investigator's original intention can be understood by the measured object, that is, whether the measured object's behavior or idea can be effectively measured by the questionnaire, or the extent to which each question item in the questionnaire accurately portrays the corresponding concept. The accuracy of the questionnaire measurement results reflects its control over the system error.

In academic research, the content validity, discriminant validity, and construct validity are usually discussed in validity evaluation. Content validity refers to the degree of consistency between contents measured by a questionnaire and its research objective; criterion validity refers to the degree of consistency between questionnaire measurement results and the validity criteria; structure validity is an index evaluating whether the correlation between observable variables is consistent with the theoretical prediction. It is generally believed that construct validity is the most powerful index for validity evaluation [8], and factor analysis a commonly-used statistical method for testing the construct validity.

In the use of factor analysis to test the construct validity, 3 indexes including communalities, cumulative contribution and factor loading are used. The process of factor analysis is as follows.

3.2.1   Evaluating the Appropriateness of Using Factor Analysis

In KMO and Bartlett's test scale, when the value of KMO measure is greater than 0.9, it indicates a high appropriateness of using factor analysis; when the value of KMO measure is between 0.8 and 0.9, it is appropriate to adopt factor analysis; when the value of KMO measure is between 0.7 and 0.8, it indicates that it is moderate to adopt factor analysis; when the value Of KMO measure is between 0.6 and 0.7, it is inappropriate to adopt factor analysis; when the value of KMO measure is smaller than 0.6, the correlation between variables is weak, and variables should be re-selected.

As shown in Table 5, the KMO value is 0.967; the value of Bartlett's Test of Sphericity is 16064.649; the degree of freedom is 105; the p value is about 0. This indicates that question items have communalities, so it is appropriate to adopt factor analysis on data.

Table 5. KMO & Bartlett's Test

Sampling sufficient KMO measures		0.967
Bartlett's Test of Sphericity	Chi-square	16064.649
	df	105
	Sig.	0.000

3.2.2   Communalities of Statistical Variables

According to data about communalities shown in Table 6, except that the extracted value of Q15 is smaller than 0.4, the extracted values of the other variables are quite high, indicating an excellent overall effect of factor analysis.

3.2.3   Cumulative Contribution of Statistical Factors

The cumulative contribution is used in factor analysis to determine the number of factors. When the cumulative contribution is greater than 80%, that is, in the case that the amount of information loss is lower than 20%, the number of factors is considered to be sufficient.

As shown in Table 7, the cumulative contribution of the first 3 factors is 81%, which meets the requirement for the number of factors. Therefore, it is acceptable to choose 3 factors to make loading matrixes.

Table 6. Variables' communalities and rotating component matrixes

	Variable communalities		Rotating components		
Variable	Original value	Extracted value	1	2	3
Q1	1.000	0.840	**0.904**	0.119	0.096
Q2	1.000	0.860	**0.908**	0.157	0.107
Q3	1.000	0.888	**0.928**	0.133	0.098
Q4	1.000	0.879	**0.925**	0.126	0.085
Q5	1.000	0.868	**0.922**	0.115	0.081
Q6	1.000	0.846	**0.907**	0.093	0.118
Q7	1.000	0.867	**0.915**	0.109	0.134
Q8	1.000	0.855	**0.917**	0.080	0.090
Q9	1.000	0.850	**0.908**	0.141	0.085
Q10	1.000	0.856	**0.910**	0.119	0.120
Q11	1.000	0.734	**0.819**	0.171	0.189
Q12	1.000	0.441	0.108	**0.857**	-0.099
Q13	1.000	0.671	0.205	**0.714**	0.360
Q14	1.000	0.566	0.102	**0.561**	0.502
Q15	1.000	0.366	0.171	0.104	**0.878**

Table 7. Total variance explained

Component	Initial Eigenvalues			Squared extraction and loading			Squared rotation and loading		
	Total	Variance %	Total %	Total	Variance %	Total %	Total	Variance %	Total %
1	9.689	64.593	64.593	9.689	64.593	64.593	9.072	60.480	60.480
2	1.699	11.326	75.919	1.699	11.326	75.919	2.316	15.439	75.919
3	0.787	5.245	81.165						
4	0.713	4.756	85.921						
5	0.463	3.087	89.008						
6	0.328	2.188	91.196						
7	0.274	1.826	93.022						
8	0.210	1.398	94.419						
9	0.163	1.084	95.503						
10	0.146	0.974	96.477						
11	0.133	0.886	97.363						
12	0.117	0.778	98.141						
13	0.110	0.735	98.876						
14	0.095	0.632	99.508						
15	0.074	0.492	100.000						

3.2.4   Factor Loading Matrix After Rotation

As shown in Table 6, the first 11 items are variables with greater loadings on Factor 1, which indicates that the 11 variables share common characteristics. Variables with close relations with Factor 1 happen to fall into the dimension of "students' evaluation of teachers' professional qualification". Variables with greater loadings on Factor 2 are about attitudes towards mathematical contest, elective courses and knowledge. All the 3 variables are closely related to students' interest in mathematics. Whether the QR code in textbook exercises is conducive to students' mathematics learning is the only variable with greater loading on Factor 3. This means that this variable has a weak correlation with other variables and needs to be modified and improved.

4   Conclusions

The above research shows that the existing evaluation system for mathematics teaching in this college has good reliability and validity in general. To supplement and improve some question items in Dimension 2 can improve the validity and reliability of the evaluation results.

In the process of factor analysis, three latent factors that affect the quality of classroom teaching are detected. According to the correlation coefficient between factor indexes, indexes can be roughly classified into 3 modules: teachers' professional qualification, students' interest in learning and students' after-class performance. The variance contribution of each latent factor is the weight coefficient affecting the results of classroom teaching quality evaluation, which can help to effectively determine the weight coefficient of each module in the evaluation index system, thus reducing the blindness and randomness in practice [8].

Factors affecting teachers' classroom teaching quality and outcome are large in number. It is relatively difficult to evaluate the quality of mathematics teaching in higher vocational colleges. Evaluation results usually cannot fully reflect the actual situation of mathematics teaching, so the evaluation of the mathematics teaching quality tends to become a formalized operation with meaningless practice. To truly implement evaluation results, it is necessary to establish supporting reform rules and incentive measures in the follow-up work, so that the effectiveness of classroom teaching quality evaluation can be enhanced.

Acknowledgement. Thanks to associate professor Xia Liu, correspondent of this paper.

References

1. Song Zhengfu, F., Research Team.: Research and practice of teaching quality evaluation system in higher vocational colleges. J. Chongqing Ind. Trade Polytech. (2), 1–18 (2010)
2. Nie Li, F.: The present situation of college mathematics teaching quality and its countermeasures. J. Cap. Univ. Econ. Bus. (6), 122–124 (2014)
3. Chen Jialiang, F.: Analysis of factors affecting the satisfaction of teaching quality in higher vocational colleges. Chin. Vocat. Tech. Educ. (8), 5–9 (2017)

4. Zhang Wentong, F.: SPSS Statistical Analysis Advanced Tutorial, 1st edn. Higher Education Press, Beijing (2014)
5. Zeng Wuyi, F.: Analysis of the reliability and validity of the questionnaire. Stat. Inf. Forum 6 (20), 11–15 (2005)
6. Xia Yifan, F.: SPSS Statistical Analysis Essentials and Example Details, 1st edn. Publishing House of Electronics Industry, Beijing (2010)
7. Liang Naiwen, F.: Questionnaire of public security in hunan province and its reliability and validity analysis. J Appl. Stat. Manage. 6(31), 1039–1048 (2012)
8. Li Changxi, F.: Empirical study on the reliability and validity of classroom teaching quality evaluation system——A Case Study of Shandong University of Science and Technology. J. Shandong Univ. Sci. Tech. 6(15), 86–92 (2013)

Online Education Resource Evaluation Systems Based on MOOCs

Yan Zhang[1] and Han Cao[1,2,3(✉)]

[1] School of Computer Science, Shaanxi Normal University,
Xi'an Shaanxi 710119, China
{jasonzhang, caohan}@snnu.edu.cn
[2] Shaanxi Key Laboratory of Tourism Informatics, Shaanxi Normal University,
Xi'an Shaanxi 710119, China
[3] Shaanxi Tourism Information Engineering Laboratory,
Shaanxi Normal University, Xi'an Shaanxi 710119, China

Abstract. Massive Open Online Courses(MOOCs) have been generalized as one of wildly developed tools implemented to improve educational quality in colleges and universities. However, the education quality that performed through MOOCs could not be evaluate easily due to the lack of systemic and dynamic analysis. Thus, the paper proposes a new online education resource evaluation system based on MOOCs, which combines the subjective analytic hierarchy process (AHP) and the objective entropy weight method(EWM) to determine the weight of each evaluation index. The proposed system's evaluation of online educational resources is more accurate and objective.

Keywords: MOOCs · Evaluation system · Evaluation index
Analytic hierarchy process

1 Introduction

MOOC is an online curriculum development model and online teaching model that has emerged in recent years. As a new mode of online education in the era of information age, MOOC has been widely carried out in universities and has become an important part of higher education. Using keywords such as MOOC and online education platform, search engines Baidu searched 96 platforms related to MOOC [1]. According to MOE Research Center for Online Education data, there were 1500,000 Chinese users registered on MOOC in 2014. In October 2016, it reached 10 million people, far more than any other country.

The advantages of the current MOOC have been fully demonstrated: the optimal combination of teaching resources, no limitations of the region and learners, making lifelong education possible, short teaching time and timely feedback. However, MOOC also exposes some shortcomings. MOOC is difficult to assess the teaching quality and achieve the full function of education in the university. The analysis of MOOC learners' characteristics cannot be performed comprehensively. And there is a lack of systematic and dynamic analysis of the MOOC learners' online learning behavior. On the one hand, these issues have hampered the sustainable development of online

© Springer Nature Singapore Pte Ltd. 2018
Q. Zhou et al. (Eds.): ICPCSEE 2018, CCIS 902, pp. 605–615, 2018.
https://doi.org/10.1007/978-981-13-2206-8_51

education. On the other hand, it also highlights the contradiction between the rapid development of online education and the lack of mastery of development laws. How to evaluate MOOC learners' learning effectiveness and courses is an important issue. Whether the MOOC courses learning is effective or not, and whether the education investment has the expected educational benefits, MOOC learners and MOOC courses are both need to be evaluated. The focus of MOOC-based online education research is gradually turning to MOOC learners' learning effectiveness evaluation. Online learning evaluation methods and evaluation results have a direct impact on learners' learning attitude and learning effectiveness. Online learning assessment based on behavior analysis is to understand learners' intrinsic learning styles and preferences, which through a series of behavioral analysis. And based on this, online learning assessment carries out comprehensive evaluation of students' learning process [2]. The feedback of evaluation information can enable online learners to purposely adjust their learning strategies and learning progress. It can also provide a basis for the construction of online learning platforms and the development of online learning resources.

Therefore, this paper proposes new online education resource evaluation systems that are consistent with MOOCs education ideas and online education characteristics. Using online measurement as a means to analyze students' access to educational resources and the online education resources of MOOCs are analyzed. Through the evaluation of MOOCs learners' learning effectiveness and MOOCs courses, the application effects of online education resources are evaluated. And give a reasonable evaluation system, put forward online education resources evaluation system and online education resources optimization framework.

2   Online Education Resource Evaluation Index

The overall index of online evaluation is generally divided into three subsystems:

1. Condition index system: the conditions that the network should have when it undertaking or completing a mission, that is the hardware of the network facility.
2. Process index system: the evaluation standards for the responsibility of the network education resources and the completion of the learning task.
3. Efficiency index system: Online education resources enable learners to obtain evaluation criteria defined by the human, material, and financial resources spent on learning.

The general process of designing the evaluation index system: firstly, analyze the target, derive the characteristics of the evaluation object. Secondly, establish evaluation criteria and describe the standard. And then establish the principle of value orientation. Finally, determine the weight of the index [3]. The evaluation of online education resources is a complex process. It is necessary to translate the overall goal of evaluation into actionable and concrete indicators. The operation method of online education resource evaluation is divided into the 3 steps.

2.1 Design Online Education Resources Evaluation Scheme

1. Clear the purpose of online education resource evaluation, determine the object and scope of online education resource evaluation;
2. Determine the specific criteria for the evaluation of online education resources and choose reasonable evaluation methods;
3. Organize online education resources evaluation information materials;
4. Use scientific and reasonable methods and tools;
5. The arrangement of online education resource data and the interpretation of the results;
6. Error Analysis and Control in the Implementation of Online Education Resources Evaluation.

2.2 Design Procedures of Online Education Resources Evaluation Index System

1. Determine the object and goal of online education resource evaluation;
2. Determine the ways to evaluate online education resources;
3. Establish online education resources evaluation index system and evaluation criteria;
4. Build an index system tree structure;
5. Determine the weight factors;
6. Classification and quantification of index items;
7. Establish evaluation indicators;
8. List the evaluation system.

2.3 Data Collection, Statistics and Analysis

Evaluation of online education resources can be conducted through the Internet. By setting up survey pages and forms within the network site, the respondents can submit the survey within a specified period of time and complete the statistics. According to the purpose of the evaluation to prepare the questionnaire. A summary evaluation, formative evaluation, or diagnostic evaluation is performed on the evaluation object. Use the question bank for automated online testing and adaptive quizzes [4].

The evaluation data analysis is to organize and analyze the data in the previously collected information. Reveal the evaluation results contained in the data. At the same time, some dedicated statistical analysis tools can be used to complete data processing. According statistical methods to organize evaluation data, perform statistical statistics and data processing. Finally, the classification analysis was conducted according to the value judgment classification, and the evaluation result was obtained.

The Fig. 1 is the design of online education resource evaluation index system and Fig. 2 is the evaluation program for online education resources.

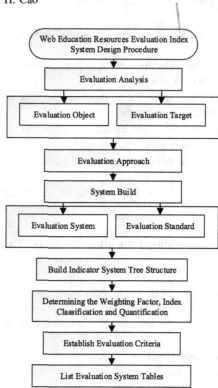

Fig. 1. Flowchart of design on evaluation index system of online education resources

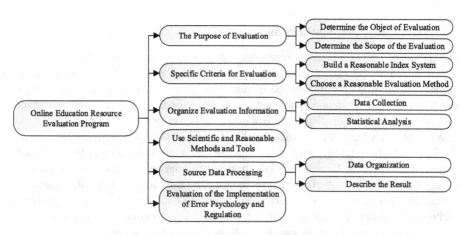

Fig. 2. Online education resource evaluation scheme

3 Establish Evaluation Model and Evaluation System

3.1 Index Assignment

The online education resource platform, the resource quality assessment system, the student online learning behavior and the effect evaluation system are mainly based on the Delphi method [5] and the analytic hierarchy process(AHP). Firstly, select a number of experts from different disciplines and use the pairwise comparison method in the analytic hierarchy process to assign values to each indicator in the model. The index's valuation is mainly based on the 9-class scale given by Saaty in AHP. The scale rule is based on the numbers 1, 3, 5, 7, 9 and 2, 4, 6, 8 combined text narrative ratings [6]. Finally, we retrospect and verify the results of the evaluation of experts. Until the results of the assignment of experts to the various factors tend to be consistent.

3.2 Determination of Index Weights

After the experts assign values to each index, the index system is evaluated using the analytic hierarchy process. In order to ensure the correctness of the decision-making results, the consistency of the judgment matrix of the index system constructed was tested. In the AHP method, the consistency test is first to build an N-order matrix according to the expert's evaluation results. The second is to normalize each column of the N-order matrix. Summing each row of the matrix yields a weight vector. The maximum eigenvalue is calculated from the weight vector. Finally, consistent results are obtained based on the maximum eigenvalue and the random consistency index. When the consistency result <= 0.1, the consistency of the matrix is within the allowable range, and the matrix is reasonable. Otherwise, the index system will be assigned again and a new judgment matrix will be constructed [7].

3.3 Handle Uncertainty Information

When evaluating the factors affecting college students' online learning behavior, the evaluation information of each index obtained must have a lot of "uncertainty". Therefore, it needs to be evaluated using methods that can process uncertain information. The cloud model can represent the randomness and ambiguity of things in the objective world. Randomness refers to the lack of decisive causality between conditions and results caused by insufficient conditions; ambiguity refers to the mutuality of boundaries [8, 9]. At the same time, the cloud model also reflects the correlation between the two uncertainties. The feature parameters of the cloud model are important concepts in the cloud model theory and can reflect the overall characteristics of the cloud. This model is a qualitative concept and quantitative numerical conversion model, which is the unity of randomness and ambiguity.

This paper use the subjective AHP method and objective entropy weight method to determine the weight of each index based on the cloud model theory, and optimize the integration of weights. In this way, when evaluating the indicators, the results are more objective and accurate. Figure 3 shows the evaluation model in the measurement.

Figure 4 shows the online educational resource platform, the quality of online educational resources and the online learning performance evaluation system of students.

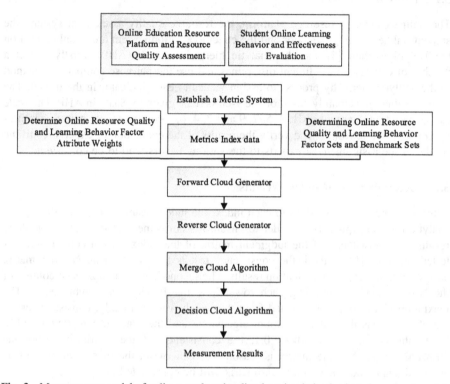

Fig. 3. Measurement model of college students' online learning behavior based on cloud model

3.4   Establish a Gauge Index System

An electronic performance support system (EPSS) was introduced to effectively evaluate online education resources and learning behaviors [10]. Compared with traditional evaluation methods, its advantages:

1. EPSS is learner-centered and supports the interaction among students, teachers, and media, making up for the teacher-centered singleness in traditional evaluation;
2. EPSS aims at performance, making up for the ambiguity of the objects in the tradition and focusing on formative evaluation;
3. EPSS implements the test modifications and then avoids the confusion of the traditional index system;
4. EPSS provides an online help system to provide immediate help for improvement after evaluation.

The three dimensions of online learning interaction quality, collaborative learning for online learning and online learning support system are evaluated. Consider that teachers, students, and computer media interact in all aspects of online learning. In each

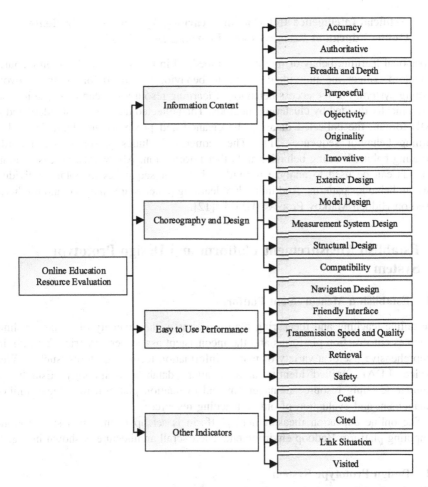

Fig. 4. The evaluation system of online educational resource platform and Students' online learning behavior

dimension, an indicator system was developed based on three factors, and a five-level evaluation method was used: 5 is strongly agree, 4 is agree, 3 is neutral, 2 is disagree, 1 is extremely disagree.

The basic steps: first step is data entry. The data is input into the EPSS as a spreadsheet, where the corresponding variable values are entered in the spreadsheet. And the scores selected by each student for all evaluation criteria are entered in corresponding spaces. Second step is analysis of data. At the beginning, reliability analysis is performed on all the evaluation criteria of interaction quality. A statistical view can be obtained in a new window, and then the reliability standard of the evaluation criteria for web-based collaborative learning is analyzed.

3.5   Artificial Intelligence and Machine Learning Approaches to Evaluate Online Education Resources and Learning Behaviors

A personal learning behavior monitor is embedded in the network education resource system to collect the individual's specific behavior and operation in the network learning system. For the access behavior of learning resources, it can be classified and hierarchically divided by clustering methods. The collected data can be standardized in XML format, the collected data can be cleaned and processed to obtain individual learning behavior sequences [11]. The sequence includes personal visit records, operating habits, learning behavior and other information. Afterwards, process mining methods can be used to analyze individual behavior sequences to obtain individual learning behavior patterns. And individual learning patterns are analyzed and predicted based on Hidden Markov Process (HMM) [12].

4   Establish Measurement Platform and Design Prototype System

4.1   Establish a Measurement Platform

The online teaching and measuring platform aims at the diversity of online teaching resources construction platforms and the inconsistent assessment criteria. Through the comprehensive use of a variety of mature information technology tools, such as Web Service, LDAP, unified identity authentication, database technology. Establish a complete teaching resource management and evaluation platform to achieve unified management and evaluation of online teaching resources.

The online education measurement platform is set up in the open source cloud computing platform Hadoop environment. The overall architecture is shown in Fig. 5.

4.2   Design Prototype System

Design a Student Behavior Analysis System Based on Hadoop Technology
Based on the Hadoop cloud platform, HDFS distributed file system is used to complete data collection, processing, and storage. Design network education resources and student learning behavior data storage schemes. The data collection framework and key algorithms for the evaluation of online education resources and student learning behaviors are designed. And the analysis algorithms are written using the MapReduce parallel programming model. Then data mining and intelligent analysis of the data. Design education resource and network learning quality assessment system, based on JAVAWEB and adopt the mainstream SSH architecture to complete the development of related modules. At the same time, the results of the analysis are graphically and intuitively displayed. In order to design a student behavior analysis system based on Hadoop technology.

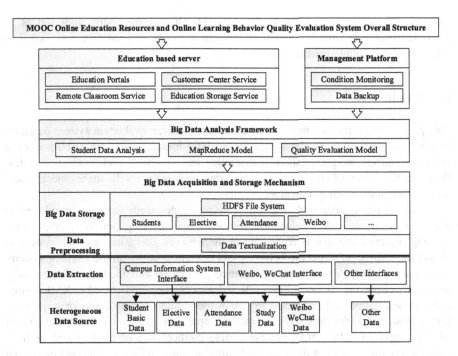

Fig. 5. Online education resources and online learning behavior quality evaluation system overall structure

Design a MOOCs Online Education Resource and Online Learning Behavior Quality Evaluation System Under a Cloud Platform Environment

For the MOOCs online teaching platform used by Shaanxi Normal University, the secondary development of the API interface provided by the MOOCs online teaching platform is carried out. Through the relevant data exchange interface and functional service interface provided by the API of the MOOC online teaching platform, the system is integrated.

Comprehensive Evaluation of Online Education Resources and Students' Online Learning Behavior

According to individual analysis, situation monitoring, comprehensive display, teaching aid, decision support and other key functions, the function of MOOCs online teaching platform is extended. The Building Blocks TM technology provided by the MOOC online teaching platform puts these functions as important value-adding capabilities in the MOOC online teaching platform. Thus, a comprehensive evaluation of online education resources and student network learning behaviors is provided, and a model tool is provided for student learning achievement evaluation in a diversified and re-learning process.

Design Education Decision Support System

Using data mining and data warehousing techniques to design data education warehouses and student learning behavior data warehouse assessment models and data warehouse data mining algorithms. Take the MOOCs online course appraisal as an example to conduct empirical research. According to the analysis data, come up with suggestions for countermeasures.

5 Conclusion

This paper first proposes a new online education resource evaluation system based on MOOCs, which combines the subjective analytic hierarchy process (AHP) and the objective entropy weight method to determine the weight of each evaluation index. Based on the theory of cloud model, this system makes the results more objective and accurate when evaluating indicators.

Artificial intelligence and machine learning model methods were introduced to effectively evaluate online education resources and learning behavior. A personal learning behavior monitor is embedded in the network education resource system to collect the individual's specific behavior and operation in the online learning system. For the access behavior of learning resources, it can be classified and hierarchically divided by clustering methods. The collected data can be standardized in XML format and can be cleaned and processed to obtain individual learning behavior sequences. The sequence includes personal visit records, operating habits, learning behavior and other information. Afterwards, process mining methods can be used to analyze individual behavior sequences to obtain individual learning behavior patterns, individual learning patterns are analyzed and predicted based on Hidden Markov Process (HMM).

Acknowledgements. This work was supported by the National Natural Science Foundation of China under grant number: 40471102.

References

1. Zhao, H., Zheng, Q.H., Chen, L.: Research on the construction and development of MOOCs in China: status quo and reflection. Distance Educ. China **010**(11), 55–62 (2017)
2. Zhai, Y.H., Yang, M.H.: Learning behavior and effectiveness under the "MOOC +" Environment. J. Heilongjiang Univ. Technol. **17**(4), 29–33 (2017). (Comprehensive Edition)
3. Sun, P.: The design of micro-lesson evaluation system based on user experience. Mod. Educ. Technol. **25**(3), 63–68 (2015)
4. Xie, H.B.: Review of the appraisal of educational network resources in universities. Distance Educ. J. **29**(4), 60–64 (2011)
5. Wang, C.Z., Si, Q.: A study of data statistical processing method of delphi method and its application. J. Inner Mongolia Univ. Finan. Econ. **09**(4), 92–96 (2011)
6. He, K.: A study on the scale of analytic hierarchy process. Syst. Eng. Theory Pract. **17**(6), 58–61 (1997)
7. Li, H.Y., Ren, L.C., Han, S.R.: Application of AHP in the teaching quality appraisal of higher education. J. Taiyuan Univ. Sci. Technol. **24**(3), 223–228 (2003)
8. Li, Z.T., Wang, Y.B.: An e—learning ecosystem model based on cloud computing. J. Henan Normal Univ. **41**(2), 64–67 (2013). (Natural Science Edition)

9. Zhang, S.B., Xu, C.X.: Study on the trust evaluation approach based on cloud model. Chin. J. Comput. **36**(2), 422–431 (2013)
10. Li, H.X.: EPSS is one choice of the training of teachers in higher school. Vocat. Tech. Educ. **25**(10), 47–49 (2004)
11. Li, S., Zhong, Y., Yu, C., Cheng, G., Wei, S.: Exploring the online learning participation behavior pattern based on behavioral sequence analysis. China Educ. Technol. **3**, 88–95 (2017)
12. Huang, Z.C.: Evaluation on learning behavior based on hidden Markov model. Comput. Appl. Softw. **31**(6), 59–62 (2014)

Analysis on Psychological Health Education of Graduate Students from the Strengths Perspective

Xiaoli Liu(✉)

Wuhan University of Science and Technology, Wuhan 430025, China
348205869@qq.com

Abstract. This paper uses the scale test to count the psychological character-istics of graduate students. Self-reporting Inventory (SCL-90) was used as a test tool for network testing and 944 valid questionnaires were received. Statistics showed that 9.96% of graduate students have any of a division factor score equal to or more than 2.5. A Comparative analysis, in particular, found that for the graduate students with a factor score of 2.5 or more, and the forcing factor scored the most. Based on the results of the above data, this paper has a deep analysis on the psychological health status of graduate students and the current shortage of postgraduate mental health education. It proposes to establish strengths evalua-tion system, set psychological courses for graduate students, play a guiding role for mentors in education and create a positive campus atmosphere according to the strengths theory, so as to achieve the goals of elite education.

Keywords: Strengths perspective · Graduate students · Psychological health

With fierce changes of social environment, psychological health status of college students have new characteristics, and thus managers gradually pay attention to sig-nificance of psychological health education of college students. The Party and national education sectors issued documents like The Basic construction standard of education for students from general institutes of higher education (trial) and The Basic require-ments of psychological health education course teaching for students from general institutes of higher education (trial), a rapid development has been got by the psy-chological health education of college students. During university, graduate students generally have received psychological health education, which has a great effect to promote their psychological health development.

At present, psychological health education courses has been set and psychological consultancy organization established in many colleges and universities, but they basically face to undergraduate students and few of them to graduate students. Meanwhile, from the perspective of practice effect, in psychological health education, colleges and universities in China generally carry out "problem mode" or "disease mode", and controlling malignant psychological incidents like suicide and self-injury of students and reducing psychological problems and psychological diseases of stu-dents are the core idea and major objective of schools to have psychological health education [1].

© Springer Nature Singapore Pte Ltd. 2018
Q. Zhou et al. (Eds.): ICPCSEE 2018, CCIS 902, pp. 616–622, 2018.
https://doi.org/10.1007/978-981-13-2206-8_52

1   The Basic Theory of Strengths Perspective

The basic assumption of the strengths perspective theory is that every individual, group, family or community has strengths, and all environments are filled with resources and even in trauma, pain and trouble, which can also be regarded as hope and seed, which will make a difference [2]. The so-called strengths not only include personal qualities and virtues, such as goodness, sense of humor, kindness, creativity and insight, but also personal skills like playing a musical instrument, having artistic creation, telling story, having manual skill, repairing furniture and writing. Strengths covers a wide range of content, almost everything in a specific atmosphere can be seen as an advantage.

From the strengths perspective, psychological health education in colleges and universities can be defined as: in the process of constructing the psychological heath education system for college students, while concerning common psychological problems of students, evaluate and analyze strengths of students through professional means, discover and encourage each student to develop and apply their own advantages to overcome their mental blocks, reduce problem behavior and realize self-growth and development [3].

2   The Existing Circumstances of Psychological Health Education of Graduate Students

2.1   Research Methods

The survey object in this study is the first-year graduate students in a comprehensive university. Self-reporting Inventory (SCL-90) was used as a test tool for network testing and 944 valid questionnaires were received.

SCL-90 score of 2.5 is set as a screening standard, equal to or greater than 2.5 means that the individual has a psychological problem in the dimension above moderate, on which future interview screenings will be based.

Statistics showed: 9.96% of graduate students have any of a division factor score equal to or more than 2.5. A Comparative analysis, in particular, found that for the graduate students with a factor score of 2.5 or more, the forcing factor scored the most (Table 1).

2.2   Psychological Health Status of Graduate Students

As special group, graduate students grow in campus and stand in the highest level in the educational structure, belonging to outstanding persons in the learning field. However, their personnel structure is relatively complex. According to the current educational system of China, full-time graduate students are generally at the age from 21 to over 30 years, and, moreover, they differ in the aspects of background, age, work experience and marital status. The university encourages graduate students to study independently, and the daily learning of graduate students takes scientific research as the main content. They often communicate with tutors and members of research

Table 1. The detection rate of each factor

Factors	The number of students scoring more than 2.5	Percentage of total (%)
Compelling	41	4.34
Interrelationship sensitivity	11	1.17
Depression	10	1.06
Anxiety	10	1.06
Paranoid	8	0.85
Hostile	5	0.53
Somatization	4	0.42
Psychosis	3	0.32
Terror	2	0.21

teammate. In addition, their learning and living environment is featured with academic property, independence and closure, thus forming some special psychological characteristics of graduate students.

Graduate students have differences in physiological development, mentality and social development. They are physiologically mature, and at the stage of going to mature rapidly but not totally mature in psychological development, but their social development is far from mature. Therefore, although they are at a period of great prosperity in intellectual development and their physical conditions have reached to the peak, but their cognitive competence and self-control ability should be improved and their psychological quality is not good enough. However, a higher requirement and expectation the society and they themselves have to them than other groups, thus causing the internal contradictions and they are easy to be frustrated. Many graduate students have radical behaviors reaction because of their difficulties in adjustment, giving rise to various extreme results which would severely affect the quality of postgraduate training and bring harms to the society, school, their family and themselves.

2.3   The Current Shortage of Postgraduate Mental Health Education

Related departments of colleges and universities lack essential concerns to psychological health of graduate students. The psychological health education of graduate students is not fully professionalized, the number of full-time staff is seriously insufficient, and the coverage and deepening degree of psychological health education of graduate students are not enough. More ideological educations and less professional psychological guidance, no timely screening of psychological problems of graduate students and inappropriate treatment, resulting in delays in the best time for mental health education [4]. Graduate students have a heavy burden in the aspects of school work, employment, economy, love and marriage, and they are faced with complex challenges. Number of graduate students who have psychological problems keeps increasing, and it had been reported that psychological problems causing malignant incidents such as failures in school work, dropping out and even suicide. The current domestic studies indicated that psychological health status of graduate students is not optimistic, graduate students have psychological problems of varying degrees, and employment pressure, economy and interpersonal relationship have a great influence on their psychological health [5, 6]. At present, psychological health education of most colleges and universities often use experience management or entrust psychological health education institutions for undergraduates, thus lacking effective carriers and having little effects.

3   Approaches to Promote Psychological Health of Graduate Students from the Strengths Perspective

In the report of the 19th National Congress of the Communist Party of China, Xi Jinping pointed out that "reinforce construction of social psychological service, and cultivate self-esteem and confident, rational and peaceful, and positive social mentality". The several opinions of the ministry of education to further strengthen and improve ideological and political education of graduate students clearly proposed to "enhance psychological health education and consultation work for graduate students, actively carry out psychological health screening, psychological health education, psychological consultation and crisis intervention for graduate students, provide individual and group counseling service projects according to characteristics of graduate students, help them to solve problems in emotion adjustment, environmental adaptation, personality development, interpersonal communication, friend making and love, ad occupation selection and employment, enhance their ability to adjust mentality and improve level of psychological health". Therefore, supported by the strengths perspective theory, psychological health work for graduate students can be carried out with the following approaches.

3.1   Evacuate Personal Advantageous Resources and Establish Strengths Evaluation System

It is the basis of psychological health education to identify and evaluate personal strengths of graduate students. Researches indicated that not a few graduate students did not understand themselves very well and the higher the grade, the higher expectation they would have. And when they were confronted with problems which could not be solved, they were easy to blame themselves or adopted the method of "self-justification" to make them feel better. That's to say, may graduate students do not know their strengths [7]. Currently, most existing evaluation systems were modified from strength classification method and evaluation systems which are widely used in foreign countries, and a typical one is the actively psychological trait table for Chinese college students of W.J. Men. It is constituted by 6 component tables, including 20 positive qualities and 62 items. Statistical test indicated that the table had favorable reliability and validity, thus it can be used for large scale evaluation [8].

The strengths perspective does not ignore personal "problems" but weaken the "problems". It makes personal energy focus on strengths and believes that strengths of human have infinite potentials, but whether they can realize it or not is related to personal changes. First, it should depend on efforts of graduate students themselves. They have s strong sense of independence, and the willingness and ability to have a happy life should be believed. On the basis of strengths evaluation, it can stimulate them to think about how to improve the current situation and pursue for future targets and power. Moreover, psychological health education departments of schools should help individuals to explore their strengths resources, develop their personality and improve themselves according to their characteristics of physical and mental development and psychological needs of graduate students so as to promote comprehensive free development of graduate students and fully reflect their concerns to individual life [9]. Furthermore, student psychological self-help organizations like psychological health association and psychological societies for graduate students to make them communicate with each other should be encouraged and guided.

3.2   Set up a Strengths Course System and Adjust Psychological Consultation Mode

At present, colleges and universities have less psychological health education courses which is specialized for graduate students, thus such a course system should be established on the basis of strengths perspective by taking recognition and cultivation of "favorable psychological qualities" and "favorable characters and morals" as the key to give full play to the driving role of internet channels in the new era and taking promotion of integrated development as the ultimate goal. For example, some psychological health optional courses and psychological health knowledge lectures for graduate students can help them to grasp some applicable self-adjustment methods and skills, and the participation in voluntary services to guide them to discover their own kind and altruistic positive qualities. On the basis of scientific evaluation of character strengths, graduate students can establish advantageous groups, encounter groups for difficulties of interpersonal communication, student-teacher groups for tense tutor-student relationships, occupation mutual-help groups for employment pressure and other groups for emotion relationship. Group working can guide graduate students to apply their advantages actively in various aspects including learning, life and employment to experience the sense of achievement brought by strengths.

Psychological counseling is an important form of psychological health education of colleges and universities. The target of traditional psychological counseling is to relieve psychological problems, while strengths perspective requires psychological counseling teachers to understand natures of psychological consultation from a positive aspect, not only providing more understanding and acceptance to visiting graduate students but also striving to discover strengths and power of them. Moreover, they should concern strengths of graduate students first in a roundabout manner, share their former success or proud experience, and help them to clear up relevant resources, regain self-confidence through life review and finally see new life goals. In addition, they should encourage graduate students to use their own strengths to solve current problems to obtain the sense of achievement and thus have better development.

3.3   Expand Environment Strength Background and Create a Positive Campus Atmosphere

The strengths perspective theory also emphasizes on environment strengths, thus psychological health education for graduate students should first be fully concerned by educational management institutions at all levels, regards psychological health education for graduate students as equal important as that for undergraduate students and enhances support in personnel, equipment, site and expenditure. Within colleges and universities, psychological health education for graduate students cannot be limited to psychological counseling center and psychological teachers but should also need cooperation and support of various functional departments and participation of all members, thus forming a join force. Moreover, tutors of graduate students, professional teachers, instructors and management and service staff should be also covered in the scope of the psychological health education system for graduate students [10].

Tutor is the first person responsible for health growth of graduate students. As the conductor of students in study and scientific research, they have more perceptual and accurate understanding to psychological status and a decisive significance to academic development of graduate students. Therefore, tutors should reinforce their sense of responsibility and help graduate students to solve problems in study, thought and life. In addition, related departments of schools should take active measures to improve study and living environment of graduate students, for example, setting up scholarship and grants and posts like graduate student assistant manager, assistant teacher and research assistant, so as to help the students with financial difficulties to finish their study successfully.

Graduate students are the group which has highest academic degree. Whether they can have a healthy development can directly affect the socialist modernization of China and realization of the great rejuvenation of the Chinese nation. Therefore, based on the strengths perspective theory, it has great and profound strategic significance to concern psychological health of graduate students and enhance psychological health education of graduate students.

References

1. Chen, H.J.: Active transfer of the centre of psychological health education for undergraduates. J. Changsha Univ. Sci. Technol. (Soc. Sci.) (1) (2014)
2. Saleebey, D., Du, L.J., Li, Y.W.: Strength Perspective – New Mode of Social Working Practice, pp. 4–7. East China University of Science and Technology Press (2004)
3. Cheng, Z.Q.: Strengths perspective: transformation of psychological health education in colleges and universities. China Educ. Light Ind. (5), 42 (2015)
4. Luan, H.Q.: Psychological health education of graduate students from colleges and universities: reflection and construction. Heilongjiang Res. High. Educ. (12), 93 (2017)
5. Fang, W., Wang, S.Y., Yang, X.F.: Analysis on psychological health status of graduate students and influence factors. Chin. J. Public Health (4), 433–435 (2009)
6. Shi, T.P., He, C.Y., Chen, R.: Epidemiological survey on psychological health status of medical graduate students. Mod. Prev. Med. (9), 1638–1642 (2006)

7. Tsai, Y.Q., Jin, S.F.: Study on negative life and self-efficiency of graduate students and solutions. Ideol. Theoret. Educ. (7), 87–88 (2014)
8. Men, W.J., Guan, Q.: Report of compilation of active psychological quality table for Chinese college students. Chin. J. Spec. Educ. (8), 71–75 (2009)
9. Zhu, M.Y.: Analysis on the current situation of psychological health education for graduate students and solutions. Heilongjiang Res. High. Educ. (3), 120 (2017)
10. Fen, R., Zhang, Y.T., Ma, X.T.: The current research situation and progress of psychological health education for graduate students from Chinese colleges and universities. Educ. Res. Grad. Stud. (2), 25 (2015)

Design and Implement of International Students' Management and Security Warning System Based on B/S Architecture

Yulu Zhang[1], Zhikun Li[2], Ya Wen[1], Jifu Wang[1], and Ruigai Li[1(✉)]

[1] College of Information and Computer Engineering,
Northeast Forestry University, Harbin, China
Lirg751@163.com
[2] School of International Education and Exchanges,
Northeast Forestry University, Harbin, China

Abstract. In recent years, the number of international students in China has been increasing, which has greatly increased the difficulty of the international students' information's management. It's a difficult problem for all the schools of international education and exchanges to manage international students successfully and normatively in order to ensure that they can complete their studies smoothly in China. Thus, Information technology can provide the necessary technical support to solve these problems. Firstly, the development of this system can realize the efficient management of international students' information. Secondly, according to the functional requirements of different users, this system has developed modules for students, student managers and administrators, realizing the students' leaving applications, security warning, the letter and other functions. It is convenient for managers to manage all kinds of information of international students effectively. The development of this system can promote international students' management in various schools, simplify the process of various work and improve the working efficiency.

Keywords: International student · Security warning · Management system

International students play an important role in promoting regional cultural internationalization and the development of students' exchanges in China can greatly improve our cultural level. Especially in recent years, with the rapid economic development in China, more and more international students choose to study in our China. However, with the increasing number of international students, the difficulty of international students' management has increased, and the security of international students has become a key issue. China pays more and more attention to the security and management of international students. Ministry of Education in China promulgates the plan of studying in china in September 21st, 2010 which explicitly point out that the total number of international students in Chinese primary schools, middle schools, colleges and universities will reach half a million until 2020, among which that in colleges and universities will be more than 150000.

However, at present, the level of International Student Management System in most universities is not high, and the management efficiency of international students is low.

© Springer Nature Singapore Pte Ltd. 2018
Q. Zhou et al. (Eds.): ICPCSEE 2018, CCIS 902, pp. 623–631, 2018.
https://doi.org/10.1007/978-981-13-2206-8_53

Therefore, we designed NEFU International Student Management and Security Warning System. The development of this system basically meets the requirements of information management and security management for international students in most universities. It's convenient to manage international students' basic information and provide warning for international security.

1 The Purpose and Significance of System Development

1.1 The Purpose of System Development

School of International Education and Exchanges in Northeast Forestry University use this website for warning. In order to assist the college's administrators better to manage the information of international students. The main functions of the system are the management and maintenance of the international students' basic information, students' supervisors' information, major information and security information, leaving application, the dormitory management, the access record and security warning, the relevant information of teachers, administrators and dormitory, and the user login and permission management. These systems implement the maintenance of leaving application and fieldwork according to the type of international student. The information can be used for scholarship security warning and management. This system can meet the basic requirements for international students' information management and security management.

1.2 Project Research Background

In recent years, our country's overall national strength is growing and our political stability has attracted a lot of international students to study in China which caused that the number of international students in China is increasing. Meanwhile, instable and disharmonious factors which may affect national security and campus stability such as differences in cultural background, religious conflict, life changes, etc. are increasing. These factors bring many challenges for the maintenance of security stability. What's more is that some students who came to study in China have little awareness of Chinese law, and some of universities are not able to manage it properly, so that some of the students can't get the proper guidance in time, which has lead to the increasing number of illegal criminal cases in China, violating school regulations, endangering social security, and even endangering the safety of international students and our citizens, and the security management in China is getting more and more difficult. In order to adapt to the new situation, paying more attentions to the research of international students' management mechanism, promoting the safety management system, improving international students' managers' abilities and working efficiency are the most important unsolved problems.

Many universities have made outstanding contributions to establishing international student's safety management systems and forming institutions. For example, Beijing Language and Culture University proposed to strengthen the law publicity and education for international students, further improved the school network information security strategy, and completes the work of sorting out and documenting university's personnel, places, events, objects, and organizations in 2015. These provide multi-dimensional support for security work. Zhejiang University actively explored the security and stability management system for international students in China and formed a three-dimensional management model with better management efficiency: strengthen regulations and discipline education, strengthen the establishment of international students' management team and improve students' working mechanism, normalize security checks, form the training mechanism for survival of the fittest, pay close attention to students, and use the news media. However, there are still many deficiencies in the safety management of international students, which are mainly shown in: (1) the management system is not perfect. The safety management of international students in most local colleges and universities is mainly carry out in the form of "problems arise - solving problems", but there is no systematic and scientific management of international students from the points of security situation analysis, early deployment, key monitoring, security early warning and real-time management. (2) Information exchange is not smooth. There is no effective linkage mechanism and enough cooperation between the international students management department, school functional management departments, local police stations, and national security agencies, etc. (3) Some international students' management personnel lack management capabilities, are not careful in their operations, and are concerned about safety man-agreement accidents, leading to the phenomenon of "don't want to deal with, don't dare to manage, have no ability to handle" of foreign-related cases.

In the research on the security management mechanism of international students, Liu [1] conducted an in-depth study on the security management issues of college students abroad and their countermeasures expounded the importance of early warning mechanisms, and at the same time dealt with the emergencies, cultures conflict and accident harm that occurred during the period when international students were studying in China to establish a security warning mechanism. The early-warning mechanism can resolve contradictions and conflicts, earn a buffer time for the outbreak of events, and prepare for the battle, narrow the scope of the outbreak of security incidents, and realize the transformation from "crisis treatment" to "crisis prevention." Zheng [2] conducted a study of the problems in the campus security management of African international(overseas) students and found that the school's response to campus security incidents lags behind, (at the same time school) lacks management experience and strategic planning is not in place. Michael Lester (crisis management expert, US.) once said: "Prevention is the best way to solve the crisis", it profoundly reveals the necessity of "emphasizing prevention." Therefore, it is urgent to analyze the characteristics and causes of emergencies, effectively prevent and properly handle the emergencies of international students coming to China and establish an international student security management system to implement a standardized, systematic, and institutionalized campus security management system.

At present, although there are many research cases in the development of international student information management automation systems, and at the same time they are all aware of the importance of security management and early warning, in the aspect of early warning of student security, there are no established systems. Therefore, in order to standardize and simplify the security management of international students, the research group has designed and developed a international student management and security early warning system, aiming at the personal security of international students by regulating the individual behavior of international students, including the comprehensive transaction process of travel, Leaving application, registration after the leave, and accordingly information fulfills the function of security warning for international students to ensure the security of international students and national security.

2 Design of the System

By researching and analyzing the requirements of the system, knowing the detailed requirements of the warning system in functions, performance and reliability and understanding the procedure of international students' management and detailed procedures, the author designed the development platform, functional modules, detailed system and database. On the basis of those, the author developed and implemented the system.

2.1 System Development Platform Design

This International Student Management and Security Early Warning System was developed in Visual 2010 with SQL Server 2008 as database management system. Visual Studio is a development environment launched by Microsoft Corporation and is currently the most popular Windows platform application development environment. Microsoft Visual Studio 2010 is a relatively comprehensive and stable program development software, which can be developed in many languages such as C, C++, C++.net, C#, VB.NET, VB. Script, etc. In Visual Studio 2010, which introduced preprocessing text templates, it has become easier to generate any type of text file from an application. It also improves the support for code integration through better integration with the production system, so that the generated source code will always be updated after any changes to the source model. The study mainly uses C# language and ASP.NET technology to develop a international student management and security early warning system based on the B/S architecture, which realizes the efficient management of basic information of international students and realizes security management and early warning for students according to "student apply for a leave and register after the leave" module.

The system database management system uses SQL Server 2008 R2, which is a relational database management system launched by Microsoft Corporation. SQL Server 2008 releases data directly from structured, semi-structured, and unstructured documents into a database. Thus, you can query, search, synchronize, report, and analyze data. Data can be stored on a variety of devices, from the largest server in the data center to desktop computers and mobile devices, and it can control data regardless of where the data is stored.

2.2 System Modules Design

The user type of NEFU international Student Management and Security Early Warning System contains administrator, manager, student and teacher four types. System function modules contain login module, security early warning module, scientific research security management module, international student module and system management module. System function modules are shown in Fig. 1.

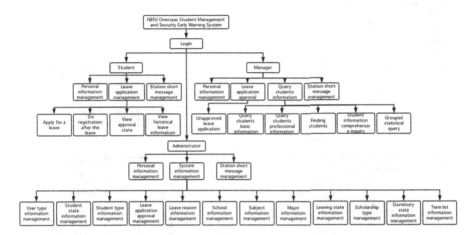

Fig. 1. System function module

2.3 Design of Main Module

(1) International students' security warning module

The International Students' Security Warning System focuses on the management and early warning of students' security. It has developed a "Leaving Application and Registration" module. Students must submit application to the college before leaving school. The application needs to be approved at all levels. The application is first checked by counselors. After approving, it will be checked by the Educational Administrator Department. Then turn to the Life Service Department. After being approved, it will be checked by the college dean. Finally, after being approved at all levels, the application takes effect. The authority of approval of each level is maintained dynamically by the system administrators in the administrator module.

When a department has more than one personnel with the authority of approval, if one of them approves the application, others needn't to approve again. The approval process is shown in Fig. 2.

After the student's holiday is completed, he must do registration after the leave. After clicking on the register button, the system will find the student's leaving application automatically and determine whether the type of the application is scientific research and then register. Then the graduate tutor or the counselor needs to check the registration. According to whether the date is late, it is confirmed as on time or as expired. Meanwhile the system will record this information to student's historical leaving records for later inquiry and management. The system scans all unregistered leave automatically every day. If one student doesn't return until the expiration date,

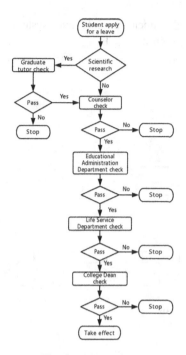

Fig. 2. Approval process

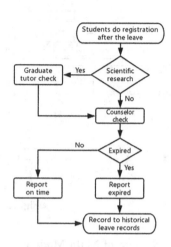

Fig. 3. Do registration process after the leave

the system will send a warning message to counselors and the college dean for reminding them to verify relevant information. So that the system can assist all levels of management personnel in approving various leave affairs more scientifically. Students do registration after the leave process is shown in Fig. 3.

(2) Scientific Research Security Management Module

Scientific research leave includes scientific field research/investigation and scientific activities such as various domestic conferences, international conferences and etc. This system monitors student's scientific research which involves sensitive areas, sensitive research, etc. Checking the research content is the responsibility of the student's school but monitoring the student's research area is the responsibility of this system. When students apply for a leave to do a scientific research, they need to provide research content, research areas, and other relevant information to make management and tracking become convenient. The process of applying for a scientific research leaving is shown in Fig. 3.

The process of applying for a scientific research leaving is mainly the same as above. The only difference is that it needs to be submitted to graduate tutors for approval before they are approved by counselors and others.

(3) Students and Manager Module

The system includes three major functional modules: student module, manager module and administrator module.

Student module: It mainly implements personal information management, apply for a leave, register after the leave and station short message management functions. Students can view and modify some personal information, add leaving application form, submit leaving application form, review leaving application form for approval progress, delete uncommitted leaving application form, do registration after the leave, and view station short messages.

Manager module: It mainly implements personal information management, student information query, student leave application management and station short message management functions. Manager can view and modify some personal information, query student information according to the conditions, approve student's leave application form, view station short messages, and send station short message. Among them, the leave application forms will be approved in the order of counselors, the educational administration department, the student life service department, and the college dean.

(4) System Basic Information Maintenance Module

The basic information of the system is maintained by the administrator module. The administrator module mainly implements personal information management, system information management, station short message management functions. Administrator can review and modify personal information. In the system information management function administrator can administer user type information, Student state information, Student type information, Leave application approval, Leave reason information, School information, Subject information, Major information, Leaving state information, Scholarship type, Dormitory state information, Term list information. The operations to the information are mainly addition, modification, and deletion. Meanwhile administrator can view station short messages and send station short message.

3   The Development and Implementation of the System

In this paper, author developed a international student management and security early warning system with C# language based on ASP.NET technology. It adopted B/S architecture and Code-Behind Model design pattern. In ASP.NET, code-behind is implemented using the new partial classes. Code-behind class is defined as a partial class meaning that it contains only part of the class definition. The remaining part of the class definition is dynamically generated by ASP.NET using the ASPX page at runtime or when the Web site is precompiled. Code files contain only the code that developers have created. The link between the code-behind file and the ASPX page is still established using the @ Page directive. It also provides for a true separation of code and content because there are no instance variable declarations in the code-behind file. All the code files in a project will be compiled into a DLL file, which can protect the code of developers.

In this system, we implemented the following system functions: student user module, teacher management module and administrator module. It not only can manage the information about international students, colleges, majorities and many other things, but also can implement students' security management and early warning

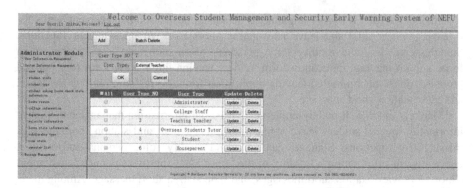

Fig. 4. Administrator module system interface

based on leaving apply and information about reporting back after leave of absence. The administrator module system interface is shown in Fig. 4.

4 Conclusion

In this paper we designed and developed an International Student Management and Security Early-Warning System. It is helpful for college staffs to regulate international students' management in order to manage international student conveniently and effectively. Meanwhile, the international students can also deal with their own personal affairs through this system such as leaving application, reporting back after leave of absence, etc. This system provides many early-warning services about personal safety problems for international student.

But there are still many deficiencies in the study. The system has no ability to manage information about state security, and lacks of models about early-warning and result evaluation as theoretical support. In the study of future, the authors will make further research in two ways. Firstly, we will improve on the system functions about early-warning about state security such as homeland security, information security and social security. Secondly, we will research on a algorithm based on probability to make a prediction about danger degree, build and implement detection methods and evaluation models to improve the performance of the system to make it more convenient for users.

Acknowledgment. This paper is supported by three projects:

1. Heilongjiang Province Higher Education Teaching Reform Project in 2017: "the Application and Practice of Teaching Aided System in the Background of Engineering Education" (SJGY20170145);

2. China Association for International Education (CSFSA) Project in 2017: Study on International Students Management Mechanism Based on the Perspective of National Security (2016-2017Z008);

3. National Undergraduate Innovation Plans in 2018: Design and Implement of international Student Management and Security Early Warning System.

References

1. Liu, F.: Research on security management strategies for international students in colleges and universities. Educ. Watch. **2**(7), 19–21 (2013)
2. Zheng, J., Hou, F., Zhang, L.: Research of security management of African students on university campus. J. Tianjin Vocat. Tech. Norm. Univ. **22**(4), 72–74 (2012)

Performance Prediction Based on Analysis of Learning Behavior

Shaowei Sun, Xiaojie Qian[(✉)], Lingling Mu, Hongying Zan,
and Qing Zhang

School of Information Engineering, Zhengzhou University, Zhengzhou, China
iexjqian@zzu.edu.cn

Abstract. With the rise of large-scale online open courses, the era of MOOC has come and it has created opportunities and challenges for higher education at home and abroad. Many colleges and universities use the combination of MOOC and classroom teaching. Through online and offline synchronization courses, the mixed teaching based on MOOC is formed. A series of learning behavior data has been generated when students participate in the online MOOC system. The resulting data facilitates learning analysis to improve teaching quality and improve learning behavior. This paper collects the learning behavior data from MOOC, uses the Pearson coefficient to select the learning behavior characteristics related to the learning effect, and establishes a learning performance classification model based on the Support Vector Machine (SVM), and predicts the learning performance according to the learning behavior data. The accuracy of the performance forecast was 95.26%.

Keywords: MOOC · Learning analysis · SVM · Forecast

1 Introduction

Massive Open Online Courses (MOOC) is emerging online courses that are based on curriculum and teaching theories as well as web and mobile communication technologies. Teaching methods based on MOOC include flipping classrooms and blended teaching. In recent years, many colleges and universities have carried out mixed teaching attempts based on MOOC [1–3]. Mixed teaching is a teaching method that combines with online and offline teaching resources. It has a lot of advantages and strong operability for a single learning subject with the same knowledge base. Teachers guide students to use online resources to study independently and organize students to discuss when watching the learning video; provide targeted counseling and answering questions which the questions are occurred in the online learning process in the offline classrooms. Teachers in the classroom to outline the content of teaching content, guide students to read textbooks and courseware, ask students to watch the online video tutorial, complete online and offline operations and participate in online discussions so that they can identify problems and solve it. Then teachers concentrate on difficulties in courses and teaching to explain and discuss issues that students cannot solve in the classroom. This teaching model combines the advantages of traditional teaching methods and E-Learning methods. It not only can play the leading role of teachers in

© Springer Nature Singapore Pte Ltd. 2018
Q. Zhou et al. (Eds.): ICPCSEE 2018, CCIS 902, pp. 632–644, 2018.
https://doi.org/10.1007/978-981-13-2206-8_54

guiding, inspiring, and monitoring the teaching process, but also can fully reflect the students' initiative, enthusiasm, and creativity as the main body of the learning process [4]. Practice shows that hybrid teaching is feasible and effective in engineering courses. Comprehensive utilization of online and offline teaching methods can accomplish teaching tasks better and achieve teaching goals.

Learning Analytics aim to understand and optimize learners' learning environments through measures, collects, analyzes, and reports learners' learning history. This can provide feedback that can be applied for learners, teachers and schools. Learning analytics play an important role when it was embedded in the learning process. It can also extract information from the learning process, provide information for teachers and schools and grasp student learning status. Nowadays, the rapid development of the Internet has driven the research of big data. More and more educational researchers are aware of the potential of data analysis in improving the learning experience. Therefore, Learning Analytics (LA) gradually separates itself from the analysis domain and absorbs lots of analysis methods such as data mining, social network analysis, statistical analysis e.g. LA regarded as an independent emerging field in 2010. Since the formal formation of the field in 2010, LA has been applied to all levels of the education system.

In the learning process based on the MOOC, many learning data and academic evaluation data will be generated and these data will be fully utilized to explore the correlation between learning behaviors and learning results. Use online learning data to provide support for teaching and learning has important practical significance. In the mixed teaching practice, the design of teaching resources for MOOC is aimed at non-discrete learners and lacks the design of experimental links. Teachers need to adjust course design and teaching content according to feedback and academic performance of students' online learning behavior. In the teaching practice, teachers generally design the teaching content by reading the speeches of the students in the discussion area, the quantity and quality of the tasks completed, and observing the students' time for watching videos to urge the students to study independently. Without analysis of learning data or the characteristics, these teaching tasks have a certain degree of blindness. With the advancement of data analysis technology and the success of machine learning algorithms, the data generated during the use of MOOC teaching can be more effective and targeted to guide teaching and learning.

Since 2016, Zhengzhou University has carried out MOOC, and Assembly Language has become one of the first MOOC courses. After analyzing the teaching mode of MOOC and "flipped classroom", the course adopted a mixed teaching method based on MOOC, and achieved a successful [5]. This paper aims at the mixed teaching in the course of the course design of the assembler language course of Zhengzhou University. Based on all the data collected by the MOOC, this paper uses the machine learning algorithm to predict the classification of the learning achievement, assists teachers in the future teaching design, improves the teaching content, and urges the students to learn.

2 Related

Learning analysis is one of the research hotspots of data mining technology in the field of education in recent years. Learning analysis aims at collecting and analyzing online learning data, digging the relationship between learning activities and learning outcomes to help educators and learners improve the teaching and learning process [6].

Various attempts have been made at home and abroad in the field of learning and analysis theory and practice. Purdue University's "course signal" system [7] uses learning analysis to predict academic performance, urges learners to focus on their own learning and enables teachers to push learning content based on individual characteristics of students. Learning analysis helps education build on the precise grasp of learners and it provides a theoretical basis for the adjustment of curriculum construction, instructional design and teaching process. Teachers can make clear instructional intentions based on learning analysis, thus designing targeted and effective teaching interventions [8]. Learning analysis can also be used to achieve better course design and adjustment [9]. It has the advantages in monitoring, predicting and adjusting the academic performance [10].

In recent years, China has also conducted research on the analysis of learning behavior. Zhang Yu et al. pointed out in 2013 that although the analysis and research of online curriculum big data is facing challenges and difficulties, it will greatly push forward the development of education measurement and learning analysis and promote the improvement of education quality [11]. After 2015, domestic scholars conducted theoretical research and practical application of the online learning platform. Li et al. [12] used Tobit and Logit quantitative analysis models based on the course data of the XuetangX to analyze the course participation and completion of MOOC learners. The study found that learners' learning data affects their learning behavior and different learning behaviors can affect their learning outcomes. Peking University analyzed and excavated the massive learning behavior data of 6 courses on Coursera with about 80,000 people involved [13]. According to the characteristics of learning behavior in Chinese MOOC, students are categorized for a deeper look at the relationship between learning behavior and learning outcomes. On this basis, through the selection of a number of typical behavioral characteristics of learners, predict their final learning outcomes. Zhao et al. [14] constructed a system for learning behavior analysis and recommendation. The system uses Web mining technology to conduct data classification mining on learning behavior from three aspects of learning content, learning path and space usage record. Through the similarity comparison with learning behaviors of excellent learner paths and path knowledge points, personalized learning for learners. Sun et al. [15] achieved a subdivision prediction of adult degree English test scores by analyzing the results and learning information of online degree education for undergraduate students in English related courses. They first used the K-means algorithm to cluster the scores of existing student English test scores, determined a more specific score distribution interval. Then, utilizing the C5.0 algorithm in data classification, Using this fractional interval as a prediction target, a subdivision forecasting rule for scores is constructed, established a score prediction system for adult degree English test, analyzed the importance of related variables in performance prediction and proposed a corresponding strategy to improve the English learning level of online education for undergraduate students and achievement of adult degree English test. After 2017, learning analysis studies for mixed teaching is occurred. Zhang et al. [16] proposed learning support services based on MOOC after researching on hybrid teaching learning support services at home and abroad, including educational technology support, environmental support, resource support, management support and interactive activity support. Then analyze every category deeply. Li et al. [17] classified online learning behaviors and used factor analysis, regression analysis and other LA techniques to quantitatively analyze the online learning representation data of mixed teaching cases and constructed a student-related online learning behavior and learning performance-related model. Ding et al. [18] adopted two different levels of teaching intervention strategies in a mixed-learning course in colleges; successfully predict the grades of the students' tasks. The results of the study show that teaching intervention has a positive effect on students' learning promotion. Tang [19] collected the data from XuetangX to design a content-aware algorithm framework using content based users' access behaviors to extract user-specific latent information to represent students' interest profile after systematic investigation. Employing the curriculum preconditions of collaborative filtering strategies, a course recommendation algorithm was developed based on the user's interests and the characteristics of the population. The algorithm has been deployed as a new feature on XuetangX.

The above research has been predicted for the achievement of learning, but in the stage of data analysis, it does not make a thorough inquiry into the stage of data analysis and feature selection, and predicts the classification of the results. On the basis of the study of the previous work, this paper, through the hybrid teaching model of the Assembly Language of Zhengzhou University,[1] uses the learning data produced by the MOOC to observe the characteristics of the data distribution, and uses the Pearson coefficient to observe the correlation between the data characteristics and the results, and select the data characteristics and establish the SVM. The model is used to predict and analyze the results of the data. Through the prediction of the classification results, the students who have poor results are intervened to improve the behavior of the students in their learning process. Through the analysis and evaluation of the learning data, we hope to provide guidance for the future hybrid teaching mode.

3   Data Analysis

This section gives a descriptive description of the data used in this article, and selects the characteristics that have an obvious impact on the academic achievement based on the relationship between the different data and the academic achievement.

---

[1] Zhengzhou University Education Teaching Reform Research and Practice Project.

3.1   Data Source and Structure

This data comes from the data of the *Assembly Language* program of Zhengzhou University, which includes 325 records of the students' learning data and students' grades from the 2015 to 2017 academic years of the course. The fields of each record are shown in Table 1.

Table 1. Feature names and meanings

Feature name	Characteristic meaning
Course visits	Total number of landing visits
Task completions	Number of course tasks completed
Percentage of mission completion	Total number of course tasks completed as a percentage of total tasks
Video completions	Course video completion quantity
Video viewing durations	Course video watch duration
Homework completions	Coursework completed
Video rumination averages	The ratio of the actual video viewing duration to the original video duration

3.2   Feature Selection

This section mainly aims at the selection of data and the selection basis of the feature in the sample data. It presents the source and distribution of data in two aspects, and calculates the correlation coefficient.

Achievement Distribution
Figure 1 is the student sample performance interpretation. According to each 10 points, the results are divided into 80–90 points, the average number is 83, the definition below 70 is ordinary, and more than 70 points are good. The distribution is shown in Fig. 1.

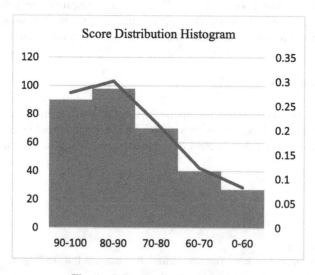

Fig. 1. Score distribution histogram

Correlation Analysis and Feature Selection

In order to build a good performance classification model, we need to select features. In this paper, there is a linear relationship between the characteristics and the performance, according to the Shapiro-Wilk test conforms to the normal distribution ($P > 0.05$), and there is no abnormal value. Pearson correlation analysis is used to evaluate the relationship between the results and the data under the various characteristics. Results as shown in Table 2, there is a moderate positive correlation between the number of courses, the number of job completion, the number of tasks, the average value of video ruminant, the completion of video, and the length of video viewing. Table 2 is the correlation coefficient.

Table 2. Main features and pearson coefficients of performance

	Course visits	Homework completions	Task completions	Video rumination averages	Video completions	Video viewing durations
Scroe	0.342**	0.445**	0.450**	0.241**	0.422**	0.241**

Tip. **. There is a significant correlation at the 0.01 level (bilateral).

Table 2 shows the specific value of the Pearson coefficient of the score and the characteristics. R indicates that r2 represents the ratio of the change of the characteristics to the score. "**" represents a moderate positive correlation between the number of courses, the number of assignments, the number of tasks, the average of the video ruminant, the video completion and the viewing length of the video. The correlation coefficients are 0.342, 0.445, 0.450, 0.241, 0.422 and 0.241, respectively. Taking the number of curriculum visits as an example, the specific meaning can be expressed as "a moderate positive correlation between the number of curriculum visits and achievements in the related factors affecting the performance, the correlation coefficient $r = 0.342$, $r2 = 0.12$, indicating that the number of course visits can explain the fluctuation of 12%".

Through the Pearson coefficient of Table 2, it is found that each feature has a moderate positive correlation with the score. Considering the relationship between the characteristics and the results, and combining with the correlation coefficient calculated in Table 2, 6 of the main feature data are selected, such as the number of courses, the completion of the job, the number of tasks, the average value of the video ruminant, the number of video, and the time of viewing the video as the feature of the prediction and classification. Among them, the "percentage of task completion" is not selected in Table 1, because there is no significant difference in the distribution of data, and it is weakly related to the classification of grades, so this feature is abandoned. Figure 2 is the data distribution of "percentage of task completion". From Fig. 2, we can see that there is no significant difference in the data of 93.6% of the data samples studied, so this feature is not selected as the influencing factor of the score.

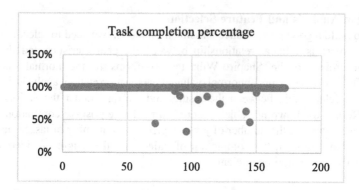

Fig. 2. Task completion percentage

4 Performance Prediction Model Based on SVM

Support Vector Machine (SVM) is first proposed in Cortes and Vapnik [20] equal to 1995. It shows many unique advantages in solving small sample, nonlinear and high dimensional pattern recognition, and can be applied to other machine learning problems such as function fitting. SVM is a classification algorithm, by finding a classification plane, separating the data on both sides of the plane, so as to achieve the purpose of classification. As shown in Fig. 3, a straight line represents a classification plane trained to separate data effectively.

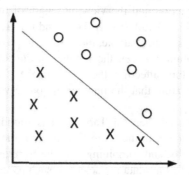

Fig. 3. SVM classification diagram

In the case of the hyperplane $\omega^T * x + b = 0$, $|\omega^T * x + b|$ indicates the distance from the point distance to the hyperplane, and the hyperplane is the two classifier. If $\omega^T * x + b > 0$, the judgement Class Y is 1, otherwise the decision is -1. This leads to the definition of function interval.

$$\hat{\gamma} = y(\omega^T * x + b) = yf(x) \tag{1}$$

Where y is a class marking value for training data, if y $(\omega^T * x + b) > 0$ shows that the value of the prediction is the same as the value of the mark, the classification is correct and the greater the value, the farther the point is from the plane, the more reliable the classification is. This is a function definition for a single sample. For the whole sample set, we need to find the minimum interval between all the samples as the interval of the entire set.

$$\hat{\gamma} = min\hat{y}i(i = 1, \ldots n) \tag{2}$$

When both ω and b are shrink or enlarge M times, the hyperplane does not change, but the function interval follows the change of ω and b. Therefore, we need to add constraints to make the function interval fixed.

Based on the distance formula from point to plane and the formula of the $\omega^T * x + b = 0$ plane, the definition of geometric interval is derived.

$$\gamma = \frac{\omega^T * x + b}{||\omega||} = \frac{f(x)}{||\omega||} \tag{3}$$

Similar to the function interval, to get the absolute value of gamma, the geometric interval is the normal partition of function interval divided by ω:

$$\tilde{\gamma} = y\gamma = \frac{\hat{\gamma}}{||\omega||} \tag{4}$$

Let the function interval equal 1 to get the original definition of the maximum margin classifier.

$$max \frac{1}{||\omega||}, s.t., y_i\left(\omega^T * x_i + b\right) \geq 1, i = 1, \ldots, n \tag{5}$$

The maximum interval classifier is the classified hyperplane, which is equal to the max (geometric interval), and the function interval is assumed to be 1, and the maximum interval hyperplane can obtain $max \frac{1}{||\omega||}$, and the constraint condition is that the interval is the minimum of the interval function of all the sample points.

$$y_i\left(\omega^T * x_i + b\right) \geq 1, i = 1, \ldots, n \tag{6}$$

Find the largest $\frac{1}{||\omega||}$.

Convert the $\frac{1}{||\omega||}$ maximum to a quadratic convex function optimization problem, solve $min \frac{1}{2}||\omega||^2$.

The solution of the maximum interval classifier can be converted to an optimization problem above.

$$min \frac{1}{2}||\omega||^2, s.t., y_i\left(\omega^T * x_i + b\right) \geq 1, i = 1, \ldots, n \tag{7}$$

Lagrange equation is used to solve the equation, as shown in formula 8.

$$\mathcal{L}(\omega, b, \alpha) = \frac{1}{2}||\omega||^2 - \sum_{i=1}^{n} \alpha_i \left(y_i \left(\omega^T * x_i + b \right) - 1 \right) \tag{8}$$

α_i means Lagrange multiplier in Eq. (8).

Using the conclusion of duality, we obtain partial derivatives of $\mathcal{L}(\omega, b, \alpha)$ on ω and b:

$$\frac{\partial \mathcal{L}}{\partial \omega} = 0 \Rightarrow \omega = \sum_{i=1}^{n} \alpha_i y_i x_i \tag{9}$$

$$\frac{\partial \mathcal{L}}{\partial b} = 0 \Rightarrow \omega = \sum_{i=1}^{n} \alpha_i y_i = 0 \tag{10}$$

By replacing the above two equations into $\mathcal{L}(\omega, b, \alpha)$ functions, it is converted into the following definition of the maximum value of α_i in this formula, no ω, b parameter solution, because ω, b can be expressed in α_i, and the value of ω and B can be solved after the calculation of α_i.

$$\max_{\alpha} \sum_{i=1}^{n} \alpha_i - \frac{1}{2} \sum_{i,j=1}^{n} \alpha_i \alpha_j y_i y_j x_i^T x_j$$
$$s.t., \alpha_i \geq 0, i = 1, \ldots, n \tag{11}$$
$$\sum_{i=1}^{n} \alpha_i y_i = 0$$

Based on the data of the learning achievement classification model of support vector machine (SVM), 6 feature data are selected after third sections of Pearson coefficient, and the data sample data are divided into training set and test set by preprocessing the selected data. According to the working principle of the above SVM, we predict and classify the data. The classification of the results is divided into two categories: "pass" and "fail". The two classification classifier is constructed with training data, and then the final two classifications of the test data are predicted. Figure 4 is a learning score classification model based on support vector machine (SVM).

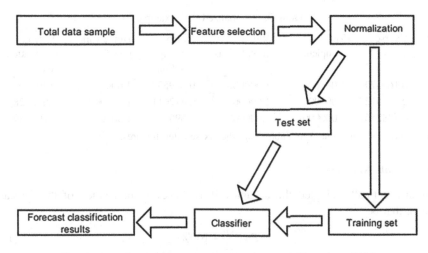

Fig. 4. Learning performance classification model based on SVM

5 Experiment

5.1 Experimental Setup

The experimental data of this paper are the data of the *Assembly Language* program design of Zhengzhou University, which includes 325 records of the students' learning data and students' achievements in the 2015 to 2017 academic years of the course. The characteristics are the number of course visits, the number of tasks completed, the number of video completion, the duration of video viewing, the number of job completion, and the ruminant average value of data. The task of this paper is to classify data as test sets and predict them. Based on the Python3.5 version, we use the SVM module in Python to classify and predict the sample data.

5.2 Data Preprocessing

The 6 types of feature data selected in this paper include 325 students learning the design of the Assemble Language. The true value of the data is uneven, which is not conducive to the analysis of the data in the process of observation. Before the experiment, normalization of data was carried out so that the range of data was distributed in (0, 1), which was convenient for observation and calculation. The normalization calculation is as follows:

$$N = \frac{fv - min(cf)}{max(cf) - min(cf)} \tag{12}$$

(*Note: N: Normalization, fv: feature value, min (cf): current feature minimum, max (cf): current feature maximum.)

Table 3 showed the normalized data examples.

Table 3. Normalized data examples

ID	Course visits	Task completions	Video completions	Video viewing durations	Homework completions	Video rumination averages
1	0.166370	1.000000	1.000000	0.409532	1.000000	0.408897
2	0.076512	1.000000	1.000000	0.472611	1.000000	0.472028
3	0.788256	1.000000	1.000000	0.698441	1.000000	0.698149

* Tip: The heads of Table 3 in Table 2 are the six selected features.

5.3   Result Analysis

This article finally adopted the accuracy P as the evaluation index of the forecast results, that is:

$$P = \frac{cn}{tn} * 100\% \tag{13}$$

(*Note: cn: correct sample number, tn: total number.)

In this paper, before the calculation of the correct rate, a pre-processing of the classification of the results of the real values is performed. By demarcating the scores, the results of the real values are classified into two types of passing and failing, and 1 is used to pass and 0 is not accepted. After that, perform the same processing for the predicted results, too. Then the number of the same 1 and 0 between the two classification results is compared to calculate the exact number of predictions, and then the ratio of the total number of samples is calculated. Get the accurate rate.

In the experiment part, according to the 6 features selected in the third part, the number of courses, the number of tasks, the number of video, the length of video viewing, the number of job completion, and the average value of data ruminant are trained and predicted. At the same time, linear regression and logistic regression were used to predict the data. The three prediction results are compared with the true values, and the accuracy is calculated. The accuracy rate is compared to Table 4.

Table 4. Comparison of model accuracy

Model	Rate (%)
Linear regression	91.69%
Logistic regression	90.15%
Support vector machine	95.26%

This article trains students in the 2015 academic year, and then predicts the student's performance category for the 2017 school year. The predicted score category is compared with the student's real performance in the 2017 academic year, reaching an accuracy of 95.26%. In addition, the linear regression algorithm and logistic regression algorithm were used to predict the data, and compared with the real results, the accuracy rate reached 91.69% and 90.15% respectively. By comparing the accuracy rate, it is not difficult to find that the accuracy of the linear regression and the accuracy of the support vector machine are significantly higher than the accuracy of the logistic regression under the same experimental data. The reason for the analysis is that in the linear regression prediction experiment, the application of the least square method reduces the loss in the linear regression iteration stage, and finally improves the accuracy of the linear regression result. In the comparison of linear regression and SVM, the accuracy of SVM is even better. The reason of the analysis is that because the feature extraction of data is more accurate in the algorithm process of the SVM, the features determined by the Pearson coefficient have higher correlation with the change of the performance, and the relationship is more closely related, so it can be more accurate in the prediction stage. Comparing the three classification prediction methods, it can be concluded that the SVM-based learning performance classification model is the most ideal for performance prediction.

6 Conclusion

Under the background of MOOC development trend, the reform of college education is imminent. The traditional lecture style teaching is too simple for student to learn. Usually students have no chance to exchange ideas and ask questions in class. Hybrid teaching has once again become a significant issue in the teaching reform of many colleges and universities. Many studies and practices have shown that the teaching effect of hybrid teaching is superior to other teaching methods. The main task of this learning analysis focuses on learners and base on the learner's behavior data collected on MOOC. Then, develop machine learning model with 6 character features to predict and classify the learners' academic performance. Finally, comparing with others', SVM is the most appropriate training model which got the highest accuracy 95.26%. This system training the features of leaners to make a prediction and classification. The results will help teachers to understand the study process of each student; and encourage them to adjust the teaching style for whole class or even making suitable teaching plans for each student.

However, we found that this study mainly aims on prediction and classification of learner's learning behavior data during the learning process and based on the learning data of 325 learners of Assembly Language [5], the dataset is not big enough.

In addition, there are still some problems in the mixed teaching. Curtis Bunker [16] pointed out four main issues of the problems, teacher-mixed teaching and training, teachers' work and students' learning burdens, learners' information literacy issues and the degree of difficulty in grasping the mix (Zhan Zehui, et al. 2009). The six main issues identified by Charles Graham (2005) are the issues of on-site interaction, learners' choice and self-management, mixed teaching support and training model, balance between technological innovation and output efficiency, digital divide, and cultural adaptation. This article starts from the perspective of analyzing learners and strives to apply MOOC-based hybrid teaching methods in future education research. For future work, first we will collect more data from learners of other faculty. Then test other features to find the higher coefficient features; and test different learning algorithms with different hyper-parameter setting to develop the most appropriate model. Although, there still contains problems of the complex application for hybrid teaching, in future, MOOC-based hybrid teaching indubitable provide teachers a better chance for instructional design and curriculum arrangements, grasp the situation of classroom teaching; and combine with the online teaching closely to convey the teaching content better.

References

1. Li, S.: Research and practice of blending teaching model based on MOOCs in programming classes–a case study of "C ++ programming". Softw. Guide **16**(1), 189–191 (2017)
2. Fu, J., Wu, S., Zhao, Y.: Design and practice of blended teaching mode based on MOOCs. Commun. Vocat. Educ. **15**, 43–45 (2016)
3. Yuan, M.: Design and practice of mixed teaching course based on MOOCs–a case study of the course "computer application foundation". Educ. Forum **28**, 161–162 (2016)
4. He, K.: Observe the new development of educational technology theory from blending learning. China Educ. Technol. **4**, 10–15 (2004)
5. Mu, L., Qian, X.: Mixed teaching practice and thinking based on MOOCs. Comput. Educ. **9**, 82–86 (2017)
6. Zheng, X., Yang, J.: The learning analytics in higher education: progress, chal-lenges and future directions. China Educ. Technol. **2**, 1–7 (2016)
7. Arnold, K.E., Pistilli, M.D.: Course signals at Purdue: using learning analytics to increase student success. In: International Conference on Learning Analytics and Knowledge, pp. 267–270. ACM (2012)
8. Lockyer, L., Heathcote, E., Dawson, S.: Informing pedagogical action aligning learning analytics with learning design. Am. Behav. Sci. **57**(10), 1439–1459 (2013)
9. Greller, W., Drachsler, H.: Translating learning into numbers: a generic framework for learning analytics (Report). Educ. Technol. Soc. **15**(3), 42–57 (2012)
10. Dietzuhler, B., Hurn, J.E.: Using learning analytics to predict (and improve) student success: a faculty perspective. J. Interact. Online Learn. **12**(1), 17–26 (2013)
11. Zhang, Yu., Li, Y.: Learning analytics and education assessment based on big data generated by MOOCs. Tsinghua J. Educ. **34**(4), 22–26 (2013)
12. Li, M., Xu, S., Sun, M.: The analysis of learning behaviors in MOOCs–a case study of "circuit principle" course. Open Educ. Res. **21**(2), 63–69 (2015)
13. Jiang, Z., Zhang, Y., Li, X.: Learning behavior analysis and prediction based on MOOC data. J. Comput. Res. Dev. **52**(3), 614–628 (2015)
14. Zhao, Z., Li, Z., Zhou, D., Zhong, S.: Research on recommendation system and analyze learning behavior in intelligent learning space. Mod. Educ. Technol. **26**(1), 100–106 (2016)
15. Sun, L., Zhang, K., Ding, B.: Research and realization of subdivision prediction of network education based on data mining–a case study of undergraduate adult degree English test. Distance Educ. China **12**, 22–29 (2016)
16. Zhang, C., Li, L.: MOOC based learning support services in blended teaching. Distance Educ. China **2**, 66–71 (2017)
17. Li, X., Liang, Z., Zhao, N.: The influence of online learning behavior based on mixed learning performance. Mod. Educ. Technol. **2**, 79–85 (2017)
18. Ding, M., Wu, M., You, J., et al.: Research on teaching intervention based on academic performance prediction. Distance Educ. China **4**, 50–56 (2017)
19. Jing, X., Tang, J.: Guess you like: course recommendation in MOOCs. In: The International Conference, pp. 783–789 (2017)
20. Cortes, C., Vapnik, V.: Support-vector networks. Mach. Learn. **20**(3), 273–297 (1995)

Author Index

Aliaosha, Ye II-389
An, Bo I-33

Bai, Yang I-589
Bao, Zhenshan I-444, I-452
Bian, Wanhong II-240
Bian, Yiming I-599
Bin, Chenzhong II-213
Bourobou, Mickala II-273
Bukhari, Syed Umer I-347

Caifang, Tang II-176
Cao, Han II-605
Cao, Zhen I-575
Chang, Liang I-382, II-213
Chen, Chong I-444, I-452
Chen, Chunyi I-347
Chen, Huipeng II-539, II-564
Chen, Lei II-337
Chen, Ming-rui I-415, II-337
Chen, Ningjiang I-548, II-493
Chen, Wenwu I-638
Chen, Xiuying I-619
Chen, Yi I-269
Chen, Yihang I-279
Chen, Yilin I-212, I-322
Chen, Yong I-484, II-530
Chen, Yushi I-729
Chen, Zhiming II-81
Cheng, Xiaohui I-539
Cheng, Yi I-679
Chi, Lejun II-69

Dai, Qiejun II-431
Dang, Rouwen I-589
Dang, Xiaochao II-312
Deng, Wendong I-512
Di, Xiao qiang I-347
Ding, Chao II-263
Ding, Junfeng II-431
Ding, Linlin I-74
Dong, Lin I-60
Dong, Shuai I-182

Dong, Xiaoju I-560, I-599
Dong, Xinhua I-122
Dong, Yixuan II-27
Dong, Yulan I-269
Dong, Zhibin I-462
Dou, Jian II-389, II-401

Fan, Binbin II-227
Fan, Xin I-560, I-599
Fang, Jinhao I-241, I-253
Feng, Guangsheng I-628, II-467
Fu, Jia-run I-48, I-689

Gao, Hong I-332
Gao, Hui II-10
Gao, Jingyuan II-96
Gao, Xinyan II-552
Ge, Jidong I-140, II-27, II-477
Ge, Wen I-679
Gu, Bo I-512
Gu, Tianlong I-382, II-213
Guan, Yelei I-539
Guan, Zhongyang I-462
Guo, William I-8
Guo, Zhihong I-24

Hamza, Ertshag I-646
Han, Fei I-212, I-322
Han, Hai-yun II-442, II-594
Han, Xu II-552
Han, Yali II-27
Han, Yiliang II-192, II-201
Hao, Zhanjun II-312
He, Anping II-552
He, Heng I-122
He, Jianbiao II-150
He, Linchao I-212
He, Mengting I-140, II-27, II-477
He, Ping I-60
He, Tieke I-493
Hei, Xinhong I-60
Hong, Mingbo I-322
Hou, Weiyan II-1

Hu, Changyu   I-721
Hu, Shuaijun   II-455
Hu, Xue   I-619
Hu, Zeyu   I-738
Hualei, Yu   II-176
Huang, Haojie   I-140
Huang, Kui   II-330
Huang, Lan   II-240
Huang, Qian   I-484, II-530
Huang, Tao   I-24
Huang, Xiang   I-322
Hui, Jingya   I-101
Huo, Xu-lun   I-415

Iwahori, Yuji   I-729

Ji, Hong   II-431
Jia, Boxuan   I-192
Jia, Yawei   II-150
Jiamin, Cheng   II-176
Jiang, Pengyuan   I-738
Jiang, Ting   II-377

Lei, Ting   I-140, II-477
Li, Baogen   I-225
Li, Beibei   II-312
Li, Bochong   I-298
Li, Chao   I-122, II-442
Li, Chenglu   I-599
Li, Chuanyi   II-27, II-477
Li, Dong   I-394
Li, Dongsheng   II-296
Li, Hua   I-548
Li, Hui   II-128
Li, Jialin   II-240
Li, Jianzhong   I-403
Li, Jie   II-273
Li, Jinbao   I-192
Li, Jingwen   II-467
Li, Keliang   II-17
Li, Lingli   I-101
Li, Meishan   I-656
Li, Minghao   I-512
Li, Quansheng   I-359
Li, Ruigai   II-623
Li, Shu   II-455
Li, Songling   II-263
Li, Tianyu   I-525
Li, Wei   I-8

Li, Xiang   II-128
Li, Xiaofeng   I-394
Li, Xiaolong   II-192, II-201
Li, Xiaopei   II-455
Li, Xiaotang   I-611
Li, Xiujuan   I-512
Li, Ya   I-359
Li, Yaping   I-332
Li, Yaqiong   I-298
Li, Ying   II-227
Li, Yongfang   I-382
Li, Yuan-hui   II-442, II-594
Li, Zhanhuai   I-88
Li, Zhikun   II-623
Li, Zhiling   II-330
Li, Zhixin   I-426
Li, Zhongyue   II-477
Lian, Hao   I-493
Liang, Birui   I-548, II-493
Liang, Guanghui   I-279
Lin, Jinxiu   II-69
Ling, Jian   II-431
Ling, Peitang   I-140, II-477
Liu, Aili   I-74
Liu, Chao   II-1
Liu, Chenglin   II-359
Liu, Dekuan   II-27
Liu, Fudong   I-279
Liu, Haiyang   I-729
Liu, Hengzhu   I-505
Liu, Hongwei   II-539, II-545, II-564
Liu, Lei   I-426
Liu, Shengpeng   II-227
Liu, Wenbin   I-548
Liu, Xia   I-415, II-337, II-350, II-442, II-581,
       II-594
Liu, Xiaojiao   I-60
Liu, Xiaoli   II-616
Liu, Xinyu   I-738
Liu, Xu   I-347
Liu, Yang   I-628
Liu, Yanhong   II-377
Liu, Yong   II-41, II-55
Liu, Yongchuan   I-122
Liu, Yongjiao   II-150
Liu, Yuan   I-359
Liu, Zhaowen   I-1
Long, Sun   II-161
Lu, Hong   I-225
Lun, Lijun   I-371

Luo, Bin   I-140, II-27, II-477
Luo, Jizhou   I-525, II-296
Luo, Mengting   I-212
Luo, Wenjie   I-560
Luo, Yiqin   I-382
Lv, Hehe   II-17
Lv, Hongwu   I-628, II-467

Ma, Dongxiang   II-573
Ma, Jingyi   II-108
Meng, Fanshan   I-332
Mirador, Labrador   II-1
Mu, Lingling   II-632

Ni, FuChuan   I-706
Ning, Bo   I-308

Ou, Jiaxiang   II-263
Ou, Zhi-peng   II-581

Phatpicha, Yochum   II-213
Pi, Dechang   I-269

Qi, Wanhua   I-88
Qi, Yanwei   I-165, II-505
Qian, Xiaodong   I-241, I-253
Qian, Xiaojie   II-632
Qiao, Xueming   II-359, II-377
Qiao, Zhenjuan   II-359
Qin, Zemin   I-493
Qiu, Zhao   I-415, II-337
Qiu, Zishan   II-10

Rao, Guanjun   I-382
Reika, Sato   II-330
Ren, Honge   I-646
Ren, Yan   I-359
Rong, Yiping   II-377

Sayed, Elmustafa   I-646
Shan, Zheng   I-279
Shi, Jinling   II-17
Shi, Shengfei   I-525, II-296
Si, Jiarui   II-455
Song, Baoyan   I-74
Song, Hangcheng   I-493
Song, Mengjiao   II-41, II-55
Song, Nan   II-416
Song, Qing   I-666

Song, Yang   II-1
Su, Rui   I-560
Su, Wei   I-462
Su, Yang   I-638
Sun, Guoqiang   I-359
Sun, Li   II-455
Sun, Lin   II-17
Sun, Mingxin   I-74
Sun, Muzhen   II-136
Sun, Shaowei   II-632
Sun, Wenping   II-213

Tan, Yiming   II-81
Tang, Mao   I-628
Tang, Xiang   I-599
Tang, Yao   II-359
Tang, Yusi   I-548
Tao, Wei   II-120
Thomas, Serge   II-273
Tian, Long   I-212, I-322
Tian, Qihua   I-477

Wan, Fang   II-337
Wang, Aili   I-721, I-729
Wang, Bin   I-1, I-60
Wang, Chaoliang   II-401
Wang, Chun-zhi   I-48, I-689
Wang, Haiquan   I-33
Wang, Haoyong   II-108
Wang, Hongbin   I-575, I-589
Wang, Hongzhi   I-332, I-403, I-525, II-296
Wang, Huiqiang   I-628, II-467
Wang, Jianfeng   I-575, I-589
Wang, Jifu   II-623
Wang, Mingwen   II-81
Wang, Qing   II-539, II-564
Wang, Wei   II-17
Wang, Xin   I-426
Wang, Xingang   I-182
Wang, Yang   II-69
Wang, Yaogang   II-455
Wang, Yepei   II-41
Wang, Yiliang   II-359
Wang, Ying   I-721
Wang, Yu   II-545
Wang, Zhizhong   II-505
Wen, Ya   II-623
Wu, Jia   II-1
Wu, Jiehua   I-151

Wu, Xuguang   II-201
Wu, Yiqi   I-212, I-322

Xiao, Wenyan   II-81
Xie, Hu   II-431
Xie, Jun   I-738
Xie, Xiaolan   I-24
Xie, Zhaoyuan   I-512
Xin, Xueshi   I-666
Xiong, Houbo   I-122
Xu, Bing   II-108
Xu, Chunming   II-120
Xu, Hui   I-192, I-371
Xu, Lei   II-539, II-545, II-564
Xu, Li   I-679
Xu, Peiling   II-530
Xu, Pengfei   II-431
Xu, Tao   I-512
Xu, Yaoli   I-88
Xu, Youchang   II-493
Xue, Jiamei   I-656
Xue, Li   II-161
Xue, Ruini   I-462
Xv, Jingjing   II-17

Yan, Ling-yu   I-48, I-689
Yang, Bohong   I-225
Yang, Guangya   I-619
Yang, Han-Tao   II-350
Yang, Hongwu   II-416
Yang, Hongxi   II-455
Yang, Jian   I-359
Yang, Jie   II-573
Yang, Jing   I-371
Yang, Lei   I-241, I-253
Yang, Lu   I-666
Yang, Min   I-666
Yang, Muyun   II-108
Yang, Yan   II-96
Yang, Yi   I-698
Yang, Zi   I-308
Yao, Linxia   I-140
Ye, Fangbin   II-401
Ye, Feng   I-484, II-530
Yi, Lu   II-161
Yochum, Phatpicha   I-382
Yu, Haidong   I-477
Yu, Liwei   I-347
Yu, Ting   I-738

Yuan, Lu   I-48, I-689
Yuan, Rao   II-161, II-176
Yue, Tianbai   I-332

Zan, Hongying   II-632
Zang, Tianlei   I-738
Zeng, Wen   II-128
Zhan, Haomin   I-575, I-589
Zhang, Chenlin   II-81
Zhang, Chunyan   II-250
Zhang, Dejun   I-212, I-322
Zhang, Deyang   II-192, II-201
Zhang, Dongjia   I-575
Zhang, Fang   II-505
Zhang, Hanlin   II-493
Zhang, Haocheng   I-599
Zhang, Haoran   I-403
Zhang, Hong   I-656
Zhang, Hongjun   II-17
Zhang, Hong-xin   I-689
Zhang, Jian   II-10
Zhang, Jianfeng   I-505
Zhang, Jianpei   I-371
Zhang, Jintao   II-240
Zhang, Junwei   II-263
Zhang, Lili   II-545
Zhang, Liqiang   II-467
Zhang, Qing   II-632
Zhang, Shiguang   II-17
Zhang, Wei   I-638
Zhang, Weitao   II-213
Zhang, Wenbo   I-444, I-452
Zhang, Wenqiang   I-225
Zhang, Xiao   I-74
Zhang, Yan   II-296, II-455, II-605
Zhang, Yanhang   II-539, II-545, II-564
Zhang, Yasu   I-33
Zhang, Yi   I-539
Zhang, Yuekai   I-298
Zhang, Yulu   II-623
Zhang, Zhaogong   I-101
Zhang, Zhaoxin   II-69
Zhang, Zheng   I-225
Zhao, Bing   II-573
Zhao, Hua   II-120
Zhao, Jiejie   I-33
Zhao, Liang   I-706
Zhao, Qian   I-628, II-467
Zhao, Simeng   I-140
Zhao, Xingyu   II-55

Zhao, Yuan   II-17
Zhao, Zhihong   II-41, II-55
Zheng, Fang   I-706
Zheng, Haonan   II-136
Zhi, Pengpeng   II-416
Zhou, Li   I-505
Zhou, Ning   II-552
Zhou, Qinglei   II-250
Zhou, Xiaoyu   II-27
Zhou, Yan   I-48, I-689
Zhou, Yemao   II-27
Zhou, Yun   II-17

Zhu, Chaoping   I-698
Zhu, Conghui   II-108
Zhu, Dexin   II-330
Zhu, Enguo   II-401
Zhu, Jie   II-573
Zhu, Jinghua   I-298
Zhu, Tiejun   II-516
Zhu, Weijun   II-250
Zhu, Xiaolong   I-646
Zhuang, Chuhan   II-477
Zou, Rui   II-359
Zou, Ruofei   II-120

Printed in the United States
By Bookmasters